Advanced Rare Earth-Based Ceramic Nanomaterials

Elsevier Series in Advanced Ceramic Materials

Advanced Rare Earth-Based Ceramic Nanomaterials

Edited by

Sahar Zinatloo-Ajabshir

Department of Chemical Engineering, University of Bonab, Bonab, Iran

Series Editor

W.E. Bill Lee

Imperial College London, United Kingdom

Elsevier
Radarweg 29, PO Box 211, 1000 AE Amsterdam, Netherlands
The Boulevard, Langford Lane, Kidlington, Oxford OX5 1GB, United Kingdom
50 Hampshire Street, 5th Floor, Cambridge, MA 02139, United States

British Library Cataloguing-in-Publication Data
A catalogue record for this book is available from the British Library

Library of Congress Cataloging-in-Publication Data
A catalog record for this book is available from the Library of Congress

ISBN: 978-0-323-89957-4

For Information on all Elsevier publications
visit our website at https://www.elsevier.com/books-and-journals

Publisher: Matthew Deans
Acquisitions Editor: Gwen Jones
Editorial Project Manager: Leticia M. Lima
Production Project Manager: Sojan P. Pazhayattil
Cover Designer: Mark Rogers

Typeset by MPS Limited, Chennai, India

Contents

List of contributors

Mahnaz Amiri Neuroscience Research Center, Institute of Neuropharmacology, Kerman University of Medical Science, Kerman, Iran; Department of Hematology and Medical Laboratory Sciences, Faculty of Allied Medicine, Kerman University of Medical Sciences, Kerman, Iran

Omid Amiri Faculty of Chemistry, Razi University, Kermanshah, Iran; Department of Chemistry, College of Science, University of Raparin, Rania, Iraq

Mahdiyeh Esmaeili-Zare Institute of Nano Science and Nano Technology, University of Kashan, Kashan, Islamic Republic of Iran

Noshin Mir Department of Mechanical and Nuclear Engineering, Virginia Commonwealth University, Richmond, VA, United States

Saeid Pourmasud Department of Physics, University of Kashan, Kashan, Iran

Razieh Razavi Department of Chemistry, Faculty of Science, University of Jiroft, Jiroft, Iran

Hossein Safardoust-Hojaghan Young Researchers and Elite Club, Marand Branch, Islamic Azad University, Marand, Iran

Ali Salehabadi Environmental Technology Division, School of Industrial Technology, Universiti Sains Malaysia, Penang, Malaysia

Ali Sobhani-Nasab Autoimmune Diseases Research Center, Kashan University of Medical Sciences, Kashan, Iran; Core Research Lab, Kashan University of Medical Sciences, Kashan, Iran

Hakimeh Teymourinia Institute of Advanced Materials (INAM), University of Jaume I, Castellon, Spain; Trita Nanomedicine Research Center (TNRC), Trita Third Millennium Pharmaceuticals, Zanjan, Iran; Department of Science, University of Zanjan, Zanjan, Iran

Sahar Zinatloo-Ajabshir Department of Chemical Engineering, University of Bonab, Bonab, Iran

Advanced rare earth-based ceramic nanomaterials at a glance

Sahar Zinatloo-Ajabshir
Department of Chemical Engineering, University of Bonab, Bonab, Iran

1.1 Rare earth elements

Rare earth elements include a group of 17 elements from the periodic table, namely 15 lanthanides with scandium (Sc) and yttrium (Y) (Castor, 2006). Scandium and yttrium are also considered rare earth elements because they are often found in the same mineral deposits as lanthanides and also have similar chemical properties to lanthanides (Balaram, 2019). All rare earth elements have been reported to be found in nature but not in the form of pure metal. Among them, although Promethium has been reported as the rarest element, it is found only in very small amounts in natural materials because it has no stable isotopes (Castor, 2006).

Studies have indicated that in the group of lanthanides, the configuration of the valence electrons corresponding to the outermost shell is the same for all elements, so that the 4f orbitals are filled with enhancing atomic number. Screening of 4f orbitals can cause very similar chemical and physical attributes of these elements. The phenomenon of "lanthanide contraction" is also reported as another related consequence in which the ionic radius of the members belonging to the group reduces from La^{3+} (1.06 Å) to Lu^{3+} (0.85 Å) (Balaram, 2019).

Chemical studies have indicated that all lanthanide elements have very similar chemical attributes. The 4f orbital in the lanthanide elements is partially filled by electrons. It has been reported that f-f electron transfer in these elements can cause light absorption and thus alter spectral properties. Due to the shield of the 4f orbital sublayers by ($^5S_2$) as well as ($^5P_6$) electronic shells, the 4f sublayers can be less affected by the potential field caused by neighboring atoms. As a result, the absorption spectral attributes of lanthanide elements in each compound can remain similar to those of the linear spectra of free ions (Castor, 2006). Lanthanide elements have very different electronic energy levels and spectra than other elements. These elements are able to absorb or emit electromagnetic waves of different wavelengths, such as infrared or ultraviolet or visible light. Lanthanide elements are key and significant components of laser compounds as well as fluorescent materials (Chubb, Pal, Patton, & Jenkins, 1999; Clara Gonçalves, Santos, & Almeida, 2002; Strecker, Gonzaga, Ribeiro, & Hoffmann, 2000; Xu & Ai, 2001). The excited states of lanthanide ions (metastable state) have been reported to be relatively long-lived. These forbidden transitions are defined as spontaneous transitions between 4f-4 energy levels that can cause to the creation of most metastable states. They have a very

Advanced Rare Earth-Based Ceramic Nanomaterials. DOI: https://doi.org/10.1016/B978-0-323-89957-4.00011-6

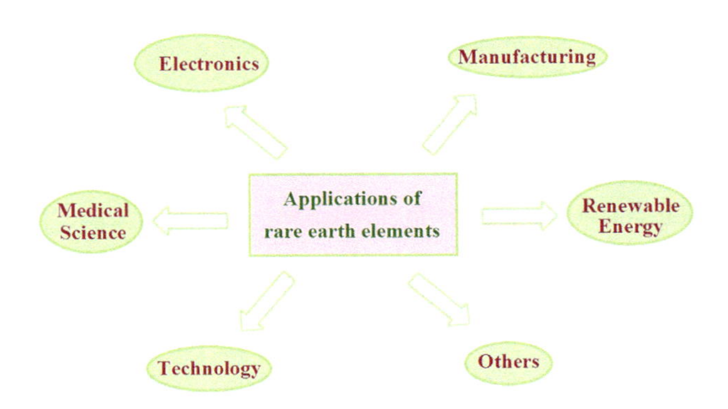

Figure 1.1 Various fields in which rare earth elements are used in the world today.

low probability of transition as well as long-lived excited states. Due to the unique properties of rare earth elements, they can be very beneficial and effective for utilization in fluorescent materials and laser compounds (Li, Xiu, & Sun, 2004; Sameshima, Ichikawa, Kawaminami, & Hirata, 1999).

Rare earth elements can exhibit excellent chemical activity. They are able to react with a variety of other elements such as oxygen, sulfur, phosphorus, and so on. Due to the unique structure of rare earth elements, the compounds containing them, both in solid and soluble form, their adsorption spectra indicate many bands. Since the 17 rare earth elements all possess their own specific wavelength range, as a result, all of these elements possess their own color (Kenyon, 2002; Li et al., 2004).

Rare earth elements have extraordinary magnetic, catalytic as well as phosphorescent attributes, so their utilization has received much attention today. The utilization of these elements in various technologies can be beneficial and effective (see Fig. 1.1). The estimated mean quantity of rare earth elements in the Earth's crust is reported to be from about 130–240 mg/g, which is actually remarkably more than other commonly utilized elements (Zepf, 2013).

1.2 Rare earth-based ceramic nanomaterials

It has been generally accepted that the strength as well as the toughness of some structural ceramics may be largely boosted with adding rare earth oxides as stabilizers or sintering aids to diminish the sintering temperature as well as diminish production costs. On the other hand, rare earth oxides may also play a very key and significant role in functional ceramics such as microwave dielectric, semiconductor recorders, and piezoelectric ceramics (Castor, 2006; Jones et al., 2005; Kenyon, 2002; Kim, Chun, Nishimura, Mitomo, & Lee, 2007; Maia, Mastelaro, Pairis, Hernandes, & Ibanez, 2007).

Oxide ceramics of the lanthanide category, along with scandium and yttrium, are all called rare earth oxides. Most of these oxide compounds are technically very key and significant and possess many applications in a variety of fields. For example: in the preparation of corrosion-resistant coatings (Bonnet, Lachkar, Larpin, & Colson, 1994; Reitz, Haetge, Suchomski, & Brezesinski, 2013), in heterogeneous catalytic processes (Mortazavi-Derazkola, Zinatloo-Ajabshir, & Salavati-Niasari, 2015; Rosynek, 1977; Zinatloo-Ajabshir, Ghasemian, Mousavi-Kamazani, & Salavati-Niasari, 2021), as phosphor/activator/host compounds (Bünzli, Comby, Chauvin, & Vandevyver, 2007; Jüstel, Nikol, & Ronda, 1998), and gate oxides with high dielectric constant (Leskelä, Kukli, & Ritala, 2006).

Recently, the utilization of rare earth elements in the ceramic industry has received much attention and has made good progress. Consumption of 30% of the total production of rare earth elements in the world has reported in the ceramic industry. Hence, the utilization of rare earths in ceramics in order to exploit the benefits of these valuable resources in order to achieve industrial superiority seems necessary (Baldacim, Santos, Silva, & Silva, 2004; Clara Gonçalves et al., 2002; Inoue, Otsu, Kominami, & Inui, 1995; Terao, Tatami, Meguro, & Komeya, 2002; Xu & Ai, 2001).

Ceramic materials owing to their unique and special attributes such as having excellent mechanical hardness and strength, good thermal and chemical stability, as well as very remarkable optical, electrical, and magnetic attributes, exhibit extraordinary performance in various fields. These areas include: in construction (Faseeva et al., 2017; Harrati, Manni, El Bouari, El Amrani El Hassani, & Sadik, 2020), energy technologies (e.g., solar cells and supercapacitors) (Bae, Kim, Choi, Son, & Shim, 2018; Chen et al., 2019; Dutta et al., 2020; Li et al., 2020; Li et al., 2020; Wang et al., 2019; Zeng et al., 2018; Zhao et al., 2019), catalytic processes (Celebanska & Opallo, 2016; Li et al., 2019; Liu, Xiao, Zhang, & Sun, 2020; Singh et al., 2018; Verma, Sinha-Ray, Sinha-Ray, & Electrospun, 2020), various chemical industries (Cheng et al., 2019), electronics (Jiang et al., 2020), medical engineering (Shu et al., 2019), machinery, aerospace (Rathod, Kumar, & Jain, 2017), hypersonic (Gopinath, Jagadeesh, & Basu, 2019), bioceramics (Reveron et al., 2017), and laser compounds (Li et al., 2017). Although ceramic materials have been reported to have specific uses in different fields, their utilization in energy-related uses has attracted the attention of many scientists and researchers. Today, we are witnessing an energy crisis with the reduction of nonrenewable energy sources such as coal, oil, and natural gas, as well as the overuse of fossil fuels and the disadvantages of using them. The utilization of ceramic compounds can be beneficial and effective in dealing with this crisis. Thus, today, in order to achieve goals such as energy saving and decreasing emissions of pollutants, as well as the principle of sustainable development, researchers continue to design and produce different advanced compounds such as advanced ceramic compounds (Zeng, Song, Shen, & Moskovits, 2021). Since ceramic compounds have unique attributes including corrosion resistance, heat shock resistance, high temperature resistance and also radiation resistance, they can be very beneficial and efficient for use in the preparation of electrode material in supercapacitors (Zeng et al., 2021). Thus a large number of ceramic electrode compounds have been designed and developed for utilization in supercapacitors, including carbide and nitride ceramics (Kim et al., 2016; Liu, Du, & Chu, 2020; Mujib et al., 2020; Pazhamalai, Krishnamoorthy, Sahoo, Mariappan,

& Kim, 2020; Wan et al., 2019; Xiao et al., 2020; Zhao et al., 2016), metal oxide ceramics (Jose et al., 2016; Knöller et al., 2018; Li, Cao, Xue, Sun, & Zhu, 2020; Liu et al., 2014; Vidyadharan et al., 2014; Xing et al., 2017; Yus, Ferrari, Sanchez-Herencia, & Gonzalez, 2020), metal hydroxide ceramics (Chang et al., 2019; Gonzalez, Ferrari, Sanchez-Herencia, Caballero, & Morales, 2016; Milne, Marques Silva, & Zhitomirsky, 2019), multielement oxide ceramics (Gu et al., 2019; Liu et al., 2019; Lu et al., 2020; McOwen et al., 2018; Xia et al., 2020; Zhang, Guo, Han, & Xiao, 2018), carbon-based ceramics (Bon et al., 2018; Liu et al., 2020; Moyano et al., 2019; Singh et al., 2019; Yus, Bravo, Sanchez-Herencia, Ferrari, & Gonzalez, 2019), and other particular ceramics (Jiang, Kurra, Alhabeb, Gogotsi, & Alshareef, 2018; Kim, Wang, & Alshareef, 2019; Pang et al., 2019; Wang et al., 2019; Zeng, Cheng, Yu, & Stucky, 2020).

Nanotechnology is generally accepted as the utilization of scientific knowledge to design, manipulate, and adjust the attributes of compounds such as particle size and their particle size distribution at the nanoscale with the aim of using attributes and phenomena dependent on size and structure. The attributes of nanostructured compounds are more distinct than those of individual atoms or bulk compounds. Nanotechnology can also offer novel opportunities in order to meet the need for everyday products to develop these products while enhancing their performance, reducing the estimated cost of their production, and also utilizing less raw materials to produce them (Gleiter, 2000; Santos et al., 2015). In addition to design and fabrication, nanotechnology describes characterization of structures, devices, or compounds that possess one or more dimensions less than 100 nm. Today, nanotechnology as a very significant, key, and helpful technology for a wide range of industrial uses has been considered by many researchers and scientists. In other words, nanotechnology has become a major priority for the development of science and technology policies (Santos et al., 2015; Sudha et al., 2018).

It has been reported that any material that has any external dimension at the nanoscale (i.e., 1—100 nm) or any material that has an internal structure as well as a surface structure at the nanoscale is referred to as a nanomaterial. Nanomaterials can be a natural substance or artificial substance comprising particles (Pokropivny & Skorokhod, 2007; Sudha et al., 2018). It has been demonstrated that by going to the nanoscale, each material can exhibit distinct and different attributes compared to the same material on a larger nonnanoscale scale. Depending on how many dimensions of each material are at the nanoscale, different classes of nanomaterials have been introduced. In general, nanomaterials are of three classes in terms of dimensions: either they have three nanoscale dimensions (such as nanoparticles) or they have two nanoscale dimensions (such as mineral nanotubes) or they have only one nanoscale dimension (such as thin films) (Aversa, Modarres, Cozzini, Ciancio, & Chiusole, 2018; Sudha et al., 2018).

Among the inorganic nanomaterials are nanoceramics, which can display very significant optical, structural, and electronic attributes as their particle size approaches the nanoscale (Saleh, 2020; Sudha et al., 2018). Nanoceramics are a kind of nanocompounds consisting of ceramics and are mostly known as heat-resistant, mineral, and nonmetallic compounds that can be prepared from nonmetallic and metal components with dimensions less than 100 nm (Saleh, 2020). So far, different forms including amorphous or crystalline, dense or porous, as well as hollow ceramic nanomaterials have been prepared (Sigmund et al., 2006). Today, ceramic nanostructures have been considered by many research teams owing to their unique attributes. They can be employed in a variety of fields including catalytic processes, heterogeneous photocatalytic reactions, environmental modification, bone repair, energy supply and storage, and construction, as well as imaging applications (Thomas, Harshita, Mishra, & Talegaonkar, 2015).

Ceramic nanomaterials based on rare earths can exhibit significant or new distinctive attributes compared to bulk ceramic materials. Advanced ceramic

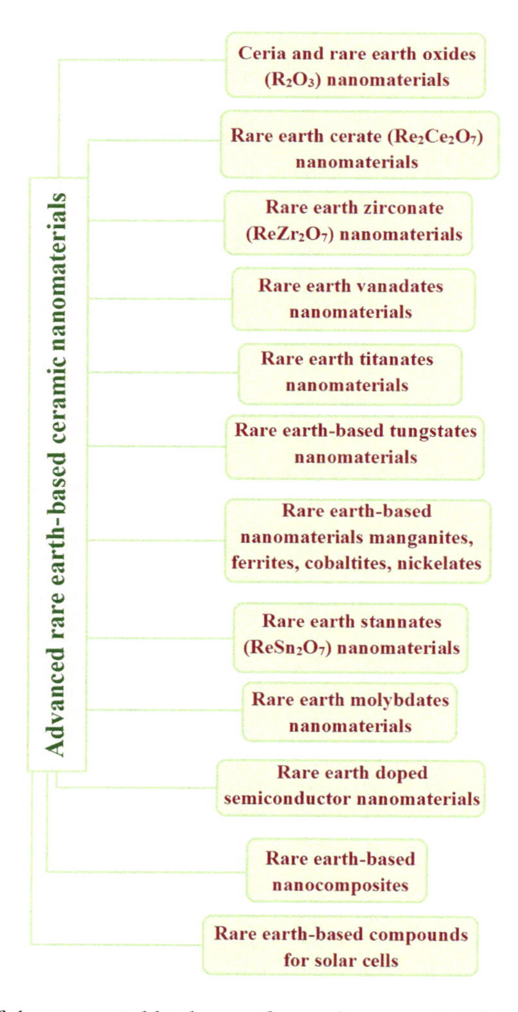

Figure 1.2 Some of the most notable classes of ceramic nanomaterials based on rare earths.

nanomaterials can be developed by adding the rare earth elements with a certain ratio to different compounds. So far, various families of advanced ceramic nanostructures based on rare earths have been prepared and studied. Some of the most notable classes of ceramic nanomaterials based on rare earths that have recently attracted the attention of various researchers and scientists are shown in Fig. 1.2.

In this book, we focus on recent advances in the preparation methods and applications of advanced rare earth-based ceramic nanomaterials. Different approaches to the synthesis of each family of ceramic nanomaterials based on rare earths with their significant advantages or disadvantages as well as their applications in different fields are presented in separate chapters of the book. It is generally accepted that attributes such as dimensions, purity in terms of chemical composition and particle size distribution, as well as the performance of nanostructured ceramic materials are largely dependent on the approach used to fabricate them. Thus choosing a very appropriate approach for large-scale production of ceramic nanomaterials with desirable attributes at a low cost is very important. This book can be considered as a valuable reference source for researchers and students and cover comprehensive and useful information about ceramic nanomaterials based on rare earths as shown in Fig. 1.2.

References

Aversa, R., Modarres, M. H., Cozzini, S., Ciancio, R., & Chiusole, A. (2018). The first annotated set of scanning electron microscopy images for nanoscience. *Scientific Data*, *5*, 180172.

Bae, K., Kim, D. H., Choi, H. J., Son, J.-W., & Shim, J. H. (2018). High-performance protonic ceramic fuel cells with 1 μm thick Y:Ba(Ce, Zr)O_3 electrolytes. *Advanced Energy Materials*, *8*, 1801315.

Balaram, V. (2019). Rare earth elements: A review of applications, occurrence, exploration, analysis, recycling, and environmental impact. *Geoscience Frontiers*, *10*, 1285–1303.

Baldacim, S. A., Santos, C., Silva, O. M. M., & Silva, C. R. M. (2004). Ceramics composites Si_3N_4–SiC(w) containing rare earth concentrate (CRE) as sintering aids. *Materials Science and Engineering: A*, *367*, 312–316.

Bon, C. Y., Mohammed, L., Kim, S., Manasi, M., Isheunesu, P., Lee, K. S., & Ko, J. M. (2018). Flexible poly(vinyl alcohol)-ceramic composite separators for supercapacitor applications. *Journal of Industrial and Engineering Chemistry*, *68*, 173–179.

Bonnet, G., Lachkar, M., Larpin, J. P., & Colson, J. C. (1994). Organometallic chemical vapor deposition of rare earth oxide thin films. Application for steel protection against high temperature oxidation. *Solid State Ionics*, *72*, 344–348.

Bünzli, J.-C. G., Comby, S., Chauvin, A.-S., & Vandevyver, C. D. B. (2007). New opportunities for lanthanide luminescence. *Journal of Rare Earths*, *25*, 257–274.

Castor, S. B. (2006). Rare earth elements. *Industrial Minerals & Rocks*, 769–792.

Celebanska, A., & Opallo, M. (2016). Layer-by-layer gold–ceramic nanoparticulate electrodes for electrocatalysis. *ChemElectroChem*, *3*, 1629–1634.

Chang, P., Mei, H., Zhao, Y., Huang, W., Zhou, S., & Cheng, L. (2019). 3D structural strengthening urchin-like $Cu(OH)_2$-based symmetric supercapacitors with adjustable capacitance. *Advanced Functional Materials*, *29*, 1903588.

Chen, G., Chen, J., Pei, W., Lu, Y., Zhang, Q., Zhang, Q., & He, Y. (2019). Bismuth ferrite materials for solar cells: Current status and prospects. *Materials Research Bulletin*, *110*, 39–49.

Cheng, Z., Ye, F., Liu, Y., Qiao, T., Li, J., Qin, H., ... Zhang, L. (2019). Mechanical and dielectric properties of porous and wave-transparent Si_3N_4-Si_3N_4 composite ceramics fabricated by 3D printing combined with chemical vapor infiltration. *Journal of Advanced Ceramics*, *8*, 399–407.

Chubb, D. L., Pal, A. T., Patton, M. O., & Jenkins, P. P. (1999). Rare earth doped high temperature ceramic selective emitters. *Journal of the European Ceramic Society*, *19*, 2551–2562.

Clara Gonçalves, M., Santos, L. F., & Almeida, R. M. (2002). Rare-earth-doped transparent glass ceramics. *Comptes Rendus Chimie*, *5*, 845–854.

Dutta, P., Sikdar, A., Majumdar, A., Borah, M., Padma, N., Ghosh, S., & Maiti, U. N. (2020). Graphene aided gelation of MXene with oxidation protected surface for supercapacitor electrodes with excellent gravimetric performance. *Carbon*, *169*, 225–234.

Faseeva, G. R., Nafikov, R. M., Lapuk, S. E., Zakharov, Y. A., Novik, A. A., Vjuginova, A. A., ... Garipov, L. N. (2017). Ultrasound-assisted extrusion of construction ceramic samples. *Ceramics International*, *43*, 7202–7210.

Gleiter, H. (2000). Nanostructured materials: Basic concepts and microstructure. *Acta Materialia*, *48*, 1–29.

Gonzalez, Z., Ferrari, B., Sanchez-Herencia, A. J., Caballero, A., & Morales, J. (2016). Use of polyelectrolytes for the fabrication of porous NiO films by electrophoretic deposition for supercapacitor electrodes. *Electrochimica Acta*, *211*, 110–118.

Gopinath, N. K., Jagadeesh, G., & Basu, B. (2019). Shock wave-material interaction in ZrB_2–SiC based ultra high temperature ceramics for hypersonic applications. *Journal of the American Ceramic Society*, *102*, 6925–6938.

Gu, R., Yu, K., Wu, L., Ma, R., Sun, H., Jin, L., ... Wei, X. (2019). Dielectric properties and I-V characteristics of $Li_{0.5}La_{0.5}TiO_3$ solid electrolyte for ceramic supercapacitors. *Ceramics International*, *45*, 8243–8247.

Harrati, A., Manni, A., El Bouari, A., El Amrani El Hassani, I.-E., & Sadik, C. (2020). Elaboration and thermomechanical characterization of ceramic-based on Moroccan geomaterials: Application in construction. *Materials Today: Proceedings*, *30*, 876–882.

Inoue, M., Otsu, H., Kominami, H., & Inui, T. (1995). Glycothermal synthesis of rare earth aluminium garnets. *Journal of Alloys and Compounds*, *226*, 146–151.

Jiang, C.-S., Dunlap, N., Li, Y., Guthrey, H., Liu, P., Lee, S.-H., & Al-Jassim, M. M. (2020). Nonuniform ionic and electronic transport of ceramic and polymer/ceramic hybrid electrolyte by nanometer-scale operando imaging for solid-state battery. *Advanced Energy Materials*, *10*, 2000219.

Jiang, Q., Kurra, N., Alhabeb, M., Gogotsi, Y., & Alshareef, H. N. (2018). All pseudocapacitive MXene-RuO_2 asymmetric supercapacitors. *Advanced Energy Materials*, *8*, 1703043.

Jones, A. C., Aspinall, H. C., Chalker, P. R., Potter, R. J., Kukli, K., Rahtu, A., ... Leskelä, M. (2005). Recent developments in the MOCVD and ALD of rare earth oxides and silicates. *Materials Science and Engineering: B*, *118*, 97–104.

Jose, R., Krishnan, S. G., Vidyadharan, B., Misnon, I. I., Harilal, M., Aziz, R. A., ... Yusoff, M. M. (2016). Supercapacitor electrodes delivering high energy and power densities. *Materials Today: Proceedings*, *3*, S48–S56.

Jüstel, T., Nikol, H., & Ronda, C. (1998). New developments in the field of luminescent materials for lighting and displays. *Angewandte Chemie International Edition*, *37*, 3084–3103.

Kenyon, A. J. (2002). Recent developments in rare-earth doped materials for optoelectronics. *Progress in Quantum Electronics*, *26*, 225–284.

Kim, H., Wang, Z., & Alshareef, H. N. (2019). MXetronics: Electronic and photonic applications of MXenes. *Nano Energy*, *60*, 179–197.

Kim, S.-J., Kim, M.-C., Han, S.-B., Lee, G.-H., Choe, H.-S., Kwak, D.-H., ... Park, K.-W. (2016). 3D flexible Si based-composite (Si@Si_3N_4)/CNF electrode with enhanced cyclability and high rate capability for lithium-ion batteries. *Nano Energy*, *27*, 545–553.

Kim, Y.-W., Chun, Y.-S., Nishimura, T., Mitomo, M., & Lee, Y.-H. (2007). High-temperature strength of silicon carbide ceramics sintered with rare-earth oxide and aluminum nitride. *Acta Materialia*, *55*, 727–736.

Knöller, A., Kilper, S., Diem, A. M., Widenmeyer, M., Runčevski, T., Dinnebier, R. E., ... Burghard, Z. (2018). Ultrahigh damping capacities in lightweight structural materials. *Nano Letters*, *18*, 2519–2524.

Leskelä, M., Kukli, K., & Ritala, M. (2006). Rare-earth oxide thin films for gate dielectrics in microelectronics. *Journal of Alloys and Compounds*, *418*, 27–34.

Li, D., Shen, Z.-Y., Li, Z., Luo, W., Song, F., Wang, X., ... Li, Y. (2020). Optimization of polarization behavior in (1 − x)BSBNT−xNN ceramics for pulsed power capacitors. *Journal of Materials Chemistry C*, *8*, 7650–7657.

Li, D., Shen, Z.-Y., Li, Z., Luo, W., Wang, X., Wang, Z., ... Li, Y. (2020). P-E hysteresis loop going slim in Ba0.3Sr0.7TiO3-modified $Bi_{0.5}Na_{0.5}TiO_3$ ceramics for energy storage applications. *Journal of Advanced Ceramics*, *9*, 183–192.

Li, T., Cao, Y., Xue, W., Sun, B., & Zhu, D. (2020). Self-assembly of graphene-based planar micro-supercapacitor with selective laser etching-induced superhydrophobic/superhydrophilic pattern. *SN Applied Sciences*, *2*, 206.

Li, W., Huang, H., Mei, B., Song, J., Yi, G., & Guo, X. (2017). Fabrication and characterization of polycrystalline Ho:CaF_2 transparent ceramics for 2.0μm laser application. *Materials Letters*, *207*, 37–40.

Li, X. K., Xiu, Z. M., & Sun, X. D. (2004). Effect of rare earths on mechanical properties and microstructure of TiCN/Al_2O_3 ceramics. *Journal of Rare Earths*, *22*, 132–135.

Li, Z., Liu, S., Song, S., Xu, W., Sun, Y., & Dai, Y. (2019). Porous ceramic nanofibers as new catalysts toward heterogeneous reactions. *Composites Communications*, *15*, 168–178.

Liu, H., Du, B., & Chu, Y. (2020). Synthesis of the ternary metal carbide solid-solution ceramics by polymer-derived-ceramic route. *Journal of the American Ceramic Society*, *103*, 2970–2974.

Liu, M., Turcheniuk, K., Fu, W., Yang, Y., Liu, M., & Yushin, G. (2020). Scalable, safe, high-rate supercapacitor separators based on the Al_2O_3 nanowire Polyvinyl butyral nonwoven membranes. *Nano Energy*, *71*, 104627.

Liu, X., Xiao, L., Zhang, Y., & Sun, H. (2020). Significantly enhanced piezo-photocatalytic capability in $BaTiO_3$ nanowires for degrading organic dye. *Journal of Materiomics*, *6*, 256–262.

Liu, Y., Ata, M. S., Shi, K., Zhu, Gz, Botton, G. A., & Zhitomirsky, I. (2014). Surface modification and cathodic electrophoretic deposition of ceramic materials and composites using celestine blue dye. *RSC Advances*, *4*, 29652–29659.

Liu, Y., Ying, Y., Fei, L., Liu, Y., Hu, Q., Zhang, G., ... Huang, H. (2019). Valence engineering via selective atomic substitution on tetrahedral sites in spinel oxide for highly enhanced oxygen evolution catalysis. *Journal of the American Chemical Society, 141*, 8136–8145.

Lu, D.-L., Zhao, R.-R., Wu, J.-L., Ma, J.-M., Huang, M.-L., Yao, Y.-B., ... Lu, S.-G. (2020). Investigations on the properties of $Li_{3x}La_{2/3-x}TiO_3$ based all-solid-state supercapacitor: Relationships between the capacitance, ionic conductivity, and temperature. *Journal of the European Ceramic Society, 40*, 2396–2403.

Maia, L. J. Q., Mastelaro, V. R., Pairis, S., Hernandes, A. C., & Ibanez, A. (2007). A sol–gel route for the development of rare-earth aluminum borate nanopowders and transparent thin films. *Journal of Solid State Chemistry, 180*, 611–618.

McOwen, D. W., Xu, S., Gong, Y., Wen, Y., Godbey, G. L., Gritton, J. E., ... Wachsman, E. D. (2018). 3D-printing electrolytes for solid-state batteries. *Advanced Materials, 30*, 1707132.

Milne, J., Marques Silva, R., & Zhitomirsky, I. (2019). Surface modification and dispersion of ceramic particles using liquid-liquid extraction method for application in supercapacitor electrodes. *Journal of the European Ceramic Society, 39*, 3450–3455.

Mortazavi-Derazkola, S., Zinatloo-Ajabshir, S., & Salavati-Niasari, M. (2015). Novel simple solvent-less preparation, characterization and degradation of the cationic dye over holmium oxide ceramic nanostructures. *Ceramics International, 41*, 9593–9601.

Moyano, J. J., Mosa, J., Aparicio, M., Pérez-Coll, D., Belmonte, M., Miranzo, P., & Osendi, M. I. (2019). Strong and light cellular silicon carbonitride – Reduced graphene oxide material with enhanced electrical conductivity and capacitive response. *Additive Manufacturing, 30*, 100849.

Mujib, S. B., Cuccato, R., Mukherjee, S., Franchin, G., Colombo, P., & Singh, G. (2020). Electrospun SiOC ceramic fiber mats as freestanding electrodes for electrochemical energy storage applications. *Ceramics International, 46*, 3565–3573.

Pang, J., Mendes, R. G., Bachmatiuk, A., Zhao, L., Ta, H. Q., Gemming, T., ... Rummeli, M. H. (2019). Applications of 2D MXenes in energy conversion and storage systems. *Chemical Society Reviews, 48*, 72–133.

Pazhamalai, P., Krishnamoorthy, K., Sahoo, S., Mariappan, V. K., & Kim, S.-J. (2020). Carbothermal conversion of siloxene sheets into silicon-oxy-carbide lamellae for high-performance supercapacitors. *Chemical Engineering Journal, 387*, 123886.

Pokropivny, V. V., & Skorokhod, V. V. (2007). Classification of nanostructures by dimensionality and concept of surface forms engineering in nanomaterial science. *Materials Science and Engineering: C, 27*, 990–993.

Rathod, V. T., Kumar, J. S., & Jain, A. (2017). Polymer and ceramic nanocomposites for aerospace applications. *Applied Nanoscience, 7*, 519–548.

Reitz, C., Haetge, J., Suchomski, C., & Brezesinski, T. (2013). Facile and general synthesis of thermally stable ordered mesoporous rare-earth oxide ceramic thin films with uniform mid-size to large-size pores and strong crystalline texture. *Chemistry of Materials, 25*, 4633–4642.

Reveron, H., Fornabaio, M., Palmero, P., Fürderer, T., Adolfsson, E., Lughi, V., ... Chevalier, J. (2017). Towards long lasting zirconia-based composites for dental implants: Transformation induced plasticity and its consequence on ceramic reliability. *Acta Biomaterialia, 48*, 423–432.

Rosynek, M. P. (1977). Catalytic properties of rare earth oxides. *Catalysis Reviews, 16*, 111–154.

Saleh, T. A. (2020). Nanomaterials: Classification, properties, and environmental toxicities. *Environmental Technology & Innovation*, *20*, 101067.

Sameshima, S., Ichikawa, T., Kawaminami, M., & Hirata, Y. (1999). Thermal and mechanical properties of rare earth-doped ceria ceramics. *Materials Chemistry and Physics*, *61*, 31–35.

Santos, C. S. C., Gabriel, B., Blanchy, M., Menes, O., García, D., Blanco, M., ... Neto, V. (2015). Industrial applications of nanoparticles – A prospective overview. *Materials Today: Proceedings*, *2*, 456–465.

Shu, L., Liang, R., Rao, Z., Fei, L., Ke, S., & Wang, Y. (2019). Flexoelectric materials and their related applications: A focused review. *Journal of Advanced Ceramics*, *8*, 153–173.

Sigmund, W., Yuh, J., Park, H., Maneeratana, V., Pyrgiotakis, G., Daga, A., ... Nino, J. C. (2006). Processing and Structure Relationships in Electrospinning of Ceramic Fiber Systems. *Journal of the American Ceramic Society*, *89*, 395–407.

Singh, P., Kaur, G., Singh, K., Singh, B., Kaur, M., Kaur, M., ... Kumar, A. (2018). Specially designed B4C/SnO_2 nanocomposite for photocatalysis: Traditional ceramic with unique properties. *Applied Nanoscience*, *8*, 1–9.

Singh, R., Chakravarty, A., Mishra, S., Prajapati, R. C., Dutta, J., Bhat, I. K., ... Muraleedharan, K. (2019). AlN–SWCNT metacomposites having tunable negative permittivity in radio and microwave frequencies. *ACS Applied Materials & Interfaces*, *11*, 48212–48220.

Strecker, K., Gonzaga, R., Ribeiro, S., & Hoffmann, M. J. (2000). Substitution of Y_2O_3 by a rare earth oxide mixture as sintering additive of Si_3N_4 ceramics. *Materials Letters*, *45*, 39–42.

Sudha, P. N., Sangeetha, K., Vijayalakshmi, K., & Barhoum, A. (2018). Chapter 12 - Nanomaterials history, classification, unique properties, production and market. In A. Barhoum, & A. S. H. Makhlouf (Eds.), *Emerging applications of nanoparticles and architecture nanostructures* (pp. 341–384). Elsevier.

Terao, R., Tatami, J., Meguro, T., & Komeya, K. (2002). Fracture behavior of AlN ceramics with rare earth oxides. *Journal of the European Ceramic Society*, *22*, 1051–1059.

Thomas, S. C., Harshita., Mishra, P. K., & Talegaonkar, S. (2015). Ceramic nanoparticles: Fabrication methods and applications in drug delivery. *Current Pharmaceutical Design*, *21*, 6165–6188.

Verma, S., Sinha-Ray, S., & Sinha-Ray, S. (2020). Electrospun CNF supported ceramics as electrochemical catalysts for water splitting and fuel cell: A review. *Polymers*, *12*, 238.

Vidyadharan, B., Aziz, R. A., Misnon, I. I., Anil Kumar, G. M., Ismail, J., Yusoff, M. M., & Jose, R. (2014). High energy and power density asymmetric supercapacitors using electrospun cobalt oxide nanowire anode. *Journal of Power Sources*, *270*, 526–535.

Wan, P., Li, M., Xu, K., Wu, H., Chang, K., Zhou, X., ... Huang, Q. (2019). Seamless joining of silicon carbide ceramics through an sacrificial interlayer of $Dy_3Si_2C_2$. *Journal of the European Ceramic Society*, *39*, 5457–5462.

Wang, Q.-W., Zhang, H.-B., Liu, J., Zhao, S., Xie, X., Liu, L., ... Yu, Z.-Z. (2019). Multifunctional and water-resistant MXene-decorated polyester textiles with outstanding electromagnetic interference shielding and joule heating performances. *Advanced Functional Materials*, *29*, 1806819.

Wang, X., Zhai, H., Qie, B., Cheng, Q., Li, A., Borovilas, J., ... Yang, Y. (2019). Rechargeable solid-state lithium metal batteries with vertically aligned ceramic nanoparticle/polymer composite electrolyte. *Nano Energy*, *60*, 205–212.

Xia, W., Liu, Y., Wang, G., Li, J., Cao, C., Hu, Q., ... Wang, D. (2020). Frequency and temperature independent $(Nb_{0.5}Ga_{0.5})_x(Ti_{0.9}Zr_{0.1})_{1-x}O_2$ ceramics with giant dielectric permittivity and low loss. *Ceramics International*, *46*, 2954–2959.

Xiao, Z., Lei, C., Yu, C., Chen, X., Zhu, Z., Jiang, H., & Wei, F. (2020). $Si@Si_3N_4@C$ composite with egg-like structure as high-performance anode material for lithium ion batteries. *Energy Storage Materials*, *24*, 565–573.

Xing, Y., Guo, X., Wu, D., Liu, Z., Fang, S., & Suib, S. L. (2017). Construction of macroscopic 3D foams of metastable manganese oxides via a mild templating route: Effects of atmosphere and calcination. *Journal of Alloys and Compounds*, *719*, 22–29.

Xu, C., & Ai, X. (2001). Particle dispersed ceramic composite reinforced with rare earth additions. *International Journal of Refractory Metals and Hard Materials*, *19*, 85–88.

Yus, J., Bravo, Y., Sanchez-Herencia, A. J., Ferrari, B., & Gonzalez, Z. (2019). Electrophoretic deposition of RGO-NiO core-shell nanostructures driven by heterocoagulation method with high electrochemical performance. *Electrochimica Acta*, *308*, 363–372.

Yus, J., Ferrari, B., Sanchez-Herencia, A. J., & Gonzalez, Z. (2020). Understanding the effects of different microstructural contributions in the electrochemical response of Nickel-based semiconductor electrodes with 3D hierarchical networks shapes. *Electrochimica Acta*, *335*, 135629.

Zeng, X., Shui, J., Liu, X., Liu, Q., Li, Y., Shang, J., ... Yu, R. (2018). Single-atom to single-atom grafting of Pt1 onto $Fe-N_4$ center: $Pt_1@Fe-N-C$ multifunctional electrocatalyst with significantly enhanced properties. *Advanced Energy Materials*, *8*, 1701345.

Zeng, X., Cheng, X., Yu, R., & Stucky, G. D. (2020). Electromagnetic microwave absorption theory and recent achievements in microwave absorbers. *Carbon*, *168*, 606–623.

Zeng, X., Song, H., Shen, Z.-Y., & Moskovits, M. (2021). Progress and challenges of ceramics for supercapacitors. *Journal of Materiomics*.

Zepf, V. (2013). Rare earth elements: What and where they are. In V. Zepf (Ed.), *Rare earth elements: A new approach to the nexus of supply, demand and use: Exemplified along the use of neodymium in permanent magnets* (pp. 11–39). Berlin, Heidelberg: Springer Berlin Heidelberg.

Zhang, Y., Guo, Z., Han, Z., & Xiao, X. (2018). Effect of rare earth oxides doping on $MgAl_2O_4$ spinel obtained by sintering of secondary aluminium dross. *Journal of Alloys and Compounds*, *735*, 2597–2603.

Zhao, Y., Kang, W., Li, L., Yan, G., Wang, X., Zhuang, X., & Cheng, B. (2016). Solution blown silicon carbide porous nanofiber membrane as electrode materials for supercapacitors. *Electrochimica Acta*, *207*, 257–265.

Zhao, Y., Yan, J., Cai, W., Lai, Y., Song, J., Yu, J., & Ding, B. (2019). Elastic and well-aligned ceramic LLZO nanofiber based electrolytes for solid-state lithium batteries. *Energy Storage Materials*, *23*, 306–313.

Zinatloo-Ajabshir, S., Ghasemian, N., Mousavi-Kamazani, M., & Salavati-Niasari, M. (2021). Effect of zirconia on improving NOx reduction efficiency of $Nd_2Zr_2O_7$ nanostructure fabricated by a new, facile and green sonochemical approach. *Ultrasonics Sonochemistry*, *71*, 105376.

Ceria and rare earth oxides (R_2O_3) ceramic nanomaterials

2

Sahar Zinatloo-Ajabshir
Department of Chemical Engineering, University of Bonab, Bonab, Iran

2.1 General introduction

Ceria (CeO_2) and rare earth oxides (R_2O_3) nanomaterials with appropriate features have been attracting ever-growing research attention in a variety of fields. This chapter focuses on the recent progress in investigating the preparation approaches, and summarizing their significant applications in various areas. Also, this chapter of the book briefly discusses the advantages and disadvantages of each method used to prepare ceria and R_2O_3 nanomaterials in order to help readers choose the appropriate method for preparing this category of ceramic nanomaterials.

2.1.1 Ceria (CeO_2)

Among members of the rare earth family, cerium (Ce) is the most plentiful element reported. The amount of cerium in the earth's crust (66.5 ppm) is reported to be higher than copper (60 ppm) or even tin (2.3 ppm) (Lide, 2004; Sun, Li, & Chen, 2012). Ce has the electron configuration of [Xe] $4f^2 6s^2$ and two various valence state, namely cerium(III) and cerium(IV) (Gnanam & Rajendran, 2011). It can exist as a free metal or oscillate between two oxidation states (Ce^{III} and Ce^{IV}) (Ravishankar, Ramakrishnappa, Nagaraju, & Rajanaika, 2015; Sreekanth, Nagajyothi, Reddy, Shim, & Yoo, 2019).

Ceria (CeO_2) is an oxide material from a rare earth family. It contains cations with a capacity of four Ce^{4+}. It shows a cubic fluorite structure with space group *Fm3m* (Gnanam & Rajendran, 2011). Nanostructured cerium oxide can also hop between CeIII and CeIV valence states. It comprises oxygen vacancies that allow the nanostructure to act as regenerative catalyst (Sreekanth et al., 2019). Ceria does not exhibit any known crystallographic alteration from room temperature up to its melting point (Li, Ikegami, & Mori, 2004).

It has been shown that the photocatalytic performance of a nanostructured catalyst substantially depend on structure, size, and shape of its particles. Cerium dioxide has the experimental band gap about 3.2 eV owing to the O 2p→Ce 4f transition, which quantity is near to those of titanium dioxide and zinc oxide (Corma, Atienzar, García, & Chane-Ching, 2004). Such narrowed band gap denotes it can absorb the wavelength less than ~390 nm (UV region). In order to improve

Advanced Rare Earth-Based Ceramic Nanomaterials. DOI: https://doi.org/10.1016/B978-0-323-89957-4.00002-5

photocatalytic efficiency, it is significant to boost the light absorption of the cerium dioxide (Xie et al., 2017).

Another significant intrinsic feature of the cerium dioxide is it can show a continuum of oxygen-deficient, CeO_{2-x} oxides with defects of O vacancies in a reducing conditions (Körner, Ricken, Nölting, & Riess, 1989). Goris et al. observed that the surface facets possess a key influence on the cerium reduction because the [10] facets (Goris, Turner, Bals, & Van, 2014) facets exhibit lower surface reduction than that of [11] facets (Lang, Chen, & Zhao, 2014) facets (Goris et al., 2014). Besides, oxygen vacancies may be formed through exchange of lower valent cations or oxygen deletion in cerium dioxide lattice, leading to two left electrons and the creation of Ce(III) adjacent to the defects. Such distinctive structures of cerium dioxide with quantities of Ce^{3+} and O vacancies possess key influence on different catalytic reactions. Cerium dioxide is well known superior oxygen storage substance because the cerium dioxide has remarkable Ce^{3+} and O vacancies. In an oxygen-deficient or rich conditions, cerium dioxide may incorporate the content of the O through forming O vacancies or supplying O, and these characters can be always relevant to the diffusion of O vacancies and surface feature of cerium dioxide (Campbell & Peden, 2005; Doornkamp & Ponec, 2000; Goris et al., 2014).

In other hand, the surface atoms of cerium dioxide may interact simply by electron-rich heteroatoms (like, carbon or nitrogen) through weak coordination, which is advantageous for catalytic reactions and suggests cerium dioxide as desirable supports for other cocatalysts (Lang et al., 2014; Mullins, 2015).

Furthermore, the cerium dioxide surface may change the dispersion of cocatalysts owing to these interaction between the cocatalysts and the surface defects. It is generally accepted that the dispersion thermal stability of cocatalysts may be boosted by cerium dioxide as the support, compared to other metal oxide supports (Kominami, Tanaka, & Hashimoto, 2010; Tanaka, Hashimoto, & Kominami, 2011). The cerium dioxide may improve the oxidative and reductive performance of CeO_2-based photocatalysts owing to the quick redox capability among the Ce(III) and Ce(IV) as well as remarkable metal–support interaction impacts (Ge, Zang, & Chen, 2015; Lu et al., 2011; Xu & Wang, 2012).

Ceria has many distinctive and interesting properties such as the specific ability to absorb ultraviolet radiation (Malleshappa et al., 2015; Tsunekawa, Fukuda, & Kasuya, 2000), proper hardness index (Ji et al., 2018; Li et al., 2012), good stability at high temperatures, optical properties (Chiu & Lai, 2010) as well as its reactivity. It has caused a great deal of attention in recent decades (Seal et al., 2020). Since the nanostructured ceria displays improved electronic, optical, and structural characteristics owing to the increment in surface-to-volume ratio as well as the effect of quantum confinement, its study has received much attention (Marques et al., 2017).

Various advantages such as excellent chemical stability, good abundance, high oxygen storage capacity and proper oxygen mobility, less toxicity, appropriate cost, and distinctive redox property (Han, Kim, Yoon, & Lee, 2011; Jiang et al., 2012; Kang, Sun, & Li, 2012; Soykal, Sohn, & Ozkan, 2012) have made CeO_2 widely utilized in the various fields including polishing agents (Feng et al., 2006), solid oxide fuel cells (Hassanzadeh-Tabrizi, Mazaheri, Aminzare, & Sadrnezhaad, 2010;

Logothetidis, Patsalas, & Charitidis, 2003; Sangsefidi, Salavati-Niasari, Mazaheri, & Sabet, 2017; Yahiro, Baba, Eguchi, & Arai, 1988), photocatalytic processes (Bamwenda & Arakawa, 2000; Tian et al., 2013), ultraviolet absorbers (Fang & Liang, 2012; Yamashita et al., 2002), optoelectronics (Tago, Tashiro, Hashimoto, Wakabayashi, & Kishida, 2003; Wang et al., 2011), redox catalytic reactions (Trovarelli, Boaro, Rocchini, de Leitenburg, & Dolcetti, 2001), NO removal (Di Monte, Fornasiero, Graziani, & Kašpar, 1998), gas sensors (Garzon, Mukundan, & Brosha, 2000), biotechnology (Ji et al., 2012), and absorbent for water treatment (Zhang & Chen, 2016).

It has been shown that the nanostructured ceria has the oxygen storage capacity, and so can capture, store, and release oxygen from its surface. This capability will be useful for oxidation/reduction reactions that are key to various applications (Kaneko et al., 2007). For example, the use of nanostructured ceria to release oxygen to catalyze the conversion of CO to CO_2, as well as the simultaneous reduction of harmful NOx emissions, has been reported in clean-air technologies (Wang et al., 2018). Besides, the mobility of oxygen within the nanostructured cerium dioxide is made possible through diffusion utilizing oxygen extraction and the creation of vacancies (Frayret et al., 2010). This could be owing to the alteration in the oxidation state of Ce between $+3$ and $+4$, in various reducing or oxidizing conditions. For example, it has been reported that the catalytic performance of cerium dioxide structures may depend on their capability to adsorb oxygen species on their surface, as well as the easy extraction of lattice oxygen that creates vacancies (Kopia, Kowalski, Chmielowska, & Leroux, 2008; Liang, Xiao, Chen, & Li, 2010; Valechha et al., 2011). High ionic conductivity has also been observed to be very suitable for usage in fuel cells (Coduri, Checchia, Longhi, Ceresoli, & Scavini, 2018).

Strong redox capability, long-term stability, nontoxicity, and low cost are some of the significant reasons that make nanostructured ceria a suitable photocatalyst for environmental uses (Nair, Wachtel, Lubomirsky, Fleig, & Maier, 2003).

The nanostructured cerium dioxide has been examined as free-radical scavenger for its capability in providing protection against biological, chemical, and radiological insults (Brunner et al., 2006; Schubert, Dargusch, Raitano, & Chan, 2006).

As the particle size reduces, there is usually a high density of interfaces in the nanocrystalline structures. The energy for defect creation may be significantly diminished in nanocrystalline oxides, causing to a significant increment in electronic carrier production and levels of nonstoichiometry (Tuller, 2000). So, cerium dioxide with nanostructure compared to bulk materials, owing to its better redox characteristics and transport properties, as well as higher surface-to-volume ratio, has attracted the interest of many researchers (Sun et al., 2012).

It has been shown that the rare earth ions-activated oxide powders can be significant materials in various fields such as lighting, scintillator systems, display, and biosensors (Höppe, 2009; Wu et al., 2014). However, the utilization of cerium dioxide nanostructure in high-performance luminescence systems as well as biomedical fields is very limited owing to its low or weak luminescence properties. Therefore improving the optical emission of nanostructured cerium dioxide seems to be essential for their potential use in solid-state lighting and other areas (He, Su, Lanhong, & Shi, 2015). One of the effective solutions proposed for this purpose is doping with rare earth ions. For example, Wang's et al. fabricated (Eu^{3+}, Sm^{3+}, or Tb^{3+})-doped CeO_2 nanostructures via a nonhydrolytic solution way and examined their photoluminescent features (Zhenling, 2007). They found that the characteristic emissions of lanthanide ions in cerium dioxide nanostructure were not seen. The emission band around 500 nm was observed, which its intensity was gradually enhanced with the increment of oxygen vacancy concentration. Also, they observed that the absorption spectrum of the nanostructured $CeO_2:Eu^{3+}$ exhibited a red shift as compared with that of net CeO_2, implying that the blocking range for UV is boosted through adding lanthanide ions.

2.1.2 Rare earth oxides (R_2O_3)

Rare earth oxides (Re_2O_3) are well known and the most stable rare earth substances. They contain rare earth trivalent ions (Xu, Gao, & Liu, 2002). R_2O_3 compounds exhibit three phases at ambient conditions, namely, the hexagonal phase, cubic phase, and monoclinic phase. It has been reported that the most stable phase for each rare earth elements can alter along with the cationic radii. The hexagonal phase can be most stable for larger ions, the cubic phase can be most stable for smaller ions, and the monoclinic phase can be most stable for some medium ones (Yan et al., 2011). Use of rare earth oxides has been reported as promoter in catalytic reaction. They can cause an enhanced dispersion of supported metallic nanostructures as well as incorporation of nano-phase oxidized species into the catalyst (Huang, Chang, & Yeh, 2006). For example, adding La_2O_3 into Ni catalyst fabricated a more uniform distribution of Ni ion in carbon support; however, it exhibited inhibitor influence upon the carbon oxidation reactivity (Zhou, He, Wang, Chen, & Sun, 2009).

These types of oxides have distinctive chemical, optical, and electronic features owing to their 4f electrons. The use of this type of significant oxides has been reported in the fields of luminescence devices, biochemical probes, optical transmission, and medical studies (Tang et al., 2003).

Rare earth oxides can be advantageous for luminescent phosphors. Also, Re_2O_3 compounds have very desirable thermal, electronic, and optical features as well as relatively low cut-off phonon energy (around 600 cm^{-1}) (Barrera et al., 2011; Guyot et al., 2016; Meza et al., 2014; Xiao, Li, Jia, & Zhang, 2009). They can be appropriate to serve as host matrices for upconversion nanostructures and can be utilized for the optical high-temperature sensing. Besides, Re_2O_3 compounds can be beneficial in remarkable fields like dielectric formulation for multilayer ceramic capacitors and catalytic processes (Kishi, Mizuno, & Chazono, 2003; Marques et al., 2017).

Upconversion (UC) luminescent compounds are those excited through a long wavelength light source and can exhibit a short wavelength emission. The rare earth-based UC compounds, such as those codoped by Yb^{3+} and Er^{3+}, denote low cytotoxicity and desirable photo stability, which are beneficial candidates for medical uses (Morgan & Mitchell, 2007). A range of tunable colors can be emitted by the upconversion luminescent rare-earth compounds (Glaspell, Anderson, Wilkins, & El-Shall, 2008). Glaspell et al. prepared UC Y_2O_3 nanostructures by utilizing laser vaporization/condensation way. Yttrium(III) oxide nanostructures doped by some rare-earth ions (Yb^{3+}, Er^{3+}, Ho^{3+}, and Tm^{3+}) were produced and could emit various lights under 980 nm.

It has been shown that the surface modification of luminescent rare earth nanostructures not only can cause the nanostructures to be dispersible in proper solvents as colloidal solutions, but also alter the surface chemistry and assist to improve the luminescence yield remarkably. Peng et al. utilized a silane coupling agent of the dimethyldichlorosilane for the surface modification of Nd_2O_3 nanostructures (Yu et al., 2007). They found that the modified nanostructures display appropriate fluorescent features and could be interspersed as a liquid medium.

Fluorescent labeling of molecules can be significant for biological studies. It has been shown that rare earth oxide-based nanostructures can provide an appealing route for bioimaging probes. For example, Kennedy et al. utilized Gd_2O_3:Eu nanostructures to visualize some proteins (Dosev et al., 2005). They coated the nanostructures by avidin via physical adsorption. Then, they could observe the protein linked to biotin owing to the affinity of biotin and avidin.

2.2 Fabrication methods

The two main stages of nucleation and growth take place in all approaches of preparing nanostructures (Beshkar, Zinatloo-Ajabshir, & Salavati-Niasari, 2015; Burda, Chen, Narayanan, & El-Sayed, 2005). Since the method of preparation of each compound can greatly affect the morphology of the particles and their size, it is necessary to choose a very suitable method that can have good control over these two substantial factors. It is generally accepted that the morphology of particles and their size can be highly influential in determining the properties and performance of nanostructures. Thus many efforts have been made to fabricate cerium dioxide and Re_2O_3 nanostructures with the aim of controlling the shape of particles and their dimensions (Han, Li, & Wang, 2012). In this part of this chapter of the book, the methods utilized for production are discussed and also their advantages and disadvantages will be mentioned.

Solid-state reaction strategy for making cerium dioxide and Re_2O_3 nanostructures mainly uses high temperature for a long time, which is very costly (Abu-Zied & Soliman, 2008; Mortazavi-Derazkola, Zinatloo-Ajabshir, & Salavati-Niasari, 2015a). On the other hand, heterogeneous product with very large particle size is produced frequently. Lu and Fang (2006) prepared CeO_2 particles with near-spherical shape through a solid-state mechanochemical way. They utilized an organic base and produced metal-ion-free cerium dioxide nanostructure. They found that altering the annealing temperature could have a key impact on the mean particle sizes and lattice distortion (Lu & Fang, 2006). Mortazavi-Derazkola, Zinatloo-

Ajabshir, & Salavati-Niasari (2015b) synthesized Ho_2O_3 ceramic nanostructures via heat treatment of the precursor in air at 600°C within 300 minutes. They utilized [Ho $L(NO_3)_2]NO_3$ (L = Schiff base ligand) as precursor, which was formed through a solvent-free solid–solid reaction from various molar ratios of Schiff base ligand and holmium nitrate. They observed that the molar ratio of holmium source and Schiff base ligand could possess a substantial effect upon the shape and particle size of the Ho_2O_3 (see Fig. 2.1). It was demonstrated that huge ligands with appropriate steric hindrance (like Schiff base ligand) may be employed instead of usual capping agents for modification of the shape and particle size (Mortazavi-Derazkola, Zinatloo-Ajabshir, & Salavati-Niasari, 2015b).

In other work, Mortazavi-Derazkola et al. utilized sodium dodecyl sulfate (SDS) (Mortazavi-Derazkola, Zinatloo-Ajabshir, & Salavati-Niasari, 2015c) to control shape and size of Nd_2O_3 through a solvent-free solid-state reaction. They employed $[NdL(NO_3)_2]$ NO_3 as a novel precursor. Their results displayed that by utilizing various amounts of SDS, diverse oxide structures (see Fig. 2.2) can be prepared and the dose of SDS is very efficient in determining the shape and size of Nd_2O_3 (Mortazavi-Derazkola, Zinatloo-Ajabshir, & Salavati-Niasari, 2015c). Then, the formed powders were heated and Nd_2O_3 samples were fabricated. Besides, they observed that the temperature for thermal treatment could have a noteworthy impact

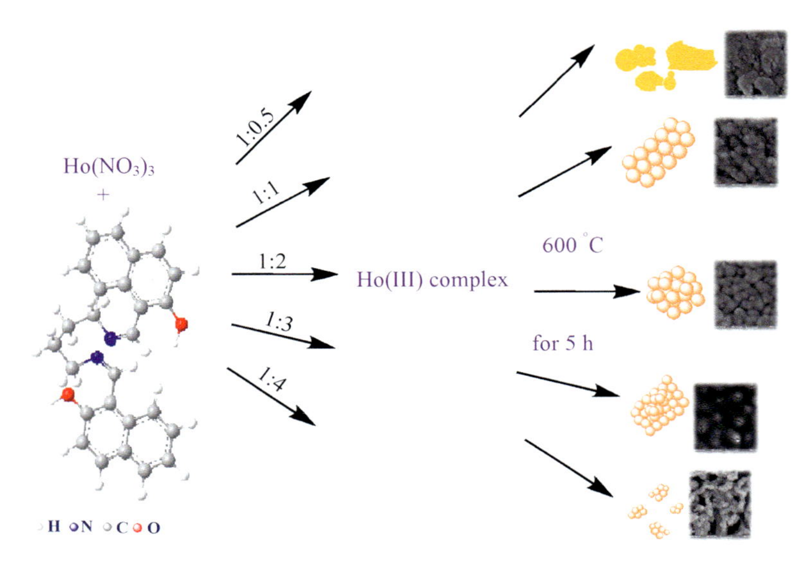

Figure 2.1 Schematic diagram illustrating the preparation of Ho_2O_3 samples by different molar ratios of $Ho(NO_3)_3 \cdot 6H_2O$ to (L) (Mortazavi-Derazkola, Zinatloo-Ajabshir, & Salavati-Niasari, 2015b).
Source: Reprinted with permission from Mortazavi-Derazkola, S., Zinatloo-Ajabshir, S., Salavati-Niasari, M. (2015b). Novel simple solvent-less preparation, characterization and degradation of the cationic dye over holmium oxide ceramic nanostructures. *Ceramics International*, *41*, 9593–9601, Copyright 2015 Elsevier.

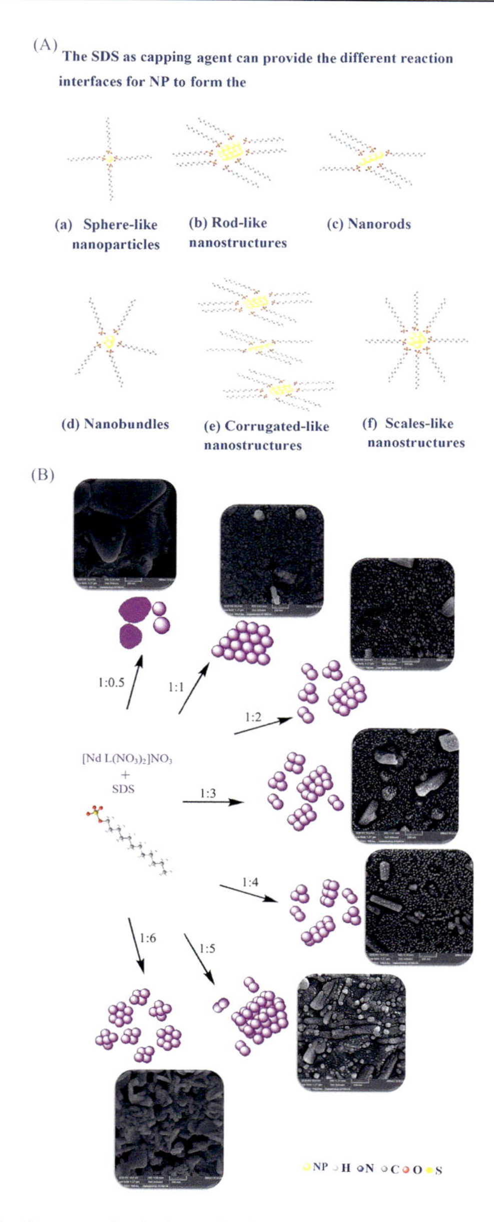

Figure 2.2 Schematic diagram depicting (A) the role of the amount of SDS on the reaction interfaces for neodymium precursor and (B) the production of neodymium oxide samples at various conditions (Mortazavi-Derazkola, Zinatloo-Ajabshir, & Salavati-Niasari, 2015c). *Source*: Reproduced from Mortazavi-Derazkola, S., Zinatloo-Ajabshir, S., Salavati-Niasari, M. (2015c). New sodium dodecyl sulfate-assisted preparation of Nd_2O_3 nanostructures via a simple route. *RSC Advances*, *5*, 56666–56676 with permission from The Royal Society of Chemistry.

on the purity of the nanostructured sample. 900°C was the proper temperature to achieve pure hexagonal Nd_2O_3.

The hydrothermal method has been reported as a simple and environmentally friendly approach for a wide range of oxide nanostructures (Beshkar, Zinatloo-Ajabshir, Bagheri, & Salavati-Niasari, 2017; Razi, Zinatloo-Ajabshir, & Salavati-Niasari, 2017b; Zhang et al., 2009). A stainless steel autoclave is a significant tool for preparing nanostructures utilizing the hydrothermal way (Sun et al., 2013; Tok, Boey, Dong, & Sun, 2007; Zhao et al., 2013). The dissolution/recrystallization mechanism for the hydrothermal process is generally accepted (Hirano & Kato, 1996). It has been reported that temperature and time for hydrothermal reaction along with the concentration of reactants and acidity as well as the utilization of capping agents can be very effective and key in regulating the shape, particle size, and purity of the resulting nanostructure (Hirano & Kato, 1996; Lin et al., 2011; Yan, Yu, Chen, & Xing, 2008). Guo, Du, Li, and Cui (2006) prepared $Ce(OH)CO_3$ triangular microplates as precursor through a hydrothermal way employing carbamide, cerium(III) nitrate, and CTAB as cationic capping agent at 150°C. The microplates had hexagonal phase. Then, the cerium dioxide triangular microplates were produced by heating the precursor at 650°C within 420 minutes.

In other work, Sun et al. (2006) fabricated cerium(IV) oxide microspheres with flowerlike shape under hydrothermal condition as well as subsequent calcination. The prepared microsphere consisted of nanosheets with a thickness of 20–30 nm. It possesses an open three-dimensional structure and therefore had a very proper large surface area, which can be very beneficial as a support for catalytic uses. Mortazavi-Derazkola, Zinatloo-Ajabshir, and Salavati-Niasari (2017b) prepared Ho $(OH)_3$ nanostructures by hydrothermal way utilizing tetraethylenepentamine, Schiff base compound as capping agent, and holmium nitrate. Then, the created $Ho(OH)_3$ nanostructure was heated in 600°C within 4 hours, and pure Ho_2O_3 nanostructure was produced (see Fig. 2.3). Hu et al. produced Lu_2O_3:Yb^{3+}/Er^{3+}/Tm^{3+} nanocubes with good uniformity via a hydrothermal way with subsequent calcination (Hu, Yang, Li, & Lin, 2012). They proposed the mechanism of forming the cubic structure as an anisotropic growth along three directions with identical velocity (Hu et al., 2012).

The precipitation approach using cerium source and alkaline agent as precursors is accepted as a simple way that has advantages such as low cost, easy control, and gentle synthesis conditions (Chang & Chen, 2005). Yao, Wang, Luo, and Size-Controllable (2017) prepared cerium dioxide nanoparticles with mean sizes of 7–12 nm utilizing a membrane dispersion microreactor. They employed cerium nitrate and NH_3 solution as starting materials. Rajan, Vilas, Rajan, John, and Philip (2020) employed oxalic acid ($C_2H_2O_4$) and polyethylene glycol (PEG) for the fabrication of nanostructured CeO_2 via a precipitation way. The resulting polymer-capped nanoparticles had a mean particle size of about 15 nm. The produced cerium dioxide nanostructure exhibited remarkable bactericidal performance against *Escherichia coli* and *Staphylococcus aureus* owing to the direct interaction of cerium dioxide nanostructure on the membrane surface of the microorganisms (Rajan et al., 2020). The utilization of urea (Gu & Soucek, 2007) or

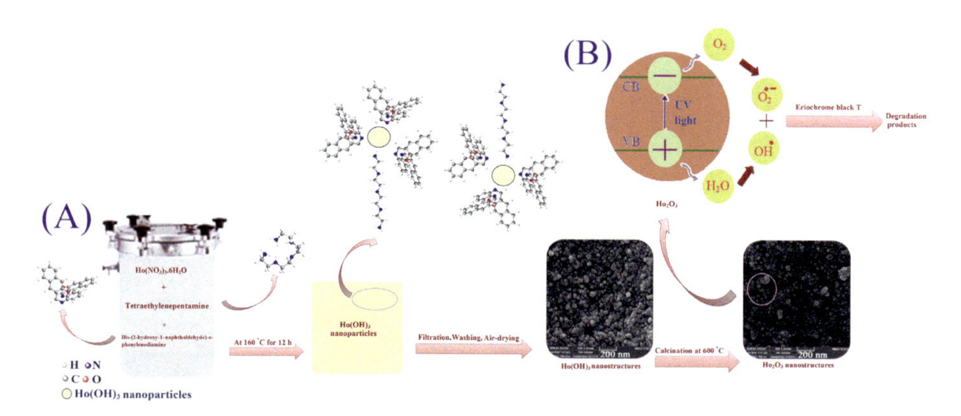

Figure 2.3 Schematic diagram of the fabrication of $Ho(OH)_3$ and Ho_2O_3 nanostructures (A) and reaction mechanism of eriochrome black T photodegradation over Ho_2O_3 nanostructures under UV light (B) (Mortazavi-Derazkola et al., 2017b).
Source: Reprinted with permission from Mortazavi-Derazkola, S., Zinatloo-Ajabshir, S., Salavati-Niasari, M. (2017b). Facile hydrothermal and novel preparation of nanostructured Ho_2O_3 for photodegradation of eriochrome black T dye as water pollutant. *Advanced Powder Technology*, *28* 747–754, Copyright 2017 Elsevier.

hexamethylenetetramine (Chen & Chen, 1993) ($C_6H_{12}N_4$) to prepare nanostructured oxides through precipitation route has also been reported.

In the precipitation way, various capping agents have been utilized to produce nanostructured oxides, both to regulate the particle size and their shape (Hassanzadeh-Tabrizi et al., 2010; Razi, Zinatloo-Ajabshir, & Salavati-Niasari, 2017a; Zinatloo-Ajabshir & Salavati-Niasari, 2015). For example, Zinatloo-Ajabshir, Mortazavi-Derazkola, and Salavati-Niasari (2017e) fabricated the nanostructured holmium oxide by a precipitation approach employing triethylenetetramine (precipitant). They examined the role of various capping agents upon the grain size, shape of holmium oxide sample (see Fig. 2.4). They observed that CTAB with proper steric hindrance influence could act as the appropriate capping agent, and control growth of nanoparticles. The results displayed that without the capping agent, nonhomogeneous nanostructures were produced.

Another approach for the preparation of nanostructured oxides is the sol–gel way, the advantages of which include achieving products with controlled stoichiometry as well as good purity (Yuan et al., 2007). Ye et al. (2020) fabricated Ni/CeO_2 nanocatalyst via an easy sol–gel way in the presence of citric acid as binding ligands, and employed for CO_2 methanation. The prepared nanocatalyst could be advantageous for CO_2 methanation at low temperature (Ye et al., 2020). Xiao et al. (2009) prepared Sm_2O_3, Dy_2O_3, and Gd_2O_3 nanostructures through a simple benzyl alcohol-assisted sol–gel way followed by calcination. They employed the rare earth acetylacetonates as metal sources. The prepared Sm_2O_3 (Xiao et al., 2009) showed the stacks of ultrathin nanodisks, and the formed Gd_2O_3 and Dy_2O_3 displayed the hierarchical microspheres. Zhao (Zhao et al., 2013) synthesized Eu^{3+} doped Lu_2O_3

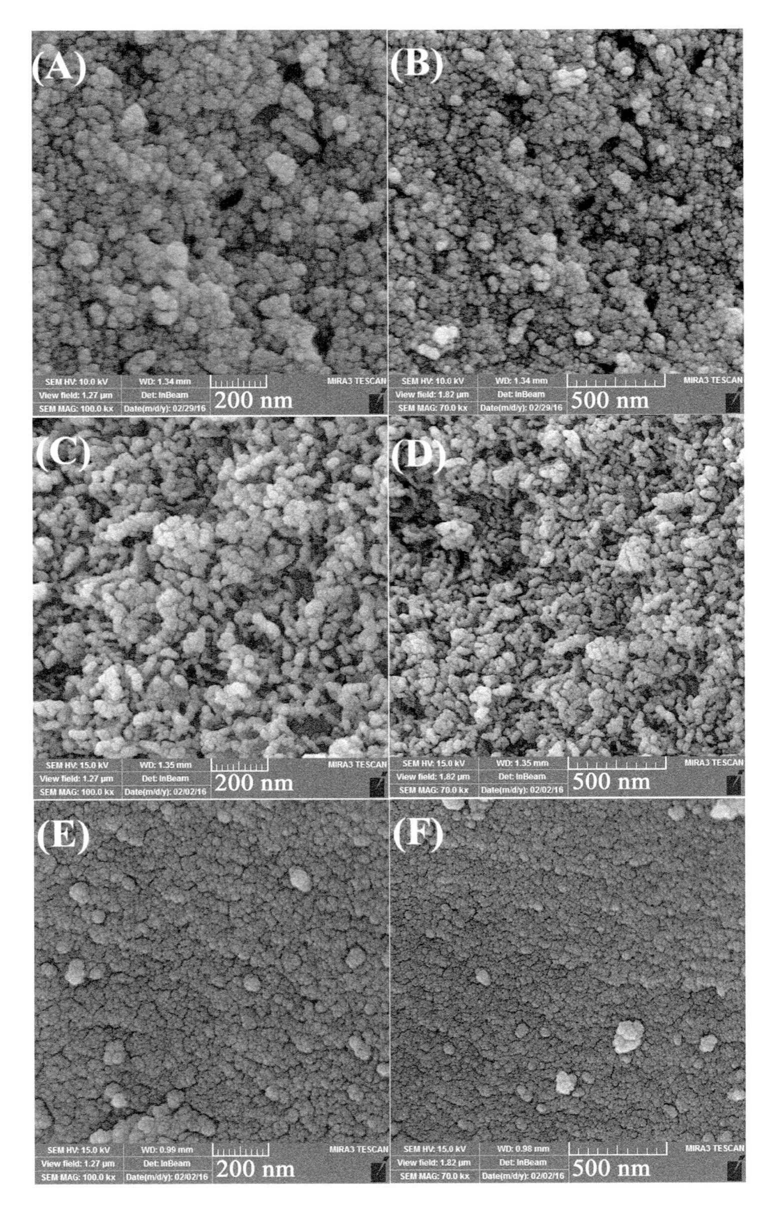

Figure 2.4 FESEM images of the holmium oxide samples prepared by applying Schiff base compound (A and B), SDBS (C and D) and CTAB (E and F) (Zinatloo-Ajabshir et al., 2017e).

Source: Reprinted with permission from Zinatloo-Ajabshir, S., Mortazavi-Derazkola, S., Salavati-Niasari, M. (2017e). Preparation, characterization and photocatalytic degradation of methyl violet pollutant of holmium oxide nanostructures prepared through a facile precipitation method. *Journal of Molecular Liquids*, *231*, 306–313, Copyright 2017 Elsevier.

samples via a surfactant-free hydrothermal route. They observed that the pH value could hs a remarkable impact on structural and luminescent features.

The combustion approach, which is accepted as a fast, low-cost, and versatile way for producing a wide range of nanostructures, actually involves a sustained reaction in a homogeneous mixture with the content of various fuels and oxidants in proper molar ratio (Aruna & Mukasyan, 2008; Zinatloo-Ajabshir, Salehi, Amiri, & Salavati-Niasari, 2019b; Zinatloo-Ajabshir, Morassaei, Amiri, Salavati-Niasari, & Foong, 2020). The advantages of this fast approach are that it is performed at a relatively low temperature and also the possibility of achieving uniform doping of low doses of dopants in a single step can be mentioned. Jamshidijam, Mangalaraja, Akbari-Fakhrabadi, Ananthakumar, and Chan (2014) synthesized rare earth (Re)-doped cerium dioxide structures through a citric acid–nitrate combustion way. They examined the role of dopant upon the structural features of nanostructured ceria.

In the modified combustion way, various capping agents have been employed to fabricate nanostructured metal oxides (Wu et al., 2014). Wu et al. (2014) produced Sm^{3+}-activated cerium dioxide nano/submicroscale structures through a combustion reaction in the presence of polyvinyl alcohol. They observed that utilization of polyvinyl alcohol could cause the formation of particles with good uniformity. The prepared structures could be efficaciously excited by 370 nm. They emitted strong orange-red light after optimization, which was proper for the high-performance UV LEDs.

The sonochemical way, which has been accepted as a low-cost and fast approach for preparing metal oxide nanostructures, is based on creating the chemical effects of ultrasound waves in a solution comprising reactants called acoustic cavitation (Mousavi-Kamazani & Azizi, 2019; Prasad, Pinjari, Pandit, & Mhaske, 2010; Zinatloo-Ajabshir & Salavati-Niasari, 2014). Explosive collapse of bubbles leads to large amounts of temperature and pressure, which can be a very effective driving force in performing chemical reactions to create the target nanostructure (Mousavi-Kamazani & Ashrafi, 2020; Mousavi-Kamazani, Zinatloo-Ajabshir, & Ghodrati, 2020; Pinjari & Pandit, 2010). Pinjari and Pandit (2011) synthesized cerium dioxide nanostructure via a sonochemically assisted precipitation way. They utilized cerium nitrate and NaOH as staring agents without any surfactants. The prepared sample had a crystal size of about 30 nm (Pinjari & Pandit, 2011). Their results displayed that the sonochemical way may be introduced as an approach that saves energy and time. Zinatloo-Ajabshir, Mortazavi-Derazkola, and Salavati-Niasari (2017f) produced the nanostructured Nd_2O_3 samples via a sonochemical process in the presence of 2,2-dimethyl-1,3 propanediamine as a new precipitant. The prepared sample consisted of homogeneous spherical nanoparticles (see Fig. 2.5). They observed that irregular microstructures/bulk structures were prepared without the utilization of sonication, and by magnetic stirring for half an hour (longer time) (Zinatloo-Ajabshir et al., 2017f). Their results displayed that the usage of ultrasound waves can play a very meaningful role in regulating growth and thus in achieving homogeneous oxide nanoparticles.

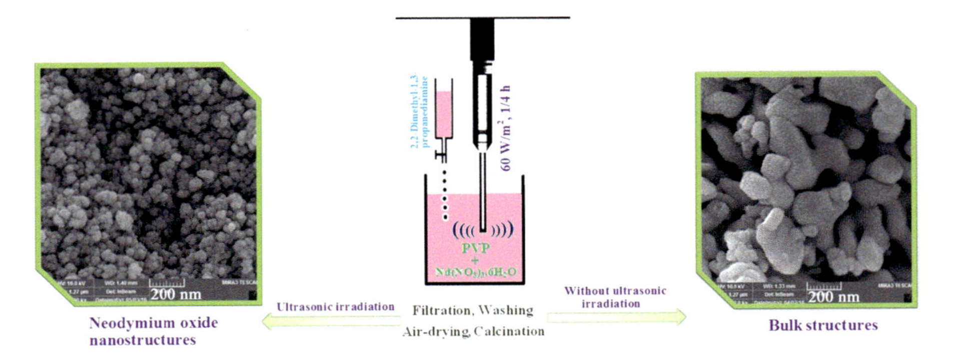

Figure 2.5 Schematic diagram of preparation of neodymium oxide nanostructures and role of utilizing of ultrasonic irradiation on the shape and particle size (Zinatloo-Ajabshir et al., 2017f).
Source: Reprinted with permission from Zinatloo-Ajabshir, S., Mortazavi-Derazkola, S., Salavati-Niasari, M. (2017). Sonochemical synthesis, characterization and photodegradation of organic pollutant over Nd_2O_3 nanostructures prepared via a new simple route. *Separation and Purification Technology*, *178*, 138–146, Copyright 2017 Elsevier.

The utilization of various capping agents as a potential solution to regulate grain size and morphology in the fabrication of nanostructured metal oxides through sonochemistry has also been studied (Karami, Ghanbari, Amiri, Ghiyasiyan-Arani, & Salavati-Niasari, 2021; Monsef, Ghiyasiyan-Arani, & Salavati-Niasari, 2021; Zinatloo-Ajabshir, Baladi, & Salavati-Niasari, 2021). Zinatloo-Ajabshir et al. (2017f) utilized three various types of capping agents to prepare neodymium oxide samples (see Fig. 2.6). They observed that all three capping agents, with their different steric inhibitory effects, could be adsorbed on the primary nuclei, preventing particles from extra aggregation (Zinatloo-Ajabshir et al., 2017f). They also cause the formation of various shapes of neodymium oxide by inducing growth in specific directions. Among the capping agents, Polyvinylpyrrolidone (PVP) is more effective and thus leads to the creation of spherical oxide nanoparticles with good uniformity (Zinatloo-Ajabshir et al., 2017f). Wang et al. (2002) prepared nanocrystalline cerium dioxide structures by a sonochemical way. They employed hexamine, ceric ammonium nitrate, and poly (ethylene glycol)-19000. The prepared samples consisted of homogeneous particles with a narrow size distribution (Wang et al., 2002).

The microemulsion way has also been utilized to fabricate metal oxide nanostructures (Gu & Soucek, 2007). Although this method has been accepted as a useful approach for the preparation of monodispersed metal oxide particles (He, Yang, & Cheng, 2003; Masui et al., 1997), its utilization on a commercial scale is limited because it is very expensive, on the other hand owing to the use of organic solvents, it is not biocompatible. Sathyamurthy, Leonard, Dabestani, and Paranthaman (2005) produced the nanostructured ceria via a microemulsion way. They utilized *n*-octane (C_8H_{18}) as the oil phase, NaOH as the alkaline agent, cerium nitrate as the metal source, and CTAB as capping agent in the presence of 1-butanol (C_4H_9OH).

Figure 2.6 FESEM images of neodymium oxide samples prepared with the aid of Schiff base structure (A and B), Titriplex (C and D) and PVP (E and F) (Zinatloo-Ajabshir et al., 2017f).
Source: Reprinted with permission from Zinatloo-Ajabshir, S., Mortazavi-Derazkola, S., Salavati-Niasari, M. (2017f). Sonochemical synthesis, characterization and photodegradation of organic pollutant over Nd_2O_3 nanostructures prepared via a new simple route. *Separation and Purification Technology*, *178*, 138–146, Copyright 2017 Elsevier.

The prepared sample had particles with a mean size of about 3.7 nm (Sathyamurthy et al., 2005), and the particles formed had a polyhedral shape.

2.3 Applications

This section describes the applications of cerium dioxide and rare earth oxides (R_2O_3) nanomaterials nanostructures in various fields. It has been reported that cerium dioxide nanostructure can be very beneficial as a support or additive in heterogeneous catalytic processes because this oxide compound has extraordinary features such as good thermal structural stability, proper catalytic efficiency, and desirable chemical selectivity (Ma et al., 2018). It has been reported that the reversible Ce^{3+}/Ce^{4+} redox pair as well as the surface acid–base features can be reasons for the highly desirable intrinsic catalytic ability of cerium dioxide as well as evidence of the improved catalytic phenomenon in a wide range of reactions (Ma et al., 2018).

In particular, the cerium dioxide nanostructure contains a large number of surface-bound defects such as oxygen vacancies, which can be advantageous as active catalytic sites on the surface to perform various reactions (Ma et al., 2018). The utilization of cerium dioxide for CO oxidation has been reported as a very desirable candidate due to its special properties (Chang et al., 2012; He et al., 2016; Liu, Wang, & Zhang, 2012; Song, Wang, & Zhang, 2015). The CO oxidation reaction on cerium dioxide can be performed according to the Mars–van Krevelen mechanism (Wu, Li, & Overbury, 2012). It is accepted that the catalytic performance of cerium dioxide for CO oxidation is highly dependent on the surface features. The nanocatalysts with great specific surface area, large amount of surface defects as well as appropriate mobility of lattice oxygen may be very advantageous for CO oxidation (Ma et al., 2018). The usage of nanostructured cerium dioxide, which due to its surface acid–base features, as well as its extraordinary ability to adsorb a variety of functional groups and carbon dioxide, has attracted the attention of many scientists (Tamura, Honda, Nakagawa, & Tomishige, 2014). This metal oxide can exhibit remarkable catalytic performance and good selectivity in reactions to produce a variety of organic compounds utilizing carbon dioxide (Tamura et al., 2014).

Numerous studies have been performed on the utilization of multifunctional cerium dioxide compound in various organic transformations, including oxidation, substitution, and so on (Liu, Luo, Gao, & Huang, 2016; Vivier & Duprez, 2010; You et al., 2017; Zhang, You, Li, Cao, & Huang, 2017). This oxide catalyst is also considered as an appropriate option in this type of catalytic reactions owing to its unique features.

The usage of cerium dioxide nanostructures to mimic several natural enzymes such as catalase, superoxide dismutate (SOD), and oxidase (Heckert, Karakoti, Seal, & Self, 2008; Heckman et al., 2013; Karakoti, Singh, Dowding, Seal, & Self, 2010; Das et al., 2013; Tian et al., 2015; Xu & Qu, 2014) has been widely studied. Due to its multiple enzymatic features, this nanostructured metal oxide can be very advantageous for a variety of applications, including biosensing, clinical diagnosis, drug delivery, and antioxidants (Celardo, Pedersen, Traversa, & Ghibelli, 2011; Kim et al., 2012; Lee et al., 2013; Pagliari et al., 2012; Wu et al., 2017; Zhang et al., 2016). One of the reasons for the usage of cerium dioxide nanostructure as synthetic nanoenzyme can be the reversible mutation of its surface in oxidation states between Ce^{3+} and Ce^{4+} (Heckert et al., 2008; Karakoti et al., 2010; Korsvik, Patil, Seal, & Self, 2007), as well as its biocompatibility properties and nontoxic nature (Pierscionek et al., 2009; Xia et al., 2008). It is reported that surface defects of cerium dioxide can normally act as catalytic active sites. Thus regulating its surface by various solutions can have a remarkable effect on the catalytic efficiency of cerium dioxide nanocatalysts as synthetic enzymes (Heckert et al., 2008).

Nesakumar, Sethuraman, Krishnan, and Rayappan (2013) synthesized the nanostructured cerium dioxide with a particle size of about 30 nm and utilized it in the preparation of an electrochemical biosensor to determine lactate quantity. Their results indicated that the formed biosensor has appropriate sensitivity and fast response time, and thus can be employed as a suitable candidate in the lactate detection system (Nesakumar et al., 2013).

Alili et al. (2013) examined the role of nanostructured cerium dioxide with a particle size of about 5 nm upon human melanoma cells in vivo. They observed that quantities of polymer-capped nanoparticles being harmless for stromal cells exhibited a cytotoxic and antiinvasive activity upon melanoma cells (Alili et al., 2013). The results displayed that tumor weight and volume could be reduced after treatment with cerium dioxide nanostructure. Thus the nanostructured cerium dioxide can be advantageous in the remedy and prevention of cancers (Alili et al., 2013).

Studies have demonstrated that cerium dioxide and rare earth oxides as nanocatalysts can play a very advantageous role in performing some photoredox reactions, including in the photodecomposition of organic pollution or water splitting or selective oxidation (Xie et al., 2017). Features of nanocatalysts with outstanding performance include excellent ability to absorb visible light, great catalytic active sites, and high isolation of photogenerated carriers (Xie et al., 2017).

The introduction of technologies with sustainable development and excellent performance as a beneficial solution to eliminate environmental pollution has received much attention from scientists in recent years (Beshkar, Zinatloo-Ajabshir, & Salavati-Niasari, 2015; Dutta et al., 2020). One of these helpful solutions is advanced oxidation processes. In this process, large amounts of oxidizing agents ($OH\cdot$) can be produced. With visible light, the electrons in the nanophotocatalyst valance band can be transferred to its conduction band (CB), resulting in electron–hole pairs. The holes in the nanocatalyst valence band can decompose H_2O to $OH\cdot$. On the other hand, electrons can react with O_2 molecules in the nanocatalyst CB to form superoxide anion ($O_2^{-}\cdot$). To perform photocatalytic oxidative degradation reactions, three key oxidizing species have been reported, namely h^+, $O_2^{-}\cdot$, and $OH\cdot$, which can be useful in the elimination of organic pollution (Dutta et al., 2020; Singh, Raizada, Pathania, Kumar, & Thakur, 2013). The mechanism of photodegradation of organic pollutants may be as follows (Mortazavi-Derazkola, Zinatloo-Ajabshir, & Salavati-Niasari, 2015c) (see Fig. 2.7):

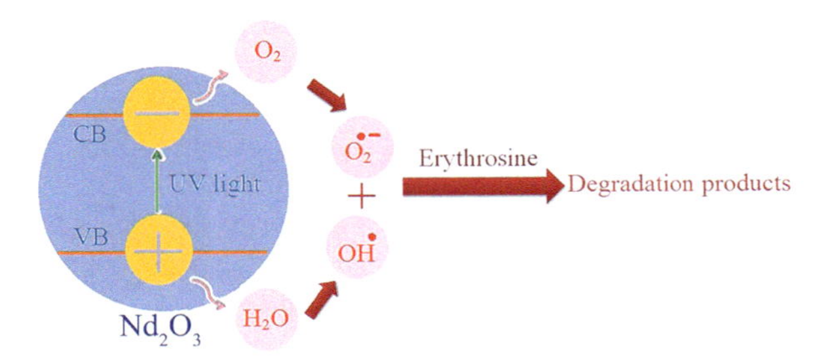

Figure 2.7 Reaction mechanism of organic pollutant photodegradation over Nd_2O_3 under UV light illumination (Mortazavi-Derazkola, Zinatloo-Ajabshir, & Salavati-Niasari, 2015c). *Source*: Reproduced from Mortazavi-Derazkola, S., Zinatloo-Ajabshir, S., Salavati-Niasari, M. (2015c). New sodium dodecyl sulfate-assisted preparation of Nd_2O_3 nanostructures via a simple route. *RSC Advances*, *5*, 56666–56676 with permission from The Royal Society of Chemistry.

$$\text{Nanocatalyst} + h\nu \rightarrow \text{Nanocatalyst}^* + e^- + h^+$$
$$h^+ + H_2O \rightarrow OH$$
$$e^- + O_2 \rightarrow \bullet O_2^-$$
$$OH\bullet \text{ or } O_2^- + \text{organic pollutants} \rightarrow \text{Degradation products}$$

Nanostructured cerium dioxide has attracted the attention of many scientists in recent years owing to its extraordinary and special features such as chemical inertness, biocompatibility, low cost and strong oxidation in wastewater treatment, and removal of pollutants through photodegradation (Bao, Zhang, Hua, & Huang, 2014; Dai, Wang, Yu, Chen, & Chen, 2014; Khan et al., 2014; Lin et al., 2012). It has been reported that the morphological features of cerium dioxide nanostructure, such as shape, dimension, and interface between particles, can greatly affect its photocatalytic performance. Thus many efforts have been made to fabricate samples of cerium dioxide with various morphologies (Chen, Cao, & Jia, 2011; Xu et al., 2011; Zhang et al., 2014). Sifontes, Rosales, and Méndez (2013) produced nanostructured cerium dioxide by utilizing chitosan as template. The formed cerium dioxide sample had large surface area, and thus displayed good photodegradation, so it was able to display appropriate performance in removal of Congo Red under visible light (Sifontes et al., 2013). Zhang et al. (2014) prepared cerium dioxide nanorods and nanowires via an electrochemical way. Both types of prepared samples exhibit much better performance in photodegradation of methyl orange owing to their hierarchical one-dimensional nanostructures as well as very suitable specific surface area compared to commercial cerium dioxide nanoparticles (Zhang et al., 2014). The results displayed that one-dimensional nanostructures could provide shorter transfer lengths as well as more efficient separation of photogenerated carriers compared to cerium dioxide nanoparticles.

The utilization of various kinds of cerium dioxide and rare earth oxide nanostructures has been reported to eliminate various pollutants (see Table 2.1).

Table 2.1 Cerium dioxide and rare earth oxide nanostructures for degradation of different pollutants.

Photocatalyst	Preparation way	Light source	Degradation rate (%)	Ref.
Ceria hierarchical nanorods	Electrochemical synthesis	White	98.2% of methyl orange after 180 min	Zhang et al. (2014)
CeO_2 nanoparticles	Solution combustion way	UV	About 100% of Trypan blue after 135 min	Ravishankar et al. (2015)
Ceria nanoparticles	Modified precipitation route	-	90% of acid orange 7 after 61.2 min	Yao et al. (2017)
CeO_2 nanoparticles	Bio-mediated way	Sun	80% of rhodamine-B after 90 min	Malleshappa et al. (2015)
Nd_2O_3 nanostructures	Solid-state route	UV	77% of rhodamine-B after 90 min	Mortazavi-Derazkola et al. (2015a)
Holmium oxide nanostructures	Solid-state reaction	UV	57% of rhodamine-B after 90 min	Mortazavi-Derazkola et al. (2015b)
Ho_2O_3/CNT nanocomposite	Solvotherml way	UV	98% of tetracycline after 60 min	Jiang et al. (2020)
La_2O_3/Mg-Al composites	Hydrothermal way	Xe lamp	99.87% of tetracycline hydrochloride after 110 min	Wu, Liu, Wen, Liu, and Zheng (2021)
Nd_2O_3 coupled tubular g-C_3N_4 composites	Solvent evaporation and high-temperature calcination way	Visible	32.32% of NO after 30 min	Tan et al. (2021)
Nd_2O_3 nanostructure	Hydrothermal way	UV	79% of eriochrome black T after 100 min	Zinatloo-Ajabshir, Mortazavi-Derazkola, and Salavati-Niasari (2017a)
Ho_2O_3 nanostructure	Sonochemical route	UV	92.3% of erythrosine after 80 min	Zinatloo-Ajabshir, Mortazavi-Derazkola, and Salavati-Niasari (2017b)

(*Continued*)

Table 2.1 (Continued)

Photocatalyst	Preparation way	Light source	Degradation rate (%)	Ref.
Nd_2O_3 nanostructure	Precipitation route	UV	89.6% of methylene blue after 90 min	Zinatloo-Ajabshir, Mortazavi-Derazkola, and Salavati-Niasari (2017c)
Au/CeO_2	Photodeposition route	Visible	56% of propylene after 60 min	Jiang, Wang, Sun, Zhang, and Zheng (2015)
CeO_2-MgAl LDH (OH^-)	Cerium impregnation way	UV	50% of phenol after 420 min	Valente, Tzompantzi, and Prince (2011)
Holmium oxide nanostructures	Precipitation route	UV	82.8% of methyl violet after 90 min	Zinatloo-Ajabshir et al. (2017e)
CeO_2/ZnO nanodisks	Chemical reaction process	Sun	93% of direct blue after 300 min	Lamba, Umar, Mehta, and Kansal (2015)
Ho_2O_3 nanostructure	Solid-state route	UV	90% of eosin Y after 120 min	Mortazavi-Derazkola, Zinatloo-Ajabshir, and Salavati-Niasari (2017a)
Nanostructured Nd_2O_3	Solid-state reaction	UV	78% of erythrosine after 120 min	Mortazavi-Derazkola et al. (2015c)
$BiVO_4/CeO_2$(6:4)	Coupling a precipitation way with hydrothermal method	Visible	80% of methylene blue after 30 min	Wetchakun et al. (2012)
Ho_2O_3-SiO_2 nanocomposites	Sonochemical way	UV	94.9% of methylene blue after 70 min	Zinatloo-Ajabshir, Mortazavi-Derazkola, and Salavati-Niasari (2017d)

Salavati-Niasari et al. fabricated the nanostructured Nd_2O_3 via a sonochemical route and employed to remove eosin Y contaminant (Zinatloo-Ajabshir et al., 2017f). They examined the role of the utilization of sonication as well as kind of capping agent on the photocatalytic behavior of neodymium oxide (see Fig. 2.8A and B) (Zinatloo-Ajabshir et al., 2017f). They observed that the neodymium oxide nanostructure made utilizing ultrasound as well as the PVP capping agent could

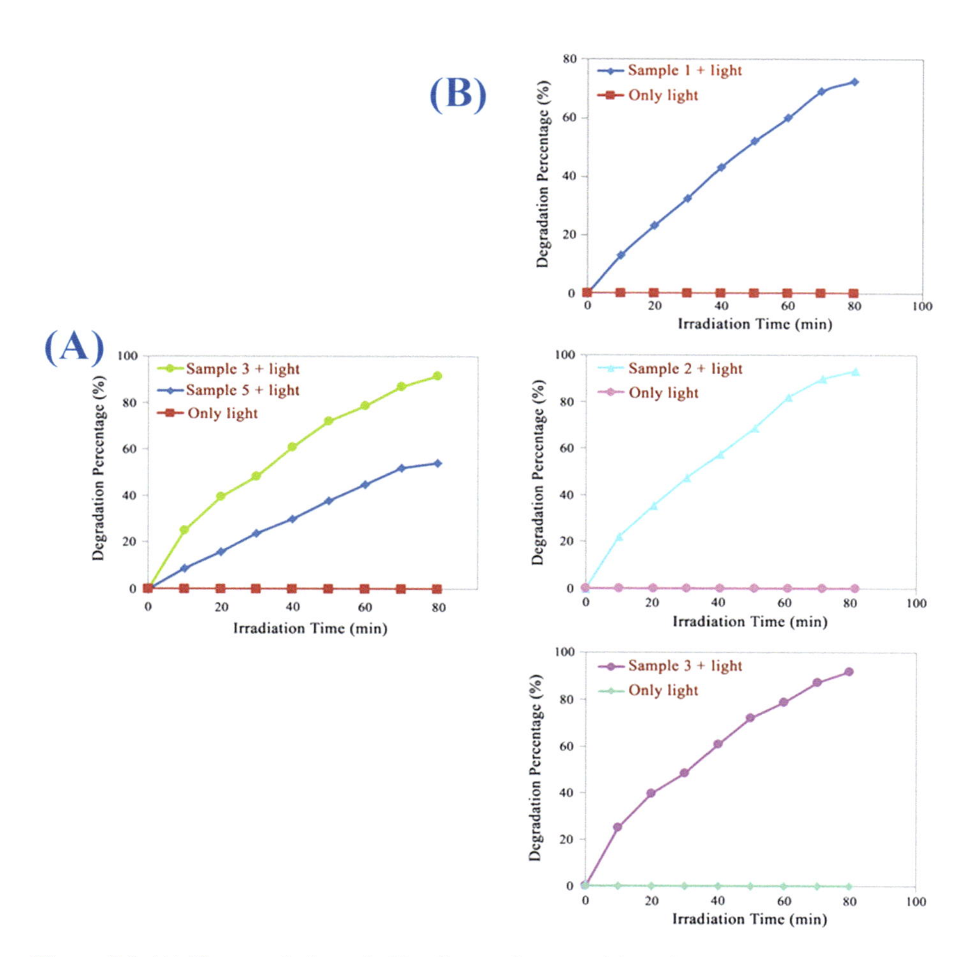

Figure 2.8 (A) Photocatalytic eosin Y pollutant decomposition of neodymium oxide structures produced with and without sonication (samples 3 and 5) and (B) Photocatalytic eosin Y pollutant decomposition of neodymium oxide structures prepared with the aid of Schiff base structure, Titriplex and PVP (samples 1–3) (Zinatloo-Ajabshir et al., 2017f). *Source*: Reprinted with permission from Zinatloo-Ajabshir, S., Mortazavi-Derazkola, S., Salavati-Niasari, M. (2017f). Sonochemical synthesis, characterization and photodegradation of organic pollutant over Nd_2O_3 nanostructures prepared via a new simple route. *Separation and Purification Technology*, *178*, 138–146, Copyright 2017 Elsevier.

display better photocatalytic performance than other samples because it had a smaller particle size than other samples (Zinatloo-Ajabshir et al., 2017f).

Salavati-Niasari et al. prepared Nd_2O_3-SiO_2 nanocomposites via a quick sonochemical way employing putrescine as a novel basic agent (Zinatloo-Ajabshir, Mortazavi-Derazkola, & Salavati-Niasari, 2018). They utilized the prepared nanocomposites to eliminate methyl violet contaminant and examined the role of

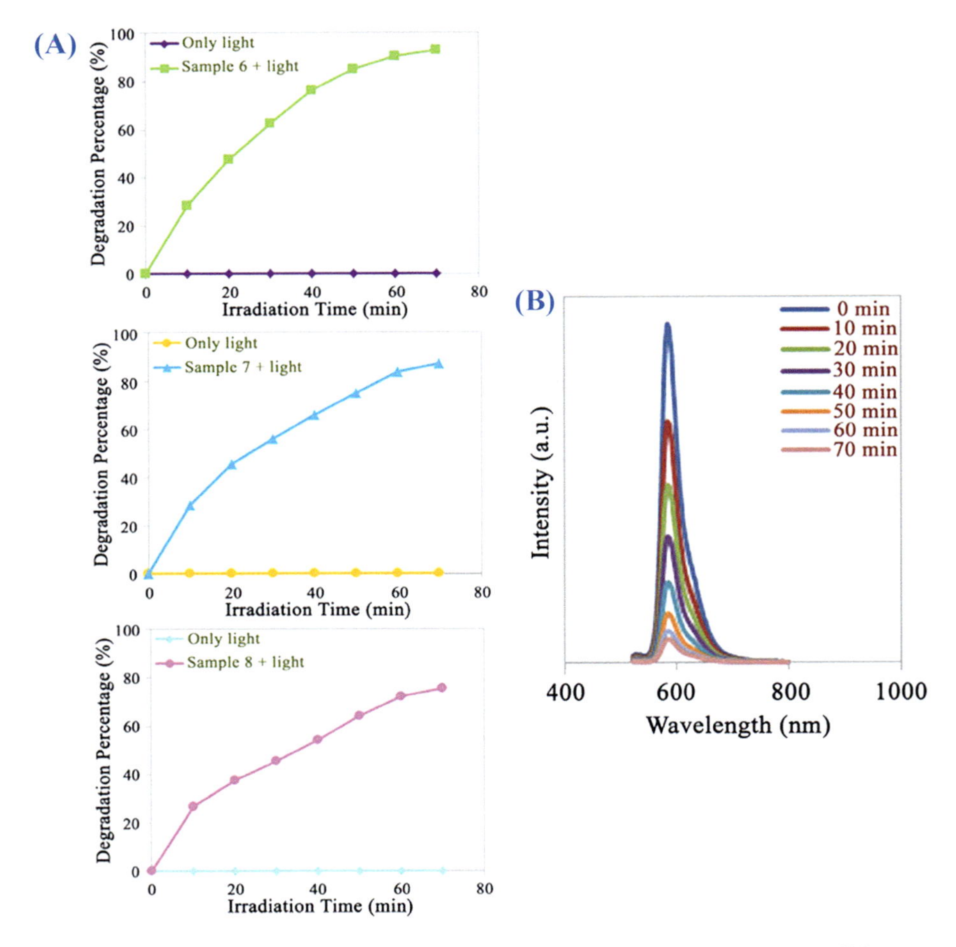

Figure 2.9 (A) Photocatalytic methyl violet contaminant decomposition of Nd_2O_3-SiO_2 nanocomposites at 40, 60, and 80 W (samples 6–8) and (B) Fluorescence spectral time scan of methyl violet contaminant illuminated at 365 nm with Nd_2O_3-SiO_2 nanocomposites (sample 6) (Zinatloo-Ajabshir et al., 2018).
Source: Reprinted with permission from Zinatloo-Ajabshir, S., Mortazavi-Derazkola, S., Salavati-Niasari, M. (2018). Nd_2O_3-SiO_2 nanocomposites: A simple sonochemical preparation, characterization and photocatalytic activity. *Ultrasonics Sonochemistry*, *42*, 171–182, Copyright 2018 Elsevier.

ultrasonic power on the photocatalytic performance of nanocomposites. They observed that the nanocomposite prepared at 60 W was able to degrade 93.1% of the methyl violet pollutant after 70 minutes under ultraviolet radiation and displayed the best photocatalytic yield compared to other samples. The best nanophotocatalyst consisted of spherical particles with appropriate uniformity (see Fig. 2.9A and B) (Zinatloo-Ajabshir et al., 2018). Their results exhibited that the removal of methyl violet pollutant with the help of Nd_2O_3-SiO_2 nanocomposite sample under UV light occurs continuously (Zinatloo-Ajabshir et al., 2018).

Hydrogen has been accepted as one of the most suitable sources of clean and environmentally friendly energy and has a very high energy capacity (Maeda et al., 2006; Zinatloo-Ajabshir, Salehi, Amiri, & Salavati-Niasari, 2019a). Utilizing solar energy to produce hydrogen through the photocatalytic water splitting in the presence of nanocatalysts is known as a very proper approach and has been considered by many scientists (Wolcott, Smith, Kuykendall, Zhao, & Zhang, 2009; Zhang, Maeda, Takata, & Domen, 2010; Zhang, Zuo, & Feng, 2010). Zeng et al. (2014) produced CeO_2/ZnO nanotube array films through an easy two-step electrodeposition way and employed for photocatalytic hydrogen evolution. The results exhibited that the prepared nanostructured films could denote remarkable photocatalytic performance with a hydrogen evolution rate of 2.7 μmol/cm^2 h beneath white light (Zeng et al., 2014). They also had good stability.

Studies have demonstrated that in the process of selective photocatalytic oxidation utilizing nanocatalysts, sunlight is utilized as a source of energy and oxygen molecules as an oxidant, so it is done in more moderate and environmentally friendly conditions (Yurdakal, Palmisano, Loddo, Augugliaro, & Palmisano, 2008; Yurdakal et al., 2009). Tanaka, Hashimoto, and Kominami (2012) fabricated Au/CeO_2 structures with various Au dosages via the multistep (Ms) photodeposition way and utilized for the photocatalytic oxidation of benzyl alcohol under visible light. They found that in the photocatalytic oxidation of benzyl alcohol possessing an amino group, the Au/CeO_2 nanostructured photocatalyst displayed good chemoselectivity toward the OH group of alcohol (99% yield) (Tanaka et al., 2012).

2.4 Conclusion and outlook

In short, cerium dioxide and rare earth oxides (R_2O_3) ceramic nanostructures have attracted the interest of many scientists in recent years owing to their unique properties that are very beneficial and noteworthy for various applications. In this chapter, the interesting and distinctive features of this type of special oxide ceramic nanostructures were mentioned. Also, the most common approaches of preparing these widely used ceramic nanostructures in recent years have been stated and the most obvious disadvantages or advantages of each approach have been pointed out. Studies exhibit that the properties as well as applications of cerium dioxide and rare earth oxides (R_2O_3) ceramic nanostructures are remarkably affected by particle size and particle shape as well as their degree of purity. Thus many efforts have been made to prepare ceramic oxide nanostructures with good purity and high uniformity. Despite the efforts made, there is still a very serious need to design and provide a suitable, low-cost, and environmentally friendly approach for the production of these widely used ceramic nanostructures on a large scale. Besides, the applications of this type of ceramic nanostructures in various fields were mentioned. Studies have indicated that depending on each specific application, ceramic nanostructures must have specific and desirable structural and morphological properties as well as a specific chemical composition in order to exhibit proper performance. Hence, during the preparation of these ceramic nanostructures, special control of these properties should be considered. Despite the efforts made, there is still a serious need to offer ceramic nanostructures with controlled properties.

References

Alili, L., Sack, M., von Montfort, C., Giri, S., Das, S., Carroll, K. S., ... Brenneisen, P. (2013). Downregulation of tumor growth and invasion by redox-active nanoparticles. *Antioxidants & Redox Signaling*, *19*, 765–778.

Abu-Zied, B. M., & Soliman, S. A. (2008). Thermal decomposition of praseodymium acetate as a precursor of praseodymium oxide catalyst. *Thermochimica Acta*, *470*, 91–97.

Aruna, S. T., & Mukasyan, A. S. (2008). Combustion synthesis and nanomaterials. *Current Opinion in Solid State and Materials Science*, *12*, 44–50.

Bamwenda, G. R., & Arakawa, H. (2000). Cerium dioxide as a photocatalyst for water decomposition to O_2 in the presence of ${Ce_{aq}}^{4+}$ and ${Fe_{aq}}^{3+}$ species. *Journal of Molecular Catalysis A: Chemical*, *161*, 105–113.

Bao, H., Zhang, Z., Hua, Q., & Huang, W. (2014). Compositions, structures, and catalytic activities of CeO_2@Cu_2O nanocomposites prepared by the template-assisted method. *Langmuir: The ACS Journal of Surfaces and Colloids*, *30*, 6427–6436.

Barrera, E. W., Pujol, M. C., Díaz, F., Choi, S. B., Rotermund, F., Park, K. H., ... Cascales, C. (2011). Emission properties of hydrothermal Yb^{3+}, Er^{3+} and Yb^{3+}, Tm^{3+}-codoped Lu_2O_3 nanorods: Upconversion, cathodoluminescence and assessment of waveguide behavior. *Nanotechnology*, *22*, 075205.

Beshkar, F., Zinatloo-Ajabshir, S., & Salavati-Niasari, M. (2015). Preparation and characterization of the $CuCr_2O_4$ nanostructures via a new simple route. *Journal of Materials Science: Materials in Electronics*, *26*, 5043–5051.

Beshkar, F., Zinatloo-Ajabshir, S., & Salavati-Niasari, M. (2015). Simple morphology-controlled fabrication of nickel chromite nanostructures via a novel route. *Chemical Engineering Journal*, *279*, 605–614.

Beshkar, F., Zinatloo-Ajabshir, S., Bagheri, S., & Salavati-Niasari, M. (2017). Novel preparation of highly photocatalytically active copper chromite nanostructured material via a simple hydrothermal route. *PLoS One*, *12*, e0158549.

Brunner, T. J., Wick, P., Manser, P., Spohn, P., Grass, R. N., Limbach, L. K., ... Stark, W. J. (2006). In vitro cytotoxicity of oxide nanoparticles: Comparison to asbestos, silica, and the effect of particle solubility. *Environmental Science & Technology*, *40*, 4374–4381.

Burda, C., Chen, X., Narayanan, R., & El-Sayed, M. A. (2005). Chemistry and properties of nanocrystals of different shapes. *Chemical Reviews*, *105*, 1025–1102.

Campbell, C. T., & Peden, C. H. F. (2005). Oxygen vacancies and catalysis on ceria surfaces. *Science (New York, N.Y.)*, *309*, 713–714.

Celardo, I., Pedersen, J. Z., Traversa, E., & Ghibelli, L. (2011). Pharmacological potential of cerium oxide nanoparticles. *Nanoscale*, *3*, 1411–1420.

Chang, H.-Y., & Chen, H.-I. (2005). Morphological evolution for CeO_2 nanoparticles synthesized by precipitation technique. *Journal of Crystal Growth*, *283*, 457–468.

Chang, S., Li, M., Hua, Q., Zhang, L., Ma, Y., Ye, B., & Huang, W. (2012). Shape-dependent interplay between oxygen vacancies and Ag–CeO_2 interaction in Ag/CeO_2 catalysts and their influence on the catalytic activity. *Journal of Catalysis*, *293*, 195–204.

Chen, F., Cao, Y., & Jia, D. (2011). Preparation and photocatalytic property of CeO_2 lamellar. *Applied Surface Science*, *257*, 9226–9231.

Chen, P.-L., & Chen, I. W. (1993). Reactive cerium(IV) oxide powders by the homogeneous precipitation method. *Journal of the American Ceramic Society*, *76*, 1577–1583.

Chiu, F.-C., & Lai, C.-M. (2010). Optical and electrical characterizations of cerium oxide thin films. *Journal of Physics D: Applied Physics*, *43*, 075104.

Coduri, M., Checchia, S., Longhi, M., Ceresoli, D., & Scavini, M. (2018). Rare earth doped ceria: the complex connection between structure and properties. *Frontiers in Chemistry*, *6*.

Corma, A., Atienzar, P., García, H., & Chane-Ching, J. Y. (2004). Hierarchically mesostructured doped CeO_2 with potential for solar-cell use. *Nature Materials*, *3*, 394–397.

Dai, Q., Wang, J., Yu, J., Chen, J., & Chen, J. (2014). Catalytic ozonation for the degradation of acetylsalicylic acid in aqueous solution by magnetic CeO_2 nanometer catalyst particles. *Applied Catalysis B: Environmental*, *144*, 686–693.

Das, S., Dowding, J. M., Klump, K. E., McGinnis, J. F., Self, W., & Seal, S. (2013). Cerium oxide nanoparticles: Applications and prospects in nanomedicine. *Nanomedicine: Nanotechnology, Biology, and Medicine*, *8*, 1483–1508.

Di Monte, R., Fornasiero, P., Graziani, M., & Kašpar, J. (1998). Oxygen storage and catalytic NO removal promoted by CeO_2-containing mixed oxides. *Journal of Alloys and Compounds, 275–*, *277*, 877–885.

Doornkamp, C., & Ponec, V. (2000). The universal character of the Mars and Van Krevelen mechanism. *Journal of Molecular Catalysis A: Chemical*, *162*, 19–32.

Dosev, D., Nichkova, M., Liu, M., Guo, B., Liu, G.-y., Xia, Y., ... Kennedy, I. (2005). Application of luminescent Eu:Gd_2O_3 nanoparticles to the visualization of protein micropatterns. *Journal of Biomedical Optics*, *10*, 064006.

Dutta, V., Sharma, S., Raizada, P., Kumar, R., Thakur, V. K., Nguyen, V.-H., ... Singh, P. (2020). Recent progress on bismuth-based Z-scheme semiconductor photocatalysts for energy and environmental applications. *Journal of Environmental Chemical Engineering*, *8*, 104505.

Fang, Q., & Liang, X. (2012). CeO_2–Al_2O_3, CeO_2–SiO_2, CeO_2–TiO_2 core-shell spheres: Formation mechanisms and UV absorption. *RSC Advances*, *2*, 5370–5375.

Feng, X., Sayle, D. C., Wang, Z. L., Paras, M. S., Santora, B., Sutorik, A. C., ... Her, Y.-S. (2006). Converting ceria polyhedral nanoparticles into single-crystal nanospheres. *Science (New York, N.Y.)*, *312*, 1504–1508.

Frayret, C., Villesuzanne, A., Pouchard, M., Mauvy, F., Bassat, J.-M., & Grenier, J.-C. (2010). Identifying doping strategies to optimize the oxide ion conductivity in ceria-based materials. *The Journal of Physical Chemistry C*, *114*, 19062–19076.

Garzon, F. H., Mukundan, R., & Brosha, E. L. (2000). Solid-state mixed potential gas sensors: Theory, experiments and challenges. *Solid State Ionics*, *136–137*, 633–638.

Ge, L., Zang, C., & Chen, F. (2015). The enhanced Fenton-like catalytic performance of PdO/CeO_2 for the degradation of acid orange 7 and salicylic acid. *Chinese Journal of Catalysis*, *36*, 314–321.

Glaspell, G., Anderson, J., Wilkins, J. R., & El-Shall, M. S. (2008). Vapor phase synthesis of upconverting Y_2O_3 nanocrystals doped with Yb^{3+}, Er^{3+}, Ho^{3+}, and Tm^{3+} to generate red, green, blue, and white light. *The Journal of Physical Chemistry C*, *112*, 11527–11531.

Gnanam, S., & Rajendran, V. (2011). Synthesis of CeO_2 or α–Mn_2O_3 nanoparticles via sol–gel process and their optical properties. *Journal of Sol-Gel Science and Technology*, *58*, 62–69.

Goris, B., Turner, S., Bals, S., & Van Tendeloo, G. (2014). Three-dimensional valency mapping in ceria nanocrystals. *ACS Nano*, *8*, 10878–10884.

Gu, H., & Soucek, M. D. (2007). Preparation and characterization of monodisperse cerium oxide nanoparticles in hydrocarbon solvents. *Chemistry of Materials*, *19*, 1103–1110.

Guo, Z., Du, F., Li, G., & Cui, Z. (2006). Synthesis and characterization of single-crystal Ce (OH)CO_3 and CeO_2 triangular microplates. *Inorganic Chemistry*, *45*, 4167–4169.

Guyot, Y., Guzik, M., Alombert-Goget, G., Pejchal, J., Yoshikawa, A., Ito, A., ... Boulon, G. (2016). Assignment of Yb^{3+} energy levels in the C2 and C3i centers of Lu_2O_3 sesquioxide either as ceramics or as crystal. *Journal of Luminescence*, *170*, 513–519.

Han, J., Kim, H. J., Yoon, S., & Lee, H. (2011). Shape effect of ceria in Cu/ceria catalysts for preferential CO oxidation. *Journal of Molecular Catalysis A: Chemical*, *335*, 82–88.

Han, X., Li, L., & Wang, C. (2012). Synthesis of monodisperse CeO_2 octahedra assembled by nanosheets with exposed {001} facets and catalytic property. *CrystEngComm*, *14*, 1939–1941.

Hassanzadeh-Tabrizi, S. A., Mazaheri, M., Aminzare, M., & Sadrnezhaad, S. K. (2010). Reverse precipitation synthesis and characterization of CeO_2 nanopowder. *Journal of Alloys and Compounds*, *491*, 499–502.

He, H., Yang, P., Li, J., Shi, R., Chen, L., Zhang, A., & Zhu, Y. (2016). Controllable synthesis, characterization, and CO oxidation activity of CeO_2 nanostructures with various morphologies. *Ceramics International*, *42*, 7810–7818.

He, L., Su, Y., Lanhong, J., & Shi, S. (2015). Recent advances of cerium oxide nanoparticles in synthesis, luminescence and biomedical studies: A review. *Journal of Rare Earths*, *33*, 791–799.

He, Y., Yang, B., & Cheng, G. (2003). Controlled synthesis of CeO_2 nanoparticles from the coupling route of homogenous precipitation with microemulsion. *Materials Letters*, *57*, 1880–1884.

Heckert, E. G., Karakoti, A. S., Seal, S., & Self, W. T. (2008). The role of cerium redox state in the SOD mimetic activity of nanoceria. *Biomaterials*, *29*, 2705–2709.

Heckman, K. L., DeCoteau, W., Estevez, A., Reed, K. J., Costanzo, W., Sanford, D., ... Erlichman, J. S. (2013). Custom cerium oxide nanoparticles protect against a free radical mediated autoimmune degenerative disease in the brain. *ACS Nano*, *7*, 10582–10596.

Hirano, M., & Kato, E. (1996). Hydrothermal synthesis of cerium(IV) oxide. *Journal of the American Ceramic Society*, *79*, 777–780.

Hu, S., Yang, J., Li, C., & Lin, J. (2012). Synthesis and up-conversion white light emission of RE^{3+}-doped lutetium oxide nanocubes as a single compound. *Materials Chemistry and Physics*, *133*, 751–756.

Huang, S.-Y., Chang, C.-M., & Yeh, C.-T. (2006). Promotion of platinum–ruthenium catalyst for electro-oxidation of methanol by ceria. *Journal of Catalysis*, *241*, 400–406.

Höppe, H. A. (2009). Recent developments in the field of inorganic phosphors. *Angewandte Chemie International Edition*, *48*, 3572–3582.

Jamshidijam, M., Mangalaraja, R. V., Akbari-Fakhrabadi, A., Ananthakumar, S., & Chan, S. H. (2014). Effect of rare earth dopants on structural characteristics of nanoceria synthesized by combustion method. *Powder Technology*, *253*, 304–310.

Ji, R., Peng, G., Zhang, S., Li, Z., Li, J., Fang, T., ... Wu, J. (2018). The fabrication of a CeO_2 coating via cathode plasma electrolytic deposition for the corrosion resistance of AZ31 magnesium alloy. *Ceramics International*, *44*, 19885–19891.

Ji, Z., Wang, X., Zhang, H., Lin, S., Meng, H., Sun, B., ... Zink, J. I. (2012). Designed synthesis of CeO_2 nanorods and nanowires for studying toxicological effects of high aspect ratio nanomaterials. *ACS Nano*, *6*, 5366–5380.

Jiang, D., Wang, W., Sun, S., Zhang, L., & Zheng, Y. (2015). Equilibrating the plasmonic and catalytic roles of metallic nanostructures in photocatalytic oxidation over Au-modified CeO_2. *ACS Catalysis*, *5*, 613–621.

Jiang, L., Yao, M., Liu, B., Li, Q., Liu, R., Lv, H., ... Wågberg, T. (2012). Controlled synthesis of CeO_2/graphene nanocomposites with highly enhanced optical and catalytic properties. *The Journal of Physical Chemistry C*, *116*, 11741–11745.

Jiang, Z., Feng, L., Zhu, J., Li, X., Chen, Y., & Khan, S. (2020). MOF assisted synthesis of a Ho_2O_3/CNT nanocomposite photocatalyst for organic pollutants degradation. *Ceramics International*, *46*, 19084–19091.

Kaneko, K., Inoke, K., Freitag, B., Hungria, A. B., Midgley, P. A., Hansen, T. W., ... Adschiri, T. (2007). Structural and morphological characterization of cerium oxide nanocrystals prepared by hydrothermal synthesis. *Nano Letters*, *7*, 421–425.

Kang, Y., Sun, M., & Li, A. (2012). Studies of the catalytic oxidation of CO Over Ag/CeO_2 catalyst. *Catalysis Letters*, *142*, 1498–1504.

Karakoti, A., Singh, S., Dowding, J. M., Seal, S., & Self, W. T. (2010). Redox-active radical scavenging nanomaterials. *Chemical Society Reviews*, *39*, 4422–4432.

Karami, M., Ghanbari, M., Amiri, O., Ghiyasiyan-Arani, M., & Salavati-Niasari, M. (2021). Sonochemical synthesis, characterization and investigation of the electrochemical hydrogen storage properties of $TlPbI_3/Tl_4PbI_6$ nanocomposite. *International Journal of Hydrogen Energy*, *46*, 6648–6658.

Khan, M. M., Ansari, S. A., Ansari, M. O., Min, B. K., Lee, J., & Cho, M. H. (2014). Biogenic fabrication of Au@CeO_2 nanocomposite with enhanced visible light activity. *The Journal of Physical Chemistry C*, *118*, 9477–9484.

Kim, C. K., Kim, T., Choi, I.-Y., Soh, M., Kim, D., Kim, Y.-J., ... Hyeon, T. (2012). Back Cover: Ceria nanoparticles that can protect against ischemic stroke (Angew. Chem. Int. (Ed.) 44/2012). *Angewandte Chemie International Edition*, *51*, 11172.

Kishi, H., Mizuno, Y., & Chazono, H. (2003). Base-metal electrode-multilayer ceramic capacitors: Past present and future perspectives. *Japanese Journal of Applied Physics*, *42*, 1–15.

Kominami, H., Tanaka, A., & Hashimoto, K. (2010). Mineralization of organic acids in aqueous suspensions of gold nanoparticles supported on cerium(iv) oxide powder under visible light irradiation. *Chemical Communications*, *46*, 1287–1289.

Kopia, A., Kowalski, K., Chmielowska, M., & Leroux, C. (2008). Electron microscopy and spectroscopy investigations of $CuO_x–CeO_{2-\delta}$/Si thin films. *Surface Science*, *602*, 1313–1321.

Korsvik, C., Patil, S., Seal, S., & Self, W. T. (2007). Superoxide dismutase mimetic properties exhibited by vacancy engineered ceria nanoparticles. *Chemical Communications*, 1056–1058.

Körner, R., Ricken, M., Nölting, J., & Riess, I. (1989). Phase transformations in reduced ceria: Determination by thermal expansion measurements. *Journal of Solid State Chemistry*, *78*, 136–147.

Kumar, A., Babu, S., Karakoti, A. S., Schulte, A., & Seal, S. (2009). Luminescence properties of europium-doped cerium oxide nanoparticles: Role of vacancy and oxidation states. *Langmuir: The ACS Journal of Surfaces and Colloids*, *25*, 10998–11007.

Lamba, R., Umar, A., Mehta, S. K., & Kansal, S. K. (2015). CeO_2ZnO hexagonal nanodisks: Efficient material for the degradation of direct blue 15 dye and its simulated dye bath effluent under solar light. *Journal of Alloys and Compounds*, *620*, 67–73.

Lang, X., Chen, X., & Zhao, J. (2014). Heterogeneous visible light photocatalysis for selective organic transformations. *Chemical Society Reviews*, *43*, 473–486.

Lee, S. S., Song, W., Cho, M., Puppala, H. L., Nguyen, P., Zhu, H., ... Colvin, V. L. (2013). Antioxidant properties of cerium oxide nanocrystals as a function of nanocrystal diameter and surface coating. *ACS Nano*, *7*, 9693–9703.

Li, J.-G., Ikegami, T., & Mori, T. (2004). Low temperature processing of dense samarium-doped CeO_2 ceramics: Sintering and grain growth behaviors. *Acta Materialia*, *52*, 2221–2228.

Li, L., He, Y., Fang, Z., Ding, Z., Wang, L., Qu, Q., & Yuan, R. (2012). Enhanced corrosion resistance of AZ31B magnesium alloy by cooperation of rare earth cerium and stannate conversion coating. *International Journal of Electrochemical Science*, *7*, 12690–12705.

Liang, X., Xiao, J., Chen, B., & Li, Y. (2010). Catalytically stable and active CeO_2 mesoporous spheres. *Inorganic Chemistry*, *49*, 8188–8190.

Lide, D. R. (2004). *CRC handbook of chemistry and physics* (85th ed.). Taylor & Francis.

Lin, F., Hoang, D. T., Tsung, C.-K., Huang, W., Lo, S. H.-Y., Wood, J. B., ... Yang, P. (2011). Catalytic properties of Pt cluster-decorated CeO_2 nanostructures. *Nano Research*, *4*, 61–71.

Lin, S., Su, G., Zheng, M., Ji, D., Jia, M., & Liu, Y. (2012). Synthesis of flower-like $Co_3O_4-CeO_2$ composite oxide and its application to catalytic degradation of 1,2,4-trichlorobenzene. *Applied Catalysis B: Environmental*, *123–124*, 440–447.

Liu, K., Wang, A., & Zhang, T. (2012). Recent advances in preferential oxidation of co reaction over platinum group metal catalysts. *ACS Catalysis*, *2*, 1165–1178.

Liu, Y., Luo, L., Gao, Y., & Huang, W. (2016). CeO_2 morphology-dependent NbO_x-CeO_2 interaction, structure and catalytic performance of NbO_x/CeO_2 catalysts in oxidative dehydrogenation of propane. *Applied Catalysis B: Environmental*, *197*, 214–221.

Logothetidis, S., Patsalas, P., & Charitidis, C. (2003). Enhanced catalytic activity of nanostructured cerium oxide films. *Materials Science and Engineering: C*, *23*, 803–806.

Lu, J., & Fang, Z. Z. (2006). Synthesis and characterization of nanoscaled cerium (IV) oxide via a solid-state mechanochemical method. *Journal of the American Ceramic Society*, *89*, 842–847.

Lu, X., Zhai, T., Cui, H., Shi, J., Xie, S., Huang, Y., ... Tong, Y. (2011). Redox cycles promoting photocatalytic hydrogen evolution of CeO_2 nanorods. *Journal of Materials Chemistry*, *21*, 5569–5572.

Ma, Y., Gao, W., Zhang, Z., Zhang, S., Tian, Z., Liu, Y., ... Qu, Y. (2018). Regulating the surface of nanoceria and its applications in heterogeneous catalysis. *Surface Science Reports*, *73*, 1–36.

Maeda, K., Teramura, K., Lu, D., Takata, T., Saito, N., Inoue, Y., & Domen, K. (2006). Photocatalyst releasing hydrogen from water. *Nature*, *440*, 295.

Malleshappa, J., Nagabhushana, H., Sharma, S. C., Vidya, Y. S., Anantharaju, K. S., Prashantha, S. C., ... Surendra, B. S. (2015). Leucas aspera mediated multifunctional CeO_2 nanoparticles: Structural, photoluminescent, photocatalytic and antibacterial properties. *Spectrochimica Acta Part A: Molecular and Biomolecular Spectroscopy*, *149*, 452–462.

Marques, T. M. F., Strayer, M. E., Ghosh, A., Silva, A., Ferreira, O. P., Fujisawa, K., ... Viana, B. C. (2017). Homogeneously dispersed CeO_2 nanoparticles on exfoliated hexaniobate nanosheets. *Journal of Physics and Chemistry of Solids*, *111*, 335–342.

Masui, T., Fujiwara, K., Machida, K.-i., Adachi, G.-y., Sakata, T., & Mori, H. (1997). Characterization of cerium(IV) oxide ultrafine particles prepared using reversed micelles. *Chemistry of Materials*, *9*, 2197–2204.

Meza, O., Villabona-Leal, E. G., Diaz-Torres, L. A., Desirena, H., Rodríguez-López, J. L., & Pérez, E. (2014). Luminescence concentration quenching mechanism in Gd_2O_3:Eu^{3+}. *The Journal of Physical Chemistry. A*, *118*, 1390–1396.

Monsef, R., Ghiyasiyan-Arani, M., & Salavati-Niasari, M. (2021). Design of magnetically recyclable ternary Fe_2O_3/$EuVO_4$/g-C_3N_4 nanocomposites for photocatalytic and electrochemical hydrogen storage. *ACS Applied Energy Materials*, *4*, 680–695.

Morgan, C. G., & Mitchell, A. C. (2007). Prospects for applications of lanthanide-based upconverting surfaces to bioassay and detection. *Biosensors and Bioelectronics*, *22*, 1769–1775.

Mortazavi-Derazkola, S., Zinatloo-Ajabshir, S., & Salavati-Niasari, M. (2015a). Preparation and characterization of Nd_2O_3 nanostructures via a new facile solvent-less route. *Journal of Materials Science: Materials in Electronics*, *26*, 5658–5667.

Mortazavi-Derazkola, S., Zinatloo-Ajabshir, S., & Salavati-Niasari, M. (2015b). Novel simple solvent-less preparation, characterization and degradation of the cationic dye over holmium oxide ceramic nanostructures. *Ceramics International*, *41*, 9593–9601.

Mortazavi-Derazkola, S., Zinatloo-Ajabshir, S., & Salavati-Niasari, M. (2015c). New sodium dodecyl sulfate-assisted preparation of Nd_2O_3 nanostructures via a simple route. *RSC Advances*, *5*, 56666–56676.

Mortazavi-Derazkola, S., Zinatloo-Ajabshir, S., & Salavati-Niasari, M. (2017a). New facile preparation of Ho_2O_3 nanostructured material with improved photocatalytic performance. *Journal of Materials Science: Materials in Electronics*, *28*, 1914–1924.

Mortazavi-Derazkola, S., Zinatloo-Ajabshir, S., & Salavati-Niasari, M. (2017b). Facile hydrothermal and novel preparation of nanostructured Ho_2O_3 for photodegradation of eriochrome black T dye as water pollutant. *Advanced Powder Technology*, *28*, 747–754.

Mousavi-Kamazani, M., & Ashrafi, S. (2020). Single-step sonochemical synthesis of Cu_2O-CeO_2 nanocomposites with enhanced photocatalytic oxidative desulfurization. *Ultrasonics Sonochemistry*, *63*, 104948.

Mousavi-Kamazani, M., & Azizi, F. (2019). Facile sonochemical synthesis of Cu doped CeO_2 nanostructures as a novel dual-functional photocatalytic adsorbent. *Ultrasonics Sonochemistry*, *58*, 104695.

Mousavi-Kamazani, M., Zinatloo-Ajabshir, S., & Ghodrati, M. (2020). One-step sonochemical synthesis of $Zn(OH)_2$/ZnV_3O_8 nanostructures as a potent material in electrochemical hydrogen storage. *Journal of Materials Science: Materials in Electronics*, *31*, 17332–17338.

Mullins, D. R. (2015). The surface chemistry of cerium oxide. *Surface Science Reports*, *70*, 42–85.

Nair, J. P., Wachtel, E., Lubomirsky, I., Fleig, J., & Maier, J. (2003). Anomalous expansion of CeO_2 nanocrystalline membranes. *Advanced Materials*, *15*, 2077–2081.

Nesakumar, N., Sethuraman, S., Krishnan, U. M., & Rayappan, J. B. B. (2013). Fabrication of lactate biosensor based on lactate dehydrogenase immobilized on cerium oxide nanoparticles. *Journal of Colloid and Interface Science*, *410*, 158–164.

Pagliari, F., Mandoli, C., Forte, G., Magnani, E., Pagliari, S., Nardone, G., ... Traversa, E. (2012). Cerium oxide nanoparticles protect cardiac progenitor cells from oxidative stress. *ACS Nano*, *6*, 3767–3775.

Pierscionek, B. K., Li, Y., Yasseen, A. A., Colhoun, L. M., Schachar, R. A., & Chen, W. (2009). Nanoceria have no genotoxic effect on human lens epithelial cells. *Nanotechnology*, *21*, 035102.

Pinjari, D. V., & Pandit, A. B. (2010). Cavitation milling of natural cellulose to nanofibrils. *Ultrasonics Sonochemistry*, *17*, 845–852.

Pinjari, D. V., & Pandit, A. B. (2011). Room temperature synthesis of crystalline CeO_2 nanopowder: Advantage of sonochemical method over conventional method. *Ultrasonics Sonochemistry*, *18*, 1118–1123.

Prasad, K., Pinjari, D. V., Pandit, A. B., & Mhaske, S. T. (2010). Phase transformation of nanostructured titanium dioxide from anatase-to-rutile via combined ultrasound assisted sol–gel technique. *Ultrasonics Sonochemistry*, *17*, 409–415.

Rajan, A. R., Vilas, V., Rajan, A., John, A., & Philip, D. (2020). Synthesis of nanostructured CeO_2 by chemical and biogenic methods: Optical properties and bioactivity. *Ceramics International*, *46*, 14048–14055.

Ravishankar, T. N., Ramakrishnappa, T., Nagaraju, G., & Rajanaika, H. (2015). Synthesis and characterization of CeO_2 nanoparticles via solution combustion method for photocatalytic and antibacterial activity studies. *ChemistryOpen*, *4*, 146–154.

Razi, F., Zinatloo-Ajabshir, S., & Salavati-Niasari, M. (2017). Preparation and characterization of HgI_2 nanostructures via a new facile route. *Materials Letters*, *193*, 9–12.

Razi, F., Zinatloo-Ajabshir, S., & Salavati-Niasari, M. (2017). Preparation, characterization and photocatalytic properties of Ag_2ZnI_4/AgI nanocomposites via a new simple hydrothermal approach. *Journal of Molecular Liquids*, *225*, 645–651.

Sangsefidi, F. S., Salavati-Niasari, M., Mazaheri, S., & Sabet, M. (2017). Controlled green synthesis and characterization of CeO_2 nanostructures as materials for the determination of ascorbic acid. *Journal of Molecular Liquids*, *241*, 772–781.

Sathyamurthy, S., Leonard, K. J., Dabestani, R. T., & Paranthaman, M. P. (2005). Reverse micellar synthesis of cerium oxide nanoparticles. *Nanotechnology*, *16*, 1960–1964.

Schubert, D., Dargusch, R., Raitano, J., & Chan, S.-W. (2006). Cerium and yttrium oxide nanoparticles are neuroprotective. *Biochemical and Biophysical Research Communications*, *342*, 86–91.

Seal, S., Jeyaranjan, A., Neal, C. J., Kumar, U., Sakthivel, T. S., & Sayle, D. C. (2020). Engineered defects in cerium oxides: Tuning chemical reactivity for biomedical, environmental, & energy applications. *Nanoscale*, *12*, 6879–6899.

Sifontes, A. B., Rosales, M., Méndez, F. J., Oviedo, O., & Zoltan, T. (2013). Effect of calcination temperature on structural properties and photocatalytic activity of ceria nanoparticles synthesized employing chitosan as template. *Journal of Nanomaterials*, *2013*, 265797.

Singh, P., Raizada, P., Pathania, D., Kumar, A., & Thakur, P. (2013). Preparation of BSA-$ZnWO_4$ nanocomposites with enhanced adsorptional photocatalytic activity for methylene blue degradation. *International Journal of Photoenergy*, *2013*, 726250.

Song, S., Wang, X., & Zhang, H. (2015). CeO_2-encapsulated noble metal nanocatalysts: Enhanced activity and stability for catalytic application. *NPG Asia Materials*, *7*, e179.

Soykal, I. I., Sohn, H., & Ozkan, United States (2012). Effect of support particle size in steam reforming of ethanol over Co/CeO_2 catalysts. *ACS Catalysis*, *2*, 2335–2348.

Sreekanth, T. V. M., Nagajyothi, P. C., Reddy, G. R., Shim, J., & Yoo, K. (2019). Urea assisted ceria nanocubes for efficient removal of malachite green organic dye from aqueous system. *Scientific Reports*, *9*, 14477.

Sun, C., Sun, J., Xiao, G., Zhang, H., Qiu, X., Li, H., & Chen, L. (2006). Mesoscale organization of nearly monodisperse flowerlike ceria microspheres. *The Journal of Physical Chemistry. B*, *110*, 13445–13452.

Sun, C., Li, H., & Chen, L. (2012). Nanostructured ceria-based materials: Synthesis, properties, and applications. *Energy & Environmental Science*, *5*, 8475–8505.

Sun, Y., Liu, Q., Gao, S., Cheng, H., Lei, F., Sun, Z., ... Xie, Y. (2013). Pits confined in ultrathin cerium(IV) oxide for studying catalytic centers in carbon monoxide oxidation. *Nature Communications*, *4*, 2899.

Tago, T., Tashiro, S., Hashimoto, Y., Wakabayashi, K., & Kishida, M. (2003). Synthesis and optical properties of SiO_2-coated CeO_2 nanoparticles. *Journal of Nanoparticle Research*, *5*, 55–60.

Tamura, M., Honda, M., Nakagawa, Y., & Tomishige, K. (2014). Direct conversion of CO_2 with diols, aminoalcohols and diamines to cyclic carbonates, cyclic carbamates and cyclic ureas using heterogeneous catalysts. *Journal of Chemical Technology & Biotechnology*, *89*, 19–33.

Tan, Y., Wei, S., Liu, X., Pan, B., Liu, S., Wu, J., ... He, Y. (2021). Neodymium oxide (Nd_2O_3) coupled tubular g-C3N4, an efficient dual-function catalyst for photocatalytic hydrogen production and NO removal. *Science of The Total Environment*, *773*, 145583.

Tanaka, A., Hashimoto, K., & Kominami, H. (2011). Gold and copper nanoparticles supported on cerium(IV) Oxide—a photocatalyst mineralizing organic acids under red light irradiation. *ChemCatChem*, *3*, 1619–1623.

Tanaka, A., Hashimoto, K., & Kominami, H. (2012). Preparation of Au/CeO_2 exhibiting strong surface plasmon resonance effective for selective or chemoselective oxidation of alcohols to aldehydes or ketones in aqueous suspensions under irradiation by green light. *Journal of the American Chemical Society*, *134*, 14526–14533.

Tang, Q., Liu, Z., Li, S., Zhang, S., Liu, X., & Qian, Y. (2003). Synthesis of yttrium hydroxide and oxide nanotubes. *Journal of Crystal Growth*, *259*, 208–214.

Tian, J., Sang, Y., Zhao, Z., Zhou, W., Wang, D., Kang, X., ... Huang, H. (2013). Enhanced photocatalytic performances of CeO_2/TiO_2 nanobelt heterostructures. *Small (Weinheim an der Bergstrasse, Germany)*, *9*, 3864–3872.

Tian, Z., Li, J., Zhang, Z., Gao, W., Zhou, X., & Qu, Y. (2015). Highly sensitive and robust peroxidase-like activity of porous nanorods of ceria and their application for breast cancer detection. *Biomaterials*, *59*, 116–124.

Tok, A. I. Y., Boey, F. Y. C., Dong, Z., & Sun, X. L. (2007). Hydrothermal synthesis of CeO_2 nano-particles. *Journal of Materials Processing Technology*, *190*, 217–222.

Trovarelli, A., Boaro, M., Rocchini, E., de Leitenburg, C., & Dolcetti, G. (2001). Some recent developments in the characterization of ceria-based catalysts. *Journal of Alloys and Compounds, 323–*, *324*, 584–591.

Tsunekawa, S., Fukuda, T., & Kasuya, A. (2000). Blue shift in ultraviolet absorption spectra of monodisperse CeO_{2-x} nanoparticles. *Journal of Applied Physics*, *87*, 1318–1321.

Tuller, H. L. (2000). Ionic conduction in nanocrystalline materials. *Solid State Ionics*, *131*, 143–157.

Valechha, D., Lokhande, S., Klementova, M., Subrt, J., Rayalu, S., & Labhsetwar, N. (2011). Study of nano-structured ceria for catalytic CO oxidation. *Journal of Materials Chemistry*, *21*, 3718–3725.

Valente, J. S., Tzompantzi, F., & Prince, J. (2011). Highly efficient photocatalytic elimination of phenol and chlorinated phenols by CeO_2/MgAl layered double hydroxides. *Applied Catalysis B: Environmental*, *102*, 276–285.

Vivier, L., & Duprez, D. (2010). Ceria-based solid catalysts for organic chemistry. *ChemSusChem*, *3*, 654–678.

Wang, H., Zhu, J.-J., Zhu, J.-M., Liao, X.-H., Xu, S., Ding, T., & Chen, H.-Y. (2002). Preparation of nanocrystalline ceria particles by sonochemical and microwave assisted heating methods. *Physical Chemistry Chemical Physics*, *4*, 3794–3799.

Wang, Y., Chen, Z., Han, P., Du, Y., Gu, Z., Xu, X., & Zheng, G. (2018). Single-atomic Cu with multiple oxygen vacancies on ceria for electrocatalytic CO_2 reduction to CH_4. *ACS Catalysis*, *8*, 7113–7119.

Wang, Z.-L., Li, G.-R., Ou, Y.-N., Feng, Z.-P., Qu, D.-L., & Tong, Y.-X. (2011). Electrochemical deposition of Eu^{3+}-doped CeO_2 nanobelts with enhanced optical properties. *The Journal of Physical Chemistry C*, *115*, 351–356.

Wetchakun, N., Chaiwichain, S., Inceesungvorn, B., Pingmuang, K., Phanichphant, S., Minett, A. I., & Chen, J. (2012). $BiVO_4$/CeO_2 nanocomposites with high visible-light-induced photocatalytic activity. *ACS Applied Materials & Interfaces*, *4*, 3718–3723.

Wolcott, A., Smith, W. A., Kuykendall, T. R., Zhao, Y., & Zhang, J. Z. (2009). Photoelectrochemical study of nanostructured ZnO thin films for hydrogen generation from water splitting. *Advanced Functional Materials*, *19*, 1849–1856.

Wu, H., Liu, X., Wen, J., Liu, Y., & Zheng, X. (2021). Rare-earth oxides modified Mg-Al layered double oxides for the enhanced adsorption-photocatalytic activity. *Colloids and Surfaces A: Physicochemical and Engineering Aspects*, *610*, 125933.

Wu, J., Shi, S., Wang, X., Li, J., Zong, R., & Chen, W. (2014). Controlled synthesis and optimum luminescence of Sm^{3+}-activated nano/submicroscale ceria particles by a facile approach. *Journal of Materials Chemistry C*, *2*, 2786–2792.

Wu, J., Shi, S., Wang, X., Song, H., Luo, M., & Chen, W. (2014). Self-rising synthesis and luminescent properties of Eu^{3+}-doped nanoceria. *Journal of Luminescence*, *152*, 142–144.

Wu, X., Zhang, Y., Lu, Y., Pang, S., Yang, K., Tian, Z., ... Pei, Z. (2017). Synergistic and targeted drug delivery based on nano-CeO_2 capped with galactose functionalized pillar[5]arene via host–guest interactions. *Journal of Materials Chemistry B*, *5*, 3483–3487.

Wu, Z., Li, M., & Overbury, S. H. (2012). On the structure dependence of CO oxidation over CeO_2 nanocrystals with well-defined surface planes. *Journal of Catalysis*, *285*, 61–73.

Xia, T., Kovochich, M., Liong, M., Mädler, L., Gilbert, B., Shi, H., ... Nel, A. E. (2008). Comparison of the mechanism of toxicity of zinc oxide and cerium oxide nanoparticles based on dissolution and oxidative stress properties. *ACS Nano*, *2*, 2121–2134.

Xiao, H., Li, P., Jia, F., & Zhang, L. (2009). General nonaqueous sol – gel synthesis of nanostructured Sm_2O_3, Gd_2O_3, Dy_2O_3, and Gd_2O_3:Eu^{3+} phosphor. *The Journal of Physical Chemistry C*, *113*, 21034–21041.

Xie, S., Wang, Z., Cheng, F., Zhang, P., Mai, W., & Tong, Y. (2017). Ceria and ceria-based nanostructured materials for photoenergy applications. *Nano Energy*, *34*, 313–337.

Xu, A.-W., Gao, Y., & Liu, H.-Q. (2002). The preparation, characterization, and their photocatalytic activities of rare-earth-doped TiO_2 nanoparticles. *Journal of Catalysis*, *207*, 151–157.

Xu, C., & Qu, X. (2014). Cerium oxide nanoparticle: A remarkably versatile rare earth nanomaterial for biological applications. *NPG Asia Materials*, *6*, e90.

Xu, L., & Wang, J. (2012). Magnetic nanoscaled Fe_3O_4/CeO_2 composite as an efficient fenton-like heterogeneous catalyst for degradation of 4-chlorophenol. *Environmental Science & Technology*, *46*, 10145–10153.

Xu, M., Xie, S., Lu, X.-H., Liu, Z.-Q., Huang, Y., Zhao, Y., ... Tong, Y.-X. (2011). Controllable electrochemical synthesis and photocatalytic activity of CeO_2 octahedra and nanotubes. *Journal of the Electrochemical Society*, *158*, E41.

Yahiro, H., Baba, Y., Eguchi, K., & Arai, H. (1988). High temperature fuel cell with ceria-yttria solid electrolyte. *Journal of the Electrochemical Society*, *135*, 2077–2080.

Yamashita, M., Kameyama, K., Yabe, S., Yoshida, S., Fujishiro, Y., Kawai, T., & Sato, T. (2002). Synthesis and microstructure of calcia doped ceria as UV filters. *Journal of Materials Science*, *37*, 683–687.

Yan, C.-H., Yan, Z.-G., Du, Y.-P., Shen, J., Zhang, C., & Feng, W. (2011). Chapter 251 - Controlled synthesis and properties of rare earth nanomaterials. In K. A. Gschneidner, J.-C. G. Bünzli, & V. K. Pecharsky (Eds.), *Handbook on the physics and chemistry of rare earths* (pp. 275–472). Elsevier.

Yan, L., Yu, R., Chen, J., & Xing, X. (2008). Template-free hydrothermal synthesis of CeO_2 nano-octahedrons and nanorods: investigation of the morphology evolution. *Crystal Growth & Design*, *8*, 1474–1477.

Yao, H., Wang, Y., Luo, G., & Size-Controllable, A. (2017). Precipitation method to prepare CeO_2 nanoparticles in a membrane dispersion microreactor. *Industrial & Engineering Chemistry Research*, *56*, 4993–4999.

Ye, R.-P., Li, Q., Gong, W., Wang, T., Razink, J. J., Lin, L., ... Yao, Y.-G. (2020). High-performance of nanostructured Ni/CeO_2 catalyst on CO_2 methanation. *Applied Catalysis B: Environmental*, *268*, 118474.

You, R., Zhang, X., Luo, L., Pan, Y., Pan, H., Yang, J., ... Huang, W. (2017). NbO_x/CeO_2-rods catalysts for oxidative dehydrogenation of propane: Nb–CeO_2 interaction and reaction mechanism. *Journal of Catalysis*, *348*, 189–199.

Yu, R. B., Yu, K. H., Wei, W., Xu, X. X., Qiu, X. M., Liu, S., ... Peng, B. (2007). Nd_2O_3 nanoparticles modified with a silane-coupling agent as a liquid laser medium. *Advanced Materials*, *19*, 838–842.

Yuan, Q., Liu, Q., Song, W.-G., Feng, W., Pu, W.-L., Sun, L.-D., ... Yan, C.-H. (2007). Ordered mesoporous $Ce_{1-x}Zr_xO_2$ solid solutions with crystalline walls. *Journal of the American Chemical Society*, *129*, 6698–6699.

Yurdakal, S., Palmisano, G., Loddo, V., Augugliaro, V., & Palmisano, L. (2008). Nanostructured rutile TiO_2 for selective photocatalytic oxidation of aromatic alcohols to aldehydes in water. *Journal of the American Chemical Society*, *130*, 1568–1569.

Yurdakal, S., Palmisano, G., Loddo, V., Alagöz, O., Augugliaro, V., & Palmisano, L. (2009). Selective photocatalytic oxidation of 4-substituted aromatic alcohols in water with rutile TiO_2 prepared at room temperature. *Green Chemistry*, *11*, 510–516.

Zeng, C.-h., Xie, S., Yu, M., Yang, Y., Lu, X., & Tong, Y. (2014). Facile synthesis of large-area CeO_2/ZnO nanotube arrays for enhanced photocatalytic hydrogen evolution. *Journal of Power Sources*, *247*, 545–550.

Zhang, C., Zhang, X., Wang, Y., Xie, S., Liu, Y., Lu, X., & Tong, Y. (2014). Facile electrochemical synthesis of CeO_2 hierarchical nanorods and nanowires with excellent photocatalytic activities. *New Journal of Chemistry*, *38*, 2581–2586.

Zhang, C., Bu, W., Ni, D., Zhang, S., Li, Q., Yao, Z., ... Shi, J. (2016). Synthesis of iron nanometallic glasses and their application in cancer therapy by a localized Fenton reaction. *Angewandte Chemie International Edition*, *55*, 2101–2106.

Zhang, F., Maeda, K., Takata, T., & Domen, K. (2010). Modification of oxysulfides with two nanoparticulate cocatalysts to achieve enhanced hydrogen production from water with visible light. *Chemical Communications*, *46*, 7313–7315.

Zhang, W., & Chen, D. (2016). Preparation and performance of CeO_2 hollow spheres and nanoparticles. *Journal of Rare Earths*, *34*, 295–299.

Zhang, X., You, R., Li, D., Cao, T., & Huang, W. (2017). Reaction sensitivity of ceria morphology effect on Ni/CeO_2 catalysis in propane oxidation reactions. *ACS Applied Materials & Interfaces*, *9*, 35897–35907.

Zhang, Y., Han, K., Yin, X., Wang, Z., Xu, Z., & Zhu, W. (2009). Synthesis and characterization of single-crystalline $PrCO_3OH$ dodecahedral microrods and its thermal conversion to Pr6O11. *Journal of Crystal Growth*, *311*, 3883–3888.

Zhang, Z., Zuo, F., & Feng, P. (2010). Hard template synthesis of crystalline mesoporous anatase TiO_2 for photocatalytic hydrogen evolution. *Journal of Materials Chemistry*, *20*, 2206–2212.

Zhao, Q., Guo, N., Jia, Y., Lv, W., Shao, B., Jiao, M., & You, H. (2013). Facile surfactant-free synthesis and luminescent properties of hierarchical europium-doped lutetium oxide phosphors. *Journal of Colloid and Interface Science*, *394*, 216–222.

Zhenling, W. Zewei, Q. & Jun, L. (2007)., Remarkable Changes in the Optical Properties of CeO2 Nanocrystals Induced by Lanthanide Ions Doping. *Inorganic Chemistry, 46(13)*, 5237–5242. Available from https://doi.org/10.1021/ic0701256.

Zhou, J., He, J., Wang, T., Chen, X., & Sun, D. (2009). Synergistic effect of RE_2O_3 (RE = Sm, Eu and Gd) on Pt/mesoporous carbon catalyst for methanol electro-oxidation. *Electrochimica Acta*, *54*, 3103–3108.

Zinatloo-Ajabshir, S., & Salavati-Niasari, M. (2014). Synthesis of pure nanocrystalline ZrO_2 via a simple sonochemical-assisted route. *Journal of Industrial and Engineering Chemistry*, *20*, 3313–3319.

Zinatloo-Ajabshir, S., & Salavati-Niasari, M. (2015). Nanocrystalline Pr_6O_{11}: Synthesis, characterization, optical and photocatalytic properties. *New Journal of Chemistry*, *39*, 3948–3955.

Zinatloo-Ajabshir, S., Mortazavi-Derazkola, S., & Salavati-Niasari, M. (2017a). Schiff-base hydrothermal synthesis and characterization of Nd_2O_3 nanostructures for effective photocatalytic degradation of eriochrome black T dye as water contaminant. *Journal of Materials Science: Materials in Electronics*, *28*, 17849–17859.

Zinatloo-Ajabshir, S., Mortazavi-Derazkola, S., & Salavati-Niasari, M. (2017b). Sono-synthesis and characterization of Ho_2O_3 nanostructures via a new precipitation way for photocatalytic degradation improvement of erythrosine. *International Journal of Hydrogen Energy*, *42*, 15178–15188.

Zinatloo-Ajabshir, S., Mortazavi-Derazkola, S., & Salavati-Niasari, M. (2017c). Nd_2O_3 nanostructures: Simple synthesis, characterization and its photocatalytic degradation of methylene blue. *Journal of Molecular Liquids*, *234*, 430–436.

Zinatloo-Ajabshir, S., Mortazavi-Derazkola, S., & Salavati-Niasari, M. (2017d). Simple sonochemical synthesis of Ho_2O_3-SiO_2 nanocomposites as an effective photocatalyst for degradation and removal of organic contaminant. *Ultrasonics Sonochemistry*, *39*, 452–460.

Zinatloo-Ajabshir, S., Mortazavi-Derazkola, S., & Salavati-Niasari, M. (2017e). Preparation, characterization and photocatalytic degradation of methyl violet pollutant of holmium oxide nanostructures prepared through a facile precipitation method. *Journal of Molecular Liquids*, *231*, 306–313.

Zinatloo-Ajabshir, S., Mortazavi-Derazkola, S., & Salavati-Niasari, M. (2017f). Sonochemical synthesis, characterization and photodegradation of organic pollutant over

Nd_2O_3 nanostructures prepared via a new simple route. *Separation and Purification Technology*, *178*, 138–146.

Zinatloo-Ajabshir, S., Mortazavi-Derazkola, S., & Salavati-Niasari, M. (2018). Nd_2O_3-SiO_2 nanocomposites: A simple sonochemical preparation, characterization and photocatalytic activity. *Ultrasonics Sonochemistry*, *42*, 171–182.

Zinatloo-Ajabshir, S., Salehi, Z., Amiri, O., & Salavati-Niasari, M. (2019a). Green synthesis, characterization and investigation of the electrochemical hydrogen storage properties of $Dy_2Ce_2O_7$ nanostructures with fig extract. *International Journal of Hydrogen Energy*, *44*, 20110–20120.

Zinatloo-Ajabshir, S., Salehi, Z., Amiri, O., & Salavati-Niasari, M. (2019b). Simple fabrication of $Pr_2Ce_2O_7$ nanostructures via a new and eco-friendly route; a potential electrochemical hydrogen storage material. *Journal of Alloys and Compounds*, *791*, 792–799.

Zinatloo-Ajabshir, S., Morassaei, M. S., Amiri, O., Salavati-Niasari, M., & Foong, L. K. (2020). $Nd_2Sn_2O_7$ nanostructures: Green synthesis and characterization using date palm extract, a potential electrochemical hydrogen storage material. *Ceramics International*, *46*, 17186–17196.

Zinatloo-Ajabshir, S., Baladi, M., & Salavati-Niasari, M. (2021). Enhanced visible-light-driven photocatalytic performance for degradation of organic contaminants using $PbWO_4$ nanostructure fabricated by a new, simple and green sonochemical approach. *Ultrasonics Sonochemistry*, *72*, 105420.

Rare earth cerate ($Re_2Ce_2O_7$) ceramic nanomaterials

Sahar Zinatloo-Ajabshir
Department of Chemical Engineering, University of Bonab, Bonab, Iran

3.1 General introduction

In recent years, the preparation and study of solid solution of cerium dioxide (CeO_2) doped with high amounts of lanthanides (Ln), that is, with a compositions containing equimolar (1:1) Ln_2O_3-CeO_2 has received much attention from scientists because these solid solutions have unique features that can be beneficial for many potential applications (Wang, Wang, Li, Cheng, & Chi, 2014). Lanthanide cerates possess a general formula $Ln_2Ce_2O_7$ in which Ln denotes the lanthanide ion (Cao et al., 2003). It has been reported that these oxide compounds have special and interesting attributes such as good thermal expansion coefficients, high melting points, desirable chemical resistance, proper thermal stability, appropriate ionic conductivity, and low thermal conductivity (Hongsong, Jianguo, Gang, Zheng, & Xinli, 2012; Khan et al., 2012; Salehi, Zinatloo-Ajabshir, & Salavati-Niasari, 2016b; Zhang, Liao, Yuan, & Guan, 2011). These attributes are very beneficial and effective for the potential uses of these oxide compounds in various fields including thermal barrier coatings (TBCs), oxygen sensors, hosts for luminescence centers, fuel cells, hosts for nuclear waste, environmental remediation and removal of various pollutants, and high temperature catalytic systems (Lin, Wang, Liu, & Meng, 2009; Mori, Drennan, Lee, Li, & Ikegami, 2002; Narayanan, Lommens, De Buysser, Vanpoucke, Huehne, Molina & Van Driessche, 2012; Raj et al., 2014; Weng, Wang, & Lee, 2013; Yamamura, Nishino, Kakinuma, & Nomura, 2003; Yan et al., 2011). This chapter discusses recent advances in ceramic lanthanide cerate compounds, as well as approaches for preparing these specific oxide compounds. The advantages and disadvantages of each preparation approach are stated to be advantageous for readers in choosing the appropriate and reliable preparation way. The uses of these special ceramic compounds are also briefly described.

3.2 Lanthanide cerates ($Ln_2Ce_2O_7$)

Lanthanide cerate is known as a solid solution of lanthanide oxide in cerium dioxide with defective fluorite structure (Cao et al., 2003). Lanthanide cerate ceramic systems with fluorite structure have attracted the attention of many researchers

Advanced Rare Earth-Based Ceramic Nanomaterials. DOI: https://doi.org/10.1016/B978-0-323-89957-4.00009-8

owing to their special electrical, catalytic, and mechanical attributes (Higashi, Sonoda, Ono, Sameshima, & Hirata, 1999; Sung Bae, Kil Choo, & Hee, 2004). Studies have shown that the replacement of Ce^{4+} ions by appropriate lanthanide trivalent ions can enhance chemical stability, improve ionic conductivity, and also suppress the reducibility of lanthanide cerate materials (O'Neill & Morris, 1999; Wang, Huang, et al., 2012). By partially substituting Ce^{4+} with lanthanide ions, a solid ceramic solution can be formed. Since the ionic radius of lanthanides is larger than that of Ce^{4+}, a lattice expansion can occur owing to the incorporation of lanthanide ions in the cerium dioxide lattice (Reddy, Katta, & Thrimurthulu, 2010). This could be a reason to reduce the activation energy for the diffusion of oxygen ions as well as to improve its performance in reduction (Yu, Zhang, & Lin, 2003).

Lanthanide cerate structures have extraordinary attributes such as desirable redox and proper transport (Singh, Kumar, & Chowdhury, 2017), so they are widely utilized in oxygen sensors (Mori et al., 2002), three-way catalysts with the usage of removing harmful exhaust gases (Matsouka, Konsolakis, Lambert, & Yentekakis, 2008), and fuel cells (Andrievskaya, Kornienko, Sameljuk, & Sayir, 2011; Mori et al., 2005; Wang, Mori, Li, & Drennan, 2005).

It has been reported that lanthanide cerate compounds can be very beneficial in TBC uses owing to their low thermal conductivity as well as large thermal expansion coefficient (Ma, Gong, Xu, & Cao, 2006). For example, the coefficient of thermal expansion (CTE) for $La_2Ce_2O_7$ compound $12.3 \times 10^{-6}\,K^{-1}$ (bulk, $300°C-1200°C$) has been reported (Singh et al., 2017).

Zhang, Zheng, Jiang, and Liu (2013) observed that lanthanide cerate lattice can be primarily stabilized with forming the anion frenkel defects. Further, they reported that lanthanide cerate can display substantial proton conductivity under reduced atmospheric conditions (Zhang et al., 2013).

The usage of TBCs on the surface of parts exposed to great temperatures, including engine blades as well as combustion chambers can be very beneficial and effective. Because it can protect special superalloys from rapid failure at high service temperatures close to their melting points (Praveen, Sravani, Alroy, Shanmugavelayutham, & Sivakumar, 2019; Ren, Wang, Cao, Shao, & Wang, 2021; Yi, Liu, Che, & Liang, 2020; Zhang, Hu, & Xu, 2010). It has been accepted that the service environment of TBCs must possess certain characteristics such as low thermal conductivity, compatible thermal expansion coefficient with the metal substratum, great sintering resistance, and sufficient phase stability (from room temperature to service temperature) along with environmental corrosion resistance (D.R.C. & Levi, 2003; Levi, 2004; Miller, 1997; Padture, Gell, & Jordan, 2002; Perepezko, 2009). Ceramic structures of lanthanide cerates are known as an appropriate alternative to yttria-stabilized zirconia and form a new generation of TBC (Darolia, 2013; Qu, Wan, & Pan, 2012; Schelling, Phillpot, & Grimes, 2004; Vassen, Cao, Tietz, Basu, & Stöver, 2000). These ceramic structures can exhibit relatively lower thermal conductivity (1.3–1.8 W/m K) with a CTE quantity close to that of yttria-stabilized zirconia (Cao et al., 2003; Xiaoge et al., 2014). Studies have shown that the greater CTE of lanthanide cerates can be ascribed to the bigger ionic radius of the Ce^{4+} ion as well as the weaker Ce-O bond energy (Ma et al., 2006; Yang,

Zhao, Zhang, Wang, & Pan, 2018). The great CTE of lanthanide cerates can cause CTE to approach that of the bonding layer and substratum, illustrating less stress mismatch in the coating as well as longer serving lifetime (Ren et al., 2021).

Studies have shown that lanthanide ions have attracted the attention of many scientists owing to their extraordinary optical attributes. They can demonstrate special functions in a wide range of industrial fields, including adjustable lasers, organic light-emitting diodes, optical communication amplifiers, and inorganic and mineral pigments (Biju, Reddy, Cowley, & Vasudevan, 2009; Bünzli & Piguet, 2005; Kido & Okamoto, 2002; Taniguchi, Kido, Nishiya, & Sasaki, 1995). In recent times, the industrial use of lanthanide-containing compounds has received much attention owing to their low toxicity. In this regard, a large number of lanthanide-containing pigments have been introduced as suitable alternatives to traditional toxic mineral pigments (Dohnalová, Šulcová, & Trojan, 2009; George, Sandhya Kumari, Vishnu, Ananthakumar, & Reddy, 2008; Martos, Julián-López, Cordoncillo, & Escribano, 2008; Martos, Julián-López, Folgado, Cordoncillo, & Escribano, 2008; Pailhé, Gaudon, & Demourgues, 2009; Petra & Trojan, 2008). Among the oxide pigments containing lanthanides, lanthanide cerates have opacity, low toxicity, and also great thermal stability and therefore have been the focus of scientists (Furukawa, Masui, & Imanaka, 2008; García, Llusar, Calbo, Tena, & Monrós, 2001; Kumari, George, Rao, & Reddy, 2008; Nobuhito, Toshiyuki, & Shinya, 2008; Šulcová & Trojan, 2004; Vishnu, George, Divya, & Reddy, 2009; Vishnu, George, & Reddy, 2010).

Solid oxide fuel cells (SOFCs) have been reported to be advanced sustainable energy conversion systems that can instantly turn chemical energy efficiently into electrical energy (Chen, Zhou, et al., 2015; Li, Wang, Wang, & Bi, 2020; Sun, Fang, Yan, & Liu, 2012; Wachsman, Marlowe, & Lee, 2012). They have been introduced as potential candidates to meet global energy requirement with substantial environmental advantages. These systems can use different fuel gases such as hydrogen, various kinds of hydrocarbons, and even carbon to generate power (Fan, Zhu, Su, & He, 2018; Gao, Mogni, Miller, Railsback, & Barnett, 2016). It has been reported that fluorite-type oxides of lanthanide cerate can display higher proton conductivity than lanthanide zirconates and can also have better sintering capability (Wang, Xie, Zhang, Liu, & Li, 2005). For example, Besikiotis, Knee, Ahmed, Haugsrud, and Norby (2012) observed that the ceramic structure of $La_2Ce_2O_7$ can be a protonic conductor at low temperature.

It has been demonstrated that when these ceramic structures are fabricated at the nanoscale, for reasons such as decreasing particle size as well as enhancing specific surface area, phenomena such as changes in phase transfer temperature, boosted catalytic performance as well as improved processability can be observed (Yan et al., 2011).

3.3 Preparation methods

Generally, these lanthanide complex oxide nanostructures may be fabricated through conventional solid-state reactions (Hongsong, Xiaochun, Gang, & Zhenjun,

2014; Zhang et al., 2020; Zhao et al., 2013), and solution-based soft chemical approaches (Hong-song, Xiao-ge, Gang, Xin-Li, & Xu-dan, 2012; Wang, Huang, et al., 2012; Wang et al., 2013) as well as green chemistry-based ways (Zinatloo-Ajabshir, Salehi, Amiri, & Salavati-Niasari, 2019a).

Solid-state reactions have been reported as an approach of preparing lanthanide cerate ceramic nanostructures involving a series of heating cycles at very high temperatures as well as frequent grinding steps. On the other hand, the powders obtained from this approach often have attributes such as low phase purity, and inhomogeneous composition as well as very high agglomeration (Dai, Zhong, Li, Meng, & Cao, 2006; Xiang et al., 2012; Xu et al., 2009, 2010; Zhang, Liao, Dang, Guan, & Zhang, 2011). Hongsong, Suran, and Shaokang (2012) fabricated $Dy_2Ce_2O_7$ ceramic powders via solid reaction way utilizing Dy_2O_3 and cerium dioxide as precursors at 1600°C within 600 minutes. They observed that the ceramic sample made had a single phase with a fluorite structure. Their results indicated that the scattering of phonons through oxygen vacancies, as well as the atomic mass difference between the substitutional atoms and the host atoms, could lead to a thermal conductivity of $Dy_2Ce_2O_7$ less than 8YSZ (Hongsong et al., 2012), and thus the ceramic compound could serve as a new and beneficial candidate compounds for TBCs. In other work, Hongsong, Shuqing, and Xiaoge (2014) prepared $(Sm_{1-x}Dy_x)_2Ce_2O_7$ solid solutions via solid reaction at 1600°C during 600 minutes in air. The prepared ceramic sample had a pure fluorite structure. They observed that the thermal conductivity as well as the thermal expansion coefficient of ceramic structures could be significantly reduced via adding dysprosium (Hongsong et al., 2014). Their results displayed that the coefficients of thermal expansion of these ceramic compounds were greater than yttria stabilized zirconia (YSZ) and also their thermal conductivity was much lower than 8YSZ. Thus these ceramic compounds with very good thermophysical attributes may be very advantageous and effective for usage in the ceramic layer in TBCs (Hongsong et al., 2014).

In contrast to solid-state reactions, the usage of soft solution-based chemical approaches for the preparation of lanthanide cerate ceramic structures can lead to chemically homogeneous products with pure phase as well as narrow particle size distributions and lower crystallization temperatures (Beshkar, Zinatloo-Ajabshir, & Salavati-Niasari, 2015; Besikiotis et al., 2012; Yan et al., 2011; Zinatloo-Ajabshir & Salavati-Niasari, 2015). It has been reported that the utilization of coprecipitation approach as one of the soft solution-based chemical approaches can lead to the preparation of pure phase product. Also, one of the most obvious advantages of this approach is its relatively simple experimental procedure, as well as its suitability for mass production of products (Chatzichristodoulou, Hendriksen, & Hagen, 2010; Fu, Chen, Tsai, & Hu, 2009; Li, Ikegami, Mori, & Wada, 2001; Moshtaghi, Zinatloo-Ajabshir, & Salavati-Niasari, 2016a; Razi, Zinatloo-Ajabshir, & Salavati-Niasari, 2017; Yan et al., 2011). Tadokoro and Muccillo (2007) prepared the solid solutions of $Ce_{1-x}(Y_{0.5}Dy_{0.5})_xO_{2-\delta}$ with $0 \leq x \leq 0.15$ through the precipitation of cerium, dysprosium, and yttrium hydroxides utilizing NH_4OH as precipitant agent. They investigated the role of cojoint additions of yttrium and dysprosium on the electrical conductivity of cerium dioxide ceramic structures via impedance spectroscopy (Tadokoro & Muccillo, 2007). They observed that the electrical conductivity of ceramic samples could vary depending on the dopant amount. The maximum conductivity was obtained for a ceramic structure with a lower dopant value. It was also observed that ceramic specimens with the same dopant value were able to exhibit similar electrical conductivity values. This phenomenon indicates that the quantity of oxygen vacancy can play a very key role in the conduction process of codoped ceramic structures (Tadokoro & Muccillo, 2007).

Besikiotis et al. (2012) produced $La_2Ce_2O_7$ ceramic structures through the citrate coprecipitation way utilizing lanthanum(III) nitrate hexahydrate and cerium(III) nitrate hexahydrate as starting materials. The prepared ceramic powders had a pure fluorite structure.

Another approach of preparing ceramic lanthanide cerate structures is sol−gel, which was discussed in the previous chapter. It has been reported that the utilization of this approach can produce pure ceramic structures that are chemically homogeneous (Fuentes & Baker, 2008; Sánchez-Bautista, Dos santos-García, Peña-Martínez, & Canales-Vázquez, 2010; Wang, Wang, et al., 2012; Weng et al., 2013). Xu et al. (2018) reported sol−gel way for synthesis of a series of $Ln_2Ce_2O_7$ structures (Ln = La, Pr, Sm, and Y) with heat treatment at 780°C within 240 minutes. They utilized the prepared samples for the oxidative of methane and compared their catalytic activity. They observed that $La_2Ce_2O_7$ could display the greatest performance among all the catalytic samples at low temperature (Xu et al., 2018).

Salehi, Zinatloo-Ajabshir, and Salavati-Niasari (2017) fabricated the nanostructured dysprosium cerate through a modified Pechini way utilizing trimesic acid as new chelating agent. They utilized PEG 600 as a novel space-filling template (see Fig. 3.1). Different kinds of chelating agents can be utilized to control the particle size and morphology of ceramic nanostructures in the modified sol−gel (Pechini) approach (Salehi et al., 2017). Thus various chelating agents were utilized to fabricate dysprosium cerate, including maleic acid, succinic acid, and trimic acid. They observed that highly agglomerated particle-like structures, uniform nanostructures with less uniformity, and highly homogeneous nanoparticles could be prepared with maleic acid, succinic acid, and trimic acid, respectively (Fig. 3.2) (Salehi et al., 2017). Their results displayed that thermic acid, which has the highest complexing effect, may better cape the surface of the formed primary nanocrystals and prevent their uncontrolled growth, and thus can better adjust the morphology and size of the ceramic nanostructured particles (Salehi et al., 2017).

It has been accepted that the utilization of hydrothermal way can produce ceramic products that are chemically homogeneous, have a narrow particle size distribution, and also possess very good crystallinity, at relatively low temperature (Cheng, Hwang, Sheu, & Hwang, 2008; Shuk & Greenblatt, 1999; Singh & Hegde, 2008; Somacescu et al., 2012; Wang et al., 2013; Yan et al., 2011). Khademinia and Behzad (2015) produced $La_2Ce_2O_7$ ceramic nanopowders through a hydrothermal reaction at 180°C during 2 days. The prepared nanopowders had a cubic structure. They observed that the morphology of $La_2Ce_2O_7$ sample could change from particle-like to rod-like by alkalizing the reaction solution (Khademinia & Behzad, 2015). Also, crystal growth was much better in alkaline medium. Therefore the pH

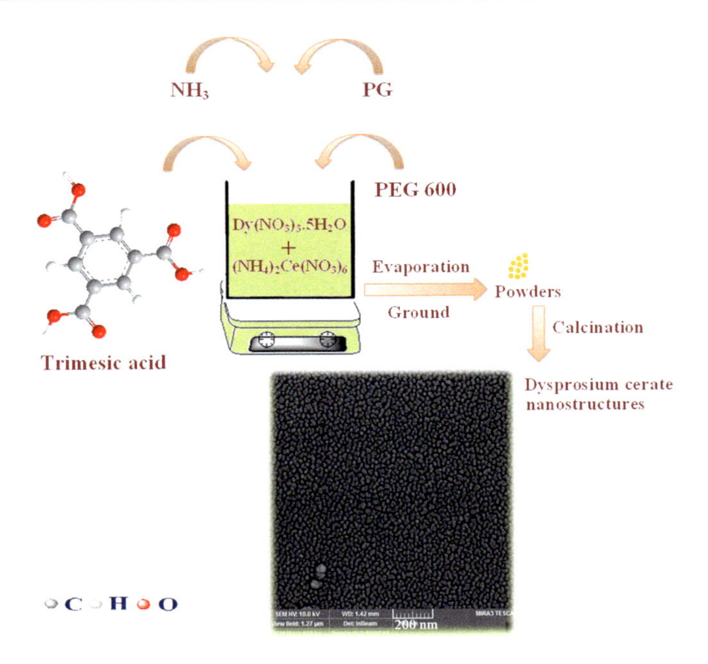

Figure 3.1 Schematic depiction for the production of dysprosium cerate nanostructure (Salehi et al., 2017).
Source: Reprinted with permission from Salehi, Z., Zinatloo-Ajabshir, S., Salavati-Niasari, M. (2017). Dysprosium cerate nanostructures: Facile synthesis, characterization, optical and photocatalytic properties. *Journal of Rare Earths*, *35*, 805–812, Copyright 2017 Elsevier.

of the reaction solution in the hydrothermal approach is very effective and key in determining the morphology of the ceramic nanostructure product as well as its crystallinity and crystalline nature.

Wang et al. (2014) reported the preparation of $La_2Ce_2O_7$ ceramic nanostructures via hydrothermal way with the help of two different types of surfactants (PEG and CTAB) at 180°C during 1 day. They observed that both samples of ceramic $La_2Ce_2O_7$ nanostructure had a cubic fluorite structure (Wang et al., 2014). Their results displayed that the kind of surfactant can play a noteworthy role in regulating the attributes of ceramic samples such as specific surface area and lattice parameter as well as their mean crystal size (Wang et al., 2014).

It has been accepted that the usage of combustion way can lead to the production of ceramic nanostructures with very fine particle sizes that are pure in chemical composition. The prepared nanostructures have high uniformity and also less agglomeration (Liu, Ouyang, & Sun, 2011; Morassaei, Zinatloo-Ajabshir, & Salavati-Niasari, 2016b; Yan et al., 2010, 2011). In this way, an aqueous solution comprising a proper fuel and metal salts (such as metal nitrate) is prepared in a certain molar ratio, and heated to boil, after which the resulting mixture is ignited and a quick combustion reaction and dry powders are prepared (Ling et al., 2013;

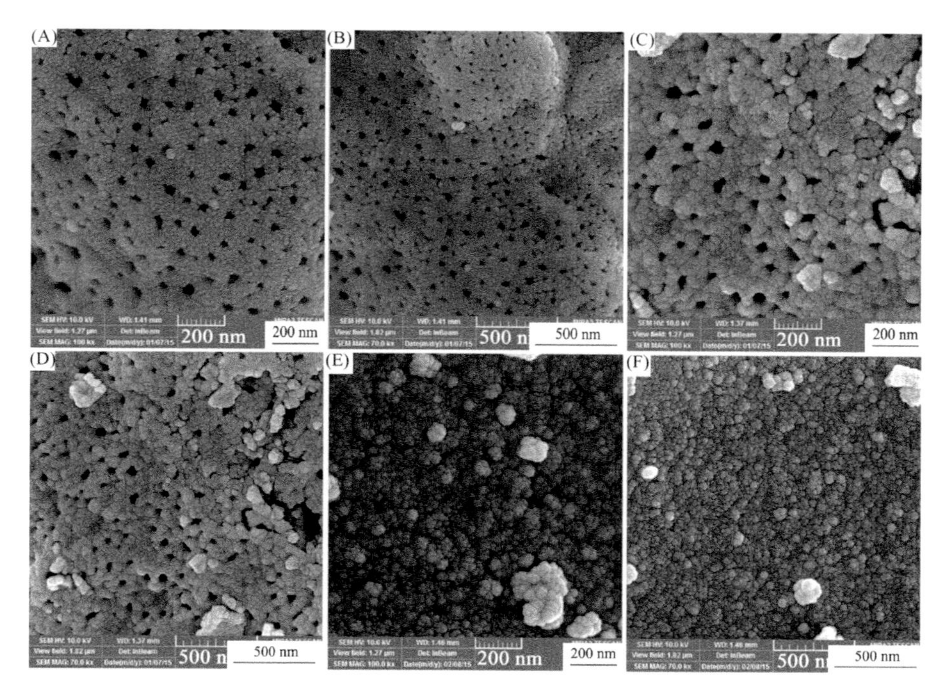

Figure 3.2 FESEM images of the samples 4–6 prepared via PG in the presence of maleic acid (A, B), succinic acid (C, D), and trimesic acid (E, F) (Salehi et al., 2017).
Source: Reprinted with permission from Salehi, Z., Zinatloo-Ajabshir, S., Salavati-Niasari, M. (2017). Dysprosium cerate nanostructures: Facile synthesis, characterization, optical and photocatalytic properties. *Journal of Rare Earths*, *35*, 805–812, Copyright 2017 Elsevier.

Morassaei, Zinatloo-Ajabshir, & Salavati-Niasari, 2016a; Yan et al., 2011). Zhong, Jiang, Lian, Song, and Peng (2020) prepared (Mg-, Ca-, Sr-, and Ba-) doped $Nd_2Ce_2O_7$ ceramic structures via a gel auto-combustion way utilizing metal nitrates and citric acid. They examined the role of addition of alkaline earth metals on the attributes of proton conductor $Nd_2Ce_2O_7$. They observed that all electrolytes displayed a cubic fluorite structure and had outstanding chemical stability at various atmospheres (Zhong et al., 2020). Their results indicated that the grain size as well as the density of ceramic structures could be remarkably enhanced and a large number of oxygen vacancies could be generated after the addition of alkaline earth metals into $Nd_2Ce_2O_7$ (Zhong et al., 2020). Since the total conductivity is largely related to the number of oxygen vacancies as well as the microstructure of the ceramic structures, the addition of magnesium dopant caused a significant change in the electrical conductivity (Zhong et al., 2020).

In another work, Salehi, Zinatloo-Ajabshir, and Salavati-Niasari (2016a) fabricated the nanostructured $Dy_2Ce_2O_7$ via a combustion route. They utilized the mixture of succinic acid and phenylalanine as fuel as well as sodium chloride salt as dispersant (see Fig. 3.3) (Salehi et al., 2016a). The prepared ceramic nanostructure

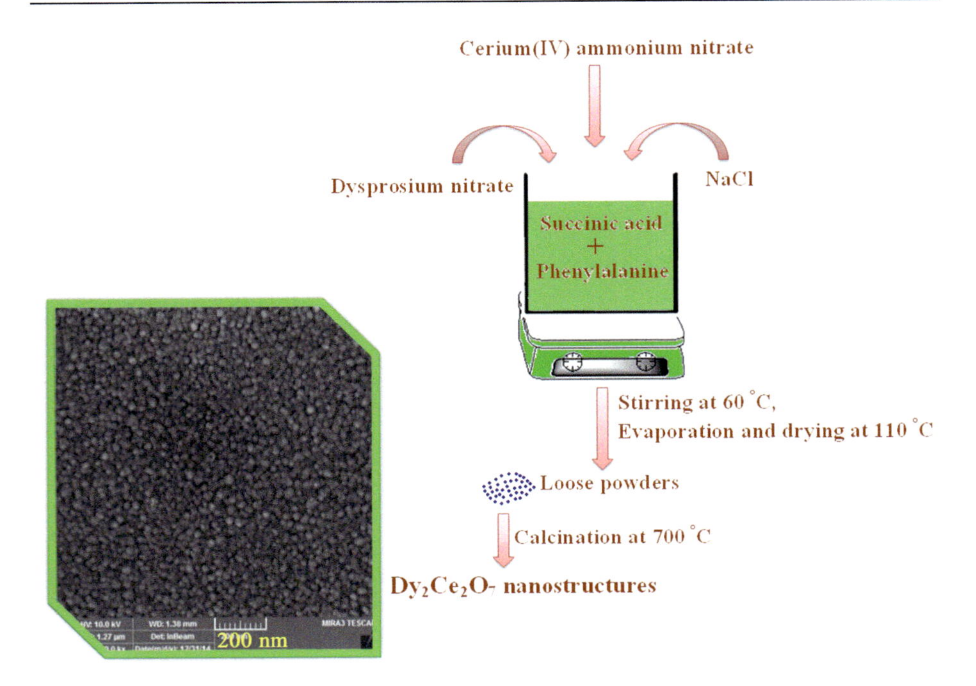

Figure 3.3 Schematic diagram of the preparation of the nanocrystalline $Dy_2Ce_2O_7$ (Salehi et al., 2016a).
Source: Reprinted with permission from Salehi, Z., Zinatloo-Ajabshir, S., Salavati-Niasari, M. (2016). New simple route to prepare $Dy_2Ce_2O_7$ nanostructures: Structural and photocatalytic studies. *Journal of Molecular Liquids*, *222*, 218–224, Copyright 2016 Elsevier.

sample had a pure fluorite structure. They observed that altering the kind of amino acid could play a very substantial role in regulating the attributes of a ceramic sample, such as particle shape and particle size (see Fig. 3.4). The amino acids employed in the preparation of ceramic samples owing to their different steric hindrance effects may bind to the surface of the formed primary particles and lead to the provision of different interfacial layers to perform the reaction and thus cause the assembly of primary particles in certain directions. As a result, different kinds of samples could be formed in terms of particle shape and particle size distribution (Salehi et al., 2016a).

One of the most harmful problems reported in the usage of chemical ways is the production of toxic and unsafe by-products during the process. After large-scale industrial production, these toxic and waste substances eventually accumulate, which can also lead to biological accumulation (Matussin, Harunsani, Tan, & Khan, 2020; Vijayaraghavan & Ashokkumar, 2017; Zinatloo-Ajabshir, Morassaei, Amiri, Salavati-Niasari, & Foong, 2020a). This can be a very serious and dangerous warning to human health as well as the environment. Thus, today, the usage of environmentally friendly, safe, and nontoxic synthesis ways based on green chemistry is

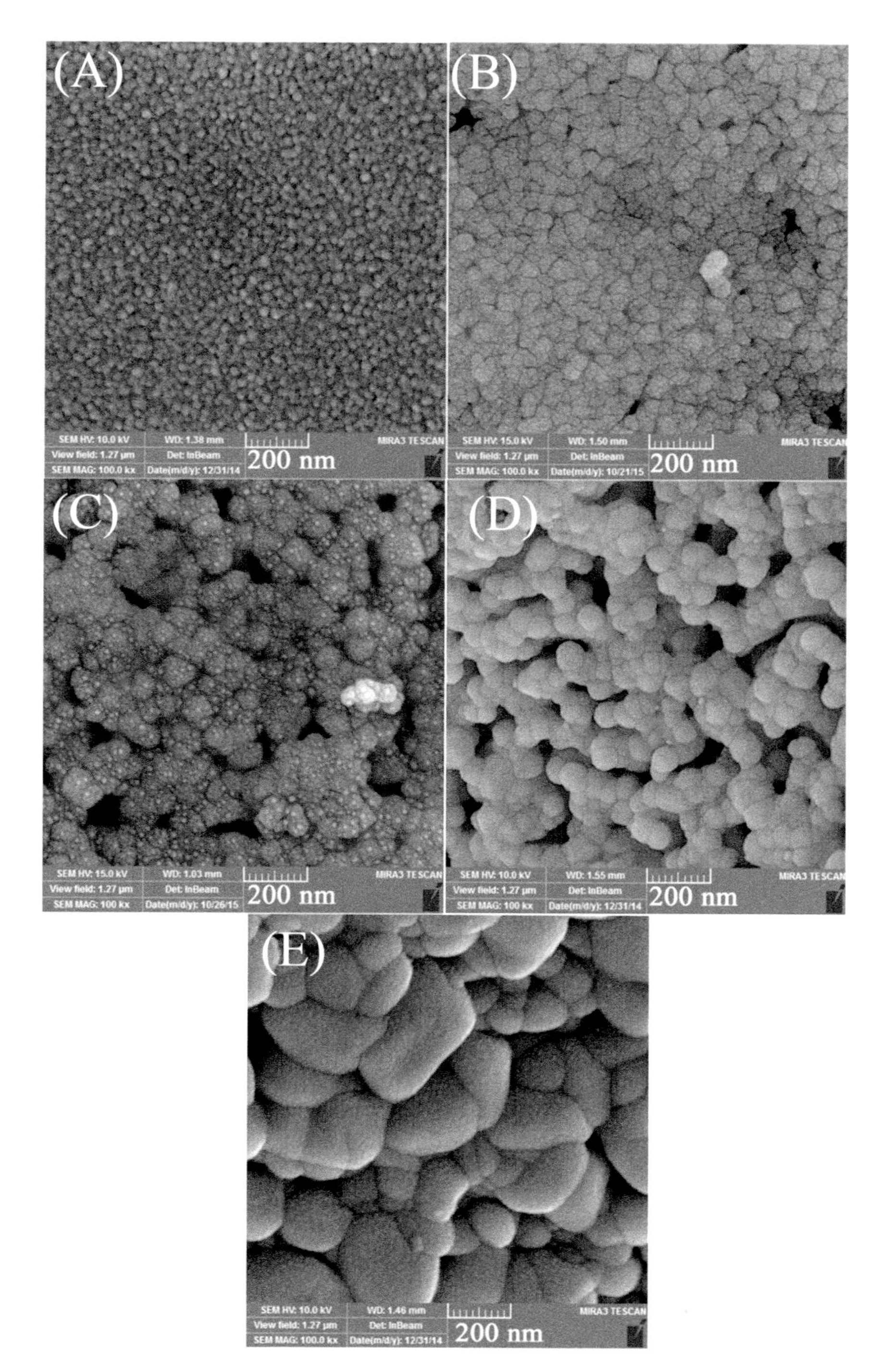

Figure 3.4 FESEM images of the samples 1, 2, 3, 4, and 5 prepared by the mixtures of succinic acid and phenylalanine (A), valine (B), serine (C), L-alanine (D), and glycine (E) as fuel (Salehi et al., 2016a).

Source: Reprinted with permission from Salehi, Z., Zinatloo-Ajabshir, S., Salavati-Niasari, M. (2016). New simple route to prepare $Dy_2Ce_2O_7$ nanostructures: Structural and photocatalytic studies. *Journal of Molecular Liquids*, *222*, 218–224, Copyright 2016 Elsevier.

preferred and their usage in the preparation of ceramic nanostructures has been considered by many scientists. Consequently, the development of environmentally friendly synthesis ways to achieve ceramic nanostructures has begun to flourish (Heidari-Asil, Zinatloo-Ajabshir, Amiri, & Salavati-Niasari, 2020; Huang et al., 2013; Lounis et al., 2018; Matussin et al., 2020; Zinatloo-Ajabshir, Morassaei, & Salavati-Niasari, 2019).

Zinatloo-Ajabshir et al. (2019a) prepared the nanostructured $Dy_2Ce_2O_7$ structures employing fig extract as new kind of fuel. They observed that the $Dy_2Ce_2O_7$ ceramic nanostructure prepared at 400°C was chemically pure and had a pure fluorite structure and also consisted of nanoparticles with very desirable uniformity and a narrow particle size distribution (see Fig. 3.5) (Zinatloo-Ajabshir et al., 2019a). They stated that fig is a fruit found in abundance in nature and comprise remarkable amounts of nontoxic and biocompatible sugars, namely fructose and glucose. The sugars in fig extract may act as a green capping agent and prevent the primary particles from overgrowth and undesired aggregation (Zinatloo-Ajabshir et al., 2019a). Their results indicated that by altering the highly effective variable, that is, production temperature, the morphological attributes of $Dy_2Ce_2O_7$ ceramic nanostructure with fluorite crystal structure can be changed. By altering the preparation temperature, they were able to fabricate different kinds of micro/nanostructures in terms of particle shape and particle size (Zinatloo-Ajabshir et al., 2019a) (Fig. 3.6).

3.4 Applications

In this section, we describe the applications of lanthanide cerate ceramic nanostructures in various fields. Today, climate change is a very serious problem for human life (Chen, Tan, Koros, & Jones, 2015; Tan et al., 2015; Yan et al., 2015; Zhu, Liu, Shen, Lou, & Ji, 2016). Solutions such as low carbon emissions as well as energy efficient technology have been proposed to solve this problem. As a result, the need to develop and provide energy technology that is environmentally friendly and efficacious, in order to resolve the energy and environmental crisis seems very necessary and serious (Zhu et al., 2016). It has been reported that hydrogen energy technology as well as fuel cells can be considered as remarkable, useful, and reliable technologies in order to achieve goals such as low carbon emissions as well as renewable energy (Chen, Liu, et al., 2015; Huang, Li, Yan, & Xing, 2014; Zhu et al., 2016). The use of proton-conducting ceramic oxide nanostructures can significantly enhance hydrogen usage, save energy, and diminish environmental contamination, and thus accelerating the development of the hydrogen economy (Medvedev et al., 2014; Tao et al., 2015; Zhu et al., 2016).

Since lanthanide cerate ceramic structures can exhibit appropriate proton conductivity, their usage as electrolytes in IT-SOFCs was investigated. For example, Tao, Bi, Fang, and Liu (2011) studied the $La_{1.95}Ca_{0.05}Ce_2O_7$ ceramic sample as an IT-SOFC electrolyte. They observed that the maximum powder density as well as the open circuit voltage were 259 mW/cm^2 and 0.832 V, correspondingly (Tao

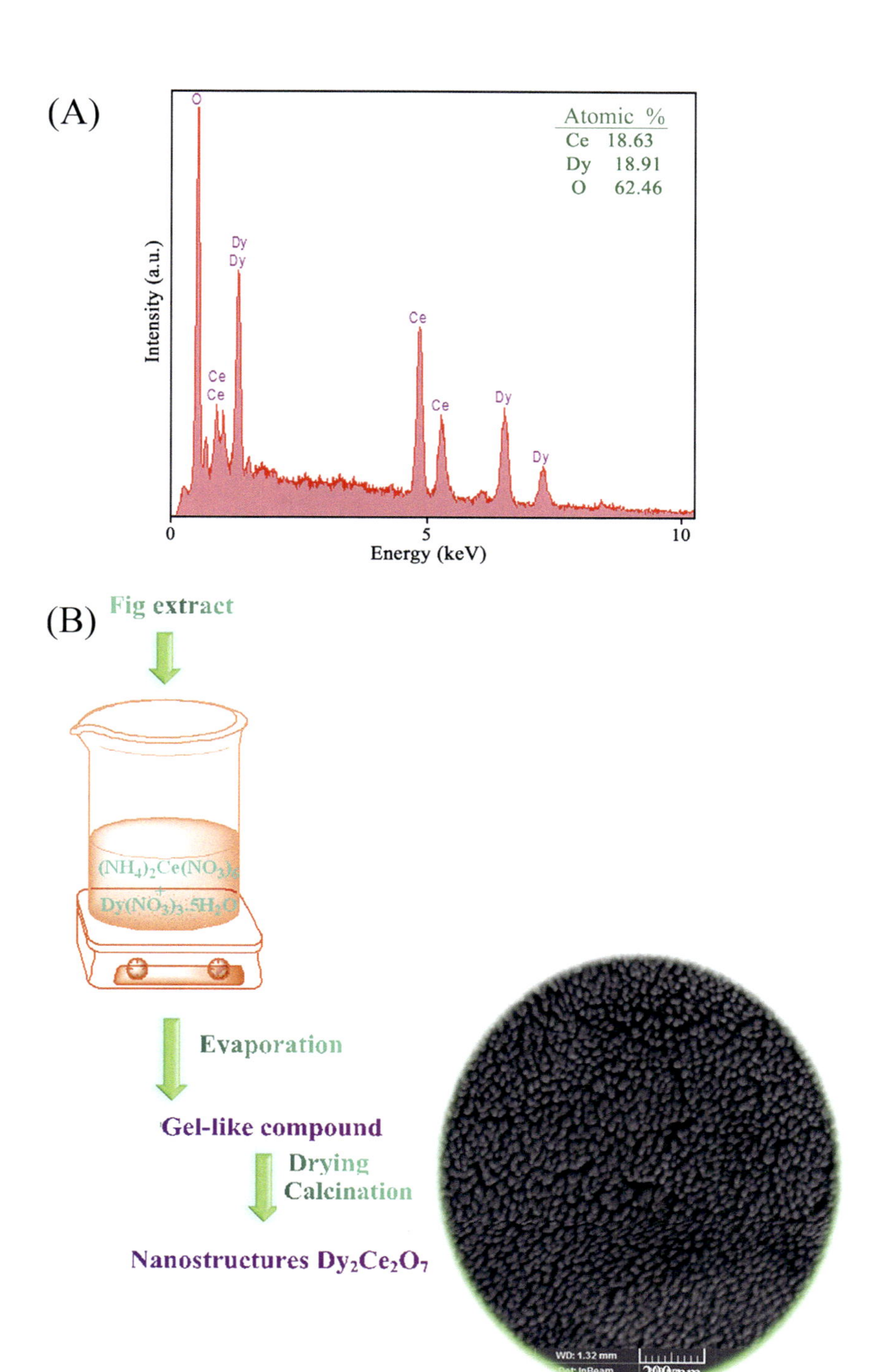

Figure 3.5 (A) EDS pattern of the produced sample in the presence of fig extract at 400°C (sample 4) and (B) schematic diagram to produce nanostructured $Dy_2Ce_2O_7$ (Zinatloo-Ajabshir et al., 2019a). *Source*: Reprinted with permission from Zinatloo-Ajabshir, S., Salehi, Z., Amiri, O., Salavati-Niasari, M. (2019). Green synthesis, characterization and investigation of the electrochemical hydrogen storage properties of $Dy_2Ce_2O_7$ nanostructures with fig extract. *International Journal of Hydrogen Energy*, *44*, 20110–20120, Copyright 2019 Elsevier.

et al., 2011). Their results indicated that the ceramic $La_{1.95}Ca_{0.05}Ce_2O_7$ sample could have acceptable stability in H_2O and CO_2 conditions (Tao et al., 2011).

The utilization of lanthanide cerate ceramic structures in the preparation of hydrogen separation membranes has also been reported (Zhu et al., 2016). In this separation approach, hydrogen is transferred from the high hydrogen pressure to the low hydrogen pressure by means of protons as carrier. It has been illustrated that the hydrogen permeation flux can largely depend on factors such as electronic conductivity, hydrogen gradient, proton conductivity, and membrane thickness (Tao et al., 2015; Zhu et al., 2016). For example, the rate of hydrogen permeability was investigated by Yan et al. (2010) using a membrane prepared on the basis of $La_2Ce_2O_7$. They found that enhancing the appropriate amount of samarium to the ceramic $La_2Ce_2O_7$ sample could play a remarkable role in increasing the hydrogen permeation flux. The results exhibited that the maximum hydrogen permeation flux could be 2.86×10^{-8} mol/cm^2 s at 900°C (Yan et al., 2010).

Recent studies have illustrated that the addition of $La_2Ce_2O_7$ ceramic nanoparticles to polybenzimidazole (PBI) membranes can enhance the hydrophilicity of the polymer and also adjust the hydrophilic/hydrophobic attributes of PLC_x (PBI/$La_2Ce_2O_7$ nanocomposite membranes, x denotes wt.% of $La_2Ce_2O_7$ nanoparticles) (Shabanikia, Javanbakht, Amoli, Hooshyari, & Enhessari, 2014; Zhu et al., 2016). This phenomenon can also enhance the conductivity of protons as well as the ion exchange capacity, so the addition of $La_2Ce_2O_7$ nanoparticles can provide possible routes for proton transport. Their results illustrated that PLC4 could exhibit a maximum power density of 0.43 W/cm^2 (Shabanikia et al., 2014; Zhu et al., 2016). Thus, the addition of $La_2Ce_2O_7$ ceramic nanoparticles as a proper proton conductor to proton polymer membranes can be very beneficial and improve the usage of proton polymer membranes in the field of high temperature proton exchange membrane fuel cells (Shabanikia et al., 2014; Zhu et al., 2016). Of course, this proton conductor can be very advantageous for hydrogen economy (Zhu et al., 2016).

The usage of lanthanide cerate structures in proton ceramic conductors as an electrolyte to produce ammonia at atmospheric pressure has also been reported. For example, the utilization of Ca-doped proton-conducting structures of $La_2Ce_2O_7$ as an electrolyte for ammonia production has been investigated (Wang et al., 2005; Zhu et al., 2016). The results indicated that during the usage of doped $La_2Ce_2O_7$ proton conductors, the maximum ammonia production rate could reach 1.3×10^{-9} mol/cm^2 s. Hence, materials based on $La_2Ce_2O_7$ ceramic structure can be introduced as a very beneficial and effective proton conductor in hydrogen economy (Wang et al., 2005; Zhu et al., 2016).

Also, the usage of ceramic nanostructures with desirable attributes for hydrogen storage as a clean source of energy, today can be very beneficial and effective in eliminating environmental pollution (Ghodrati, Mousavi-Kamazani, & Zinatloo-Ajabshir, 2020; Mousavi-Kamazani, Zinatloo-Ajabshir, & Ghodrati, 2020; Zinatloo-Ajabshir, Morassaei, et al., 2019; Zinatloo-Ajabshir, Salehi, & Salavati-Niasari, 2019; Zinatloo-Ajabshir et al., 2020a). Zinatloo-Ajabshir et al. (2019a) prepared the nanostructured $Dy_2Ce_2O_7$ utilizing fig extract at various temperatures 600°C, 500°C, and 400°C and compared their electrochemical hydrogen storage

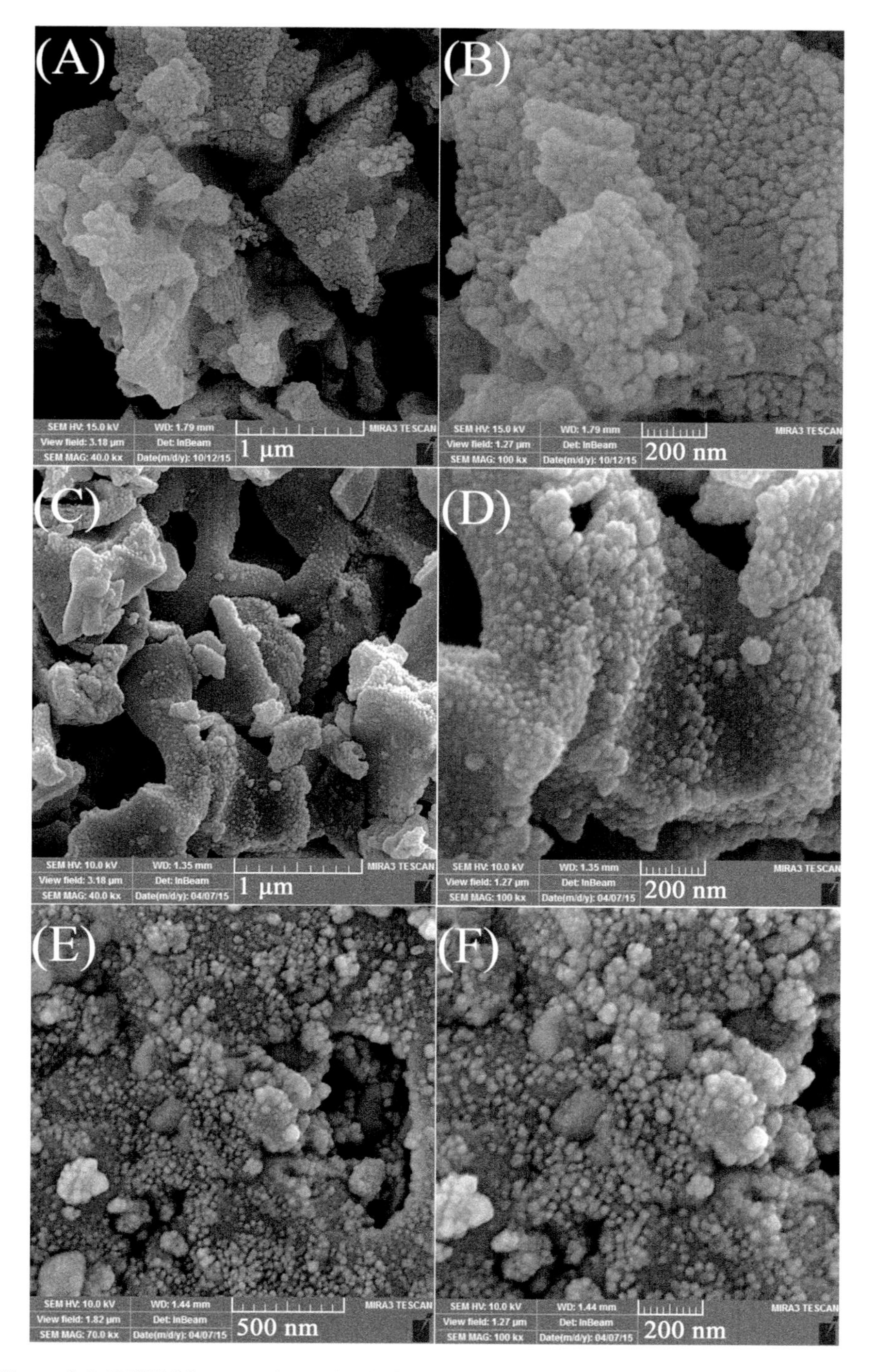

Figure 3.6 FESEM images of samples 1, 2, and 3 produced at 700°C (A and B), 600°C (C and D), and 500°C (E and F) (Zinatloo-Ajabshir et al., 2019a).
Source: Reprinted with permission from Zinatloo-Ajabshir, S., Salehi, Z., Amiri, O., Salavati-Niasari, M. (2019). Green synthesis, characterization and investigation of the electrochemical hydrogen storage properties of $Dy_2Ce_2O_7$ nanostructures with fig extract. *International Journal of Hydrogen Energy*, *44*, 20110–20120, Copyright 2019 Elsevier.

features via chronopotentiometry way at potash solution (see Fig. 3.7A–C) (Zinatloo-Ajabshir et al., 2019a). They observed that the maximum discharge capacity for samples 2, 3, and 4 could achieve to 1224, 1769, and 3070 mAh/g after 18 cycle charge/discharge, correspondingly, and sample 4 that produced at 400°C could indicate the best efficiency. Besides, their results illustrated that the discharge

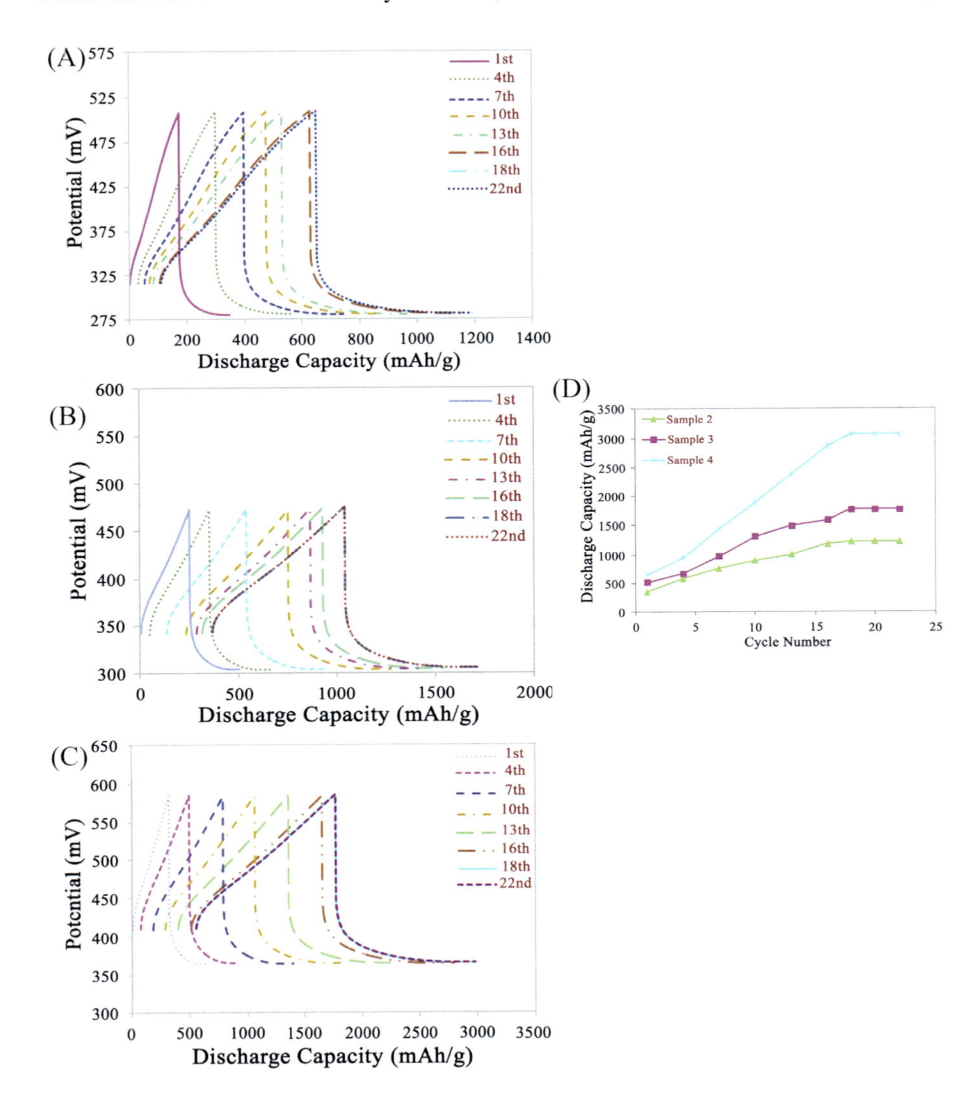

Figure 3.7 Discharge capacity curves of samples 2 (A), 3, (B), and 4 (C) under 1 mA, and (D) cycling performance of the samples 2, 3, and 4 (Zinatloo-Ajabshir et al., 2019a). *Source*: Reprinted with permission from Zinatloo-Ajabshir, S., Salehi, Z., Amiri, O., Salavati-Niasari, M. (2019). Green synthesis, characterization and investigation of the electrochemical hydrogen storage properties of $Dy_2Ce_2O_7$ nanostructures with fig extract. *International Journal of Hydrogen Energy*, *44*, 20110–20120, Copyright 2019 Elsevier.

plateau potential for all samples were about 0.28, 0.31, and 0.36 V, implying proper performance of sample 4 for electrochemical hydrogen storage (Zinatloo-Ajabshir et al., 2019a). They observed that the discharge capacity could be enhanced by increasing the number of cycles for all samples. They attributed the phenomenon to the creation of newer, more suitable sites for hydrogen uptake and desorption (see Fig. 3.7D) (Zinatloo-Ajabshir et al., 2019a).

Suitable porosity, large specific surface area, having well-uniformized particles as well as fine particle size are among the attributes that ceramic nanostructures must have in order to be able to exhibit optimal discharge capacity (Zinatloo-Ajabshir, Salehi, & Salavati-Niasari, 2019; Zinatloo-Ajabshir, Salehi, Amiri, & Salavati-Niasari, 2019b).

In recent years, owing to the rapid growth of industrial activities as well as agriculture, serious problems such as water contamination and poor quality water as well as environmental pollution have arisen. Therefore, all over the world, the very effective removal of toxic and unsafe contaminants from wastewater of various industries is still a serious and remarkable challenge (Arumugam & Choi, 2020; Kamal et al., 2019; Malathi, Arunachalam, Madhavan, Al-Mayouf, & Ghanem, 2018; Wetchakun, Wetchakun, & Sakulsermsuk, 2019; Xing et al., 2018). Thus, in order to meet the simultaneous demands such as the growth of the human population as well as the increment of industrial activities, it seems very necessary to provide a very efficient and capable solution to solve the wastewater problem. Among the various ways that have been proposed so far for the removal of organic and inorganic contaminants from aquatic environments, the photocatalysis way is very remarkable for the rapid and economical elimination of various contaminants as an environmentally friendly way. This approach, as one of the powerful and green approaches in order to eliminate organic and inorganic contaminants, has been considered by many scientists (Arumugam & Choi, 2020; Liu, Li, Du, Song, & Hou, 2014; Mao, Bao, Fang, & Yi, 2019; Zhang et al., 2019; Zinatloo-Ajabshir, Baladi, & Salavati-Niasari, 2021; Zinatloo-Ajabshir, Heidari-Asil, & Salavati-Niasari, 2021). Today, lanthanide cerate ceramic nanostructures have received a great deal of attention as a new and beneficial type of nanocatalysts owing to their excellent electrical and optical attributes and also due to their substantial photocatalytic performance (Malathi et al., 2017; Salehi et al., 2016b). Table 3.1 summarizes the photocatalytic results of lanthanide cerate ceramic nanostructures for the removal of various contaminants reported by various research groups.

Studies have indicated that the attributes of ceramic nanostructures such as chemical composition purity, particle size distribution, particle shape as well as particle uniformity and specific surface area and structural porosity can play a substantial role in regulating photocatalytic performance for removal of pollutants (Moshtaghi, Zinatloo-Ajabshir, & Salavati-Niasari, 2016b; Zinatloo-Ajabshir et al., 2018a; Zinatloo-Ajabshir, Morassaei, Amiri, & Salavati-Niasari, 2020b). Zinatloo-Ajabshir et al. (2018b) fabricated pure $Dy_2Ce_2O_7$ nanostructures utilizing juice of *Punica granatum* as fuel at several time and temperature. They compared the photocatalytic performance of different $Dy_2Ce_2O_7$ nanostructures, all of which differed in particle shape and particle size distribution and were prepared at different temperatures and times, in the degradation of erythrosine contaminant under visible

Table 3.1 Lanthanide cerate ceramic nanostructures for the removal of various contaminants.

Photocatalyst	Preparation way	Light source	Degradation rate (%)	Ref.
$Dy_2Ce_2O_7$ nanostructures	Salt-assisted combustion way	UV	80% of methyl orange after 60 min	Salehi et al. (2016b)
$CoFe_2O_4@SiO_2@Dy_2Ce_2O_7$ nanocomposites	In a three-step method	UV	90% of methyl violet after 30 min	Zinatloo-Ajabshir and Salavati-Niasari (2019)
$CoFe_2O_4@SiO_2@Dy_2Ce_2O_7$ nanocomposites	In a three-step method	UV	94.5% of methylene blue after 30 min	Zinatloo-Ajabshir and Salavati-Niasari (2019)
$CoFe_2O_4@SiO_2@Dy_2Ce_2O_7$ nanocomposites	In a three-step method	UV	98% of rhodamine B after 30 min	Zinatloo-Ajabshir and Salavati-Niasari (2019)
$Nd_2Ce_2O_7$ structures	Sol−gel way	Visible	77% of methylene blue after 180 min	Malathi et al. (2017)
$Pr_2Ce_2O_7$ structures	Sol−gel way	Visible	85% of methylene blue after 180 min	Malathi et al. (2017)
$Eu_2Ce_2O_7$ structures	Sol−gel way	Visible	56% of methylene blue after 180 min	Malathi et al. (2017)
$Pr_2Ce_2O_7$ nanostructures	Improved Pechini procedure	UV	92.1% of methyl orange after 50 min	Zinatloo-Ajabshir, Salehi, and Salavati-Niasari (2016)
Nanostructured $Dy_2Ce_2O_7$	Salt-assistant combustion route	UV	87% of eosin Y after 60 min	Salehi et al. (2016a)
Dysprosium cerate nanostructures	Modified Pechini approach	UV	91.3% of erythrosine after 60 min	Salehi et al. (2017)
Nanostructured $Dy_2Ce_2O_7$	Green method	Visible	92.8% of methylene blue after 70 min	Zinatloo-Ajabshir, Salehi, and Salavati-Niasari (2018b)
$Dy_2Ce_2O_7$ nanostructure	Green approach	Visible	96.8% of 2-naphthol after 20 min	Zinatloo-Ajabshir, Salehi, and Salavati-Niasari (2018c)
Nanoscale $Dy_2Ce_2O_7$	Green method	Visible	91.7% of eriochrome black T after 60 min	Zinatloo-Ajabshir, Salehi, and Salavati-Niasari (2018a)

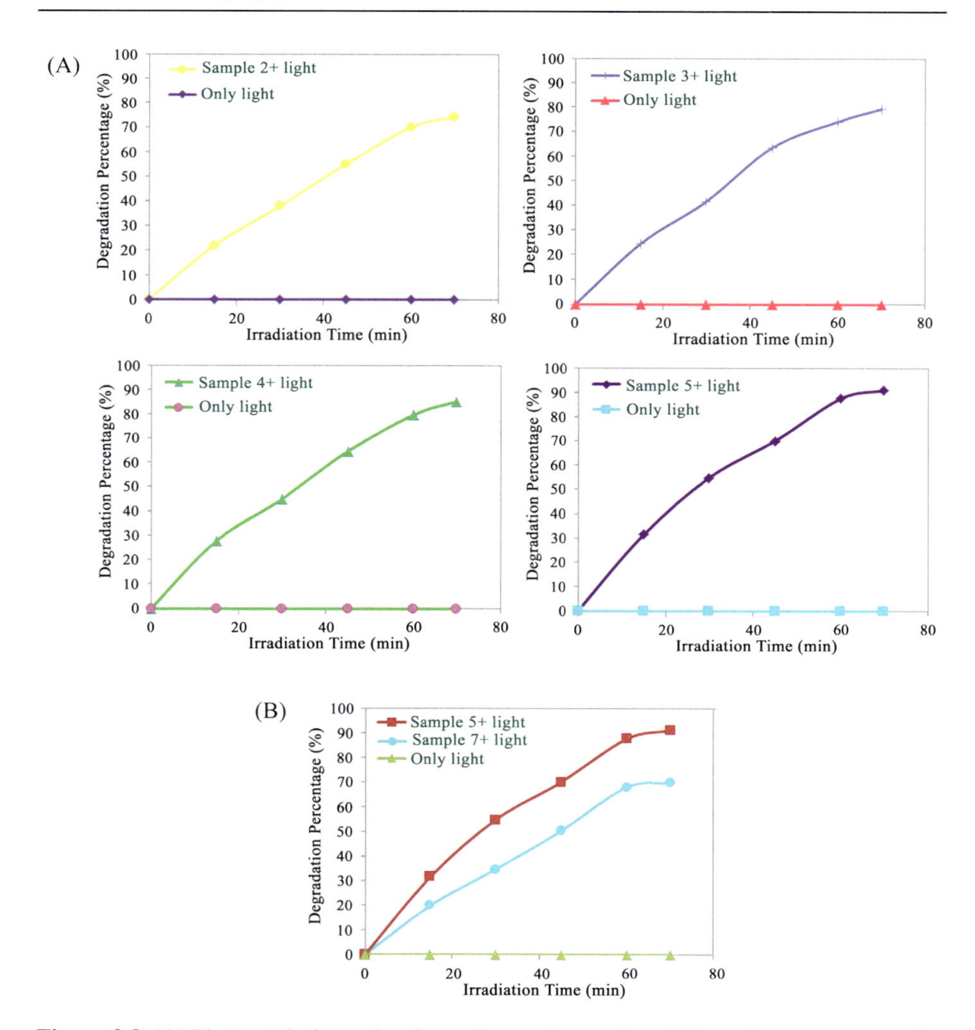

Figure 3.8 (A) Photocatalytic erythrosine pollutant destruction of the various structures of $Dy_2Ce_2O_7$ obtained at 750°C, 650°C, 550°C, and 450°C (samples 2, 3, 4, and 5) and (B) photocatalytic erythrosine pollutant destruction of the various structures of $Dy_2Ce_2O_7$ obtained within 4 and 5 h (samples 5 and 7) (Zinatloo-Ajabshir et al., 2018b).
Source: Reprinted with permission from Zinatloo-Ajabshir, S., Salehi, Z., Salavati-Niasari, M. (2018). Green synthesis of $Dy_2Ce_2O_7$ ceramic nanostructures using juice of *Punica granatum* and their efficient application as photocatalytic degradation of organic contaminants under visible light. *Ceramic International*, *44*, 3873–3883, Copyright 2018 Elsevier.

light (see Fig. 3.8) (Zinatloo-Ajabshir et al., 2018b). They observed that the ceramic sample prepared at 450°C for 4 hours was able to exhibit the highest photocatalytic efficiency in the degradation of erythrosine contaminant (91.4%). They attributed this excellent performance to the distinctive attributes of the ceramic nanostructure sample, such as fine particle size and suitable specific surface area (Zinatloo-Ajabshir et al.,

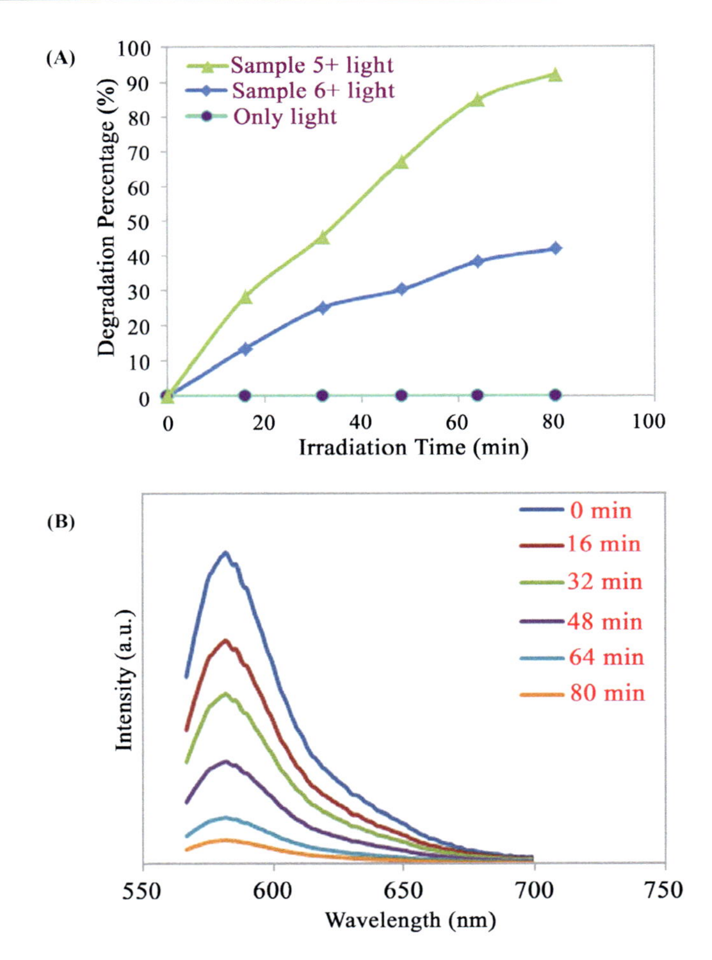

Figure 3.9 (A) Photocatalytic methyl orange decomposition of the fabricated $Dy_2Ce_2O_7$ structures (samples 5 and 6) and (B) fluorescence spectral time scan of methyl orange under visible illumination over $Dy_2Ce_2O_7$ sample fabricated with *Vitis vinifera* juice at 400°C (sample 5) (Zinatloo-Ajabshir et al., 2018c).
Source: Reprinted with permission from Zinatloo-Ajabshir, S., Salehi, Z., Salavati-Niasari, M. (2018). Green synthesis and characterization of $Dy_2Ce_2O_7$ ceramic nanostructures with good photocatalytic properties under visible light for removal of organic dyes in water. *Journal of Cleaner Production*, *192*, 678–687, Copyright 2018 Elsevier.

2018b). In another work, Zinatloo-Ajabshir et al. (2018c) prepared $Dy_2Ce_2O_7$ nanostructures employing *Vitis vinifera* juice. They observed that the ceramic nanostructure sample prepared with *V. vinifera* extract was able to display a very good photocatalytic performance in removing the methyl orange contaminant under visible light (92.4%) and the utilization of *V. vinifera* extract was very key to achieve this result. They attributed this very appropriate performance to the smaller particles of the nanostructured sample as well as its very desirable specific surface area (see Fig. 3.9A)

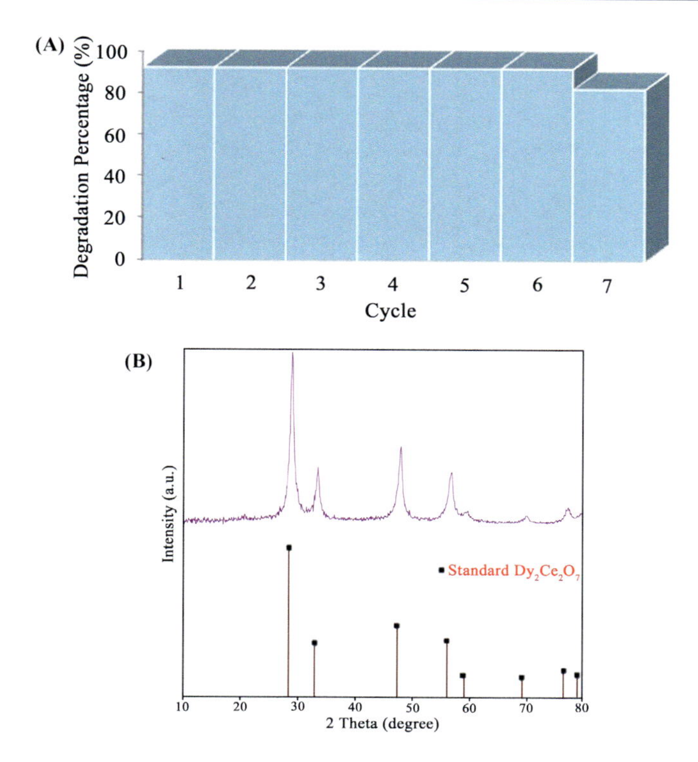

Figure 3.10 (A) Cycling runs of decomposition of methyl orange under visible illumination over $Dy_2Ce_2O_7$ nanostructures (sample 5) and (B) XRD pattern of the fabricated $Dy_2Ce_2O_7$ nanostructures (sample 5) after photocatalytic performance (Zinatloo-Ajabshir et al., 2018c). *Source*: Reprinted with permission from Zinatloo-Ajabshir, S., Salehi, Z., Salavati-Niasari, M. (2018). Green synthesis and characterization of $Dy_2Ce_2O_7$ ceramic nanostructures with good photocatalytic properties under visible light for removal of organic dyes in water. *Journal of Cleaner Production*, *192*, 678–687, Copyright 2018 Elsevier.

(Zinatloo-Ajabshir et al., 2018c). Their results indicated that the removal of methyl orange with the help of a nanostructured sample could occur continuously (see Fig. 3.9B) (Zinatloo-Ajabshir et al., 2018c).

Among the remarkable attributes of ceramic nanostructured samples as photocatalysts for the elimination of pollutants, excellent stability as well as appropriate ability to be used multiple times have been reported (Zinatloo-Ajabshir, Mortazavi-Derazkola, & Salavati-Niasari, 2017; Zinatloo-Ajabshir, Baladi, Amiri, & Salavati-Niasari, 2020). It was observed that the nanostructured $Dy_2Ce_2O_7$ sample did not display a remarkable reduction in its photocatalytic efficiency after 7 reuse for elimination of methyl orange contaminant (see Fig. 3.10) (Zinatloo-Ajabshir et al., 2018c). Also, by comparing its XRD outcomes, before and after photocatalytic tests, it was observed that $Dy_2Ce_2O_7$ nanostructure sample has a very appropriate stability.

3.5 Conclusion and outlook

In summary, lanthanide cerate ceramic nanostructures have been the focus of attention in recent years owing to their unique attributes. The utilization of this kind of nanostructures in various fields has received much attention in recent years because these kinds of nanostructures are very beneficial and effective. In this chapter of the book, the most common approaches that have recently been employed to prepare ceramic lanthanide cerate nanostructures are mentioned, and the most obvious attribute of each approach is summarized so that the book readers can choose the best approach to prepare these ceramic nanostructures. Studies have shown that the attributes and thus the performance of lanthanide cerate ceramic nanostructures depend to a large extent on purity in terms of chemical composition, particle shape as well as particle size distribution. On the other hand, the utilization of expensive and uneconomical approaches employing toxic substances can limit the usage of these momentous and widely used nanostructures. Despite the great efforts that have been made to provide approaches for the preparation of lanthanide cerate ceramic nanostructures, there is still a serious and significant need to provide low-cost, simple, and environmentally friendly approaches for the preparation of these nanostructures. Moreover, the uses of lanthanide cerate ceramic nanostructures in various fields in recent years were mentioned. Owing to the high dependence of the performance of ceramic nanostructures on their specific attributes, including structural and morphological attributes, it can be said that there is still a serious and significant need to regulate these attributes during the preparation of nanostructures to achieve ideal performance.

References

Andrievskaya, E. R., Kornienko, O. A., Sameljuk, A. V., & Sayir, A. (2011). Phase relation studies in the $CeO_2-La_2O_3$ system at 1100–1500°C. *Journal of the European Ceramic Society*, *31*, 1277–1283.

Arumugam, M., & Choi, M. Y. (2020). Recent progress on bismuth oxyiodide (BiOI) photocatalyst for environmental remediation. *Journal of Industrial and Engineering Chemistry*, *81*, 237–268.

Beshkar, F., Zinatloo-Ajabshir, S., & Salavati-Niasari, M. (2015). Simple morphology-controlled fabrication of nickel chromite nanostructures via a novel route. *Chemical Engineering Journal*, *279*, 605–614.

Besikiotis, V., Knee, C. S., Ahmed, I., Haugsrud, R., & Norby, T. (2012). Crystal structure, hydration and ionic conductivity of the inherently oxygen-deficient $La_2Ce_2O_7$. *Solid State Ionics*, *228*, 1–7.

Biju, S., Reddy, M. L. P., Cowley, A. H., & Vasudevan, K. V. (2009). 3-Phenyl-4-acyl-5-isoxazolonate complex of Tb^{3+} doped into poly-β-hydroxybutyrate matrix as a promising light-conversion molecular device. *Journal of Materials Chemistry*, *19*, 5179–5187.

Bünzli, J.-C. G., & Piguet, C. (2005). Taking advantage of luminescent lanthanide ions. *Chemical Society Reviews*, *34*, 1048–1077.

Cao, X., Vassen, R., Fischer, W., Tietz, F., Jungen, W., & Stöver, D. (2003). Lanthanum–cerium oxide as a thermal barrier-coating material for high-temperature applications. *Advanced Materials*, *15*, 1438–1442.

Chatzichristodoulou, C., Hendriksen, P. V., & Hagen, A. (2010). Defect chemistry and thermomechanical properties of $Ce_{0.8}Pr_xTb_{0.2-x}O_{2-\delta}$. *Journal of the Electrochemical Society*, *157*, B299.

Chen, G., Tan, S., Koros, W. J., & Jones, C. W. (2015). Metal organic frameworks for selective adsorption of t-butyl mercaptan from natural gas. *Energy & Fuels*, *29*, 3312–3321.

Chen, Y., Liu, B., Chen, J., Tian, L., Huang, L., Tu, M., & Tan, S. (2015). Structure design and photocatalytic properties of one-dimensional SnO_2-TiO_2 composites. *Nanoscale Research Letters*, *10*, 200.

Chen, Y., Zhou, W., Ding, D., Liu, M., Ciucci, F., Tade, M., & Shao, Z. (2015). Advances in cathode materials for solid oxide fuel cells: complex oxides without alkaline earth metal elements. *Advanced Energy Materials*, *5*, 1500537.

Cheng, M.-Y., Hwang, D.-H., Sheu, H.-S., & Hwang, B.-J. (2008). Formation of $Ce_{0.8}Sm_{0.2}O_{1.9}$ nanoparticles by urea-based low-temperature hydrothermal process. *Journal of Power Sources*, *175*, 137–144.

D.R.C., & Levi, C. G. (2003). Materials design for the next generation thermal barrier coatings. *Annual Review of Materials Research*, *33*, 383–417.

Dai, H., Zhong, X., Li, J., Meng, J., & Cao, X. (2006). Neodymium–cerium oxide as new thermal barrier coating material. *Surface and Coatings Technology*, *201*, 2527–2533.

Darolia, R. (2013). Thermal barrier coatings technology: Critical review, progress update, remaining challenges and prospects. *International Materials Reviews*, *58*, 315–348.

Dohnalová, Ž., Šulcová, P., & Trojan, M. (2009). Effect of Er^{3+} substitution on the quality of Mg–Fe spinel pigments. *Dyes and Pigments*, *80*, 22–25.

Fan, L., Zhu, B., Su, P.-C., & He, C. (2018). Nanomaterials and technologies for low temperature solid oxide fuel cells: Recent advances, challenges and opportunities. *Nano Energy*, *45*, 148–176.

Fuentes, R. O., & Baker, R. T. (2008). Synthesis and properties of Gadolinium-doped ceria solid solutions for IT-SOFC electrolytes. *International Journal of Hydrogen Energy*, *33*, 3480–3484.

Furukawa, S., Masui, T., & Imanaka, N. (2008). New environment-friendly yellow pigments based on CeO_2–ZrO_2 solid solutions. *Journal of Alloys and Compounds*, *451*, 640–643.

Fu, Y.-P., Chen, S.-H., Tsai, F.-Y., & Hu, S.-H. (2009). Aqueous tape casting and crystallization kinetics of Ce0.8La0.2O1.9 powder. *Ceramics International*, *35*, 609–615.

Gao, Z., Mogni, L. V., Miller, E. C., Railsback, J. G., & Barnett, S. A. (2016). A perspective on low-temperature solid oxide fuel cells. *Energy & Environmental Science*, *9*, 1602–1644.

García, A., Llusar, M., Calbo, J., Tena, M. A., & Monrós, G. (2001). Low-toxicity red ceramic pigments for porcelainised stoneware from lanthanide–cerianite solid solutions. *Green Chemistry*, *3*, 238–242.

George, G., Sandhya Kumari, L., Vishnu, V. S., Ananthakumar, S., & Reddy, M. L. P. (2008). Synthesis and characterization of environmentally benign calcium-doped $Pr_2Mo_2O_9$ pigments: Applications in coloring of plastics. *Journal of Solid State Chemistry*, *181*, 487–492.

Ghodrati, M., Mousavi-Kamazani, M., & Zinatloo-Ajabshir, S. (2020). $Zn_3V_3O_8$ nanostructures: Facile hydrothermal/solvothermal synthesis, characterization, and electrochemical hydrogen storage. *Ceramics International*, *46*, 28894–28902.

Heidari-Asil, S. A., Zinatloo-Ajabshir, S., Amiri, O., & Salavati-Niasari, M. (2020). Amino acid assisted-synthesis and characterization of magnetically retrievable $ZnCo_2O_4-Co_3O_4$ nanostructures as high activity visible-light-driven photocatalyst. *International Journal of Hydrogen Energy*, *45*, 22761–22774.

Higashi, K., Sonoda, K., Ono, H., Sameshima, S., & Hirata, Y. (1999). Synthesis and sintering of rare-earth-doped ceria powder by the oxalate coprecipitation method. *Journal of Materials Research*, *14*, 957–967.

Hong-song, Z., Xiao-ge, C., Gang, L., Xin-Li, W., & Xu-dan, D. (2012). Influence of Gd_2O_3 addition on thermophysical properties of $La_2Ce_2O_7$ ceramics for thermal barrier coatings. *Journal of the European Ceramic Society*, *32*, 3693–3700.

Hongsong, Z., Jianguo, L., Gang, L., Zheng, Z., & Xinli, W. (2012). Investigation about thermophysical properties of $Ln_2Ce_2O_7$ (Ln = Sm, Er and Yb) oxides for thermal barrier coatings. *Materials Research Bulletin*, *47*, 4181–4186.

Hongsong, Z., Shuqing, Y., & Xiaoge, C. (2014). Preparation and thermophysical properties of fluorite-type samarium–dysprosium–cerium oxides. *Journal of the European Ceramic Society*, *34*, 55–61.

Hongsong, Z., Suran, L., & Shaokang, G. (2012). Preparation and thermal conductivity of $Dy_2Ce_2O_7$ ceramic material. *Journal of Materials Engineering and Performance*, *21*, 1046–1050.

Hongsong, Z., Xiaochun, L., Gang, L., & Zhenjun, L. (2014). Preparation, characterization and thermophysical properties of $(Sm_{1-x}Gd_x)_2Ce_2O_7$ solid solutions. *Ceramics International*, *40*, 4567–4573.

Huang, F., Pu, F., Lu, X., Zhang, H., Xia, Y., Huang, W., & Li, Z. (2013). Photoelectrochemical sensing of Cu^{2+} ions with SnO_2/CdS heterostructural films. *Sensors and Actuators B: Chemical*, *183*, 601–607.

Huang, K., Li, Y., Yan, L., & Xing, Y. (2014). Nanoscale conductive niobium oxides made through low temperature phase transformation for electrocatalyst support. *RSC Advances*, *4*, 9701–9708.

Kamal, S., Balu, S., Palanisamy, S., Uma, K., Velusamy, V., & Yang, T. C. K. (2019). Synthesis of boron doped $C_3N_4/NiFe_2O_4$ nanocomposite: An enhanced visible light photocatalyst for the degradation of methylene blue. *Results in Physics*, *12*, 1238–1244.

Khademinia, S., & Behzad, M. (2015). Lanthanum cerate ($La_2Ce_2O_7$): Hydrothermal synthesis, characterization and optical properties. *International Nano Letters*, *5*, 101–107.

Khan, Z. S., Zou, B., Chen, X., Wang, C., Fan, X., Gu, L., . . . Cao, X. (2012). Novel double ceramic coatings based on $Yb_2Si_2O_7/La_2(Zr_{0.7}Ce_{0.3})_2O_7$ by plasma spraying on Cf/SiC composites and their thermal shock behavior. *Surface and Coatings Technology*, *207*, 546–554.

Kido, J., & Okamoto, Y. (2002). Organo lanthanide metal complexes for electroluminescent materials. *Chemical Reviews*, *102*, 2357–2368.

Kumari, L. S., George, G., Rao, P. P., & Reddy, M. L. P. (2008). The synthesis and characterization of environmentally benign praseodymium-doped $TiCeO_4$ pigments. *Dyes and Pigments*, *77*, 427–431.

Levi, C. G. (2004). Emerging materials and processes for thermal barrier systems. *Current Opinion in Solid State and Materials Science*, *8*, 77–91.

Li, J., Wang, C., Wang, X., & Bi, L. (2020). Sintering aids for proton-conducting oxides – A double-edged sword? A mini review. *Electrochemistry Communications*, *112*, 106672.

Li, J.-G., Ikegami, T., Mori, T., & Wada, T. (2001). Reactive $Ce_{0.8}RE_{0.2}O_{1.9}$ (RE = La, Nd, Sm, Gd, Dy, Y, Ho, Er, and Yb) powders via carbonate coprecipitation. 1. Synthesis and characterization. *Chemistry of Materials*, *13*, 2913–2920.

Ling, Y., Chen, J., Wang, Z., Xia, C., Peng, R., & Lu, Y. (2013). New ionic diffusion strategy to fabricate proton-conducting solid oxide fuel cells based on a stable $La_2Ce_2O_7$ electrolyte. *International Journal of Hydrogen Energy*, *38*, 7430–7437.

Lin, B., Wang, S., Liu, X., & Meng, G. (2009). Stable proton-conducting Ca-doped $LaNbO_4$ thin electrolyte-based protonic ceramic membrane fuel cells by in situ screen printing. *Journal of Alloys and Compounds*, *478*, 355–357.

Liu, J., Li, H., Du, N., Song, S., & Hou, W. (2014). Synthesis, characterization, and visible-light photocatalytic activity of BiOI hierarchical flower-like microspheres. *RSC Advances*, *4*, 31393–31399.

Liu, Z. G., Ouyang, J. H., & Sun, K. N. (2011). Electrical conductivity improvement of $Nd_2Ce_2O_7$ ceramic co-doped with Gd_2O_3 and ZrO_2. *Fuel Cells*, *11*, 153–157.

Lounis, Z., Bouslama, Mh, Zegadi, C., Ghaffor, D., Gazzoul, Mh, Baizid, A., ... Ouerdane, A. (2018). Surface stoichiometry analysis by AES, EELS spectroscopy and AFM microscopy in UHV atmosphere of SnO_2 thin film. *Journal of Electron Spectroscopy and Related Phenomena*, *226*, 9–16.

Malathi, A., Arunachalam, P., Madhavan, J., Al-Mayouf, A. M., & Ghanem, M. A. (2018). Rod-on-flake α-FeOOH/BiOI nanocomposite: Facile synthesis, characterization and enhanced photocatalytic performance. *Colloids and Surfaces A: Physicochemical and Engineering Aspects*, *537*, 435–445.

Malathi, M., Sreenu, K., Gundeboina, R., Kumar, P., Ch, S., Guje, R., ... Vithal, M. (2017). Low temperature synthesis of fluorite-type Ce-based oxides of composition $Ln_2Ce_2O_7$ (Ln = Pr, Nd and Eu): Photodegradation and luminescence studies. *Journal of Chemical Sciences*, *129*.

Mao, S., Bao, R., Fang, D., & Yi, J. (2019). Fabrication of sliver/graphitic carbon nitride photocatalyst with enhanced visible-light photocatalytic efficiency through ultrasonic spray atomization. *Journal of Colloid and Interface Science*, *538*, 15–24.

Martos, M., Julián-López, B., Cordoncillo, E., & Escribano, P. (2008). Structural and spectroscopic study of a novel erbium titanate pink pigment prepared by sol – gel methodology. *The Journal of Physical Chemistry. B*, *112*, 2319–2325.

Martos, M., Julián-López, B., Folgado, J. V., Cordoncillo, E., & Escribano, P. (2008). Sol–gel synthesis of tunable cerium titanate materials. *European Journal of Inorganic Chemistry*, *2008*, 3163–3171.

Matsouka, V., Konsolakis, M., Lambert, R. M., & Yentekakis, I. V. (2008). In situ DRIFTS study of the effect of structure (CeO_2–La_2O_3) and surface (Na) modifiers on the catalytic and surface behaviour of Pt/γ-Al_2O_3 catalyst under simulated exhaust conditions. *Applied Catalysis B: Environmental*, *84*, 715–722.

Matussin, S., Harunsani, M. H., Tan, A. L., & Khan, M. M. (2020). Plant-extract-mediated SnO_2 nanoparticles: Synthesis and applications. *ACS Sustainable Chemistry & Engineering*, *8*, 3040–3054.

Ma, W., Gong, S., Xu, H., & Cao, X. (2006). On improving the phase stability and thermal expansion coefficients of lanthanum cerium oxide solid solutions. *Scripta Materialia*, *54*, 1505–1508.

Medvedev, D., Murashkina, A., Pikalova, E., Demin, A., Podias, A., & Tsiakaras, P. (2014). $BaCeO_3$: Materials development, properties and application. *Progress in Materials Science*, *60*, 72–129.

Miller, R. A. (1997). Thermal barrier coatings for aircraft engines: History and directions. *Journal of Thermal Spray Technology*, *6*, 35.

Morassaei, M. S., Zinatloo-Ajabshir, S., & Salavati-Niasari, M. (2016a). New facile synthesis, structural and photocatalytic studies of NdOCl-$Nd_2Sn_2O_7$-SnO_2 nanocomposites. *Journal of Molecular Liquids*, *220*, 902–909.

Morassaei, M. S., Zinatloo-Ajabshir, S., & Salavati-Niasari, M. (2016b). Simple salt-assisted combustion synthesis of $Nd_2Sn_2O_7-SnO_2$ nanocomposites with different amino acids as fuel: an efficient photocatalyst for the degradation of methyl orange dye. *Journal of Materials Science: Materials in Electronics*, *27*, 11698–11706.

Mori, T., Drennan, J., Lee, J.-H., Li, J.-G., & Ikegami, T. (2002). Oxide ionic conductivity and microstructures of Sm- or La-doped CeO_2-based systems. *Solid State Ionics*, *154–155*, 461–466.

Mori, T., Kobayashi, T., Wang, Y., Drennan, J., Nishimura, T., Li, J.-G., & Kobayashi, H. (2005). Synthesis and characterization of nano-hetero-structured dy doped CeO_2 solid electrolytes using a combination of spark plasma sintering and conventional sintering. *Journal of the American Ceramic Society*, *88*, 1981–1984.

Moshtaghi, S., Zinatloo-Ajabshir, S., & Salavati-Niasari, M. (2016a). Preparation and characterization of $BaSnO_3$ nanostructures via a new simple surfactant-free route. *Journal of Materials Science: Materials in Electronics*, *27*, 425–435.

Moshtaghi, S., Zinatloo-Ajabshir, S., & Salavati-Niasari, M. (2016b). Nanocrystalline barium stannate: facile morphology-controlled preparation, characterization and investigation of optical and photocatalytic properties. *Journal of Materials Science: Materials in Electronics*, *27*, 834–842.

Mousavi-Kamazani, M., Zinatloo-Ajabshir, S., & Ghodrati, M. (2020). One-step sonochemical synthesis of $Zn(OH)_2/ZnV_3O_8$ nanostructures as a potent material in electrochemical hydrogen storage. *Journal of Materials Science: Materials in Electronics*, *31*, 17332–17338.

Narayanan, V., Lommens, P., De Buysser, K., Vanpoucke, D. E. P., Huehne, R., Molina, L., ... Van Driessche, I. (2012). Aqueous CSD approach for the growth of novel, lattice-tuned $La_xCe_{1-x}O_\delta$ epitaxial layers. *Journal of Materials Chemistry*, *22*, 8476–8483.

Nobuhito, I., Toshiyuki, M., & Shinya, F. (2008). Novel nontoxic and environment-friendly inorganic yellow pigments. *Chemistry Letters*, *37*, 104–105.

O'Neill, W. M., & Morris, M. A. (1999). The defect chemistry of lanthana–ceria mixed oxides by MASNMR. *Chemical Physics Letters*, *305*, 389–394.

Padture, N. P., Gell, M., & Jordan, E. H. (2002). Thermal barrier coatings for gas-turbine engine applications. *Science (New York, N.Y.)*, *296*, 280–284.

Pailhé, N., Gaudon, M., & Demourgues, A. (2009). (Ca^{2+}, V^{5+}) co-doped $Y_2Ti_2O_7$ yellow pigment. *Materials Research Bulletin*, *44*, 1771–1777.

Perepezko, J. H. (2009). The hotter the engine, the better. *Science (New York, N.Y.)*, *326*, 1068–1069.

Petra, Š., & Trojan, M. (2008). Thermal analysis of the $(Bi_2O_3)_{1-x}(Y_2O_3)_x$ pigments. *Journal of Thermal Analysis and Calorimetry*, *91*, 151–154.

Praveen, K., Sravani, N., Alroy, R. J., Shanmugavelayutham, G., & Sivakumar, G. (2019). Hot corrosion behaviour of atmospheric and solution precursor plasma sprayed $(La_{0.9}Gd_{0.1})_2Ce_2O_7$ coatings in sulfate and vanadate environments. *Journal of the European Ceramic Society*, *39*, 4233–4244.

Qu, Z., Wan, C., & Pan, W. (2012). Thermophysical properties of rare-earth stannates: Effect of pyrochlore structure. *Acta Materialia*, *60*, 2939–2949.

Raj, A. K. V., Prabhakar Rao, P., Sreena, T. S., Sameera, S., James, V., & Renju, U. A. (2014). Remarkable changes in the photoluminescent properties of $Y_2Ce_2O_7{:}Eu^{3+}$ red phosphors through modification of the cerium oxidation states and oxygen vacancy ordering. *Physical Chemistry Chemical Physics*, *16*, 23699–23710.

Razi, F., Zinatloo-Ajabshir, S., & Salavati-Niasari, M. (2017). Preparation and characterization of HgI_2 nanostructures via a new facile route. *Materials Letters*, *193*, 9–12.

Reddy, B. M., Katta, L., & Thrimurthulu, G. (2010). Novel nanocrystalline $Ce_{1-x}La_xO_{2-\delta}$ (x = 0.2) solid solutions: Structural characteristics and catalytic performance. *Chemistry of Materials*, *22*, 467–475.

Ren, K., Wang, Q., Cao, Y., Shao, G., & Wang, Y. (2021). Multicomponent rare-earth cerate and zirconocerate ceramics for thermal barrier coating materials. *Journal of the European Ceramic Society*, *41*, 1720–1725.

Salehi, Z., Zinatloo-Ajabshir, S., & Salavati-Niasari, M. (2016a). New simple route to prepare $Dy_2Ce_2O_7$ nanostructures: Structural and photocatalytic studies. *Journal of Molecular Liquids*, *222*, 218–224.

Salehi, Z., Zinatloo-Ajabshir, S., & Salavati-Niasari, M. (2016b). Novel synthesis of $Dy_2Ce_2O_7$ nanostructures via a facile combustion route. *RSC Advances*, *6*, 26895–26901.

Salehi, Z., Zinatloo-Ajabshir, S., & Salavati-Niasari, M. (2017). Dysprosium cerate nanostructures: Facile synthesis, characterization, optical and photocatalytic properties. *Journal of Rare Earths*, *35*, 805–812.

Sánchez-Bautista, C., Dos santos-García, A. J., Peña-Martínez, J., & Canales-Vázquez, J. (2010). The grain boundary effect on dysprosium doped ceria. *Solid State Ionics*, *181*, 1665–1673.

Schelling, P. K., Phillpot, S. R., & Grimes, R. W. (2004). Optimum pyrochlore compositions for low thermal conductivity. *Philosophical Magazine Letters*, *84*, 127–137.

Shabanikia, A., Javanbakht, M., Amoli, H. S., Hooshyari, K., & Enhessari, M. (2014). Effect of $La_2Ce_2O_7$ on the physicochemical properties of phosphoric acid doped polybenzimidazole nanocomposite membranes for high temperature proton exchange membrane fuel cells applications. *Journal of the Electrochemical Society*, *161*, F1403–F1408.

Shuk, P., & Greenblatt, M. (1999). Hydrothermal synthesis and properties of mixed conductors based on $Ce_{1-x}Pr_xO_{2-\delta}$ solid solutions 1 In memory of Professor Mikhail Perfiliev.1. *Solid State Ionics*, *116*, 217–223.

Singh, P., & Hegde, M. S. (2008). Controlled synthesis of nanocrystalline CeO_2 and $Ce_{1-x}M_xO_{2-\delta}$ (M = Zr, Y, Ti, Pr and Fe) solid solutions by the hydrothermal method: Structure and oxygen storage capacity. *Journal of Solid State Chemistry*, *181*, 3248–3256.

Singh, K., Kumar, R., & Chowdhury, A. (2017). Synergistic effects of ultrasonication and ethanol washing in controlling the stoichiometry, phase-purity and morphology of rare-earth doped ceria nanoparticles. *Ultrasonics Sonochemistry*, *36*, 182–190.

Somacescu, S., Parvulescu, V., Calderon-Moreno, J. M., Suh, S.-H., Osiceanu, P., & Su, B.-L. (2012). Uniform nanoparticles building $Ce_{1-x}Pr_xO_{2-\delta}$ mesoarchitectures: Structure, morphology, surface chemistry, and catalytic performance. *Journal of Nanoparticle Research*, *14*, 885.

Šulcová, P., & Trojan, M. (2004). Thermal synthesis of the CeO_2-PrO_2-La_2O_3 pigments. *Journal of Thermal Analysis and Calorimetry*, *77*, 99.

Sun, W., Fang, S., Yan, L., & Liu, W. (2012). Investigation on proton conductivity of $La_2Ce_2O_7$ in wet atmosphere: dependence on water vapor partial pressure. *Fuel Cells*, *12*, 457–463.

Sung Bae, J., Kil Choo, W., & Hee, C. (2004). Lee, The crystal structure of ionic conductor $La_xCe_{1-x}O_{2-x/2}$. *Journal of the European Ceramic Society*, *24*, 1291–1294.

Tadokoro, S. K., & Muccillo, E. N. S. (2007). Effect of Y and Dy co-doping on electrical conductivity of ceria ceramics. *Journal of the European Ceramic Society*, *27*, 4261–4264.

Taniguchi, H., Kido, J., Nishiya, M., & Sasaki, S. (1995). Europium chelate solid laser based on morphology-dependent resonances. *Applied Physics Letters*, *67*, 1060–1062.

Tan, S., Gil, L. B., Subramanian, N., Sholl, D. S., Nair, S., Jones, C. W., ... Pendergast, J. G. (2015). Catalytic propane dehydrogenation over $In_2O_3-Ga_2O_3$ mixed oxides. *Applied Catalysis A: General*, *498*, 167–175.

Tao, Z., Bi, L., Fang, S., & Liu, W. (2011). A stable $La_{1.95}Ca_{0.05}Ce_2O_{7-\delta}$ as the electrolyte for intermediate-temperature solid oxide fuel cells. *Journal of Power Sources*, *196*, 5840–5843.

Tao, Z., Yan, L., Qiao, J., Wang, B., Zhang, L., & Zhang, J. (2015). A review of advanced proton-conducting materials for hydrogen separation. *Progress in Materials Science*, *74*, 1–50.

Vassen, R., Cao, X., Tietz, F., Basu, D., & Stöver, D. (2000). Zirconates as new materials for thermal barrier coatings. *Journal of the American Ceramic Society*, *83*, 2023–2028.

Vijayaraghavan, K., & Ashokkumar, T. (2017). Plant-mediated biosynthesis of metallic nanoparticles: A review of literature, factors affecting synthesis, characterization techniques and applications. *Journal of Environmental Chemical Engineering*, *5*, 4866–4883.

Vishnu, V. S., George, G., Divya, V., & Reddy, M. L. P. (2009). Synthesis and characterization of new environmentally benign tantalum-doped $Ce_{0.8}Zr_{0.2}O_2$ yellow pigments: Applications in coloring of plastics. *Dyes and Pigments*, *82*, 53–57.

Vishnu, V. S., George, G., & Reddy, M. L. P. (2010). Effect of molybdenum and praseodymium dopants on the optical properties of $Sm_2Ce_2O_7$: Tuning of band gaps to realize various color hues. *Dyes and Pigments*, *85*, 117–123.

Wachsman, E. D., Marlowe, C. A., & Lee, K. T. (2012). Role of solid oxide fuel cells in a balanced energy strategy. *Energy & Environmental Science*, *5*, 5498–5509.

Wang, C., Huang, W., Wang, Y., Cheng, Y., Zou, B., Fan, X., ... Cao, X. (2012). Synthesis of monodispersed $La_2Ce_2O_7$ nanocrystals via hydrothermal method: A study of crystal growth and sintering behavior. *International Journal of Refractory Metals and Hard Materials*, *31*, 242–246.

Wang, C., Wang, Y., Zhang, A., Cheng, Y., Chi, F., & Yu, Z. (2013). The influence of ionic radii on the grain growth and sintering-resistance of $Ln_2Ce_2O_7$ (Ln = La, Nd, Sm, Gd). *Journal of Materials Science*, *48*, 8133–8139.

Wang, C., Wang, Y., Fan, X., Huang, W., Zou, B., & Cao, X. (2012). Preparation and thermophysical properties of $La_2(Zr_{0.7}Ce_{0.3})_2O_7$ ceramic via sol–gel process. *Surface and Coatings Technology*, *212*, 88–93.

Wang, J.-D., Xie, Y.-H., Zhang, Z.-F., Liu, R.-Q., & Li, Z.-J. (2005). Protonic conduction in Ca^{2+}-doped $La_2M_2O_7$ (M = Ce, Zr) with its application to ammonia synthesis electrochemically. *Materials Research Bulletin*, *40*, 1294–1302.

Wang, Y., Mori, T., Li, J.-G., & Drennan, J. (2005). Synthesis, characterization, and electrical conduction of 10mol% Dy_2O_3-doped CeO_2 ceramics. *Journal of the European Ceramic Society*, *25*, 949–956.

Wang, Y., Wang, C., Li, C., Cheng, Y., & Chi, F. (2014). Influence of different surfactants on crystal growth behavior and sinterability of $La_2Ce_2O_7$ solid solution. *Ceramics International*, *40*, 4305–4310.

Weng, S.-F., Wang, Y.-H., & Lee, C.-S. (2013). Autothermal steam reforming of ethanol over $La_2Ce_{2-x}Ru_xO_7$ (x = 0–0.35) catalyst for hydrogen production. *Applied Catalysis B: Environmental*, *134–135*, 359–366.

Wetchakun, K., Wetchakun, N., & Sakulsermsuk, S. (2019). An overview of solar/visible light-driven heterogeneous photocatalysis for water purification: TiO_2- and ZnO-based photocatalysts used in suspension photoreactors. *Journal of Industrial and Engineering Chemistry*, *71*, 19–49.

Xiang, J., Chen, S., Huang, J., Liang, W., Cao, Y., Wang, R., & He, Q. (2012). Synthesis kinetics and thermophysical properties of $La_2(Zr_{0.7}Ce_{0.3})_2O_7$ ceramic for thermal barrier coatings. *Journal of Rare Earths*, *30*, 228–232.

Xiaoge, C., Shusen, Y., Hongsong, Z., Gang, L., Zhenjun, L., Bo, R., ... An, T. (2014). Preparation and thermophysical properties of $(Sm_{1-x}Er_x)_2Ce_2O_7$ oxides for thermal barrier coatings. *Materials Research Bulletin*, *51*, 171–175.

Xing, Z., Zhang, J., Cui, J., Yin, J., Zhao, T., Kuang, J., ... Zhou, W. (2018). Recent advances in floating TiO_2-based photocatalysts for environmental application. *Applied Catalysis B: Environmental*, *225*, 452–467.

Xu, Z. H., He, L. M., Mu, R. D., He, S. M., Huang, G. H., & Cao, X. Q. (2010). Double-ceramic-layer thermal barrier coatings based on $La_2(Zr_{0.7}Ce_{0.3})_2O_7/La_2Ce_2O_7$ deposited by electron beam-physical vapor deposition. *Applied Surface Science*, *256*, 3661–3668.

Xu, Z., He, L., Mu, R., He, S., Zhong, X., & Cao, X. (2009). Influence of the deposition energy on the composition and thermal cycling behavior of $La_2(Zr_{0.7}Ce_{0.3})_2O_7$ coatings. *Journal of the European Ceramic Society*, *29*, 1771–1779.

Xu, J., Peng, L., Fang, X., Fu, Z., Liu, W., Xu, X., ... Wang, X. (2018). Developing reactive catalysts for low temperature oxidative coupling of methane: On the factors deciding the reaction performance of $Ln_2Ce_2O_7$ with different rare earth A sites. *Applied Catalysis A: General*, *552*, 117–128.

Yamamura, H., Nishino, H., Kakinuma, K., & Nomura, K. (2003). Crystal phase and electrical conductivity in the pyrochlore-type composition systems, $Ln_2Ce_2O_7$ (Ln = La, Nd, Sm, Eu, Gd, Y and Yb). *Journal of the Ceramic Society of Japan*, *111*, 902–906.

Yan, L., Sun, W., Bi, L., Fang, S., Tao, Z., & Liu, W. (2010). Effect of Sm-doping on the hydrogen permeation of Ni–$La_2Ce_2O_7$ mixed protonic–electronic conductor. *International Journal of Hydrogen Energy*, *35*, 4508–4511.

Yan, L., Xu, Y., Zhou, M., Chen, G., Deng, S., Smirnov, S., ... Zou, G. (2015). Porous TiO_2 conformal coating on carbon nanotubes as energy storage materials. *Electrochimica Acta*, *169*, 73–81.

Yan, C.-H., Yan, Z.-G., Du, Y.-P., Shen, J., Zhang, C., & Feng, W. (2011). Chapter 251 - Controlled synthesis and properties of rare earth nanomaterials. In K. A. Gschneidner, J.-C. G. Bünzli, & V. K. Pecharsky (Eds.), *Handbook on the physics and chemistry of rare earths* (pp. 275–472). Elsevier.

Yang, J., Zhao, M., Zhang, L., Wang, Z., & Pan, W. (2018). Pronounced enhancement of thermal expansion coefficients of rare-earth zirconate by cerium doping. *Scripta Materialia*, *153*, 1–5.

Yi, H., Liu, X., Che, J., & Liang, G. (2020). Thermochemical compatibility between $La_2(Ce_{1-x}Zr_x)_2O_7$ and 4 mol% Y2O3 stabilized zirconia after high temperature heat treatment. *Ceramics International*, *46*, 4142–4147.

Yu, J. C., Zhang, L., & Lin, J. (2003). Direct sonochemical preparation of high-surface-area nanoporous ceria and ceria–zirconia solid solutions. *Journal of Colloid and Interface Science*, *260*, 240–243.

Zhang, H. S., Hu, R. X., & Xu, Q. (2010). Forming mechanism of $Dy_2Ce_2O_7$ ceramic for thermal barrier coatings. *Advanced Materials Research*, *105–106*, 383–385.

Zhang, C., Li, Y., Shuai, D., Shen, Y., Xiong, W., & Wang, L. (2019). Graphitic carbon nitride ($g\text{-}C_3N_4$)-based photocatalysts for water disinfection and microbial control: A review. *Chemosphere*, *214*, 462–479.

Zhang, H., Liao, S., Dang, X., Guan, S., & Zhang, Z. (2011). Preparation and thermal conductivities of $Gd_2Ce_2O_7$ and $(Gd_{0.9}Ca_{0.1})_2Ce_2O_{6.9}$ ceramics for thermal barrier coatings. *Journal of Alloys and Compounds*, *509*, 1226–1230.

Zhang, H., Yuan, J., Song, W., Zhou, X., Dong, S., Duo, S., ... Cao, X. (2020). Composition, mechanical properties and thermal cycling performance of YSZ toughened $La_2Ce_2O_7$ composite thermal barrier coatings. *Ceramics International*, *46*, 6641–6651.

Zhang, H. S., Liao, S. R., Yuan, W., & Guan, S. K. (2011). Preparation and thermal conductivity of $Y_2Ce_2O_7$ ceramic material. *Advanced Materials Research*, *266*, 59–62.

Zhang, Q., Zheng, X., Jiang, J., & Liu, W. (2013). Structural stability of $La_2Ce_2O_7$ as a proton conductor: A first-principles study. *The Journal of Physical Chemistry C*, *117*, 20379–20386.

Zhao, S., Gu, L., Zhao, Y., Huang, W., Zhu, L., Fan, X., ... Cao, X. (2013). Thermal cycling behavior and failure mechanism of $La_2(Zr_{0.7}Ce_{0.3})_2O_7/Eu^{3+}$-doped 8YSZ thermal barrier coating prepared by atmospheric plasma spraying. *Journal of Alloys and Compounds*, *580*, 101–107.

Zhong, Z., Jiang, Y., Lian, Z., Song, X., & Peng, K. (2020). Exploring the effects of divalent alkaline earth metals (Mg, Ca, Sr, Ba) doped $Nd_2Ce_2O_7$ electrolyte for proton-conducting solid oxide fuel cells. *Ceramics International*, *46*, 12675–12685.

Zhu, Z., Liu, B., Shen, J., Lou, Y., & Ji, Y. (2016). $La_2Ce_2O_7$: A promising proton ceramic conductor in hydrogen economy. *Journal of Alloys and Compounds*, *659*, 232–239.

Zinatloo-Ajabshir, S., Baladi, M., Amiri, O., & Salavati-Niasari, M. (2020). Sonochemical synthesis and characterization of silver tungstate nanostructures as visible-light-driven photocatalyst for waste-water treatment. *Separation and Purification Technology*, *248*, 117062.

Zinatloo-Ajabshir, S., Baladi, M., & Salavati-Niasari, M. (2021). Enhanced visible-light-driven photocatalytic performance for degradation of organic contaminants using $PbWO_4$ nanostructure fabricated by a new, simple and green sonochemical approach. *Ultrasonics Sonochemistry*, *72*, 105420.

Zinatloo-Ajabshir, S., Heidari-Asil, S. A., & Salavati-Niasari, M. (2021). Recyclable magnetic $ZnCo_2O_4$-based ceramic nanostructure materials fabricated by simple sonochemical route for effective sunlight-driven photocatalytic degradation of organic pollution. *Ceramics International*, *47*, 8959–8972.

Zinatloo-Ajabshir, S., Morassaei, M. S., Amiri, O., Salavati-Niasari, M., & Foong, L. K. (2020a). $Nd_2Sn_2O_7$ nanostructures: Green synthesis and characterization using date palm extract, a potential electrochemical hydrogen storage material. *Ceramics International*, *46*, 17186–17196.

Zinatloo-Ajabshir, S., Morassaei, M. S., Amiri, O., & Salavati-Niasari, M. (2020b). Green synthesis of dysprosium stannate nanoparticles using *Ficus carica* extract as photocatalyst for the degradation of organic pollutants under visible irradiation. *Ceramics International*, *46*, 6095–6107.

Zinatloo-Ajabshir, S., Mortazavi-Derazkola, S., & Salavati-Niasari, M. (2017). Simple sonochemical synthesis of Ho_2O_3-SiO_2 nanocomposites as an effective photocatalyst for degradation and removal of organic contaminant. *Ultrasonics Sonochemistry*, *39*, 452–460.

Zinatloo-Ajabshir, S., & Salavati-Niasari, M. (2015). Preparation and characterization of nanocrystalline praseodymium oxide via a simple precipitation approach. *Journal of Materials Science: Materials in Electronics*, *26*, 5812–5821.

Zinatloo-Ajabshir, S., & Salavati-Niasari, M. (2019). Preparation of magnetically retrievable $CoFe_2O_4@SiO_2@Dy_2Ce_2O_7$ nanocomposites as novel photocatalyst for highly efficient degradation of organic contaminants. *Composites Part B: Engineering*, *174*, 106930.

Zinatloo-Ajabshir, S., Salehi, Z., Amiri, O., & Salavati-Niasari, M. (2019a). Green synthesis, characterization and investigation of the electrochemical hydrogen storage properties of

$Dy_2Ce_2O_7$ nanostructures with fig extract. *International Journal of Hydrogen Energy*, *44*, 20110–20120.

Zinatloo-Ajabshir, S., Salehi, Z., Amiri, O., & Salavati-Niasari, M. (2019b). Simple fabrication of $Pr_2Ce_2O_7$ nanostructures via a new and eco-friendly route; a potential electrochemical hydrogen storage material. *Journal of Alloys and Compounds*, *791*, 792–799.

Zinatloo-Ajabshir, S., Salehi, Z., & Salavati-Niasari, M. (2016). Preparation, characterization and photocatalytic properties of $Pr_2Ce_2O_7$ nanostructures via a facile procedure. *RSC Advances*, *6*, 107785–107792.

Zinatloo-Ajabshir, S., Salehi, Z., & Salavati-Niasari, M. (2018a). Green synthesis and characterization of $Dy_2Ce_2O_7$ nanostructures using *Ananas comosus* with high visible-light photocatalytic activity of organic contaminants. *Journal of Alloys and Compounds*, *763*, 314–321.

Zinatloo-Ajabshir, S., Salehi, Z., & Salavati-Niasari, M. (2018b). Green synthesis of $Dy_2Ce_2O_7$ ceramic nanostructures using juice of *Punica granatum* and their efficient application as photocatalytic degradation of organic contaminants under visible light. *Ceramics International*, *44*, 3873–3883.

Zinatloo-Ajabshir, S., Salehi, Z., & Salavati-Niasari, M. (2018c). Green synthesis and characterization of $Dy_2Ce_2O_7$ ceramic nanostructures with good photocatalytic properties under visible light for removal of organic dyes in water. *Journal of Cleaner Production*, *192*, 678–687.

Zinatloo-Ajabshir, S., Salehi, Z., & Salavati-Niasari, M. (2019). Synthesis of dysprosium cerate nanostructures using *Phoenix dactylifera* extract as novel green fuel and investigation of their electrochemical hydrogen storage and Coulombic efficiency. *Journal of Cleaner Production*, *215*, 480–487.

Zinatloo-Ajabshir, S., Zinatloo-Ajabshir, S., Morassaei, M. S., & Salavati-Niasari, M. (2019). Simple approach for the synthesis of $Dy_2Sn_2O_7$ nanostructures as a hydrogen storage material from banana juice. *Journal of Cleaner Production*, *222*, 103–110.

Rare earth zirconate ($Re_2Zr_2O_7$) ceramic nanomaterials

Hakimeh Teymourinia[1,2,3]
[1]Institute of Advanced Materials (INAM), University of Jaume I, Castellon, Spain, [2]Trita Nanomedicine Research Center (TNRC), Trita Third Millennium Pharmaceuticals, Zanjan, Iran, [3]Department of Science, University of Zanjan, Zanjan, Iran

4.1 General introduction

Rare earth zirconate ceramic materials with the chemical formula Re_2 (or Ln_2) Zr_2O_7, (where Re (Ln) is a lanthanide element or trivalent rare earth, (Dy, Gd, Nd, and La)) are a kind of ceramic materials with a generic pyrochlore structure. Recently, complex oxide structures with the chemical formula $A_2M_2O_7$, which can be in pyrochlore crystal, or cubic fluorite structure have attracted significant research interest. In $A_2M_2O_7$, A commonly is an element with an inert single pair of electrons or a rare earth element, and M is a post transition metal with a variable oxidation state or a transition metal. Generally, in pyrochlore lattice, the ratio of ionic radius value ($RR = rA^{3+}/rM^{4+}$) at room temperature is between 1.46 and 1.80 (Davis, 2004; Subramanian, Aravamudan, & Rao, 1983; Yan et al., 2011; Zhang, Guo, Jung, Li, & Knapp, 2017). Additionally, rare earth zirconate (ReZrs) pyrochlores as an exciting category of compounds display empty spaces that make them good ionic conductors (Gokul Raja, Balamurugan, & Reshma, 2021; Rushton, Grimes, Stanek, & Owens, 2004) and demonstrate the wide diversity of chemical and physical properties. The most important properties are chemical composition, oxygen vacancies, and the ordering (disordering) of the Re site and Zr cations (Zinatloo-Ajabshir, Mortazavi-Derazkola, & Salavati-Niasari, 2017). Although the predominant crystalline structure of rare earth zirconates is pyrochloride, these structures can also be formed in the fluorite phase. Fig. 4.1 demonstrates the fluorite and pyrochlore crystal structure of $A_2M_2O_7$ (Pokhrel, Alcoutlabi, & Mao, 2017). As shown in Fig. 4.1A, oxygen vacancies in fluorite form are accidentally distributed on the anion site (8c). In pyrochlore form of $Re_2Zr_2O_7$, with the face-centered cubic arrangement (FCC) which formed by Re and Zr cations, eight- and sixfold oxygen atoms coordinated Re-site and Zr-site cations, respectively (Fig. 4.1B) (Pokhrel et al., 2017).

Among the zirconate pyrochlores, there is a category, called defective pyrochlores (e.g., $Gd_2Zr_2O_7$), which show an incompletely disordered atomic array at ambient condition. Pyrochlores have shown ionic, electronic, or mixed conductors and unusual magnetic properties (Snyder, Slusky, Cava, & Schiffer, 2001; Wuensch

Advanced Rare Earth-Based Ceramic Nanomaterials. DOI: https://doi.org/10.1016/B978-0-323-89957-4.00004-9

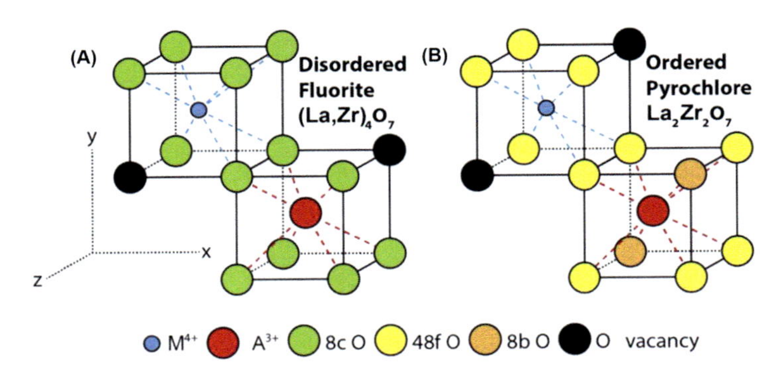

Figure 4.1 Crystal structures of $Re_2Zr_2O_7$: (A) defect fluorite structure where La^{3+} and Zr^{4+} occupy the unique cation site with sevenfold coordination; (B) ordered pyrochlore structure where La^{3+} ion occupy the A-site with eightfold coordination with oxygen. The smaller Zr^{4+} cation occupies the M-site (16c) and is a sixfold coordination with oxygen. Green, yellow, and orange spheres represent the oxygen on the 8c, 48f, and 8b sites, respectively (Pokhrel et al., 2017).
Source: Reprinted with permissions from Pokhrel, M., Alcoutlabi, M., & Mao, Y. (2017). Optical and X-ray induced luminescence from Eu^{3+} doped $La_2Zr_2O_7$ nanoparticles. *Journal of Alloys and Compounds*, *693*, 719–729.

et al., 2000). Fabrication of materials as nanoparticles (NPs) caused the reduction of particle size and expansion specific surface area, resulted various phase transition temperatures, improved catalytic activity, and developed process ability (Teymourinia, Amiri, & Salavati-Niasari, 2021; Teymourinia, Salavati-Niasari, Amiri, & Farangi, 2018; Teymourinia, Salavati-Niasari, Amiri, & Safardoust-Hojaghan, 2017; Teymourinia, Salavati-Niasari, Amiri, & Yazdian, 2019; Yan et al., 2011; Zinatloo-Ajabshir, Morassaei, & Salavati-Niasari, 2018; Zinatloo-Ajabshir, Salehi, & Salavati-Niasari, 2018a, 2018b, 2018c). Nanostructured pyrochlore $Ln_2Zr_2O_7$ (LnZ) has been achieved by conventional solid-state reactions, coprecipitation, sol-gel, hydrothermal, and self-propagation methods (AruláDhas, 1993; Chen, Gao, Liu, & Luo, 2009; Hongming & Danqing, 2008; Tang, Sun, Chen, & Jiao, 2012; Tong, Yu, Lu, Yang, & Wang, 2008).

In general, complex oxides of zirconate ceramic materials $Re_2Zr_2O_7$, with unique physical, chemical, and thermal properties and possible applications, have been noticed. The potential applications in electro/photocatalysis, magnetism, and nuclear waste storage, oxygen monitoring sensors, photoluminescent host materials, and solid electrolytes in high-temperature fuel cells correlated with distinctive properties of these materials (Hongming & Danqing, 2008; Keyvani, Mahmoudinezhad, Jahangiri, & Bahamirian, 2020; Lei et al., 2015; Zinatloo-Ajabshir, Salavati-Niasari, Sobhani, & Zinatloo-Ajabshir, 2018). In addition, these compounds display high thermal stabilities, chemical resistance, complex compositions, high thermal expansion coefficients, high melting points, low thermal conductivities, great ionic conductivity, high sintering rate, and high endurance to defects (Cao et al., 2001; Cao, Vassen, & Stöver, 2004; Padture, Gell, & Klemens, 2000; Torres-Rodriguez et al., 2019). In this chapter, the $Re_2Zr_2O_7$ materials, which are potential candidates for thermal-barrier coatings (TBCs), have been investigated. Some of the lanthanum zirconate materials as top ceramic materials (for example $Gd_2Zr_2O_7$, $Dy_2Zr_2O_7$, $Nd_2Zr_2O_7$ and $La_2Zr_2O_7$) have been noticed in recent researches as significant TBCs (Hongming & Danqing, 2008; Wu et al., 2002). Thermal conductivity of rare earth zirconates (ReZs) is closely equal. By doping or codoping of these materials with oxides and formation of defect cluster, thermal conductivity has been reduced (Cao et al., 2001, 2004). The state-of-the-art synthesis approaches, properties and applications of ReZs nanostructures will be reviewed in this chapter.

4.2 Preparation methods of $Re_2Zr_2O_7$ ceramic nanomaterials

Properties and applications of nanomaterials depend on composition, shape, size, and morphology of these materials, which are influenced by preparation method (Teymourinia et al., 2017, 2018, 2019, 2020; Zinatloo-Ajabshir, Salavati-Niasari, et al., 2018). There are several methods for preparing rare earth zirconate ceramic nanomaterials, which has been summarized in the following section.

4.2.1 Solid state reaction

For the preparation of rare earth zirconate ceramic materials by conventional solid-state reaction, ZrO_2 and lanthanum oxide (La_2O_3) powders are mechanically mixed at high temperature (T = 1773K) for very long time under an argon atmosphere for 10 hours and then followed by calcination of obtained powders for long time at high temperature (Bolech et al., 1997; Duarte, Meguekam, Colas, Vardelle, & Rossignol, 2015; Kalinkin, Vinogradov, & Kalinkina, 2021; Sedmidubský, Beneš, & Konings, 2005; Wang et al., 2014; Zhang, Guo, et al., 2017). Tong et al. has prepared $Ln_2Zr_2O_7$ by solid-state reaction method. In this way, La_2O_3 (Nd_2O_3), $K_2CO_3 \cdot 1.5H_2O$ and $ZrOCl_2 \cdot 8H_2O$ (the molar ratio is 1.6:1:2) has been ground together at 900° in air for 10 hours. They found that, by adding appropriate K_2CO_3 during reaction the $Ln_2Zr_2O_7$ nanocrystals produced at lower temperature (Teng et al., 2021). XRD patterns of $Ln_2Zr_2O_7$ precursor powders are shown in Fig. 4.2. When the temperature was increased to 700°, the compounds KCl, ZrO_2, La_2O_3 and $La_2Zr_2O_7$ existed at the same time. By sintering the precursors at 800° for 10 hours, small amount of ZrO_2 and La_2O_3 exist yet. Howbeit, when the reaction time increased to 24 hours, the impurity vanished, but the pyrochlore phase is not recognizable yet. The characteristic peaks that always, make the pyrochlore form diverse from the fluorite phase are (3 3 1) and (5 1 1) (Tong, Zhu, Lu, Wang, & Yang, 2008).

When the temperature was increased to 900°, XRD pattern showed no impure peaks, but a single phase of pure pyrochlore form of as synthesized $La_2Zr_2O_7$ (JCPDS: 73−0444), and $Nd_2Zr_2O_7$ (JCPDS: 78−1617) nanocrystals are shown

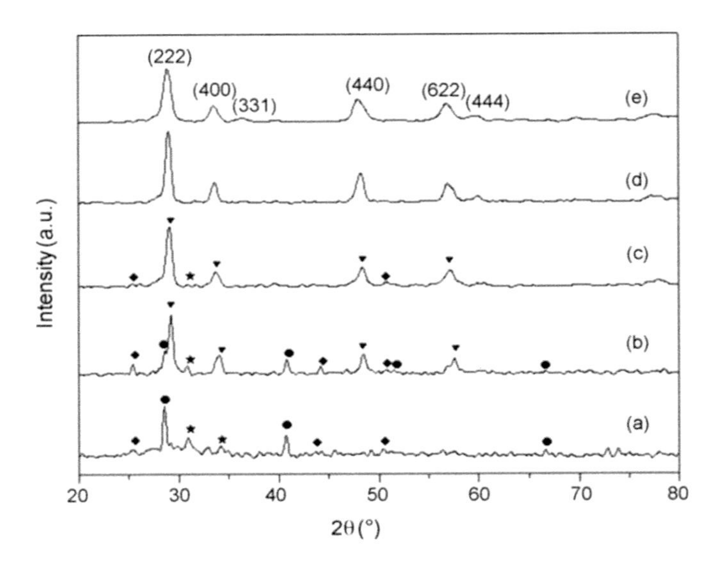

Figure 4.2 The XRD patterns of $La_2Zr_2O_7$ precursor powders at temperatures: (A) 400°C, 5 h; (B) 700°C, 5 h; (C) 800°C, 10 h; (D) 800°C, 24 h; (E) 900°C, 10 h. (♦) Denotes $La_2Zr_2O_7$; (●) denotes KCl; (□) denotes La_2O_3; (★) denotes ZrO_2 (Tong, Zhu, et al., 2008).

Source: Reprinted with permissions from Tong, Y., Zhu, J., Lu, L., Wang, X., & Yang, X. (2008). Preparation and characterization of $Ln_2Zr_2O_7$ (Ln = La and Nd) nanocrystals and their photocatalytic properties. *Journal of Alloys and Compounds*, *465*, 280–284.

(Figs. 4.2 and 4.3) (Mohr, Brezinski, Audrieth, Ritchey, & McFarlin, 1953; Tong, Zhu, et al., 2008).

In another study, Hagiwara et al. synthesized $Eu_2Zr_2O_7$ and $La_2Zr_2O_7$ with the solid-state reaction by using Eu_2O_3, ZrO_2 and La_2O_3 as precursors (Hagiwara, Nomura, & Kageyama, 2017). Quiroz et al. were prepared and studied structural and thermoelectric properties of as synthesized $A_2Zr_2O_7$ ceramic nanomaterials, with A = Pr, Nd, Sm, Gd, and Er, by solid-state reaction, in temperatures between 1000° and 1400° at ambient pressure. As synthesized $A_2Zr_2O_7$ ceramic nanomaterials, were characterized by XRD and SEM. The formation of heterogeneous grains of nanomaterials with the size that amplitudes from 0.7 to 4.7 μm, has shown by SEM images. All $A_2Zr_2O_7$ ceramic nanomaterials samples present porous surfaces (Quiroz, Chavira, Garcia-Vazquez, Gonzalez, & Abatal, 2018). The instructions for preparing the nanomaterials were as follows: in order to obtain a homogeneous powder, first the stoichiometric values of the reactants were ground in a mortar for 30 minutes (Quiroz et al., 2018). By applying a pressure of 3 tons/cm^2 for 5 minutes, under vacuum, the produced $A_2Zr_2O_7$ powders were compressed into pellets (13 mm diameter, 1.0–1.5 mm thickness) (Naik & Hasolkar, 2020; Quiroz et al., 2021). The production of $A_2Zr_2O_7$ ceramic nanomaterials was followed by heating pellets in the air from room temperature to 1400° at 10°/min, and kept at 1400° in

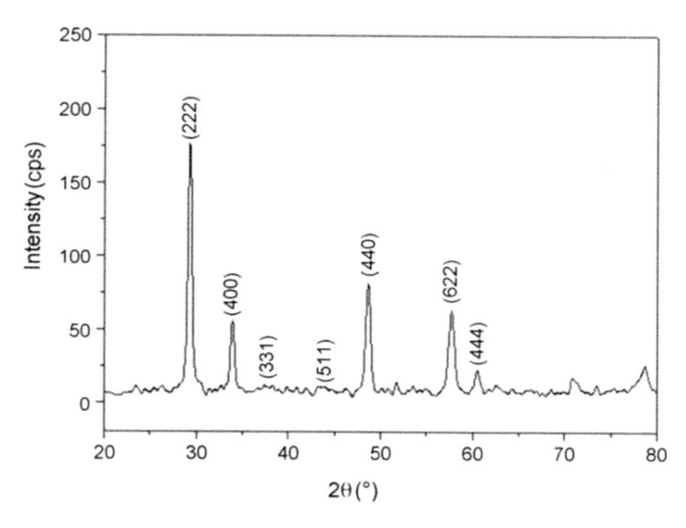

Figure 4.3 The XRD patterns of $Nd_2Zr_2O_7$ precursor powders at 900°C for 10 h (Tong, Zhu, et al., 2008).
Source: Reprinted with permissions from Tong, Y., Zhu, J., Lu, L., Wang, X., & Yang, X. (2008). Preparation and characterization of $Ln_2Zr_2O_7$ (Ln = La and Nd) nanocrystals and their photocatalytic properties. *Journal of Alloys and Compounds*, *465*, 280–284.

72 hours, and then cooled down to room temperature at 3°/min (Quiroz et al., 2021). The resultant $A_2Zr_2O_7$ were studied by XRD and SEM. The XRD patterns were obtained at ambient temperature from 10° to 90°. In addition, refinement of the XRD patterns was performed by using the MAUD refinement software to analyze diffraction data to gain the crystal structures of the samples (Ferrari & Lutterotti, 1994; Prize, 1999; Wenk, Matthies, & Lutterotti, 1994). The results indicate the formation of pyrochlore-type cubic phase with Fd3m space group (S.G. number 227). Main diffraction peaks of all samples involve the (222), (400), (331), (440), (662), (711), (800, and (662), and (840), that belongs to pyrochlorate structure (Popov et al., 2019; Quiroz et al., 2021; Wang et al., 2020). Based on the results of MAUD refinement, there is a good agreement with the experimental pattern, confirming that the main phase presented in the samples is the cubic $A_2Zr_2O_7$ phase (Quiroz et al., 2018).

4.2.2 Coprecipitation

In the coprecipitation method, hydroxide precipitates are concurrently created by adding an aqueous solution of $La(NO_3)_3 \cdot 6H_2O$ and $ZrOCl_2.8H_2O$ via ammonia solution, then followed by calcination at 900°C during 5 hours, to prepare $Ln_2Zr_2O_7$ structures (Cao & Tietz, 2000). Liu et al. prepared $Sm_xZr_{1-x}O_{2-x/2}$ ceramic powders with coprecipitation method and calcination at 1073K within 5 hours (Liu, Ouyang, Wang, Zhou, & Li, 2009). Transmission electron microscopy images of the resultant powders are shown in Fig. 4.4. The most obvious benefit of the coprecipitation method associated with the preparation of homogeneous materials is high purity at relatively low temperature. However, for large grain-sized powders obtained by coprecipitation method, a milling step is required (Duarte et al., 2015). $(Gd_{1-x}Eux)_2\ Zr_2O_7$ ($0 < x < 0.100$) powders have been prepared with fluorite phase via coprecipitation method with calcination at 800° (Hu, Liu, Wang, Wang, & Ouyang, 2012).

4.2.3 Hydrothermal

In hydrothermal synthesis of $Re_2Zr_2O_7$, the required precursors are dissolved in a suitable solvent, mixed by stirring and then the transparent solution is transferred to autoclave (Huo et al., 2019; Matsuda et al., 2020; Teymourinia et al., 2019; Trujillano, Martín, & Rives, 2016). For adjustment of pH of the mixed solution in common value (above 7), some agents like ammonium hydroxide and sodium hydroxides have been used (Sankar & Kumar, 2021; Teymourinia et al., 2017). The hydrothermal synthesis method is the only method, with lower reaction temperature (below 400°) than other methods and with a cheap, interminable and available solvent, which is water (Emadi, Salavati-Niasari, & Sobhani, 2017; Razi, Zinatloo-Ajabshir, & Salavati-Niasari, 2017; Teymourinia et al., 2017). In this method, the reaction time depends on the pH of the solution and usually has long reaction duration. Surfactants' use and microwave-assisted hydrothermal synthesis features reduced reaction time (Huo et al., 2019; Sankar & Kumar, 2021; Wang, Cheng, Li, & Jin, 2016).

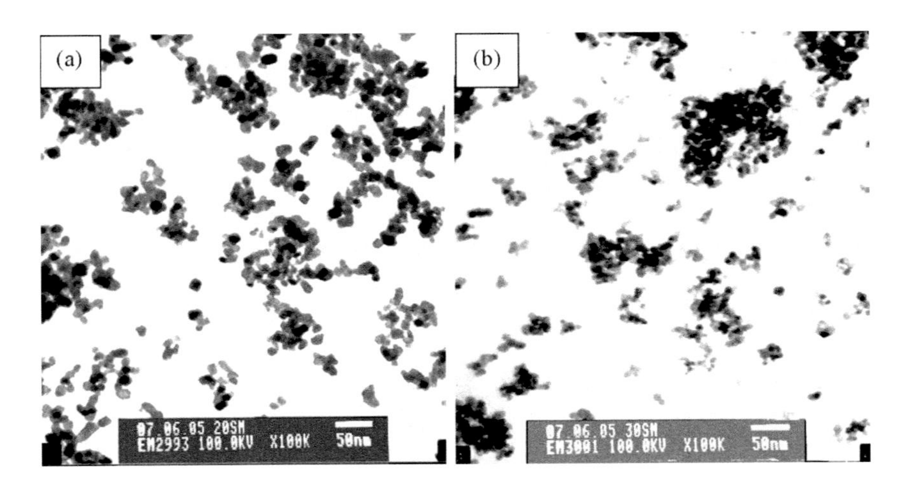

Figure 4.4 Morphologies of $Sm_xZr_{1-x}O_{2-x/2}$ (x = 0.2, 0.3) ceramic powders: (A) x = 0.2; (B) x = 0.3.
Source: Reprinted with permissions from Liu, Z. -G., Ouyang, J. -H., Wang, B. -H., Zhou, Y., & Li, J. (2009). Thermal expansion and thermal conductivity of $Sm_xZr_{1-x}O_{2-x/2}$ ($0.1 \le x \le 0.5$) ceramics. *Ceramics International*, *35*, 791–796.

Wang et al. prepared $Sm_2Zr_2O_7$ pyrochlore by using hydrothermal synthesis. First, solution of SmO_2 and zirconium (IV) oxide chloride octahydrate, that were dissolved in nitric acid and stirred for 10 minutes, were prepared. Then, pH of the above mixture adjusted by aqueous ammonia in 5–9, which resulted a white precipitate, and vigorous stirring KOH solution was added to the washed precipitates, then the mixture was moved into a teflon-lined autoclave. A hydrothermal synthesis was done at 100° and 200°C. The resultant powder was washed and dried at a vacuum at 70°C for 18 hours (Wang et al., 2016). In another investigation, Hongming and Danqing prepared lanthanum zirconates by hydrothermal approach using lanthanum and zirconate nitrate salts, and NaOH was used to tuning the pH in 11. The hydrothermal reaction was completed at 200° in 1 hours (Hongming & Danqing, 2008). Gao et al. synthesized $Gd_2Zr_2O_7:Tb^{3+}$ phosphors by hydrothermal route by employing $Gd(NO_3)_3$, $Tb(NO_3)_3$, $ZrOCl_2$ and NH_3 as starting materials at 200° for 20 hours (Gao, Zhu, Wang, & Ou, 2011; Wang et al., 2016). However, the significant benefits of the hydrothermal synthesis that has reported till now, is homogeneous nucleation processes, that causes the formation of fine particle size powders, without calcination stage (Fang et al., 2019). Preparation of $Ln_2Zr_2O_7$ with hydrothermal synthesis is complicated because of the slow reaction rate. After hydrothermal reaction, a sensible heating step at low temperature is needed for the acceleration of the solid reaction (Chen & Xu, 1998; Zinatloo-Ajabshir, Salavati-Niasari, et al., 2018).

4.2.4 Sol-gel

The sol-gel is a method to synthesize nanomaterials with high sintering capacity for the $Ln_2Zr_2O_7$ (Ln = La, Nd, Sm) systems (Shimamura, Arima, Idemitsu, & Inagaki, 2007). Also, it is one of the wet chemical synthesis methods used to prepare porous structured materials (Kong et al., 2015). The sol-gel process, which includes hydrolysis followed by condensation, results in synthesized gel-like bulk species with a polymeric matrix. Finally, drying heat treatment resulted the cumulous produced powders (Joulia, Vardelle, & Rossignol, 2013; Kong et al., 2015; Li et al., 2006). This method is simple, low-cost and nontoxic (Salehi, Zinatloo-Ajabshir, & Salavati-Niasari, 2017; Zinatloo-Ajabshir, Salehi, & Salavati-Niasari, 2016). The other benefits of this method include low temperature reactions, low cost of the process by common lab equipment (Wang, Li, Wang, & Chen, 2015). The rare earth zirconate ceramic materials preparation by sol-gel method steps are as follows: rare earth salts and chelating agents mixture, adjust the pH of the mixture to promote the polymeric chain formation, continues stirring until mixture become transparent and then gel formation, heat treatment/ sintering, pyrochlores formation (Kong et al., 2015; Sohn, Kim, & Woo, 2003). Sohn et al. prepared $Ln_2Zr_2O_7$ (Ln = Eu, Sm, Tb and Gd) powders via sol-gel approach. By utilization of zirconium isopropoxide and lanthanide metal (Ln = Sm, Eu, Gd and Tb) nitrate as precursors, also isopropanol as non-aqueous solvent at 1200°, $Ln_2Zr_2O_7$ powders has been synthesized (Sohn et al., 2003). Wang et al. prepared $La_2Zr_2O_7$ pyrochlore through the sol-gel method. After the dissolution of zirconium chloride and lanthanum nitrate in

H_2O/ethanol, citric acid (CA) was added to the prepared mixture. Then, a uniform transparent sol solution was formed by adding formamide and polyethylene glycol as a chelating agent to the above mixture, which was then stirred at room temperature for 2 hours. The resultant solution was kept at 80° for 20 hours for aging, which resulted in gel formation. The resultant gel was dried at 110° for 5 hours and calcined at 1200° for 2 hours in a muffle furnace to form homogeneous $La_2Zr_2O_7$ powders (Li et al., 2021; Wang, Li, Wang, Zhang, & Chen, 2016). Jovaní et al. prepared pigment based on Fe and Tb doped yttrium zirconate ($Y_2Zr_2O_7$) with fluorite phase by using sol-gel approach at 1300°C in 12 hours (Jovaní, Sanz, Beltrán-Mir, & Cordoncillo, 2016). Tang et al. prepared lanthanum zirconate fibers via sol-gel combined electrospinning route (Tang et al., 2012). After dissolving an equimolar ratio of $LaCl_3 \cdot 7H_2O$ in zirconium acetate ($C_8H_{12}O_8Zr$) solution, silica sol (SiO_2, pH = 3.2) was added to the above solution with stirring at 6 hours. The lanthanum zirconate gel fibers were fabricated by adding transparent sol to the absolute ethanol with PVP, under the electrostatic force, and drying and calcining at high temperatures (Tang et al., 2012). In the sol-gel method, despite the proper control of raw materials, nano-sized metal species are used. The high sensitivity of metal alkoxides as the main and costly precursors to humidity, the reaction is promoted with stabilizers, which called the Pachini approach (Lee, Sheu, & Kao, 2010; Lee, Sheu, Deng, & Kao, 2009; Morassaei, Zinatloo-Ajabshir, & Salavati-Niasari, 2017; Rao, Banu, Vithal, Swamy, & Kumar, 2002; Simonenko, Sakharov, Simonenko, Sevastyanov, & Kuznetsov, 2015; Zinatloo-Ajabshir, Morassaei, & Salavati-Niasari, 2017; Zinatloo-Ajabshir, Zinatloo-Ajabshir, Salavati-Niasari, Bagheri, & Abd Hamid, 2017). $Ln_2Zr_2O_7$ powders has been synthesized through Pechini method by using lanthanide and zirconyl nitrate as precursors at 1073K in 5 hours (Uno et al., 2006). Chen and Zhang synthesized pyrochlore $Sm_2Zr_2O_7$ and $Sm_2(Zr_{0.7}Ce_{0.3})_2O_7$ NPs by Pechini method at 900° (Chen & Zhang, 2013). Salavati-Niasari et al. research group has synthesized neodymium zirconate $Nd_2Zr_2O_7$ nanostructures by a modified Pechini method. They utilized succinic acid ($(CH_2)_2(CO_2H)_2$) as stabilization agent, propane-1,2-diol (CH_3CHCH_2OH) as a connecting agent for the existence of tetraethylene pentamine as pH corrector at 700° in 4 hours (Zinatloo-Ajabshir & Salavati-Niasari, 2017). The effect of propanol quantity and the kind of pH regulator on the morphology and size of neodymium zirconate nanostructures has been investigated. Based on the results, the investigated parameters have outstanding roles in the morphology and size of neodymium zirconate nanostructures. Fig. 4.5 displays the schematic diagram of the production of neodymium zirconate nanostructures, which prepared praseodymium zirconate nanostructures by a facile route with benzene tricarboxylic acid as stabilization agent in attendance of propylene glycol as cross linker at a lower temperature (Zinatloo-Ajabshir, Salavati-Niasari, & Zinatloo-Ajabshir, 2017). Based on the results, the purity, size, and morphology of praseodymium zirconate depend on the stabilization and alkaline agents.

Using the new compounds can open a way for controlling the size and morphology of fabricated materials. $Nd_2Zr_2O_7$-Nd_2O_3 nanocomposites prepared through a simple modified Pechini method with applying salicylic acid as a complexing agent (Zinatloo-Ajabshir, Salavati-Niasari, & Zinatloo-Ajabshir, 2016). Tong et al. prepared $Ln_2Zr_2O_7$

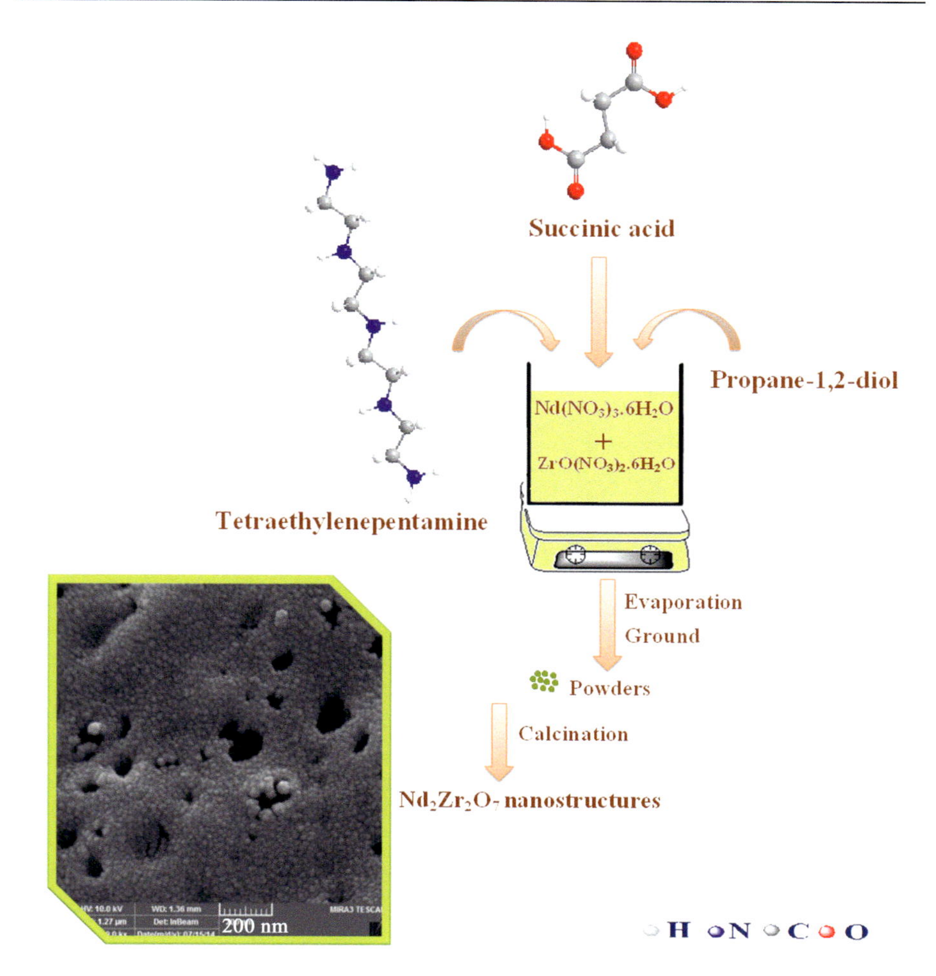

Figure 4.5 Schematic diagram of preparation of $Nd_2Zr_2O_7$ nanostructures (Zinatloo-Ajabshir, Salavati-Niasari, et al., 2017).
Source: Reprinted with permissions from Zinatloo-Ajabshir, S., & Salavati-Niasari, M. (2017). Photo-catalytic degradation of erythrosine and eriochrome black T dyes using $Nd_2Zr_2O_7$ nanostructures prepared by a modified Pechini approach. *Separation and Purification Technology*, *179*, 77–85.

nanostructures via stearic acid sol-gel method (Tong, Wang, et al., 2008; Tong, Xue, et al., 2008; Tong, Zhao, Li, & Li, 2010). The selected molar ratio of Ln: Zr was 1:1 which provided by nitrate salts as raw materials and stearic acid ($C_{17}H_{35}COOH$) as both solvent and dispersant. Produced powders were calcined at 800° within 5 hours in air condition. In comparison to other methods, Tong group used shorter time and lower reaction temperature. Kong et al. by combining the sole-gel and complex precipitation, prepared $La_2Zr_2O_7$ powders by using lanthanum nitrate hexahydrate and zirconium(IV) bis(diethyl citrate) dipropoxide (Kong et al., 2013).

4.2.5 Combustion

The combustion method and precipitation technique described above are similar; nevertheless, the fuel source and oxygen source can cause the production of auto-inflammation and manufactured pyrochlore powders. Production of $Ln_2Zr_2O_7$ powders has resulted from combustion reaction between fuel and metal precursors (oxidants) (Aruna, Sanjeeviraja, Balaji, & Manikandanath, 2013; Du, Zhou, Zhou, & Yang, 2012; Zhang, Lü, Qiu, Zhou, & Ma, 2008; Zhang, Lü, Yang, Zhou, & Zhou, 2008). Glycine, hydrazine, CA, and glycerol used as fuel chemical agents by heat treatment of pyrochlore structure powders to remove the impurities. The combustion method, which involves a simple reaction with low energy ingesting, has benefits including high efficiency, uniform morphology of resulted powders. The pyrochlore powders, which produced via combustion method, are highly porous structures (Jeyasingh, Saji, & Wariar, 2017; Jeyasingh, Saji, Kavitha, & Wariar, 2018; Jeyasingh, Vindhya, Saji, Wariar, & Kavitha, 2019; Quader et al., 2020). Lanthanum zirconate was synthesized by Matovic et al. by dissolving lanthanum nitrate and zirconium chloride in glycine, then the prepared solution was heat-treated at 950° for 120 minutes and then sintered at 1600° within 4 hours. Matovic et al. (2020). Cerium zirconate prepared through the combustion method by Venkatesh et al. by dissolving cerium nitrate and zirconium nitrate in water. Dissolved glycine in DI water sintered for 45 minutes with mild heat and then added dropwise to the cerium-zirconium mixture at a slow stirring rate until resulted transparent solution. The resulted solution was moved to an alumina crucible and reserved in a muffle furnace at 650° for 10–15 minutes, which resulted in $Ce_2Zr_2O_7$ powder that was sintered at 1100° (Venkatesh, Subramanian, & Berchmans, 2019). The transparent $Gd_2Zr_2O_7$ ceramic synthesized via the combustion synthesis method under vacuum sintering. The mixture of dissolved $Gd(NO_3)_3 \cdot 6H_2O$ and $Zr(NO_3)_4 \cdot 5H_2O$ in water and ammonia, was heated until the formation of gelatin, and then it was transported to a muffle furnace to heat at 300°C until formation a fleecy gray powder. After calcination of resultant powder at 1200° in 120 minutes and ball-milling for 24 hours with DI water, the produced slurry was dried, pelleted and sintered at 1775°, 1800°, and 1825° within 6 hours in a vacuum and then has been annealed during 5 hours at 1500° in the atmosphere (Li et al., 2020).

4.2.6 Other preparation approaches $Ln_2Zr_2O_7$ ceramic nanomaterials

4.2.6.1 Molten salt

In the molten salt technique, in order to facilitate the mixing of precursors, they can be dissolved in the molten salt state (Mao, Guo, Huang, Wang, & Chang, 2009; Pokhrel, Brik, & Mao, 2015; Wang, Zhu, & Zhang, 2010). Huang et al. fabricated $La_2Zr_2O_7$ powders with pyrochlore form via molten salt. ZrO_2 and La_2O_3 in the existence of NaF, KCl, and NaCl mixture as reaction medium have been used as precursors to preparing $La_2Zr_2O_7$ particles (Huang et al., 2016).

4.2.6.2 Co-ions complexation

The co-ions complexation method (CCM) done in three stages: solidification, freeze-drying, and calcining of resultants. The pyrochlore form of $La_2Zr_2O_7$ particles synthesized by the co-ions complexation route at 1300° by Xu et al. (2017) (Fig. 4.6).

4.2.6.3 Cathode plasma electrolysis

The fluorite and pyrochlore structures of lanthanum zirconate microspheres has been prepared by cathode plasma electrolysis (CPE) method in the electrolyte of lanthanum and zirconium (IV) nitrate. Fig. 4.7 demonstrated the schematic diagram of this method to synthesize $Ln_2Zr_2O_7$ microspheres (Liu, Zhang, Deng, Wang, & He, 2016).

4.2.6.4 Complex precipitation

Gadolinium zirconate and neodymium zirconate nanostructures have been synthesized by Kong et al. (2013) through complex precipitation method. This method is a molecular-level combinatorial method that provides the production of uniform oxides at comparatively less temperatures. Zirconium (IV) bis (diethyl citrate) dipropoxide has been utilized as a zirconate source. After stirring and heating the

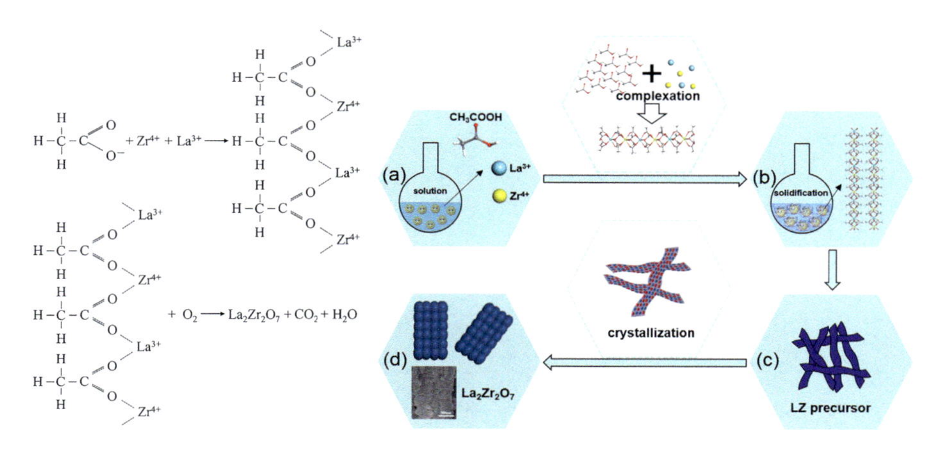

Figure 4.6 (Left) The actual reaction equations in CCM. (Right) Schematic of the complexation and crystallization in CCM (Xu et al., 2017).
Source: Reprinted with permissions from Xu, C., Jin, H., Zhang, Q., Huang, C., Zou, D., He, F., & Hou, S. (2017). A novel Co-ions complexation method to synthesize pyrochlore $La_2Zr_2O_7$. *Journal of the European Ceramic Society*, *37*, 2871–2876.

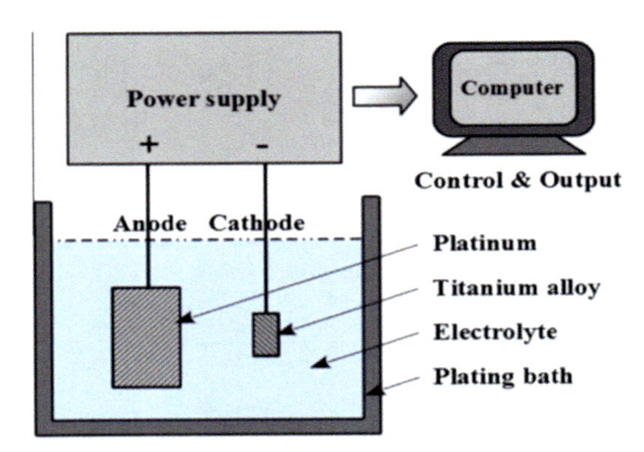

Figure 4.7 Schematic diagram of CPE device to prepare $La_2Zr_2O_7$ microspheres (Liu et al., 2016).
Source: Reprinted with permissions from Liu, C., Zhang, J., Deng, S., Wang, P., & He, Y. (2016). Direct preparation of $La_2Zr_2O_7$ microspheres by cathode plasma electrolysis. *Journal of Colloid and Interface Science*, *474*, 146–150.

prepared aqueous solution of precursor, the solution of lanthanide nitrate was added to that. Then pH of solutions was controlled by NH_3. Finally, calcination and sintering stages of products done. As synthesized $Gd_2Zr_2O_7$ and $Nd_2Zr_2O_7$ by complex precipitation method calcined at 1400° and 1200°, respectively (Kong et al., 2013).

4.2.6.5 Precursor approach

$Nd_2Zr_2O_7$ ceramics were fabricated with a low-temperature precursor route by Payne (Payne, Tucker, & Evans, 2013). After stirring the mixed aqueous solution of $Nd(NO_3)_3.6H_2O$ and $ZrO(NO_3)_2 \cdot xH_2O$, ammonium hydroxide was added and gel created. The stirring, washing, drying, and then annealing of the resulted gel have been done. In the precursor route, annealing at high temperatures terminates the phase conversion of fluorite form to the pyrochlore phase (Payne et al., 2013).

4.2.6.6 Floating zone

The floating zone method results in the preparation of crystals of the lanthanide zirconate (Ciomaga Hatnean, Decorse, Lees, Petrenko, & Balakrishnan, 2016; Hatnean, Lees, & Balakrishnan, 2015; Koohpayeh, Wen, Trump, Broholm, & McQueen, 2014). polycrystalline form of $Ln_2Zr_2O_7$ pyrochlore oxides was prepared by Hatnean et al. (2015). After calcination and then sintering of stoichiometric quantities of milled Ln_2O_3 and ZrO_2 powders, the polycrystalline rods were obtained. The produced rods with polycrystalline structure used as seeds for further growth of the crystals (Hatnean et al., 2015).

4.3 Applications of $Re_2Zr_2O_7$ ceramic nanomaterials

4.3.1 *Photocatalytic applications*

4.3.1.1 *Photocatalytic performance of $Re_2Zr_2O_7$ nanocomposites*

Rare earth zirconate nanostructures have been used significantly as heterogeneous photocatalysts because of their great stability and optical reactivity (Litter, 1999; Stylidi, Kondarides, & Verykios, 2003; Zhong et al., 2012). Heterogeneous photocatalysts as a large group of catalysts are widely used in water and wastewater treatment (Han et al., 2021; Vasseghian, Berkani, Almomani, & Dragoi, 2021). The photocatalytic efficiency of rare earth zirconate ceramics is influenced by crystalline, size, shape, and surface. Tong et al. employed $Re_2Zr_2O_7$ nanostructures as photocatalyst for degradation of the methyl orange pollution under UV illumination (Tong, Zhu, et al., 2008). Based on findings of this research, the efficiency of photocatalytic performance of $La_2Zr_2O_7$ nanostructures in 60 minutes (67.3%) is higher than $Nd_2Zr_2O_7$ (59.1%) nanostructures in decomposition of methyl orange pollutant, which is because of great surface area and high crystallinity $La_2Zr_2O_7$ nanostructures (Tong, Zhu, et al., 2008; Zinatloo-Ajabshir, Heidari-Asil, & Salavati-Niasari, 2021). Tong et al. investigated the photocatalytic properties of yttrium zirconate nanostructures prepared at various temperatures, by the demolition of dye under UV illumination (Tong, Xue, et al., 2008). The findings indicated that enhancing the calcination temperature caused the reduction of surface area and photocatalytic yield (Kaliyaperumal, Sankarakumar, & Paramasivam, 2020). In another study, the photocatalytic activity of $(LaxFe_{1-x})_2Zr_2O_7$ nanostructures has been surveyed by Bai, Lu, and Bao (2011) with the demolition of dye under UV illumination in 120 minutes. Based on outcomes, the photocatalytic yield of $La_2Zr_2O_7$ could be enhanced with iron doping. In $(La_{0.9}Fe_{0.1})_2Zr_2O_7$ structure, favorable quantity of doped iron due to higher lattice imperfections operates as the best photocatalyst.

4.3.1.2 *Photocatalytic performance of $Re_2Zr_2O_7$ nanocomposites*

Tong et al. investigated $TiO_2/La_2Zr_2O_7$ composites in the destruction of methyl orange under UV illumination (Tong, Zhao, Huo, & Yang, 2011). Based on results, the synergistic catalytic reaction of TiO_2 and $La_2Zr_2O_7$ particles in $TiO_2/La_2Zr_2O_7$ composites could operate as the best photocatalyst, in comparison to pure TiO_2 and $Re_2Zr_2O_7$. In another study, Yuping et al. perused $TiO_2/Nd_2Zr_2O_7$ composites for the demolition of methyl orange dye in UV light. $TiO_2/Nd_2Zr_2O_7$ composites indicated high efficiency in demolition of dye under UV illumination, in comparison to pure neodymium zirconate (Tong, Qian, Zhao, & Lu, 2013).

4.3.2 *Catalytic activity of $Re_2Zr_2O_7$ nanomaterials*

$Ln_2Zr_2O_7$ ceramic nanomaterials have been prepared from green tea extract (GrTeEx). These nanostructures can act as effective fuel in the propane-selective catalytic reduction of the NO_x process (propane-SCR-NO_x), which studied via various kinds of techniques. The results indicated that as prepared $Ln_2Zr_2O_7$ (Ln = Nd, Pr) nanostructures can act as a potential catalyst for propane-SCR-NO_x. Based on this study, $Pr_2Zr_2O_7$ is better than neodymium zirconate for NOx reduction. The efficiency of propane-SCR-NO_x to N_2 by $Pr_2Zr_2O_7$ was 67% and by neodymium zirconate, was 56%. In addition, the concentration of produced CO which is an unavoidable product in the SCR-NO_x process by $Pr_2Zr_2O_7$ was lower than $Nd_2Zr_2O_7$ (Zinatloo-Ajabshir, Ghasemian, & Salavati-Niasari, 2020). In another study, the synthesized neodymium zirconate ceramic nanostructure via the eco-friendly sonochemical pathway used as potential nanocatalysts for propane-SCR-NO_x. $Nd_2Zr_2O_7$-ZrO_2 nanocomposite as the best sample converted NO_x to N_2 in high amount (70%). Moreover, the quantity of CO as an unavoidable product, in the case of $Nd_2Zr_2O_7$-ZrO_2 nanocomposite, was lower than the other ceramics (Kato, Akiyama, Ogasawara, Wakabayashi, & Nakahara, 2014).

$Re_2Zr_2O_7$ with heavy rare earth metals, especially yttrium zirconate (Fang et al., 2021; Kato et al., 2014; Tong, Xue, et al., 2008; Xie et al., 2021; Zhang, Fang, et al., 2017) with fluorite structure, presented a great catalytic performance in discriminating dehydration of 1,4-butanediol to 3-buten-1-ol and also in 1,3-butanediol and formation of and 3-buten-2-ol (Karthick, Karati, & Murty, 2020), whereas the other zirconate catalysts are weak in the dehydration of 2,3-butanediol (Matsuda et al., 2020). The catalytic performance of yttrium zirconate affected by the calcination temperature as the main factor in the dehydration of 1,4-butanediol. The selective formation of unsaturated alcohols has occurred at 900° or higher temperature, which is the temperature of calcining of yttrium zirconate. The high crystallinity of yttrium zirconate demonstrates the top efficiency of 3-buten-1-ol from 1,4-butanediol production (Matsuda et al., 2020).

4.3.3 $Re_2Zr_2O_7$ materials as thermal barrier coatings

TBCs with multilayer coating systems are utilized as thermal insulators and protectors against the corrosive and hot gas flow in the ingredients of gas turbine power plants, particularly turbine blades (Clarke & Levi, 2003; Clarke, Oechsner, & Padture, 2012; Padture, Gell, & Jordan, 2002; Weber, Lein, Grande, & Einarsrud, 2013). TBCs generally act to safeguard the gas turbine engines and other hot sections where hot combustion gases with direct interaction (Ling, Qiang, Fuchi, & Hongsong, 2008; Liu et al., 2019; Schmitt, Stokes, Rai, Schwartz, & Wolfe, 2019). The basic structure of TBCs involves four layers: (1) super alloy substrate; (2) bond coat; (3) thermally grown oxide (TGO); and (4) ceramic topcoat (Vassen, Cao, Tietz, Basu, & Stöver, 2000; Vaßen, Jarligo, Steinke, Mack, & Stöver, 2010; Wright & Evans, 1999). The unique mechanical, physical and thermal properties of $Re_2Zr_2O_7$ ceramic materials make them a good candidate for TBCs (Weber et al., 2013), which will be discussed in the following part.

4.3.3.1 Physical properties of $Re_2Zr_2O_7$ ceramic nanomaterials based TBCs

4.3.3.1.1 Porosity and coating density

The theoretical density of $Re_2Zr_2O_7$ structures computed by utilizing the number of formula units per elementary cell and the molecular weight (Lehmann, Pitzer, Pracht, Vassen, & Stöver, 2003). The theoretical density of $Ln_2Zr_2O_7$ material calculated by Lehmann et al. is 6050 kg/m^3. The porous $Ln_2Zr_2O_7$ coating density can be calculated based on Archimedes' principle ASTM standard B328-96 (ASTM B328, 2003). The porosity of the $Re_2Zr_2O_7$ coatings differs by the deposition parameters, included substrate standoff interval, deposition power and powder feed velocity (Myoung et al., 2010).

4.3.3.1.2 Sintering behavior

The sintering behavior of (APS) deposited $Re_2Zr_2O_7$ material has been air plasma spraying investigated by utilizing a high-temperature dilatometer and mercury porosimetry by Vassen research group. The coating with high porosity (20%) is sintered at a high rate and are densified forcefully during the annealing process at 1250° (Vassen et al., 2000; Vassen, Cao, Tietz, Kerkhoff, & Stoever, 1999).

4.3.3.1.3 Cracks and pores morphology

Zhang et al. investigated the crack morphology of the APS deposited $Re_2Zr_2O_7$ coatings. The findings of this research showed that the cracks in the top and middle regions of the cross-section area were horizontal, while the majority of cracks in the bottom region are vertical. The crack and pore shape of TBCs is one of the important parameters that influenced the mechanical and thermal properties of the coatings (Weber, Lein, Grande, & Einarsrud, 2014).

4.3.3.2 Mechanical properties of $Re_2Zr_2O_7$ ceramic nanomaterials based TBCs

4.3.3.2.1 Elastic modulus

Elastic moduli involve bulk modulus (K), Young's modulus (E), Poisson's ratio (ν), and shear modulus (G) which can be calculated by a depth-sensing indentation technique or an ultrasound pulse-echo technique (Shimamura et al., 2007; Vassen et al., 2000). Himamura showed that in $Ln_2Zr_2O_7$ pyrochlore form, the elastic moduli apart related to the atomic radius of Re-elements. Increasing the atomic radius causes a larger E, K, and G amount (Shimamura et al., 2007).

4.3.3.2.2 Fracture toughness and hardness

The hardness of lanthanum ziconate materials differs owing to the sample density change. By increasing coating density, the hardness will increase.

4.3.3.2.3 Erosion resistance

The erosion of air plasma spraying deposited YSZ and $Ln_2Zr_2O_7$ coatings investigated. According to the findings of this research, the erosion rate of the coatings depends on impact angle, various abrasive particles, coating porosity, velocity, and amount of porosity. The porosity of coatings increased the erosion speeding. By enhancing the abrasive particle's velocity, the erosion rate will be increased (Ramachandran, Balasubramanian, & Ananthapadmanabhan, 2013).

4.3.3.3 Thermal properties of $Re_2Zr_2O_7$ ceramic nanomaterials based TBCs

4.3.3.3.1 Melting point temperature

Thermal analysis is the usual method used to discover the melting point. This measurement completed in sealed tungsten crucibles by monitoring the temperature of samples with spectro-pyrometer. The melting point of the rare earth zirconates, which measured experimentally, is about 2523K−2573K (2250°−2300°). The value of the melting point of $Re_2Zr_2O_7$ is lesser than that of YSZ (2953K, 2680°), which make it a sufficient case for most applications such as TBCs (Hong, 2015; Vassen et al., 1999).

4.3.3.3.2 Thermal conductivity

Rare earth zirconate ceramics recognized in the category of low-thermal-conductive materials, which are good cases for TBCs. Padture et al. studied the thermal conductivities of the following hot-pressed ceramics: $Gd_2Zr_2O_7$, $Gd_2Zr_2O_7$, $Gd_{2.58}Zr_{1.57}O_7$, $Nd_2Zr_2O_7$, and $Sm_2Zr_2O_7$. Results indicated almost identical thermal conductivities for these materials (Wu et al., 2002).

Thermal conductivity of MCrAlM' as an intermetallic alloy, wherein M is an element from Co, Fe, Ni and the mixture of them, and M' is one of the ytterbium, hafnium, zirconium, yttrium elements or a mixture thereof, utilized as a bond coat in TBCs (Bennett, 1986; Taylor, 2007, 2011). Deposition of TBCs on the substrate have been performed by various techniques, such as APS, EB-PVD, HVOF, vacuum and low-pressure plasma spraying, and diffusion bond method (Chen, Glynn, Ott, Hufnagel, & Hemker, 2003; Fauchais, Heberlein, & Boulos, 2014; Nelson & Orenstein, 1997). $Re_2Zr_2O_7$ nanostructures compared to zirconia partially stabilized with yttria (YSZ), which has low thermal conductivity, low coefficient of thermal expansion, poor sintering ability, and absence of toughness. $Re_2Zr_2O_7$ nanostructures are composed with standard TBCs YSZ to reach incorporate advantages of multilayer architecture (Mathanbabu, Thirumalaikumarasamy, Thirumal, & Ashokkumar, 2021). $Re_2Zr_2O_7$ nanostructures are more beneficial as TBCs due to: (1) no phase transformation from ambient temperature to its melting point (2) notable sintering resistance, (3) low thermal conductivity, and (4) less oxygen ion diffusivity, which supports the bond coat and the substrate from oxidation (Cao & Tietz, 2000; Jiang, Wang, Wang, Liu, & Han, 2019; Lyu et al., 2017; Mathanbabu et al., 2021; Vassen et al., 2000; Zhang et al., 2019). $Re_2Zr_2O_7$ with pyrochlore oxides structure are emerging TBCs material, in order to their low thermal conductivity and improved strength at raised operational temperatures (Dudnik et al., 2021; Mathanbabu et al., 2021; Song, Song, Paik, Lyu, & Jung, 2020; Zhu et al., 2021) (Fig. 4.8).

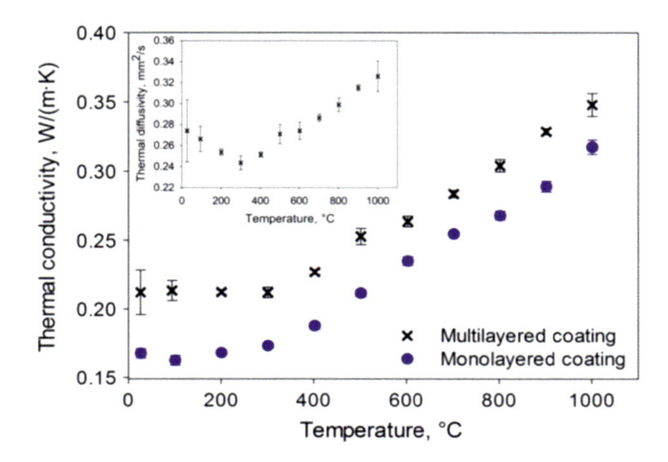

Figure 4.8 Thermal conductivity of nanostructured $Re_2Zr_2O_7$ mono- and multi-layered coatings as a function of temperature. Inset: thermal diffusivity of nanostructured $Re_2Zr_2O_7$ multilayered coating (Weber et al., 2013).
Source: Reprinted with permissions from Weber, S. B., Lein, H. L., Grande, T., & Einarsrud, M. -A. (2013). $Re_2Zr_2O_7$ thermal barrier coatings deposited by spray pyrolysis. *Surface and Coatings Technology*, *227*, 10–14.

Muhammet karabaş prepared TBCs of $(La_{0.7}(Nd, Dy)_{0.3})_2Zr_2O_7$ compound by the plasma spray route as a single and double layer with YSZ. Characterization, thermal conductivity, tensile adhesion strengths, and thermal cycle tests at 1250°–1300° of produced coatings achieved. The finding of this study showed thermal conductivity of the single and double layers of YSZ were lesser than the un-doped $Ln_2Zr_2O_7$ coatings, and this parameter will be improved by raising the temperature. Results showed the single-layer of Dy, Nd-doped, LnZ-based coatings were perverted at 163 and 300 cycles, respectively (Karabaş, 2020). In another research, Wang et al. have evidenced that TBCs have low thermal conductivity and great abidance to oxidation at elevated temperatures. They prepared TBCs consisting of a YSZ layer and of a LnZ + YSZ composite structure, on a nickel-based high-temperature substrate by APS route. The commonly used YSZ cannot operate at temperatures more than 1200°C for a prolonged-term of time. The results demonstrated that the $Re_2Zr_2O_7$-based coating by oxygen sequestration and great efficient thermal insulation, compared to the YSZ layer, decreases the growth of TGO in TBCs, therewith cause the postponement of coating failure (Wang et al., 2019).

In another study, Eu-doped $Re_2Zr_2O_7$ in pyrochlore structure was synthesized by the coprecipitation-calcination route, and then thermophysical properties, bond strength, grain growth kinetics, phase structure of the doped-$Re_2Zr_2O_7$ were investigated. The results demonstrate nano-sized doped LnZ seeds produced under elevated temperature, immediately, and by increasing Eu^{3+} ions, more imperfects, changes of bond strength, and slight disorder of the lattice have occurred. By increasing Eu^{3+} ions, The CTE of Eu-doped $La_2Zr_2O_7$ seriously meliorated. Based on the findings of this study, the thermal conductivity declines at the primary period and raises subsequently that gains the lowermost quantity for the composition of $(La_{0.6}Eu_{0.4})_2Zr_2O_7$ (Mathanbabu et al., 2021; Zhu et al., 2018).

4.4 Conclusion and outlook

Rare earth zirconate ceramic nanomaterials and developed researches pertaining to the synthesis, properties, and applications of these materials have received major interest in recent years. These materials have two different crystal structures: pyrochlore and fluorite. The unique physical, mechanical, and thermal properties of these materials made them very applicant materials in recent researches. The potential applications in electro/photocatalysis, magnetism, nuclear waste storage, oxygen monitoring sensors, photoluminescent host materials, and solid electrolytes in high-temperature fuel cells correlated with distinctive properties of these materials. This chapter contains preparation methods, properties, and applications of rare earth zirconates. The final part of this chapter covers the physical, mechanical, and thermal properties of rare earth zirconate ceramic nanomaterials based on TBCs.

References

AruláDhas, N. (1993). Combustion synthesis and properties of fine-particle rare earth-metal zirconates, $Ln_2Zr_2O_7$. *Journal of Materials Chemistry*, *3*, 1289–1294.

Aruna, S., Sanjeeviraja, C., Balaji, N., & Manikandanath, N. (2013). Properties of plasma sprayed $La_2Zr_2O_7$ coating fabricated from powder synthesized by a single-step solution combustion method. *Surface and Coatings Technology*, *219*, 131–138.

ASTM B328. (2003). Standard test method for density, oil content, and interconnected porosity of sintered metal structural parts and oil-impregnated bearing. ASTM International.

Bai, Y., Lu, L., & Bao, J. (2011). Synthesis and characterization of lanthanum zirconate nanocrystals doped with iron ions by a salt-assistant combustion method. *Journal of Inorganic and Organometallic Polymers and Materials*, *21*, 590–594.

Bennett, A. (1986). Properties of thermal barrier coatings. *Materials Science and Technology*, *2*, 257–261.

Bolech, M., Cordfunke, E., Van Genderen, A., Van Der Laan, R., Janssen, F., & Van Miltenburg, J. (1997). The heat capacity and derived thermodynamic functions of $La_2Zr_2O_7$ and $Ce_2Zr_2O_7$ from 4 to 1000K. *Journal of Physics and Chemistry of Solids*, *58*, 433–439.

Cao, X., & Tietz, F. (2000). Zirconates as new materials for thermal barrier coatings. *Journal of the American Ceramic Society*, *83*, 2023–2028.

Cao, X., Vassen, R., Jungen, W., Schwartz, S., Tietz, F., & Stöver, D. (2001). Thermal stability of lanthanum zirconate plasma-sprayed coating. *Journal of the American Ceramic Society*, *84*, 2086–2090.

Cao, X., Vassen, R., & Stöver, D. (2004). Ceramic materials for thermal barrier coatings. *Journal of the European Ceramic Society*, *24*, 1–10.

Chen, D., & Xu, R. (1998). Hydrothermal synthesis and characterization of $La_2M_2O_7$ (M = Ti, Zr) powders. *Materials Research Bulletin*, *33*, 409–417.

Chen, H., Gao, Y., Liu, Y., & Luo, H. (2009). Coprecipitation synthesis and thermal conductivity of $La_2Zr_2O_7$. *Journal of Alloys and Compounds*, *480*, 843–848.

Chen, M., Glynn, M., Ott, R., Hufnagel, T., & Hemker, K. (2003). Characterization and modeling of a martensitic transformation in a platinum modified diffusion aluminide bond coat for thermal barrier coatings. *Acta Materialia*, *51*, 4279–4294.

Chen, X. G., & Zhang, H. S. (2013). *Preparation and their fluorescent property of* $Sm_2Zr_2O_7$ *and* $Sm_2(Zr_{0.7}Ce_{0.3})_2O_7$ *powders. Advanced Materials Research* (pp. 22–25). Trans Tech Publ.

Ciomaga Hatnean, M., Decorse, C., Lees, M. R., Petrenko, O. A., & Balakrishnan, G. (2016). Zirconate pyrochlore frustrated magnets: Crystal growth by the floating zone technique. *Crystals*, *6*, 79.

Clarke, D., & Levi, C. (2003). Materials design for the next generation thermal barrier coatings. *Annual Review of Materials Research*, *33*, 383–417.

Clarke, D. R., Oechsner, M., & Padture, N. P. (2012). Thermal-barrier coatings for more efficient gas-turbine engines. *MRS Bulletin*, *37*, 891–898.

Davis, J. R. (2004). *Handbook of thermal spray technology*. ASM international.

Du, Q., Zhou, G., Zhou, H., & Yang, Z. (2012). Novel multiband luminescence of $Y_2Zr_2O_7$: Eu^{3+}, R^{3+} (R = Ce, Bi) orange–red phosphors via a sol–gel combustion approach. *Optical Materials*, *35*, 257–262.

Duarte, W., Meguekam, A., Colas, M., Vardelle, M., & Rossignol, S. (2015). Effects of the counter-cation nature and preparation method on the structure of $La_2Zr_2O_7$. *Journal of Materials Science*, *50*, 463–475.

Dudnik, O., Lakiza, S., Grechanyuk, I., Red'ko, V., Glabay, M., Shmibelsky, V., ... Grechanyuk, M. (2021). High-entropy ceramics for thermal barrier coatings produced from ZrO_2 doped with rare earth metal oxides. *Powder Metallurgy and Metal Ceramics*, *59*, 556–563.

Emadi, H., Salavati-Niasari, M., & Sobhani, A. (2017). Synthesis of some transition metal (M: 25Mn, 27Co, 28Ni, 29Cu, 30Zn, 47Ag, 48Cd) sulfide nanostructures by hydrothermal method. *Advances in Colloid and Interface Science*, *246*, 52–74.

Fang, X., Xia, L., Li, S., Hong, Z., Yang, M., Xu, X., ... Wang, X. (2021). Superior 3DOM $Y_2Zr_2O_7$ supports for Ni to fabricate highly active and selective catalysts for CO_2 methanation. *Fuel*, *293*, 120460.

Fang, X., Xia, L., Peng, L., Luo, Y., Xu, J., Xu, L., ... Wang, X. (2019). $Ln_2Zr_2O_7$ compounds (Ln = La, Pr, Sm, Y) with varied rare earth A sites for low temperature oxidative coupling of methane. *Chinese Chemical Letters*, *30*, 1141–1146.

Fauchais, P. L., Heberlein, J. V., & Boulos, M. I. (2014). *Thermal spray fundamentals: From powder to part*. Springer Science & Business Media.

Ferrari, M., & Lutterotti, L. (1994). Method for the simultaneous determination of anisotropic residual stresses and texture by x-ray diffraction. *Journal of Applied Physics*, *76*, 7246–7255.

Gao, L., Zhu, H., Wang, L., & Ou, G. (2011). Hydrothermal synthesis and photoluminescence properties of $Gd_2Zr_2O_7$: Tb^{3+} phosphors. *Materials Letters*, *65*, 1360–1362.

Gokul Raja, T., Balamurugan, S., & Reshma, A. (2021). Synthesis, different characterizations and pigment application of nanoceramic zirconate powder. *Journal of Nanoscience and Nanotechnology*, *21*, 2212–2220.

Hagiwara, T., Nomura, K., & Kageyama, H. (2017). Crystal structure analysis of $Ln_2Zr_2O_7$ (Ln = Eu and La) with a pyrochlore composition by high-temperature powder X-ray diffraction. *Journal of the Ceramic Society of Japan*, *125*, 65–70.

Han, S., Li, Z., Ma, S., Zhi, Y., Xia, H., Chen, X., & Liu, X. (2021). Bandgap engineering in benzotrithiophene-based conjugated microporous polymers: A strategy for screening metal-free heterogeneous photocatalysts. *Journal of Materials Chemistry A*, *9*, 3333–3340.

Hatnean, M. C., Lees, M., & Balakrishnan, G. (2015). Growth of single-crystals of rare-earth zirconate pyrochlores, $Ln_2Zr_2O_7$ (with Ln = La, Nd, Sm, and Gd) by the floating zone technique. *Journal of Crystal Growth*, *418*, 1–6.

Hong, Q.-J. (2015). *Methods for melting temperature calculation*. California Institute of Technology.

Hongming, Z., & Danqing, Y. (2008). Effect of rare earth doping on thermo-physical properties of lanthanum zirconate ceramic for thermal barrier coatings. *Journal of Rare Earths*, *26*, 770–774.

Hu, K.-J., Liu, Z.-G., Wang, J.-Y., Wang, T., & Ouyang, J.-H. (2012). Synthesis and photoluminescence properties of Eu^{3+}-doped $Gd_2Zr_2O_7$. *Materials Letters*, *89*, 276–278.

Huang, Z., Li, F., Jiao, C., Liu, J., Huang, J., Lu, L., ... Zhang, S. (2016). Molten salt synthesis of $La_2Zr_2O_7$ ultrafine powders. *Ceramics International*, *42*, 6221–6227.

Huo, Y.-f, Qin, N., Liao, C.-z, Feng, H.-f, Gu, Y.-y, & Cheng, H. (2019). Hydrothermal synthesis and energy storage performance of ultrafine $Ce_2Sn_2O_7$ nanocubes. *Journal of Central South University*, *26*, 1416–1425.

Jeyasingh, T., Saji, S., Kavitha, V., & Wariar, P. (2018). *Frequency dependent dielectric properties of combustion synthesized $Dy_2Ti_2O_7$ pyrochlore oxide. AIP conference proceedings* (p. 030107) AIP Publishing LLC.

Jeyasingh, T., Saji, S., & Wariar, P. (2017). *Synthesis of nanocrystalline $Gd_2Ti_2O_7$ by combustion process and its structural, optical and dielectric properties. AIP conference proceedings* (p. 020016) AIP Publishing LLC.

Jeyasingh, T., Vindhya, P., Saji, S., Wariar, P., & Kavitha, V. (2019). Structural and magnetic properties of combustion synthesized $A_2Ti_2O_7$ (A = Gd, Dy and Y) pyrochlore oxides. *Bulletin of Materials Science*, *42*, 1–7.

Jiang, D., Wang, Y., Wang, S., Liu, R., & Han, J. (2019). Thermal conductivity of air plasma sprayed yttrium heavily-doped lanthanum zirconate thermal barrier coatings. *Ceramics International*, *45*, 3199–3206.

Joulia, A., Vardelle, M., & Rossignol, S. (2013). Synthesis and thermal stability of $Re_2Zr_2O_7$, (Re = La, Gd) and $La_2(Zr_{1-x}Ce_x)_2O_{7-\delta}$ compounds under reducing and oxidant atmospheres for thermal barrier coatings. *Journal of the European Ceramic Society*, *33*, 2633–2644.

Jovaní, M., Sanz, A., Beltrán-Mir, H., & Cordoncillo, E. (2016). New red-shade environmental-friendly multifunctional pigment based on Tb and Fe doped $Y_2Zr_2O_7$ for ceramic applications and cool roof coatings. *Dyes and Pigments*, *133*, 33–40.

Kalinkin, A., Vinogradov, V. Y., & Kalinkina, E. (2021). Solid-state synthesis of nanocrystalline gadolinium zirconate using mechanical activation. *Inorganic Materials*, *57*, 178–185.

Kaliyaperumal, C., Sankarakumar, A., & Paramasivam, T. (2020). Grain size engineering in nanocrystalline $Y_2Zr_2O_7$: A detailed study on the grain size correlated electrical properties. *Journal of Alloys and Compounds*, *831*, 154782.

Karabaş, M. (2020). Production and characterization of Nd and Dy doped lanthanum zirconate-based thermal barrier coatings. *Surface and Coatings Technology*, *394*, 125864.

Karthick, G., Karati, A., & Murty, B. (2020). Low temperature synthesis of nanocrystalline $Y_2Ti_2O_7$, $Y_2Zr_2O_7$, $Y_2Hf_2O_7$ with exceptional hardness by reverse co-precipitation technique. *Journal of Alloys and Compounds*, *837*, 155491.

Kato, S., Akiyama, S., Ogasawara, M., Wakabayashi, T., & Nakahara, Y. (2014). NO reduction activity of pyrochlore-type $Ln_2Sn_{2-x}Zr_xO_7$ (Ln: La, Nd, Y)-supported Rh catalysts. *Bulletin of the Chemical Society of Japan*, *87*, 1216–1223.

Keyvani, A., Mahmoudinezhad, P., Jahangiri, A., & Bahamirian, M. (2020). Synthesis and characterization of $((La_{1-x}Gd_x)_2Zr_2O_7$; $x = 0, 0.1, 0.2, 0.3, 0.4, 0.5, 1$) nanoparticles for advanced TBCs. *Journal of the Australian Ceramic Society*, *56*, 1543–1550.

Kong, L., Karatchevtseva, I., Aughterson, R. D., Davis, J., Zhang, Y., Lumpkin, G. R., & Triani, G. (2015). New pathway for the preparation of pyrochlore $Nd_2Zr_2O_7$ nanoparticles. *Ceramics International*, *41*, 7618–7625.

Kong, L., Karatchevtseva, I., Gregg, D. J., Blackford, M. G., Holmes, R., & Triani, G. (2013). A novel chemical route to prepare $La_2Zr_2O_7$ pyrochlore. *Journal of the American Ceramic Society*, *96*, 935–941.

Koohpayeh, S., Wen, J.-J., Trump, B., Broholm, C., & McQueen, T. (2014). Synthesis, floating zone crystal growth and characterization of the quantum spin ice $Pr_2Zr_2O_7$ pyrochlore. *Journal of Crystal Growth*, *402*, 291–298.

Lee, Y., Sheu, H., Deng, J., & Kao, H.-C. (2009). Preparation and fluorite–pyrochlore phase transformation in $Gd_2Zr_2O_7$. *Journal of Alloys and Compounds*, *487*, 595–598.

Lee, Y., Sheu, H., & Kao, H.-C. (2010). Preparation and characterization of $Nd_2Zr_2O_7$ nanocrystals by a polymeric citrate precursor method. *Materials Chemistry and Physics*, *124*, 145–149.

Lehmann, H., Pitzer, D., Pracht, G., Vassen, R., & Stöver, D. (2003). Thermal conductivity and thermal expansion coefficients of the lanthanum rare-earth-element zirconate system. *Journal of the American Ceramic Society*, *86*, 1338–1344.

Lei, M., Weimin, M., Xudong, S., Jianan, L., Lianyong, J., & Han, S. (2015). Structure properties and sintering densification of $Gd_2Zr_2O_7$ nanoparticles prepared via different acid combustion methods. *Journal of Rare Earths*, *33*, 195–201.

Li, J., Dai, H., Li, Q., Zhong, X., Ma, X., Meng, J., & Cao, X. (2006). Lanthanum zirconate nanofibers with high sintering-resistance. *Materials Science and Engineering: B*, *133*, 209–212.

Li, L., Lei, L., Zhao, G., Deng, B., Yan, F., & Li, C. (2021). Highly epitaxial $YBa_2Cu_3O_{7-\delta}$ films grown on gradient $La_{2-x}Gd_xZr_2O_7$-buffered NiW-RABiTS using all sol–gel process. *Superconductor Science and Technology*, *34*, 045004.

Li, W., Zhang, K., Xie, D., Deng, T., Luo, B., Zhang, H., & Huang, X. (2020). Characterizations of vacuum sintered $Gd_2Zr_2O_7$ transparent ceramics using combustion synthesized nanopowder. *Journal of the European Ceramic Society*, *40*, 1665–1670.

Ling, L., Qiang, X., Fuchi, W., & Hongsong, Z. (2008). Thermophysical properties of complex rare-earth zirconate ceramic for thermal barrier coatings. *Journal of the American Ceramic Society*, *91*, 2398–2401.

Litter, M. I. (1999). Heterogeneous photocatalysis: Transition metal ions in photocatalytic systems. *Applied Catalysis B: Environmental*, *23*, 89–114.

Liu, B., Liu, Y., Zhu, C., Xiang, H., Chen, H., Sun, L., ... Zhou, Y. (2019). Advances on strategies for searching for next generation thermal barrier coating materials. *Journal of Materials Science & Technology*, *35*, 833–851.

Liu, C., Zhang, J., Deng, S., Wang, P., & He, Y. (2016). Direct preparation of $La_2Zr_2O_7$ microspheres by cathode plasma electrolysis. *Journal of Colloid and Interface Science*, *474*, 146–150.

Liu, Z.-G., Ouyang, J.-H., Wang, B.-H., Zhou, Y., & Li, J. (2009). Thermal expansion and thermal conductivity of $Sm_xZr_{1-x}O_{2-x/2}$ $(0.1 \le x \le 0.5)$ ceramics. *Ceramics International*, *35*, 791–796.

Lyu, G., Kim, B. G., Lee, S.-S., Jung, Y.-G., Zhang, J., Choi, B.-G., & Kim, I.-S. (2017). Fracture behavior and thermal durability of lanthanum zirconate-based thermal barrier coatings with buffer layer in thermally graded mechanical fatigue environments. *Surface and Coatings Technology*, *332*, 64–71.

Mao, Y., Guo, X., Huang, J. Y., Wang, K. L., & Chang, J. P. (2009). Luminescent nanocrystals with $A_2B_2O_7$ composition synthesized by a kinetically modified molten salt method. *The Journal of Physical Chemistry C*, *113*, 1204–1208.

Mathanbabu, M., Thirumalaikumarasamy, D., Thirumal, P., & Ashokkumar, M. (2021). Study on thermal, mechanical, microstructural properties and failure analyses of lanthanum zirconate based thermal barrier coatings: A review. *Materials Today: Proceedings*.

Matovic, B., Maletaskic, J., Zagorac, J., Pavkov, V., Maki, R. S., Yoshida, K., & Yano, T. (2020). Synthesis and characterization of pyrochlore lanthanide (Pr, Sm) zirconate ceramics. *Journal of the European Ceramic Society*, *40*, 2652–2657.

Matsuda, A., Matsumura, Y., Nakazono, K., Sato, F., Takahashi, R., Yamada, Y., & Sato, S. (2020). Dehydration of biomass-derived butanediols over rare earth zirconate. *Catalysts, Catalysts*, *10*, 1392.

Mohr, E. B., Brezinski, J., Audrieth, L., Ritchey, H., & McFarlin, R. (1953). Carbohydrazide. *Inorganic Syntheses*, *4*, 32–35.

Morassaei, M. S., Zinatloo-Ajabshir, S., & Salavati-Niasari, M. (2017). $Nd_2Sn_2O_7$ nanostructures: New facile Pechini preparation, characterization, and investigation of their photocatalytic degradation of methyl orange dye. *Advanced Powder Technology*, *28*, 697–705.

Myoung, S.-W., Kim, J.-H., Lee, W.-R., Jung, Y.-G., Lee, K.-S., & Paik, U. (2010). Microstructure design and mechanical properties of thermal barrier coatings with layered top and bond coats. *Surface and Coatings Technology*, *205*, 1229–1235.

Naik, P. P., & Hasolkar, S. S. (2020). Consequence of B-site substitution of rare earth (Gd^{+3}) on electrical properties of manganese ferrite nanoparticles. *Journal of Materials Science: Materials in Electronics*, *31*, 13434–13446.

Nelson, W. A., & Orenstein, R. M. (1997). TBC experience in land-based gas turbines. *Journal of Thermal Spray Technology*, *6*, 176–180.

Padture, N. P., Gell, M., & Jordan, E. H. (2002). Thermal barrier coatings for gas-turbine engine applications. *Science (New York, N.Y.)*, *296*, 280–284.

Padture, N. P., Gell, M., & Klemens, P. G. (2000). Ceramic materials for thermal barrier coatings. *Google Patents*.

Payne, J. L., Tucker, M. G., & Evans, I. R. (2013). From fluorite to pyrochlore: Characterisation of local and average structure of neodymium zirconate, $Nd_2Zr_2O_7$. *Journal of Solid State Chemistry*, *205*, 29–34.

Pokhrel, M., Alcoutlabi, M., & Mao, Y. (2017). Optical and X-ray induced luminescence from Eu^{3+} doped $La_2Zr_2O_7$ nanoparticles. *Journal of Alloys and Compounds*, *693*, 719–729.

Pokhrel, M., Brik, M. G., & Mao, Y. (2015). Particle size and crystal phase dependent photoluminescence of $La_2Zr_2O_7$: Eu^{3+} nanoparticles. *Journal of the American Ceramic Society*, *98*, 3192–3201.

Popov, V., Menushenkov, A., Ivanov, A., Gaynanov, B., Yastrebtsev, A., d'Acapito, F., ... Zheleznyi, M. (2019). Comparative analysis of long-and short-range structures features in titanates $Ln_2Ti_2O_7$ and zirconates $Ln_2Zr_2O_7$ ($Ln = Gd$, Tb, Dy) upon the crystallization process. *Journal of Physics and Chemistry of Solids*, *130*, 144–153.

Prize, E. (1999). International union of crystallography report of the executive committee for 1997. *Foundations of Crystallography*, *55*, 565–600.

Quader, A., Mustafa, G. M., Abbas, S. K., Ahmad, H., Riaz, S., Naseem, S., & Atiq, S. (2020). Efficient energy storage and fast switching capabilities in Nd-substituted $La_2Sn_2O_7$ pyrochlores. *Chemical Engineering Journal*, *396*, 125198.

Quiroz, A., Chavira, E., Garcia-Vazquez, V., Gonzalez, G., & Abatal, M. (2018). Structural, electrical and magnetic properties of the pyrochlorate $Er_{2-x}Sr_xRu_2O_7$ ($0 \leq x \leq 0.10$) system. *Revista Mexicana de Fı́sica*, *64*, 222–227.

Quiroz, A., Garcia-Vazquez, V., Aguilar, C. G., Agustin-Serrano, R., Chavira, E., & Abatal, M. (2021). Electrical and thermal conductivities of rare-earth $A_2Zr_2O_7$ (A = Pr, Nd, Sm, Gd, and Er). *Revista Mexicana de Fı́sica*, *67*, 255–262.

Ramachandran, C., Balasubramanian, V., & Ananthapadmanabhan, P. (2013). Erosion of atmospheric plasma sprayed rare earth oxide coatings under air suspended corundum particles. *Ceramics International*, *39*, 649–672.

Rao, K. K., Banu, T., Vithal, M., Swamy, G., & Kumar, K. R. (2002). Preparation and characterization of bulk and nano particles of $La_2Zr_2O_7$ and $Nd_2Zr_2O_7$ by sol–gel method. *Materials Letters*, *54*, 205–210.

Razi, F., Zinatloo-Ajabshir, S., & Salavati-Niasari, M. (2017). Preparation and characterization of HgI_2 nanostructures via a new facile route. *Materials Letters*, *193*, 9–12.

Rushton, M., Grimes, R. W., Stanek, C., & Owens, S. (2004). Predicted pyrochlore to fluorite disorder temperature for $A_2Zr_2O_7$ compositions. *Journal of Materials Research*, *19*, 1603–1604.

Salehi, Z., Zinatloo-Ajabshir, S., & Salavati-Niasari, M. (2017). Dysprosium cerate nanostructures: Facile synthesis, characterization, optical and photocatalytic properties. *Journal of Rare Earths*, *35*, 805–812.

Sankar, J., & Kumar, S. (2021). Synthesis of rare earth based pyrochlore structured ($A_2B_2O_7$) materials for thermal barrier coatings (TBCs)—A review. *Current Applied Science and Technology*, 601–617.

Schmitt, M. P., Stokes, J. L., Rai, A. K., Schwartz, A. J., & Wolfe, D. E. (2019). Durable aluminate toughened zirconate composite thermal barrier coating (TBC) materials for high temperature operation. *Journal of the American Ceramic Society*, *102*, 4781–4793.

Sedmidubský, D., Beneš, O., & Konings, R. (2005). High temperature heat capacity of $Nd_2Zr_2O_7$ and $La_2Zr_2O_7$ pyrochlores. *The Journal of Chemical Thermodynamics*, *37*, 1098–1103.

Shimamura, K., Arima, T., Idemitsu, K., & Inagaki, Y. (2007). Thermophysical properties of rare earth-stabilized zirconia and zirconate pyrochlores as surrogates for actinide-doped zirconia. *International Journal of Thermophysics*, *28*, 1074–1084.

Simonenko, N., Sakharov, K., Simonenko, E., Sevastyanov, V., & Kuznetsov, N. (2015). Glycol–citrate synthesis of ultrafine lanthanum zirconate. *Russian Journal of Inorganic Chemistry*, *60*, 1452–1458.

Snyder, J., Slusky, J., Cava, R., & Schiffer, P. (2001). How 'spin ice' freezes. *Nature*, *413*, 48–51.

Sohn, J. M., Kim, M. R., & Woo, S. I. (2003). The catalytic activity and surface characterization of $Ln_2B_2O_7$ (Ln = Sm, Eu, Gd and Tb; B = Ti or Zr) with pyrochlore structure as novel CH_4 combustion catalyst. *Catalysis Today*, *83*, 289–297.

Song, D., Song, T., Paik, U., Lyu, G., & Jung, Y.-G. (2020). Hot corrosion behavior in thermal barrier coatings with heterogeneous splat boundary. *Corrosion Science*, *163*, 108225.

Stylidi, M., Kondarides, D. I., & Verykios, X. E. (2003). Pathways of solar light-induced photocatalytic degradation of azo dyes in aqueous TiO_2 suspensions. *Applied Catalysis B: Environmental*, *40*, 271–286.

Subramanian, M., Aravamudan, G., & Rao, G. S. (1983). Oxide pyrochlores—A review. *Progress in Solid State Chemistry*, *15*, 55–143.

Tang, H., Sun, H., Chen, D., & Jiao, X. (2012). Fabrication and characterization of nanostructured $La_2Zr_2O_7$ fibers. *Materials Letters*, *70*, 48–50.

Taylor, T. (2007). Low thermal expansion bondcoats for thermal barrier coatings. *Google Patents*.

Taylor, T. A. (2011). Low thermal expansion bondcoats for thermal barrier coatings. *Google Patents*.

Teng, Z., Tan, Y., Zeng, S., Meng, Y., Chen, C., Han, X., & Zhang, H. (2021). Preparation and phase evolution of high-entropy oxides $A_2B_2O_7$ with multiple elements at A and B sites. *Journal of the European Ceramic Society*, *41*, 3614–3620.

Teymourinia, H., Amiri, O., & Salavati-Niasari, M. (2021). Synthesis and characterization of cotton-silver-graphene quantum dots (cotton/Ag/GQDs) nanocomposite as a new antibacterial nanopad. *Chemosphere*, *267*, 129293.

Teymourinia, H., Darvishnejad, M. H., Amiri, O., Salavati-Niasari, M., Reisi-Vanani, A., Ghanbari, E., & Moayedi, H. (2020). GQDs/Sb_2S_3/TiO_2 as a co-sensitized in DSSs: Improve the power conversion efficiency of DSSs through increasing light harvesting by using as-synthesized nanocomposite and mirror. *Applied Surface Science*, *512*, 145638.

Teymourinia, H., Salavati-Niasari, M., Amiri, O., & Farangi, M. (2018). Facile synthesis of graphene quantum dots from corn powder and their application as down conversion effect in quantum dot-dye-sensitized solar cell. *Journal of Molecular Liquids*, *251*, 267–272.

Teymourinia, H., Salavati-Niasari, M., Amiri, O., & Safardoust-Hojaghan, H. (2017). Synthesis of graphene quantum dots from corn powder and their application in reduce charge recombination and increase free charge carriers. *Journal of Molecular Liquids*, *242*, 447–455.

Teymourinia, H., Salavati-Niasari, M., Amiri, O., & Yazdian, F. (2019). Application of green synthesized TiO_2/Sb_2S_3/GQDs nanocomposite as high efficient antibacterial agent against E. coli and Staphylococcus aureus. *Materials Science and Engineering: C*, *99*, 296–303.

Tong, Y., Qian, X., Zhao, W., & Lu, L. (2013). Synthesis and catalytic peoperties of TiO_2/$Nd_2Zr_2O_7$ nanocomposites. *Journal of The Chinese Ceramic Society*, *41*, 34–37.

Tong, Y., Wang, Y., Yu, Z., Wang, X., Yang, X., & Lu, L. (2008). Preparation and characterization of pyrochlore $La_2Zr_2O_7$ nanocrystals by stearic acid method. *Materials Letters*, *62*, 889–891.

Tong, Y., Xue, P., Jian, F., Lu, L., Wang, X., & Yang, X. (2008). Preparation and characterization of $Y_2Zr_2O_7$ nanocrystals and their photocatalytic properties. *Materials Science and Engineering: B*, *150*, 194–198.

Tong, Y., Yu, Z., Lu, L., Yang, X., & Wang, X. (2008). Rapid preparation and characterization of $Dy_2Zr_2O_7$ nanocrystals. *Materials Research Bulletin*, *43*, 2736–2741.

Tong, Y., Zhu, J., Lu, L., Wang, X., & Yang, X. (2008). Preparation and characterization of $Ln_2Zr_2O_7$ (Ln = La and Nd) nanocrystals and their photocatalytic properties. *Journal of Alloys and Compounds*, *465*, 280–284.

Tong, Y. P., Zhao, S. B., Li, F. L., & Li, C. Y. (2010). *Characterization and photocatalytic activity of $La_{1.6}Ln_{0.4}Zr_2O_7$ (Ln = La, Nd, Dy, Er) nanocrystals by stearic acid method. Advanced Materials Research* (pp. 631–634). Trans Tech Publ.

Tong, Y. P., Zhao, Y. Q., Huo, H. Y., & Yang, H. (2011). *Synthesis, structure and catalytic activity of TiO_2/$La_2Zr_2O_7$ nanocomposites. Advanced Materials Research* (pp. 2878–2881). Trans Tech Publ.

Torres-Rodriguez, J., Gutierrez-Cano, V., Menelaou, M., Kaštyl, J., Cihlář, J., Tkachenko, S., ... Lázár, I. N. (2019). *Rare-earth zirconate $Ln_2Zr_2O_7$ (ln: La, Nd, Gd, and Dy) powders, xerogels, and aerogels: Preparation, structure, and properties, . Inorganic Chemistry* (58, pp. 14467−14477). .

Trujillano, R., Martín, J. A., & Rives, V. (2016). Hydrothermal synthesis of $Sm_2Sn_2O_7$ pyrochlore accelerated by microwave irradiation. A comparison with the solid state synthesis method. *Ceramics International*, *42*, 15950−15954.

Uno, M., Kosuga, A., Okui, M., Horisaka, K., Muta, H., Kurosaki, K., & Yamanaka, S. (2006). Photoelectrochemical study of lanthanide zirconium oxides, $Ln_2Zr_2O_7$ (Ln = La, Ce, Nd and Sm). *Journal of Alloys and Compounds*, *420*, 291−297.

Vasseghian, Y., Berkani, M., Almomani, F., & Dragoi, E.-N. (2021). Data mining for pesticide decontamination using heterogeneous photocatalytic processes. *Chemosphere*, *270*, 129449.

Vassen, R., Cao, X., Tietz, F., Basu, D., & Stöver, D. (2000). Zirconates as new materials for thermal barrier coatings. *Journal of the American Ceramic Society*, *83*, 2023−2028.

Vassen, R., Cao, X., Tietz, F., Kerkhoff, G., & Stoever, D. (1999). *La2Zr2O7—A new candidate for thermal barrier coatings. Proceedings of the united thermal spray conference-UTSC'99* (pp. 830−834). Düsseldorf, Germany: DVS-Verlag.

Vaßen, R., Jarligo, M. O., Steinke, T., Mack, D. E., & Stöver, D. (2010). Overview on advanced thermal barrier coatings. *Surface and Coatings Technology*, *205*, 938−942.

Venkatesh, G., Subramanian, R., & Berchmans, L. J. (2019). Phase analysis and microstructural investigations of $Ce_2Zr_2O_7$ for high-temperature coatings on Ni-Base superalloy substrates. *High Temperature Materials and Processes*, *38*, 773−782.

Wang, L., Di, Y., Wang, H., Li, X., Dong, L., & Liu, T. (2019). Effect of lanthanum zirconate on high temperature resistance of thermal barrier coatings. *Transactions of the Indian Ceramic Society*, *78*, 212−218.

Wang, Q., Cheng, X., Li, J., & Jin, H. (2016). Hydrothermal synthesis and photocatalytic properties of pyrochlore $Sm_2Zr_2O_7$ nanoparticles. *Journal of Photochemistry and Photobiology A: Chemistry*, *321*, 48−54.

Wang, S., Li, W., Wang, S., & Chen, Z. (2015). Synthesis of nanostructured $La_2Zr_2O_7$ by a non-alkoxide sol−gel method: From gel to crystalline powders. *Journal of the European Ceramic Society*, *35*, 105−112.

Wang, S., Li, W., Wang, S., Zhang, J., & Chen, Z. (2016). Deposition of SiC/$La_2Zr_2O_7$ multi-component coating on C/SiC substrate by combining sol-gel process and slurry. *Surface and Coatings Technology*, *302*, 383−388.

Wang, X., Zhu, Y., & Zhang, W. (2010). Preparation of lanthanum zirconate nano-powders by molten salts method. *Journal of Non-Crystalline Solids*, *356*, 1049−1051.

Wang, Y., Gao, B., Wang, Q., Li, X., Su, Z., & Chang, A. (2020). $A_2Zr_2O_7$ (A = Nd, Sm, Gd, Yb) zirconate ceramics with pyrochlore-type structure for high-temperature negative temperature coefficient thermistor. *Journal of Materials Science*, *55*, 15405−15414.

Wang, Z., Zhou, G., Qin, X., Yang, Y., Zhang, G., Menke, Y., & Wang, S. (2014). Transparent $La_{2-x}Gd_xZr_2O_7$ ceramics obtained by combustion method and vacuum sintering. *Journal of Alloys and Compounds*, *585*, 497−502.

Weber, S. B., Lein, H. L., Grande, T., & Einarsrud, M.-A. (2013). Lanthanum zirconate thermal barrier coatings deposited by spray pyrolysis. *Surface and Coatings Technology*, *227*, 10−14.

Weber, S. B., Lein, H. L., Grande, T., & Einarsrud, M.-A. (2014). Thermal and mechanical properties of crack-designed thick lanthanum zirconate coatings. *Journal of the European Ceramic Society*, *34*, 975−984.

Wenk, H., Matthies, S., & Lutterotti, L. (1994). *Texture analysis from diffraction spectra. Materials Science Forum* (pp. 473–480). Trans Tech Publ.

Wright, P., & Evans, A. G. (1999). Mechanisms governing the performance of thermal barrier coatings. *Current Opinion in Solid State and Materials Science*, *4*, 255–265.

Wu, J., Wei, X., Padture, N. P., Klemens, P. G., Gell, M., García, E., ... Osendi, M. I. (2002). Low-thermal-conductivity rare-earth zirconates for potential thermal-barrier-coating applications. *Journal of the American Ceramic Society*, *85*, 3031–3035.

Wuensch, B. J., Eberman, K. W., Heremans, C., Ku, E. M., Onnerud, P., Yeo, E. M., ... Jorgensen, J. D. (2000). Connection between oxygen-ion conductivity of pyrochlore fuel-cell materials and structural change with composition and temperature. *Solid State Ionics*, *129*, 111–133.

Xie, Y., Wang, L., Peng, Y., Ma, D., Zhu, L., Zhang, G., & Wang, X. (2021). High temperature and high strength $Y_2Zr_2O_7$ flexible fibrous membrane for efficient heat insulation and acoustic absorption. *Chemical Engineering Journal*, *416*, 128994.

Xu, C., Jin, H., Zhang, Q., Huang, C., Zou, D., He, F., & Hou, S. (2017). A novel Co-ions complexation method to synthesize pyrochlore $La_2Zr_2O_7$. *Journal of the European Ceramic Society*, *37*, 2871–2876.

Yan, C.-H., Yan, Z.-G., Du, Y.-p, Shen, J., Zhang, C., & Feng, W. (2011). *Controlled synthesis and properties of rare earth nanomaterials. Handbook on the physics and chemistry of rare earths* (pp. 275–472). Elsevier.

Zhang, A., Lü, M., Qiu, Z., Zhou, Y., & Ma, Q. (2008). Multiband luminescence of Eu^{3+} based on $Y_2Zr_2O_7$ nanocrystals. *Materials Chemistry and Physics*, *109*, 105–108.

Zhang, A., Lü, M., Yang, Z., Zhou, G., & Zhou, Y. (2008). Systematic research on $RE_2Zr_2O_7$ (RE = La, Nd, Eu and Y) nanocrystals: Preparation, structure and photoluminescence characterization. *Solid State Sciences*, *10*, 74–81.

Zhang, J., Guo, X., Jung, Y.-G., Li, L., & Knapp, J. (2017). Lanthanum zirconate based thermal barrier coatings: A review. *Surface and Coatings Technology*, *323*, 18–29.

Zhang, J., Guo, X., Zhang, Y., Lu, Z., Choi, H.-H., Jung, Y.-G., & Kim, I.-S. (2019). Mechanical properties of lanthanum zirconate-based composite thermal barrier coatings. *Advances in Applied Ceramics*, *118*, 257–263.

Zhang, X., Fang, X., Feng, X., Li, X., Liu, W., Xu, X., ... Zhou, W. (2017). $Ni/Ln_2Zr_2O_7$ (Ln = La, Pr, Sm and Y) catalysts for methane steam reforming: The effects of A site replacement. *Catalysis Science & Technology*, *7*, 2729–2743.

Zhong, J. b, Li, J. z, Feng, F. m, Lu, Y., Zeng, J., Hu, W., & Tang, Z. (2012). Improved photocatalytic performance of SiO_2-TiO_2 prepared with the assistance of SDBS. *Journal of Molecular Catalysis A: Chemical*, *357*, 101–105.

Zhu, J., Meng, X., Zhang, P., Li, Z., Xu, J., Reece, M. J., & Gao, F. (2021). Dual-phase rare-earth-zirconate high-entropy ceramics with glass-like thermal conductivity. *Journal of the European Ceramic Society*, *41*, 2861–2869.

Zhu, R., Zou, J., Wang, D., Zou, K., Gao, D., Mao, J., & Liu, M. (2018). X-ray diffractional, spectroscopic and thermo-physical properties analyses on Eu-doped lanthanum zirconate ceramic for thermal barrier coatings. *Journal of Alloys and Compounds*, *746*, 62–67.

Zinatloo-Ajabshir, S., Ghasemian, N., & Salavati-Niasari, M. (2020). Green synthesis of $Ln_2Zr_2O_7$ (Ln = Nd, Pr) ceramic nanostructures using extract of green tea via a facile route and their efficient application on propane-selective catalytic reduction of NOx process. *Ceramics International*, *46*, 66–73.

Zinatloo-Ajabshir, S., Heidari-Asil, S. A., & Salavati-Niasari, M. (2021). Simple and eco-friendly synthesis of recoverable zinc cobalt oxide-based ceramic nanostructure as high-performance photocatalyst for enhanced photocatalytic removal of organic contamination under solar light. *Separation and Purification Technology*, *267*, 118667.

Zinatloo-Ajabshir, S., Morassaei, M. S., & Salavati-Niasari, M. (2017). Facile fabrication of $Dy_2Sn_2O_7$-SnO_2 nanocomposites as an effective photocatalyst for degradation and removal of organic contaminants. *Journal of Colloid and Interface Science*, *497*, 298–308.
Zinatloo-Ajabshir, S., Morassaei, M. S., & Salavati-Niasari, M. (2018). $Nd_2Sn_2O_7$ nanostructures as highly efficient visible light photocatalyst: Green synthesis using pomegranate juice and characterization. *Journal of Cleaner Production*, *198*, 11–18.
Zinatloo-Ajabshir, S., Mortazavi-Derazkola, S., & Salavati-Niasari, M. (2017). Sonochemical synthesis, characterization and photodegradation of organic pollutant over Nd_2O_3 nanostructures prepared via a new simple route. *Separation and Purification Technology*, *178*, 138–146.
Zinatloo-Ajabshir, S., & Salavati-Niasari, M. (2017). Photo-catalytic degradation of erythrosine and eriochrome black T dyes using $Nd_2Zr_2O_7$ nanostructures prepared by a modified Pechini approach. *Separation and Purification Technology*, *179*, 77–85.
Zinatloo-Ajabshir, S., Salavati-Niasari, M., Sobhani, A., & Zinatloo-Ajabshir, Z. (2018). Rare earth zirconate nanostructures: Recent development on preparation and photocatalytic applications. *Journal of Alloys and Compounds*, *767*, 1164–1185.
Zinatloo-Ajabshir, S., Salavati-Niasari, M., & Zinatloo-Ajabshir, Z. (2016). $Nd_2Zr_2O_7$-Nd_2O_3 nanocomposites: New facile synthesis, characterization and investigation of photocatalytic behaviour. *Materials Letters*, *180*, 27–30.
Zinatloo-Ajabshir, S., Salavati-Niasari, M., & Zinatloo-Ajabshir, Z. (2017). Facile size-controlled preparation of highly photocatalytically active praseodymium zirconate nanostructures for degradation and removal of organic pollutants. *Separation and Purification Technology*, *177*, 110–120.
Zinatloo-Ajabshir, S., Salehi, Z., & Salavati-Niasari, M. (2016). Preparation, characterization and photocatalytic properties of $Pr_2Ce_2O_7$ nanostructures via a facile procedure. *RSC Advances*, *6*, 107785–107792.
Zinatloo-Ajabshir, S., Salehi, Z., & Salavati-Niasari, M. (2018a). Green synthesis and characterization of $Dy_2Ce_2O_7$ ceramic nanostructures with good photocatalytic properties under visible light for removal of organic dyes in water. *Journal of Cleaner Production*, *192*, 678–687.
Zinatloo-Ajabshir, S., Salehi, Z., & Salavati-Niasari, M. (2018b). Green synthesis and characterization of $Dy_2Ce_2O_7$ nanostructures using Ananas comosus with high visible-light photocatalytic activity of organic contaminants. *Journal of Alloys and Compounds*, *763*, 314–321.
Zinatloo-Ajabshir, S., Salehi, Z., & Salavati-Niasari, M. (2018c). Green synthesis of $Dy_2Ce_2O_7$ ceramic nanostructures using juice of Punica granatum and their efficient application as photocatalytic degradation of organic contaminants under visible light. *Ceramics International*, *44*, 3873–3883.
Zinatloo-Ajabshir, S., Zinatloo-Ajabshir, Z., Salavati-Niasari, M., Bagheri, S., & Abd Hamid, S. B. (2017). Facile preparation of $Nd_2Zr_2O_7$–ZrO_2 nanocomposites as an effective photocatalyst via a new route. *Journal of Energy Chemistry*, *26*, 315–323.

Rare earth orthovanadate ceramic nanomaterials

*Sahar Zinatloo-Ajabshir**
Department of Chemical Engineering, University of Bonab, Bonab, Iran
*Corresponding author. E-mail address: s.zinatloo@ubonab.ac.ir; s.zinatloo@gmail.com

5.1 General introduction

Rare earth orthovanadate (RVO_4), where R represents a trivalent rare earth element, have been reported to possess two different structures: tetragonal zircon kind and monoclinic monazite kind (Huang, Huang, Ou, & Pan, 2012; Jia et al., 2005; Jovanović, 2020; Yan et al., 2011; Yuan et al., 2015) (see Fig. 5.1). In the case of all RVO4, the zircon kind is the more stable phase. It has been reported that by enhancing the ionic radius, the RVO4 with monoclinic phase can be formed much more easily than the RVO_4 with tetragonal phase owing to its greater coordination number. Since La^{3+} ion has the largest radius compared to other Ln^{3+} ions, $LaVO_4$ with monoclinic phase can be thermodynamically stable (He et al., 2011).

RVO_4 are known as a substantial family of oxide compounds that possess beneficial optical, catalytic, magnetic, and electronic features (Huignard, Gacoin, & Boilot, 2000; Jeong, Lee, & Byeon, 2014; Martínez-Huerta et al., 2004; Yan et al., 2011). These oxide compounds have been attracting remarkable research interest due to their extraordinary magnetic features, unique optical behaviors, and also potential usages in luminescent displays (Li et al., 2011; Tang et al., 2012), light-emitting diodes (Chen, Chen, Hu, & Liu, 2011; Liu & Y.D. Li, 2007; Wang, Wang, Wu, Li, & Ruan, 2013; Zhang, Li, Zhang, & Yan, 2015), biomedical technologies (Dhanya et al., 2016; Jeyaraman, Shukla, & Sivakumar, 2016; Shen, Sun, & Yan, 2008), sensors (Wangkhem, Singh, Singh, Singh, & Singh, 2018; Yang et al., 2018; Zhu, Ni, & Sheng, 2016), solar cells (Liu, Yao, & Li, 2006; Zahedifar, Chamanzadeh, & Hosseinpoor, 2013), polarizers, catalytic processes (Huang et al., 2009; Kalevaru, Dhachapally, & Martin, 2016; Paunović et al., 2018), luminescent substances, and laser hosts (Choi, Kang, Yi, Jang, & Jeong, 2015; Deng, Yang, Xiao, Gong, & Wang, 2008; Xia et al., 2000; Xu et al., 2010; Zhang, Ma, & Xu, 2020; Zhang, Zhou, Li, & Li, 2008). Extensive use of RVO_4 structures has also been reported owing to their special 4f−5d and 4f−4f electronic transitions in a variety of advanced technologies including photocatalytic reactions, fuel cells, and luminescence (Liu, Shao, Yin, Zhao, Shao et al., 2011). Owing to their good stability as well as nontoxicity, RVO_4 structures, especially at the nanometer scale, can be beneficial for biomedical usages (Lyadov & Kurilkin, 2016). It has been reported that proper chemical stability, specific luminescence features, and great lanthanide

Advanced Rare Earth-Based Ceramic Nanomaterials. DOI: https://doi.org/10.1016/B978-0-323-89957-4.00010-4

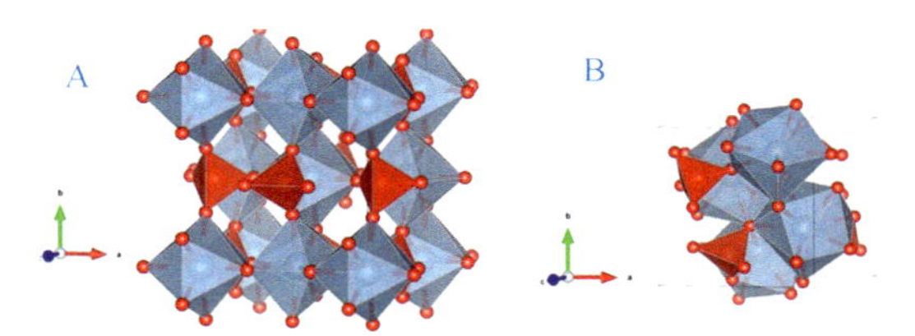

Figure 5.1 Crystal structures of AVO_4 compounds. The coordination polyhedra of V and the trivalent cation A are shown in red and blue. The oxygen atoms are shown as small red spheres: (A) zircon and (B) monazite (Errandonea & Garg, 2018).
Source: Reprinted with permission from Errandonea, D., & Garg, A. B. (2018), Recent progress on the characterization of the high-pressure behaviour of AVO_4 orthovanadates. *Progress in Materials Science 97*, 123–169, Copyright 2018, Elsevier.

ion admittances may be reasons to consider RVO_4 nanomaterials as an appropriate host kind for use in various fields such as fluorescqent probes, inorganic phosphors fabrication, and pharmaceutical carriers (Xu et al., 2010; Xu, Li, Hou, Peng, & Lin, 2011; Zhang & Liang, 2020).

A variety of methods have been utilized to prepare RVO_4 structures, including: hydrothermal, microwave-assisted approach, ultrasonic-assisted way, etc. (Liu & Y. Li, 2007; Qian et al., 2009; Wang, Meng, & Yan, 2004; Xu, Wang, Jin, & Yan, 2004). It is generally accepted that two substantial factors, namely particle size and dimensions, are very meaningful in determining the features of the oxide compounds. Hence, many efforts have been made in the field of preparing RVO_4 structures with controlled particle size and also optimizing their performance in various usages (Mialon, Gohin, Gacoin, & Boilot, 2008; Wang, Chen, Zhang, & Dai, 2008; Zhang, Shi, Tan, Wang, & Gong, 2010). So far, various kinds of RVO_4 micro/nanostructures with different morphologies and particle sizes have been prepared, such as dumbbell-like structures (Zhong & Zhao, 2015), nanorods (Weng et al., 2013), nanosheets (Zahedifar et al., 2013), hollow sphere-like structures (Yang et al., 2013), and nanoparticles (Salavati-Niasari, Saleh, Mohandes, & Ghaemi, 2014). This chapter covers recent advances in ceramic RVO_4 compounds, as well as approaches for fabricating these special oxide compounds. In order to help readers choose the most proper and reliable approach of fabrication, the advantages and disadvantages of each production approach are stated. The usages of these specific ceramic compounds are also summarized.

5.2 Fabrication methods

The solid-state reaction approach is accepted as a relatively easy approach for the fabrication of ceramic nanostructures, especially in large scaling. The steps of this approach are: mixing appropriate quantities of solid reactants (oxides or inorganic salts), grinding the powder mixture until its particle size is reduced, and finally heating the resulting precursor at the proper temperature for a certain time to react the precursors together, and thus produce the target oxide compound (Jiang, Wei, Chen, Duan, & Yin, 2013; Jovanović, 2020). The grinding stage is significant and key because it causes all the reactants to be homogeneously mixed together and also their specific surface area to be maximized. An increase in the specific surface area enhances the reaction rate (Jovanović, 2020). On the other hand, heating to high temperatures is also key and necessary. Since solid reactants cannot react with each other at ambient conditions (even after long periods of time), it is essential to heat the mixture at very high temperatures in order to react with considerable speed between the reactants (Chen, Jia et al., 2016; Gavrilović, Jovanović, & Dramićanin, 2018; Jovanović, 2020). One of the approaches employed to fabricate RVO_4 structures is solid-state reaction. Michalska et al. (2021) prepared $LaVO_4$-Eu^{3+} nanostructures by La_2O_3 and V_2O_5 as precursors via a high-energy ball-milling approach followed by heating (600°C–1200°C). Fig. 5.2 displays the flowchart of the approach for preparing $LaVO_4$-Eu^{3+} samples.

They observed that the grain size could enhance with increasing temperature. It was also observed that the shape of the grains could alter from spherical (600°C and 800°C) to irregular at high temperatures (1000°C–1200°C). Their results also illustrated that, at lower temperatures, $LaVO_4$-Eu^{3+} structures were highly agglomerated and the grain sizes could enhance with increasing temperature (see Fig. 5.3). Hence, the synthesis temperature can have a remarkable effect on controlling the morphology and grain size of $LaVO_4$-Eu^{3+} sample.

Since the solid-state reaction approach has undesirable features such as very-high-energy consumption as well as the production of structures with extensive aggregation, its use for the preparation of ceramic nanostructures in industrial scale is limited (Bishnoi & Chawla, 2017; Fan, Bu, Song, Sun, & Zhao, 2007; Grzyb et al., 2018; Yan et al., 2011). Therefore researchers have interest in utilizing wet

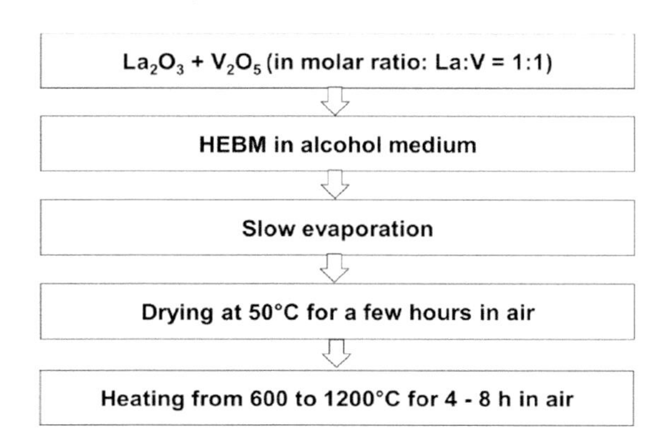

Figure 5.2 The flowchart of the preparation of $LaVO_4$-Eu^{3+} samples.
Source: Reprinted with permission from Michalska, M., Jasiński, J. B., Pavlovsky, J., Zurek-Siworska, P., Sikora, A., Gołębiewski, P., Szysiak, A., Matejka, & V., Seidlerova, J. (2021). Solid state-synthesized lanthanum orthovanadate ($LaVO_4$) Co-doped with Eu as efficient photoluminescent material. *Journal of Luminescence*, *233*, 117934, Copyright 2021, Elsevier.

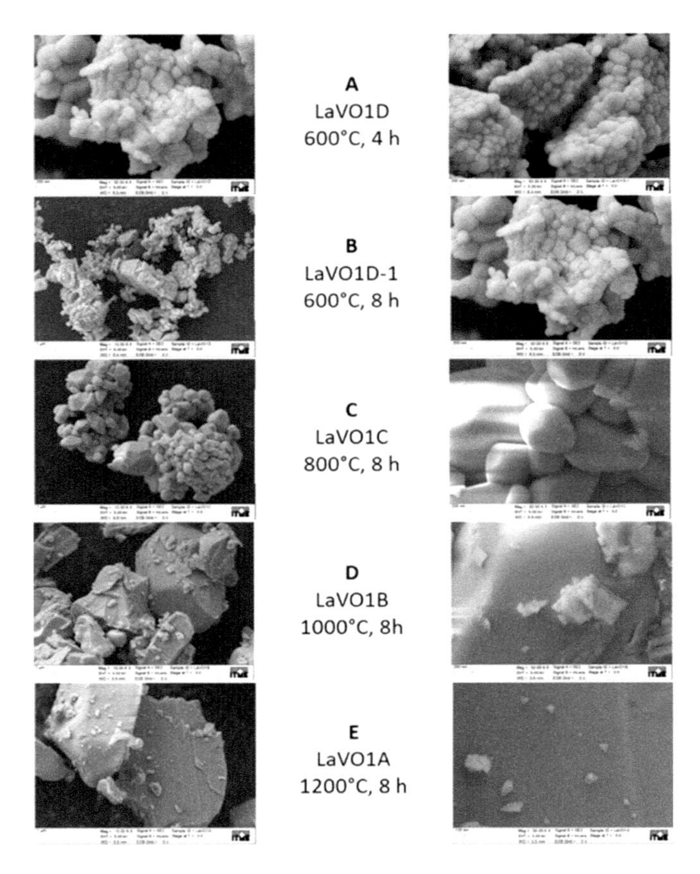

Figure 5.3 SEM images of all prepared $LaVO_4$ structures at various temperatures and time of heating.
Source: Reprinted with permission from Michalska, M., Jasiński, J. B., Pavlovsky, J., Zurek-Siworska, P., Sikora, A., Gołębiewski, P., Szysiak, A., Matejka, & V., Seidlerova, J. (2021). Solid state-synthesized lanthanum orthovanadate ($LaVO_4$) Co-doped with Eu as efficient photoluminescent material. *Journal of Luminescence*, *233*, 117934, Copyright 2021, Elsevier.

chemical approaches to fabricate ceramic nanostructures (Balakrishnan et al., 2010; Jo, Luo, Senthil, Masaki, & Yoon, 2011; Liang et al., 2017; Zheng, Sun, Su, Sun, & Qi, 2018). It has been reported that many wet chemical approaches have desirable features such as good controllability while preparing nanostructures with suitable quality as well as adjustable morphology and particle size (Jovanović, 2020; Kolesnikov et al., 2015; Zinatloo-Ajabshir, Salavati-Niasari, Sobhani, & Zinatloo-Ajabshir, 2018; Zinatloo-Ajabshir, Heidari-Asil, & Salavati-Niasari, 2021b). Because in these approaches, the reactants can be homogeneously mixed at the molecular or even ionic level, the nanostructured products fabricated are chemically pure, and their morphology and particle size are adjusted. The reaction temperature is usually low, although in some wet chemistry approaches a postheat treatment step is required to improve the crystallization of the samples, after which the morphology of the sample particles usually remains unchanged (de Sousa Filho, Gacoin, Boilot, Walton, & Serra, 2015; Grandhe et al., 2012; Jovanović, 2020; Kolesnikov, Golyeva, Kurochkin, Lähderanta, & Mikhailov, 2016; Ray, Banerjee, & Pramanik, 2009).

One of the wet chemistry approaches for preparing RVO_4 nanostructures in ambient conditions is coprecipitation approach because these nanomaterials have low solubility in water. Utilizing this approach of preparation may have an advantage such as saving energy (Takeshita, Isobe, Sawayama, & Niikura, 2011; Tamilmani, Nair, & Sreeram, 2017; Wang, Xu, Hojamberdiev Cui et al., 2009; Yan et al., 2011). It has been reported that the pH of the reaction may be a very substantial factor in the preparation of RVO_4 nanostructures by coprecipitation approach. The pH value optimized for the preparation of pure RVO_4 nanostructure is reported to be about 11 (Kumar, Sharma, Haranath, & Pandey, 2017; Yan et al., 2011).

Aqueous or organic solvents can be utilized to perform coprecipitation reactions (Beshkar, Zinatloo-Ajabshir, & Salavati-Niasari, 2015; Huignard et al., 2000; Huignard, Buissette, Franville, Gacoin, & Boilot, 2003; Huignard, Buissette, Laurent, Gacoin, & Boilot, 2002; Jovanović, 2020). Also, in order to prevent the aggregation of formed particles, to stabilize the prepared particles, as well as to achieve a narrow particle size distribution, different kinds of capping agents or dispersing agents can be utilized (Jovanović, 2020; Moshtaghi, Zinatloo-Ajabshir, & Salavati-Niasari, 2016).

Othman, Hisham Zain, Abu Haija, and Banat (2020) fabricated $CeVO_4$ nanoparticles through a facile coprecipitation approach and then employed as a nanosized catalyst for phenol removal by peroxymonosulfate activation. They applied cerium nitrate, ammonia, and NH_4VO_3 as starting materials (Othman et al., 2020). Their results demonstrated that the $CeVO_4$ sample contained rod-shaped particles with a length of about 14 nm and a width of about 5 nm and was also pure in chemical composition (see Fig. 5.4) (Othman et al., 2020).

Periša et al. (2020) prepared EVO_4 sample and $Gd_{0.75}Eu_{0.25}VO_4$ sample utilizing metal nitrates and trisodium citrate via a coprecipitation approach and examined their luminescent behavior. They observed that the samples contained nanoparticles with a mean diameter of about 2 nm and a narrow size distribution (Periša et al., 2020). It was also observed that the prepared nanoparticles possess a zircon-kind structure. Their results illustrated that Eu^{3+}-activated colloidal nanoparticles could display intense emission signals in the red spectral region (Periša et al., 2020).

Stouwdam, Raudsepp, and van Veggel (2005) fabricated Ln^{3+}-doped $LaVO_4$ nanostructures through a coprecipitation approach in ethanol/water solution. They employed dithiophosphate as capping agent to regulate the growth of $LaVO_4$ nanostructures. They observed that the sample had irregularly shaped particles about 6–10 nm in size (Stouwdam et al., 2005). Particles capped with hydrophobic ligand were also soluble in nonpolar solvents.

Jiang et al. (Jiang et al., 2013) prepared YVO_4:Yb^{3+}, Bi^{3+} structures as phosphors through a coprecipitation approach followed by calcination. Their results demonstrated that the prepared structures were highly efficient phosphors (Jiang et al., 2013).

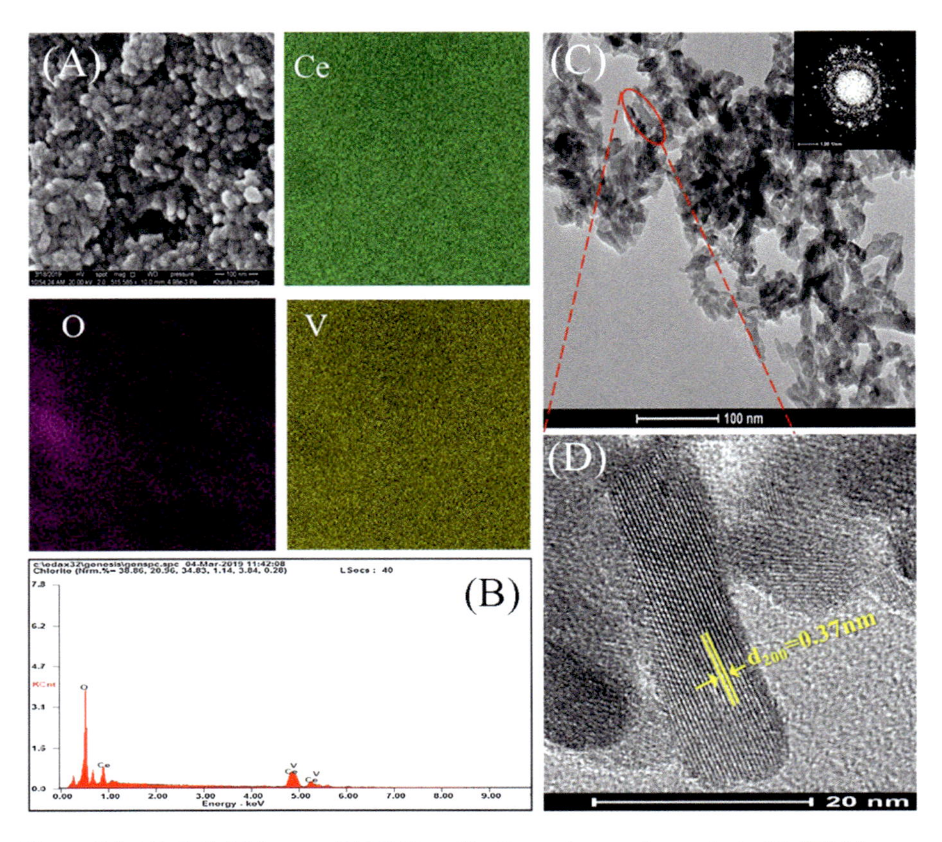

Figure 5.4 (A) FESEM image, (B) EDX profile (inset elemental mapping), (C) TEM images (inset SAED pattern) and (D) HRTEM images of as-formed $CeVO_4$ nanoparticles.
Source: Reprinted with permission from Othman, I., Zain, J. H., Haija, M. A., & Banat, F. (2020). Catalytic activation of peroxymonosulfate using $CeVO_4$ for phenol degradation: An insight into the reaction pathway. *Applied Catalysis B*, *266*, 118601, Copyright 2020, Elsevier.

It has been reported that low reaction temperature as well as precipitation rate can be very effective in regulating the crystallization and final performance of nanostructured oxide materials (Ghodrati, Mousavi-Kamazani, & Zinatloo-Ajabshir, 2020; Haase, Riwotzki, Meyssamy, & Kornowski, 2000; Riwotzki & Haase, 1998; Yan et al., 2011). Compared to traditional solid-state reactions that require very high calcination temperatures, the hydrothermal approach can provide sufficient energy to perform soluble phase reactions that can be widely utilized in the production of ceramic nanostructures (Fan et al., 2004; Oka, Yao, & Yamamoto, 2000; Yan et al., 2011; Zinatloo-Ajabshir, Mortazavi-Derazkola, & Salavati-Niasari, 2017). Under hydrothermal conditions, high pressure and temperature can significantly enhance the dissolution–reprecipitation steps, thus reducing the defect of the oxide lattice. Using this approach, nanoscale crystals can be produced with precise modulation (Razi, Zinatloo-Ajabshir, & Salavati-Niasari, 2017; Yan et al., 2011).

The hydrothermal or solvothermal approach has been reported to involve heating a Teflon-sealed container, an autoclave, comprising an aqueous or nonaqueous solution (or even suspension) of starting materials. During the hydrothermal approach, the overheated solvent as well as the autogenerated high pressure can provide a suitable and completely different microenvironment from what is common in other production systems (Jovanović, 2020; Tian et al., 2013). This preparation approach can utilize changes in solvent features as well as good solubility along with improved reactivity of precursors at great temperature and pressure, and thus can produce nanocrystals at much lower temperatures compared to traditional solid-state reactions. In this approach, various precursors can be utilized, such as hydroxides or salts of metals and so on. The autoclave heating can be done by placing it in the oven or utilizing the microwave (Beshkar, Zinatloo-Ajabshir, Bagheri, & Salavati-Niasari, 2017; Jovanović, 2020; Ma, Wu, & Ding, 2008; Shen et al., 2018). To adjust the phase composition, shape, and dimensions of the oxide structures, changes can be made to some parameters, including production temperature and reaction time, solvent kind, capping agent kind, kind and dose of precursors, as well as pH solution (Byrappa & Yoshimura, 2013; Jovanović, 2020; Li, Chao, Peng, & Chen, 2008; Shandilya, Rai, & Singh, 2016).

Tegus, Amurisana, and Zhiqiang (2019) prepared flower-like $YVO_4:Ln^{3+}$ structures through a hydrothermal approach utilizing Na_2H_2L compound as a capping agent. They observed that by adjusting the reaction time as well as the dose of Na_2H_2L, the shape, dimensions, and particle size of the oxide structure could be changed (Tegus et al., 2019). Their results demonstrated that the $YVO_4:Eu^{3+}$ sample with a flower-like architecture consisted of a primary rod-like microstructure trunk to which several secondary nails were attached (see Fig. 5.5).

Grzyb et al. (2018) prepared RVO_4 (R = Gd, Y, and La) structures doped by Eu^{3+} via a hydrothermal approach. Their results illustrated that the prepared structures consisted of spherical particles with tetragonal crystal phase (Grzyb et al., 2018).

Table 5.1 shows some rare earth orthovanadate ceramic structures prepared by hydrothermal route.

Ponnaiah and Prakash (2021) fabricated pure $CeVO_4$ nanostructures and $CeVO_4$/polypyrrole nanostructures through a hydrothermal route. Their results indicated that the prepared nanostructures as high-performance supercapacitors may be beneficial for use in energy storage devices (Ponnaiah & Prakash, 2021).

One of the approaches for preparing RVO_4 nanostructures is sol − gel, which usually follows an annealing step to achieve nanostructured products with suitable crystallization (Yan et al., 2011). Hou et al. (2008) prepared t-$YVO_4:Ln^{3+}$ microbelts (Ln = Eu, Sm, Dy) via a combination approach of sol − gel process and electrospinning. They utilized physical force to shape the sol precursors into one dimensionality and then crystallize them by calcination (Hou et al., 2008).

Chumha, Kittiwachana, Thongtem, Thongtem, and Kaowphong (2014) prepared $GdVO_4$ structures through a sol−gel approach employing malic acid, followed by calcinations at 500°C − 700°C within 180 minutes. Malic acid was utilized as a chelating agent. They observed that with an increment in calcination temperature,

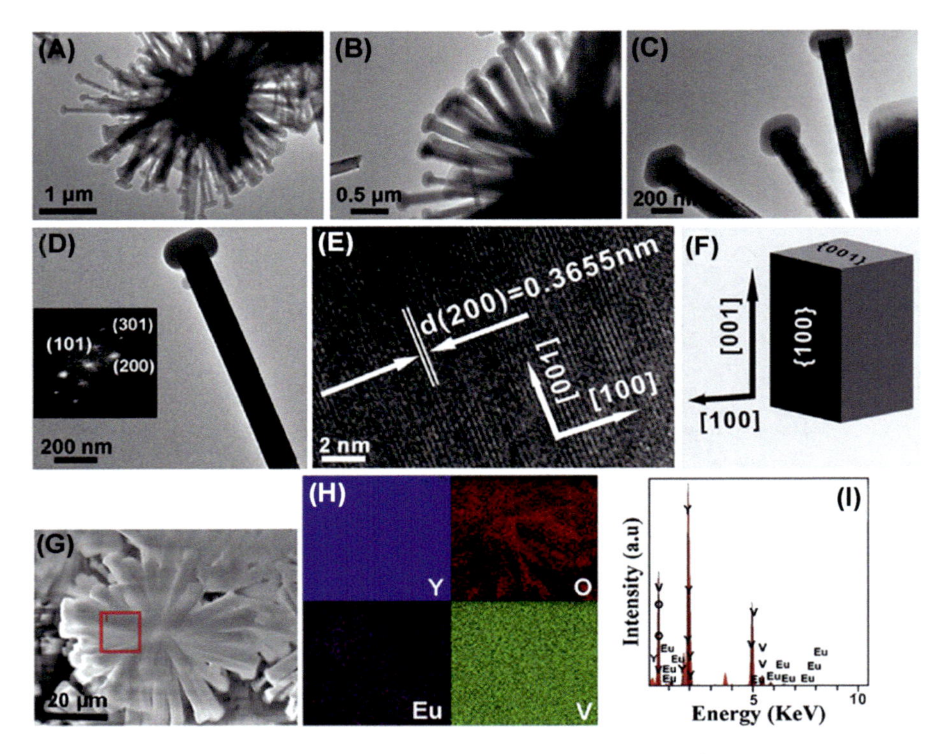

Figure 5.5 (A–C) TME images of a single flower-like YVO_4:5% Eu^{3+} hierarchitecture at different magnifications. (D) TEM, SADE (inset) image, and (E) HRTEM pattern of single nanonail with rectangular cap. (F) Structural representation of nanorod. (G) SEM image of a single flower-like YVO_4:5% Eu^{3+} hierarchitecture. (H) EDS mapping images of Y, Eu, V, and O for the square region in the (G). (I) EDX spectrum of the flower-like YVO_4:5% Eu^{3+}. *Source:* Reprinted with permission from Tegus, O., Amurisana, B., & Zhiqiang, S. (2019). Morphology-sensitive photoluminescent properties of YVO_4:Ln^{3+} (Ln^{3+} = Eu^{3+}, Sm^{3+}, Dy^{3+}, Tm^{3+}) hierarchitectures. *Journal of Luminescence 215*, 116624, Copyright 2019 Elsevier.

the particle size enhanced significantly (see Fig. 5.6) (Chumha et al., 2014). They attributed this increment in particle size to an enhancement in the surface energy of $GdVO_4$ particles, which can lead to increased particle aggregation (Chumha et al., 2014).

Li et al. (2011) fabricated $GdVO_4$:Re^{3+} (Re = Sm, Eu, Dy) nanofibers through a combination approach of sol – gel process and electrospinning. They observed that the prepared nanofibers comprised linked nanoparticles with a diameter of about 100–160 nm (Li et al., 2011).

It has been shown that the composition of the sol can be altered by adding a dispersant agent to produce a nanometer-sized oxide product. Zhang, Fu, Niu, Sun, and Xin (2004a) utilized polyacrylamide gel as the dispersing agent and were able to produce YVO_4:Eu^{3+} nanoparticles with a size of about 20 nm and with the least

Table 5.1 Several rare earth orthovanadate ceramic structures prepared by hydrothermal route.

Sample	Morphology	Precursors	Reference
$CeVO_4$	Three-dimensional (3D) hierarchical nanostructure	$Ce(NH_4)_2(NO_3)_6$, NH_4VO_3, urea	Kanna Sharma, Hwa, Santhan, and Ganguly (2021)
$SmVO_4$	Nanoparticles	$Sm(NO_3)_3$, NH_4VO_3	Baby, Sriram, Wang, and George (2021)
$GdVO_4:Eu^{3+}$	Cushion-shaped particles	Gd_2O_3, Eu_2O_3, NH_4VO_3, TTAB, NaOH	Yan et al. (2013)
t-$LaVO_4$:Eu	Nanosquares	Rare earth nitrates, $(NH_4)_2SO_4$, NH_4OH, NH_4VO_3	Wang et al. (2020)
t-$NdVO_4$	Nanorod arrays	$Nd(NO_3)_3 0.6H_2O$, NH_4VO_3, EDTA-2Na, NaOH	Tian et al. (2020)
$YVO_4:Eu^{3+}$	Trepang-like structures	$Eu(NO_3)_3$. $6H_2O$, $Y(NO_3)_3$. $6H_2O$,TETA, Na_3VO_4	Ren et al. (2019)
$NdVO_4$	Nanoparticles	EDTA-2Na, $Nd(NO_3)_3 0.6H_2O$, NH_4VO_3, NaOH	Tian et al. (2018)
$LaVO_4$:Eu	Nanorods	$LaCl_3 \cdot 7H_2O$, $Eu(NO_3)_3 \cdot 6H_2O$, catechin hydrate, sodium orthovanadate	Vairapperumal, Natarajan, Manikantan Syamala, Kalarical Janardhanan, and Balachandran (2017)
Au/$GdVO_4$:Eu	Flower-shaped nanoparticles	Trisodium citrate, $HAuCl_4$, Gd $(NO_3)_3$, $Eu(NO_3)_3$, Na_3VO_4	Chen, Wang et al. (2016)
$GdVO_4$	Octahedral microcrystals	$Gd_2(CO_3)_3.xH_2O$, $Na_3VO_4 0.2H_2O$, NaOH	Jin, Liu, and Sun (2013)
$GdVO_4:Eu^{3+}$	Ball-like structure consists of submicron-sized flakes with sharp edges	Gd_2O_3, Eu_2O_3, NH_4VO_3, HNO_3, NaOH	Yan, Hojamberdiev, Xu, Wang, and Luan (2013)
$YVO_4:Ln^{3+}$ (Ln = Eu, Dy)	Spindle-like self-assembled nanoparticles	$Y(NO_3)_3$, $Eu(NO_3)_3$, $Dy(NO_3)_3$, urea, NH_4VO_3	Zhang, You, and Yang (2012)

(*Continued*)

Table 5.1 (Continued)

Sample	Morphology	Precursors	Reference
$CeVO_4$	Nanoparticals	$Ce(NO_3)_3 \cdot 6H_2O$, EDTA, ammonium hydroxide, $Na_3VO_4 \cdot 12H_2O$	Liu, Shao, Yin, Zhao, Sun et al. (2011)
$PrVO_4$	3D nanoarchitecture	$Pr(NO_3)_3 \cdot 6H_2O$, ETDA, NH_4VO_3, NaOH	Thirumalai, Chandramohan, and Vijayan (2011)
$LaVO_4:Tb^{3+}$	Polyhedral nanocrystals and nanorods	NH_4VO_3, $La(NO_3)_3$, $TbCl_3$, Hydrochloric acid	Zhang, Wang, Peng, Li, and Chen (2010)
$YVO_4:Eu^{3+}$	Nanoparticles with polygonal shape	V_2O_5, HNO_3, Y_2O_3, Eu_2O_3, Ammonia, PEG	Choi, Moon, and Jung (2010)
Tm^{3+}-doped $GdVO_4$	Square, hexagon, and lozenge nanocrystals	$Gd(NO_3)_3$, $Tm(NO_3)_3$	Calderón-Villajos, Zaldo, and Cascales (2010)
$LaVO_4:Eu^{3+}$	Fishbone-like nanocrystals	La_2O_3, Eu_2O_3, NH_4VO_3, ammonia,	Liu, Duan, Li, and Dong (2009)
$YVO_4:Eu^{3+}$	Peanut-shaped structure	Y_2O_3, Eu_2O_3, NH_4VO_3, HNO_3, SDS	Wang, Xu, Hojamberdiev, and Zhu (2009)
$YVO_4:Er^{3+}$	Nanoparticles	$Y(NO_3)$, $Er(NO_3)$, sodium citrate, Na_3VO_4	Sun, Liu, Wang, Kong, and Zhang (2006)

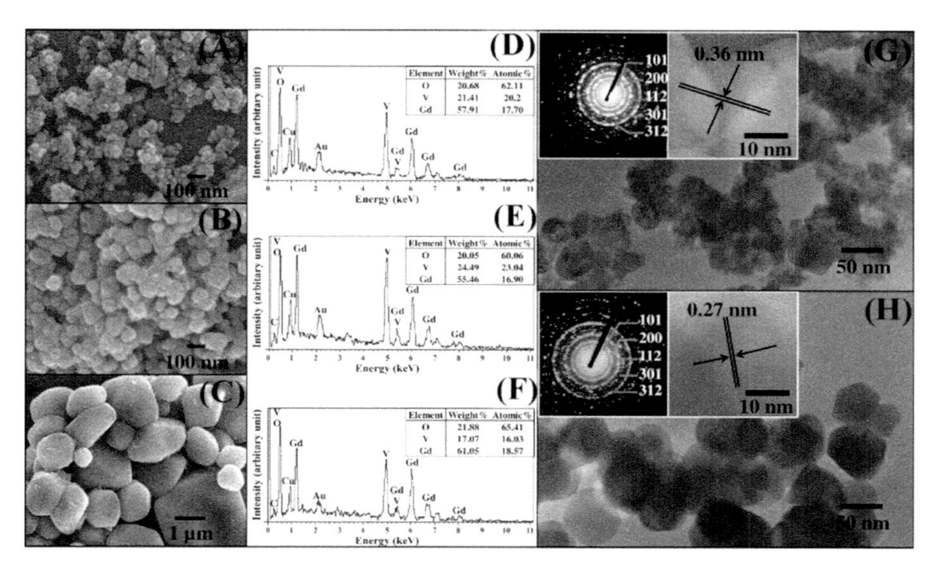

Figure 5.6 (A–C) FESEM images and (D–F) EDS spectra of the $GdVO_4$ powders calcined at 500°C, 600°C, and 700°C, respectively. TEM images of the $GdVO_4$ nanoparticles calcined at (G) 500°C and (H) 600°C. Insets: SAED patterns (left) and HRTEM images (right). *Source:* Reprinted with permission from Chumha, N., Kittiwachana, S., Thongtem, T., Thongtem, S., & Kaowphong, S. (2014). Synthesis and characterization of $GdVO_4$ nanoparticles by a malic acid-assisted sol–gel method. *Materials Letters, 136*, 18–21, Copyright 2014, Elsevier.

amount of aggregation. They added a mixture of acrylamide monomers and cross-linkers as well as polymerization initiators to a solution containing reactants (including citric acid, metal sources, and NH_4VO_3) to form a polymer gel alongside inorganic precursors (Zhang et al., 2004a). A gel was then obtained in which, as a porous framework, vanadium species and rare earth elements were well dispersed. The created static separation can significantly diminish the aggregation of particles after burning organic residues (Zhang et al., 2004a). In another work, they utilized polymeric glucose gel (Zhang, Fu, Niu, Sun, & Xin, 2004b) to fabricate the nanocrystalline YVO_4:Tm.

Zhang, Fu, Niu, and Xin (2008) fabricated the nanosized YVO_4:Re (Re = Dy, Sm) through a complex-based sol–gel approach utilizing soluble starch as a gelling agent and a chelating agent. They stated that since starch chains consist of glucose units with hydroxyl groups, they could bind to the surface of the prepared nanoparticles and thus prevent them from growing excessively (Zhang, Fu et al., 2008).

It has been reported that heat source can be considered as a significant factor in improving the reactivity of reactants. In general, the heating process can take place through three different mechanisms: radiation, convection, and heat conduction (Yan et al., 2011). Since all three heating modes are relatively inefficient because they unavoidably bring temperature gradients into the reaction medium, which makes it time-consuming to reach equilibrium. The usage of microwave and ultrasonic heating approaches has been proposed as a suitable solution to overcome these problems (Yan et al., 2011). The utilization of the microwave approach is accepted as a fast, reliable, and energetic way for heating as well as conducting chemical reactions to prepare oxide nanostructures (Ekthammathat, Thongtem, Phuruangrat, & Thongtem, 2013; Uematsu, Ochiai, Toda, & Sato, 2006; Xu, Wang, & Yan, 2007; Yan et al., 2011). Homogeneous heating by the microwave approach can be due to the direct absorption of high-penetration microwaves through resonance or relaxation by polar molecules (Liu et al., 2015; Ma & Wang, 2015; Toda, Toda, Ishigaki, Uematsu, & Sato, 2010; Wang et al., 2004; Yan et al., 2011). On the other hand, due to in situ heating as well as rapidly rising temperatures, the usage of microwaves can lead to explosive nucleation of crystalline seeds as well as quick crystallization, which leads to the formation of oxide structures with small particle size (Yan et al., 2011).

Xu, Wang, Meng, and Yan (2004) prepared YVO_4 nanoparticles via a microwave approach utilizing $Y(NO_3)_3$ and sodium metavanadate as precursors during 10 minutes. They observed that the prepared nanoparticles, depending on the pH of the reaction, had a particle size of about 5−18 nm, and the sample with the smallest particle size was prepared at pH = 7 (Xu, Wang, Meng et al., 2004).

Mahapatra, Nayak, Madras, and Guru Row (2008) prepared $ReVO_4$ (Re = Pr, Ce, Nd) nanopowders utilizing the microwave approach in a polyethylene glycol solvent. They observed that the prepared samples had a particle size in the range of 25−30 nm. Their results indicated that the $CeVO_4$ nanostructure displayed much better photocatalytic performance owing to its shorter energy gap compared to the TiO_2 sample (Mahapatra et al., 2008).

Huong et al. (2016) prepared the nanostructured $YVO_4:Eu^{3+}$ with strong emission through microwave approach employing polyethylene glycol as the solvent, europium(III) nitrate, NH_4VO_3, and yttrium(III) nitrate. They investigated the effect of microwave radiation power on morphology and dimensions; they prepared four samples under radiation with power of 300, 500, 700, and 900 W (see Fig. 5.7) (Huong et al., 2016). They observed that the sample prepared at 300 W consisted of nanoparticles with a size in the range of 15−20 nm. Their results illustrated that by changing the radiation power from 300 to 900 W, the morphology of the oxide sample could alter from a particle-like to nanowires with a diameter of about 10−20 nm and a length of 500−800 nm (Huong et al., 2016). Therefore the power of microwave radiation had a very remarkable effect on regulating the morphology and dimensions of $YVO_4:Eu^{3+}$ material (Huong et al., 2016).

It has been reported that ultrasonic radiation can heat the liquid media very quickly (most of the water system is used as the medium) (Yan et al., 2011; Yu et al., 2009). It has been stated that its heating mechanism can include the formation and explosive destruction of bubbles. The explosive destruction of bubbles may be accompanied by a high temperature of about 5000 K, after which very rapid cooling can occur (Hu & Wang, 2014; Yan et al., 2011). In other words, these bubbles can become hot spots and thus transfer heat energy continuously to the aqueous reaction medium (Mosleh & Mahinpour, 2016; Salavati-Niasari et al., 2014; Yan et al., 2011).

Figure 5.7 FESEM images of the YVO4:Eu^{3+}/PEG nanostructures with changing microwave-irradiated powers: (A) 300 W, (b) 500 W, (C) 700 W, and (D) 900 W.
Source: Reprinted with permission from Huong, T. T., Th. Vinh, L., Th. Phuong, H., Th. Khuyen, H., Anh, T. K., Tu, V. D., & Minh, L. Q. (2016). Controlled fabrication of the strong emission YVO_4: Eu^{3+} nanoparticles and nanowires by microwave assisted chemical synthesis. *Journal of Luminescence*, *173*, 89–93, Copyright 2016, Elsevier.

Zhu, J. Li et al. (2007) fabricated YVO_4:Eu^{3+} nanostructures via a sonochemical approach utilizing rare earth metal nitrates and NH_4VO_3. They observed that the utilization of ultrasound waves to produce YVO_4:Eu^{3+} material could have a very substantial effect on its morphology and dimensions because under ultrasound waves the spindle-like structures with a diameter of about 90–150 nm and a length of 250–300 nm could be produced, while without the usage of ultrasound waves, only nanoparticles could be prepared (Zhu, J. Li et al., 2007). On the other hand, their results demonstrated that spindle-like nanostructures owing to their special architecture were able to display good photoluminescence (PL) intensity compared to nanoparticles (Zhu, J. Li et al., 2007).

In another work, Zhu, Q. Li et al. (2007), utilizing various precursors, were able to prepare different $CeVO_4$ structures through sonochemical approach. They found that $CeVO_4$ nanorods with a diameter of about 5 nm and a length of 100–150 nm could be fabricated under ultrasound waves utilizing precursors including cerium (III) nitrate, NH_4VO_3, and ammonia (Zhu, Q. Li et al., 2007). While utilizing cerium(III) nitrate, V_2O_5, and sodium hydroxide, mesoporous $CeVO_4$ nanostructure can be prepared with a specific surface area of about 122 m^2/g (Zhu, Q. Li et al., 2007). Their results illustrated that the usage of ultrasound waves as well as ammonia could play a substantial role in the preparation of $CeVO_4$ nanostructures with rod-like morphology (Zhu, Q. Li et al., 2007).

Kalai Selvan, Gedanken, Anilkumar, Manikandan, and Karunakaran (2009) prepared a series of RVO_4 by sonochemical approach in the presence of two different capping agents called polyethylene glycol as well as Pluronic P123. Their results illustrated that the usage of polyethylene glycol can lead to the preparation of RVO_4 samples ($SmVO_4$, $CeVO_4$, $LaVO_4$, and $EuVO_4$) consisting of spherical nanoparticles and also lead to the formation of $GdVO_4$ and $NdVO_4$ nanorods and nanospindles (Kalai Selvan et al., 2009). Their results illustrated that the usage of P123 as a capping agent could lead to the preparation of samples with completely various morphologies, although in most cases nanospindles and nanorods were formed (Kalai Selvan et al., 2009). They observed that the nanostructured $CeVO_4$ showed the best performance in degrading methylene blue (Kalai Selvan et al., 2009).

Khorasanizadeh, Monsef, Amiri, Amiri, and Salavati-Niasari (2019) fabricated the nanostructured $HoVO_4$ via a sonochemical approach utilizing triethylenetetramine as a capping agent. They utilized probe as a source of ultrasound waves. In order to investigate the role of the duration of ultrasound radiation on the morphology and dimensions of the structure of holmium orthovanadate, samples were prepared at various times of 5, 10, 20, and 3 minutes (see Fig. 5.8) (Khorasanizadeh et al., 2019). They observed that enhancing the ultrasound time to 10 minutes could lead to a reduction in the particle size of the holmium orthovanadate structure. They attributed this decrease in particle size to the disintegration of accumulated particles by the application of energy that can be produced by the collapse of bubbles (Khorasanizadeh et al., 2019).

It was also observed that enhancing the ultrasound time to 30 minutes could lead to an increment in particle size owing to Ostwald process. According to the particle size distribution diagrams of the nanostructured samples, in the optimal time of 10 minutes, it is possible to prepare the holmium orthovanadate nanostructure with a particle size of about 30.6 nm (Khorasanizadeh et al., 2019).

Table 5.2 gives some rare earth orthovanadate ceramic structures prepared by other preparation approaches.

5.3 Applications

One of the most distinctive features of compounds based on rare earth elements is multicolored luminescence. It has been reported that lanthanide ions can have many intense emission peaks that can cover the visible and also near-infrared (NIR) region owing to the great transitions for f-orbital configurations (Yan et al., 2011).

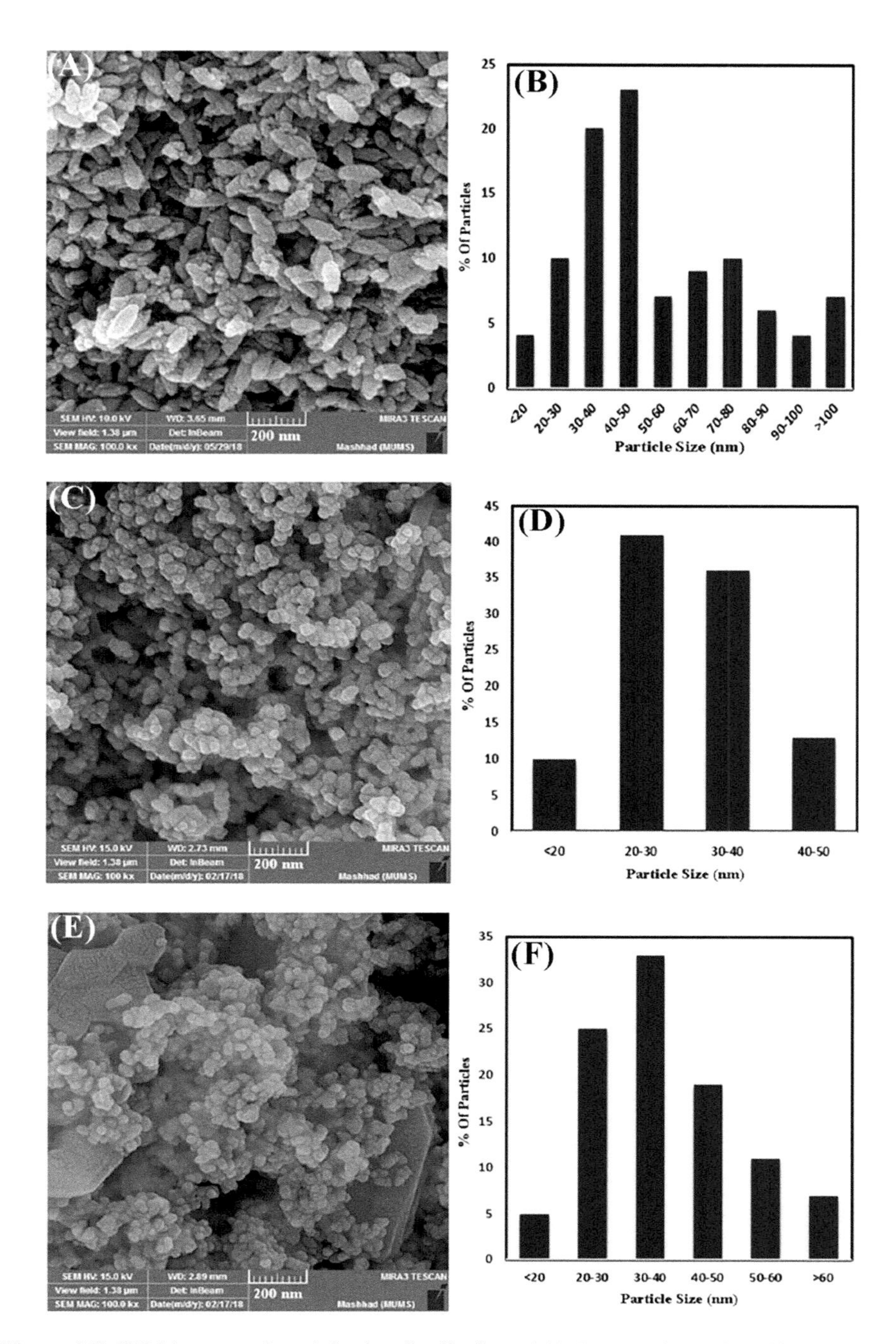

Figure 5.8 SEM image and particle size distribution of (A, B) sample produced in the presence of water after 5-min sonication, (C, D) after 10-min sonication, and (E, F) after 20-min sonication.
Source: Reprinted with permission from Khorasanizadeh, M. H., Monsef, R., Amiri, O., Amiri, M., & Salavati-Niasari, M. (2019). Sonochemical-assisted route for synthesis of spherical shaped holmium vanadate nanocatalyst for polluted waste water treatment. *Ultrasonics Sonochemistry*, *58*, 104686, Copyright 2019, Elsevier.

Table 5.2 Several rare earth orthovanadate ceramic structures prepared by other preparation approaches.

Sample	Morphology	Preparation method	Reference
$YVO_4:Eu^{3+}$	Hollow nanocrystals	Self-sacrifice template method	Yang, Peng, Zhao, and Yu (2019)
$GdVO_4:Eu^{3+}$	Nanoparticles	A protected calcination method	Zhang and Liang (2020)
$YVO_4:Eu$	Nanocrystallines with a cubic morphology	Molten salt synthesis	Wang, Liu, Zhou, Jia, and Lin (2012)
$YbVO_4$	Rod-like morphology with many bulge edges	MOF route	Vadivel et al. (2020)
$Y(V,P)O_4:Eu^{3+}$	Nearly spherical nanoparticles	Liquid-phase precursor method	Jo et al. (2011)
$YVO_4:Eu^{3+}$	Nanoparticles	Combustion method	Mentasti, Martínez, Zucchi, Santiago, and Barreto (2020)
$CeVO_4$/CNT hybrid composite	Hybrid nanocomposite	Silicone oil-bath method	Narsimulu, Kakarla, and Su Yu (2021)
YVO_4	Nanospindles	Chemical conversion route	Zheng et al. (2018)

Of course, the forbidden f−f transitions can cause narrow excitation peaks for the majority of rare earth ions. Low-adsorption cross section is a limitation for large-scale uses, thus host-sensitive emission mode is considered for rare earth phosphors. The vanadate matrix is one of the most notable candidates that can excite lanthanide ions through charge transfer energy movement (Yan et al., 2011).

Based on the kind of luminescence emission mechanism, Ln^{3+}-activated luminescent compounds are classified into two distinct categories: upconversion (UC) phosphors and also downconversion (DC) phosphors, which are generally detectable with adding different types of lanthanide ions (Jovanović, 2020). In the case of DC luminescence orthovanadates, it has been reported that energy transfer can take place from VO_4^{3-} ions to activator dopant ions located at cationic sites when excited with ultraviolet radiation. While lanthenide ions in UC compounds can absorb NIR light, phosphor can display anti-Stokes emission of PL (Jovanović, 2020).

Huong et al. (2016) prepared the nanostructured $YVO_4:Eu^{3+}$ by a microwave approach in various pH values and examined their optical features. They observed that when $YVO_4:Eu^{3+}$ structures produced in different pHs were excited utilizing ultraviolet light, they could intensely display the red luminescence with narrow signals, which could be related to the intra 4f transitions of $^5D_0-^7F_j$ ($j = 1, 2, 3$, and 4) Eu^{3+} (Huong et al., 2016). Also, the signals were seen around 594 nm ($^5D_0-^7F_1$), 619 nm ($^5D_0-^7F_2$), 652 nm ($^5D_0-^7F_3$), and 702 nm ($^5D_0-^7F_4$), among which the most intense emission was around 619 nm (see Fig. 5.9) (Huong et al., 2016).

It has been reported that double doping can be employed in the production of DC phosphors (Jovanović, 2020). Several examples of red emission have been observed through excitation with ultraviolet light (about 330 nm) owing to the DC trend in the doped Yb^{3+}/Er^{3+} $GdVO_4$ compound (Gavrilović, Jovanović, Lojpur, & Dramićanin, 2014; Gavrilović, Jovanović, Lojpur, Đorđević, & Dramićanin, 2014).

Today, cleaning the environment from all kinds of organic and dangerous contaminants that enter it through industrial effluents or domestic wastewater is one of the serious requirement for human life and other living organisms (Heidari-Asil, Zinatloo-Ajabshir, Amiri, & Salavati-Niasari, 2020; Lai & Lee, 2021; Mohd Razali et al., 2021; Zinatloo-Ajabshir, Heidari-Asil, & Salavati-Niasari, 2021a). One of the cost-effective and green solutions to degrade and eliminate these contaminants with the help of free and accessible solar energy is the usage of photocatalytic reactions, which is rapidly developing (Cani, van der Waal, & Pescarmona, 2021; Zinatloo-Ajabshir, Morassaei, Amiri, & Salavati-Niasari, 2020). The usage of RVO_4 nanostructures to remove toxic contaminants from the environment has also been considered by researchers in recent years.

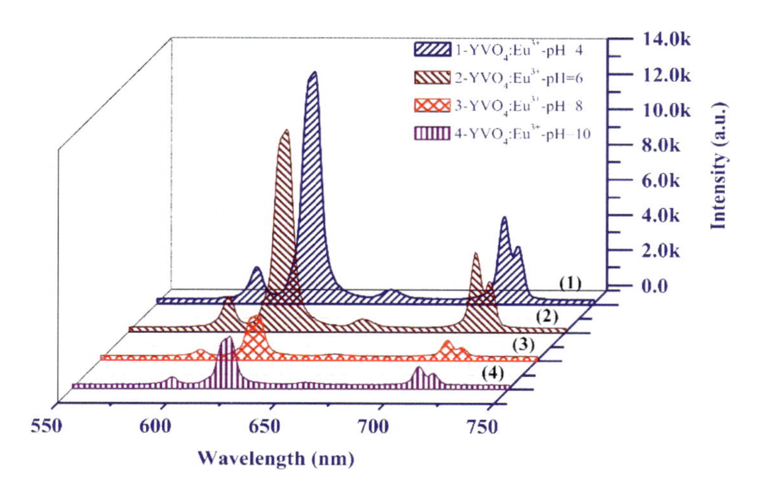

Figure 5.9 PL spectra of the $YVO_4:Eu^{3+}$ nanostructures with changing pH values pH = 4–10 at $\lambda_{exc} = 325$ nm.
Source: Reprinted with permission from Huong, T. T., Th. Vinh, L., Th. Phuong, H., Th. Khuyen, H., Anh, T. K., Tu, V. D., Minh, L. Q. (2016). Controlled fabrication of the strong emission YVO_4: Eu^{3+} nanoparticles and nanowires by microwave assisted chemical synthesis. *Journal of Luminescence*, *173*, 89–93. Copyright 2016, Elsevier.

Khorasanizadeh et al. (2019) prepared the nanostructure $HoVO_4$ with a spherical shape through a sonochemical approach and utilized it to remove methyl violet contaminant. They employed various amounts of $HoVO_4$ nanostructure to degrade methyl violet contaminant to investigate the role of the amount of nanocatalyst on photocatalytic degradation performance (see Fig. 5.10) (Khorasanizadeh et al., 2019). They observed that by altering the quantity of nanostructure to 50 mg, the contaminant degradation could enhance by 67.6%. While employing 75 mg of nanostructure, the rate of contaminant degradation reaches 43.88%. This decrement in performance was attributed to the decrease in light absorption by the nanostructure owing to the increment in its quantity (Khorasanizadeh et al., 2019). They also observed that $HoVO_4$ nanostructures irradiated with two different kinds of light, ultraviolet light and visible light, were able to degrade methyl violet contaminant, although the rate of degradation of methyl violet contaminant under visible light was relatively lower (47.8%). Also, the results of investigating the recyclability of $HoVO_4$ nanostructure in the elimination of methyl violet contaminant under ultraviolet light exhibited that the nanostructure had appropriate stability and its efficiency in pollutant elimination diminishes slightly after five consecutive periods of degradation (Khorasanizadeh et al., 2019).

Table 5.3 displays some RVO_4 nanomaterials employed for photocatalytic degradation of toxic contaminants.

5.4 Conclusion and outlook

In summary, RVO_4 ceramic nanostructures have attracted a great deal of interest from researchers and scientists owing to their unique features. Also, these ceramic nanostructures may be considered as advantageous and efficient options for utilization in various fields, especially luminescence compounds, as well as photocatalytic processes to eliminate toxic pollutants and remedy the environment. Owing to the substantial effect of some structural and morphological features as well as the purity of these ceramic nanostructures on their performance and behavior, so far many efforts have been made to prepare them by adjusting and controlling these features. Kinds of micro/nanostructures with different morphologies and various dimensions have been prepared and studied so far. This chapter of the book reviews the common approaches utilized in recent years to fabricate RVO_4 ceramic nanostructures. The distinctive features of each production approach are expressed so that it can be beneficial and instructive for readers in choosing the best, most appropriate, and most reliable approach in terms of cost, ease, and large-scale production capability. Despite the efforts made, there is still an urgent need to design and provide a low-cost, environmentally friendly, and easy-to-use approach for the fabrication of RVO_4 ceramic nanostructures with appropriate features on a large scale. In addition, the usages of RVO_4 structures in the most common fields in recent years have been summarized. Owing to the dependence of the performance and efficiency of RVO_4 nanostructures on some structural and morphological features as well as their

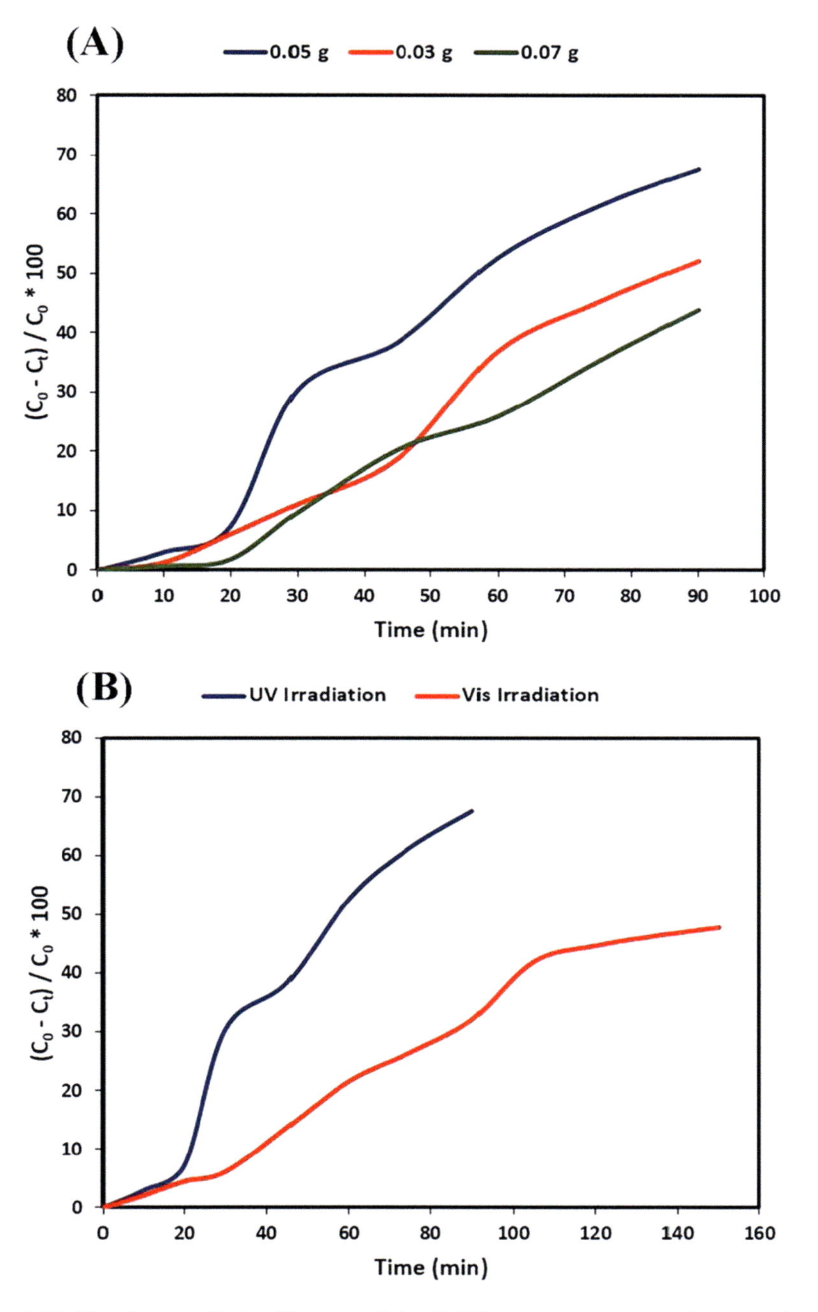

Figure 5.10 The photocatalytic efficiency of the $HoVO_4$ nanostructures produced in the presence of PG solvent: (A) influence of photocatalyst amount and (B) influence of light source.
Source: Reprinted with permission from Khorasanizadeh, M. H., Monsef, R., Amiri, O., Amiri, M., & Salavati-Niasari, M. (2019). Sonochemical-assisted route for synthesis of spherical shaped holmium vanadate nanocatalyst for polluted waste water treatment. *Ultrasonics Sonochemistry*, *58*, 104686, Copyright 2019, Elsevier.

Table 5.3 Use of some rare earth orthovanadate nanomaterials as a photocatalyst for the removal of toxic contaminants.

Photocatalyst	Target contaminant	Light source	Reference
$NdVO_4$	Malachite green, Eriochorme Black T	Ultraviolet (UV)	Monsef, Ghiyasiyan-Arani, and Salavati-Niasari (2018)
$PrVO_4$	Eriochrome Black T, erythrosine, methyl violet	UV	Monsef, Ghiyasiyan-Arani, Amiri, and Salavati-Niasari (2020)
$PrVO_4/CdO$	Erythrosine	UV and visible	Monsef et al. (2021)
$CeVO_4$	Methylene blue	UV	Phuruangrat, Kuntalue, Thongtem, and Thongtem (2016)
$YbVO_4$ and YVO_4	Methyl orange	Visible	Vadivel et al. (2020)
$CeVO_4$	Eosin yellow	Visible	Mishra et al. (2020)
$GdVO_4$	Rhodamine B	UV	Chumha et al. (2014)
$NdVO_4$	Methyl violet, eosin Y, Eriochrome Black T	UV	Monsef, Ghiyasiyan-Arani, and Salavati-Niasari (2019)

purity, it is necessary to adjust these features during the preparation step in order to achieve the most desirable and best performance in each usage.

References

Baby, J. N., Sriram, B., Wang, S.-F., & George, M. (2021). Integration of samarium vanadate/carbon nanofiber through synergy: An electrochemical tool for sulfadiazine analysis. *Journal of Hazardous Materials*, *408*, 124940.

Balakrishnan, S., Launikonis, A., Osvath, P., Swiegers, G. F., Douvalis, A. P., & Wilson, G. J. (2010). Synthesis and characterisation of optically tuneable, magnetic phosphors. *Materials Chemistry and Physics*, *120*, 649–655.

Beshkar, F., Zinatloo-Ajabshir, S., Bagheri, S., & Salavati-Niasari, M. (2017). Novel preparation of highly photocatalytically active copper chromite nanostructured material via a simple hydrothermal route. *PLoS One*, *12*, e0158549.

Beshkar, F., Zinatloo-Ajabshir, S., & Salavati-Niasari, M. (2015). Simple morphology-controlled fabrication of nickel chromite nanostructures via a novel route. *Chemical Engineering Journal*, *279*, 605–614.

Bishnoi, S., & Chawla, S. (2017). Enhancement of $GdVO_4$: Eu^{3+} red fluorescence through plasmonic effect of silver nanoprisms on Si solar cell surface. *Journal of Applied Research and Technology*, *15*, 102–109.

Byrappa, K., & Yoshimura, M. (2013). *Handbook of hydrothermal technology*. William Andrew Books, Elsevier.

Calderón-Villajos, R., Zaldo, C., & Cascales, C. (2010). Hydrothermal processes for Tm^{3+}-doped $GdVO_4$ nanocrystalline morphologies and their photoluminescence properties. *Physics Procedia*, *8*, 109–113.

Cani, D., van der Waal, J. C., & Pescarmona, P. P. (2021). Highly-accessible, doped TiO_2 nanoparticles embedded at the surface of SiO2 as photocatalysts for the degradation of pollutants under visible and UV radiation. *Applied Catalysis A: General*, *621*, 118179.

Chen, L., Chen, K.-J., Hu, S.-F., & Liu, R.-S. (2011). Combinatorial chemistry approach to searching phosphors for white light-emitting diodes in (Gd-Y-Bi-Eu)VO_4 quaternary system. *Journal of Materials Chemistry*, *21*, 3677–3685.

Chen, M., Wang, J.-H., Luo, Z.-J., Cheng, Z.-Q., Zhang, Y.-F., Yu, X.-F., ... Wang, Q.-Q. (2016). Facile synthesis of flower-shaped Au/$GdVO_4$:Eu core/shell nanoparticles by using citrate as stabilizer and complexing agent. *RSC Advances*, *6*, 9612–9618.

Chen, Z., Jia, H., Sharafudeen, K., Dai, W., Liu, Y., Dong, G., & Qiu, J. (2016). Up-conversion luminescence from single vanadate through blackbody radiation harvesting broadband near-infrared photons for photovoltaic cells. *Journal of Alloys and Compounds*, *663*, 204–210.

Choi, D. H., Kang, D. H., Yi, S. S., Jang, K., & Jeong, J. H. (2015). Up-conversion luminescent properties of $La_{(0.80-x)}VO_4:Yb_x, Er_{0.20}$ phosphors. *Materials Research Bulletin*, *71*, 16–20.

Choi, S., Moon, Y.-M., & Jung, H.-K. (2010). Luminescent properties of PEG-added nanocrystalline $YVO_4:Eu^{3+}$ phosphor prepared by a hydrothermal method. *Journal of Luminescence*, *130*, 549–553.

Chumha, N., Kittiwachana, S., Thongtem, T., Thongtem, S., & Kaowphong, S. (2014). Synthesis and characterization of $GdVO_4$ nanoparticles by a malic acid-assisted sol–gel method. *Materials Letters*, *136*, 18–21.

de Sousa Filho, P. C., Gacoin, T., Boilot, J.-P., Walton, R. I., & Serra, O. A. (2015). Synthesis and luminescent properties of $REVO_4$–$REPO_4$ (RE = Y, Eu, Gd, Er, Tm, or Yb) heteronanostructures: a promising class of phosphors for excitation from NIR to VUV. *The Journal of Physical Chemistry C*, *119*, 24062–24074.

Deng, H., Yang, S., Xiao, S., Gong, H.-M., & Wang, Q.-Q. (2008). Controlled synthesis and upconverted avalanche luminescence of cerium(III) and neodymium(III) orthovanadate nanocrystals with high uniformity of size and shape. *Journal of the American Chemical Society*, *130*, 2032–2040.

Dhanya, C. R., Jeyaraman, J., Janeesh, P. A., Shukla, A., Sivakumar, S., & Abraham, A. (2016). Bio-distribution and in vivo/in vitro toxicity profile of PEGylated polymer capsules encapsulating $LaVO_4:Tb^{3+}$ nanoparticles for bioimaging applications. *RSC Advances*, *6*, 55125–55134.

Ekthammathat, N., Thongtem, T., Phuruangrat, A., & Thongtem, S. (2013). Synthesis and characterization of $CeVO_4$ by microwave radiation method and its photocatalytic activity. *Journal of Nanomaterials*, *2013*, 434197.

Errandonea, D., & Garg, A. B. (2018). Recent progress on the characterization of the high-pressure behaviour of AVO_4 orthovanadates. *Progress in Materials Science*, *97*, 123–169.

Fan, W., Bu, Y., Song, X., Sun, S., & Zhao, X. (2007). Selective synthesis and luminescent properties of monazite- and zircon-type $LaVO_4$:Ln (Ln = Eu, Sm, and Dy) nanocrystals. *Crystal Growth & Design*, *7*, 2361–2366.

Fan, W., Zhao, W., You, L., Song, X., Zhang, W., Yu, H., & Sun, S. (2004). A simple method to synthesize single-crystalline lanthanide orthovanadate nanorods. *Journal of Solid State Chemistry*, *177*, 4399–4403.

Gavrilović, T. V., Jovanović, D. J., & Dramićanin, M. D. (2018). Chapter 2 - synthesis of multifunctional inorganic materials: From micrometer to nanometer dimensions. In B. A. Bhanvase, V. B. Pawade, S. J. Dhoble, S. H. Sonawane, & M. Ashokkumar (Eds.), *Nanomaterials for Green Energy* (pp. 55–81). Elsevier.

Gavrilović, T. V., Jovanović, D. J., Lojpur, V., & Dramićanin, M. D. (2014). Multifunctional Eu^{3+}- and Er^{3+}/Yb^{3+}-doped $GdVO_4$ nanoparticles synthesized by reverse micelle method. *Scientific Reports*, *4*, 4209.

Gavrilović, T. V., Jovanović, D. J., Lojpur, V. M., Đorđević, V., & Dramićanin, M. D. (2014). Enhancement of luminescence emission from $GdVO_4:Er^{3+}/Yb^{3+}$ phosphor by Li^+ co-doping. *Journal of Solid State Chemistry*, *217*, 92–98.

Ghodrati, M., Mousavi-Kamazani, M., & Zinatloo-Ajabshir, S. (2020). $Zn_3V_3O_8$ nanostructures: Facile hydrothermal/solvothermal synthesis, characterization, and electrochemical hydrogen storage. *Ceramics International*, *46*, 28894–28902.

Grandhe, B. K., Bandi, V. R., Jang, K., Ramaprabhu, S., Lee, H.-S., Shin, D.-S., ... Jeong, J.-H. (2012). Multi wall carbon nanotubes assisted synthesis of $YVO_4:Eu^{3+}$ nanocomposites for display device applications. *Composites Part B: Engineering*, *43*, 1192–1195.

Grzyb, T., Szczeszak, A., Shyichuk, A., Moura, R. T., Neto, A. N. C., Andrzejewska, N., ... Lis, S. (2018). Comparative studies of structure, spectroscopic properties and intensity parameters of tetragonal rare earth vanadate nanophosphors doped with Eu(III). *Journal of Alloys and Compounds*, *741*, 459–472.

Haase, M., Riwotzki, K., Meyssamy, H., & Kornowski, A. (2000). Synthesis and properties of colloidal lanthanide-doped nanocrystals. *Journal of Alloys and Compounds*, *303–304*, 191–197.

He, F., Yang, P., Wang, D., Niu, N., Gai, S., Li, X., & Zhang, M. (2011). Hydrothermal synthesis, dimension evolution and luminescence properties of tetragonal $LaVO_4$:Ln (Ln = Eu^{3+}, Dy^{3+}, Sm^{3+}) nanocrystals. *Dalton Transactions*, *40*, 11023–11030.

Heidari-Asil, S. A., Zinatloo-Ajabshir, S., Amiri, O., & Salavati-Niasari, M. (2020). Amino acid assisted-synthesis and characterization of magnetically retrievable $ZnCo_2O_4-Co_3O_4$ nanostructures as high activity visible-light-driven photocatalyst. *International Journal of Hydrogen Energy*, *45*, 22761–22774.

Hou, Z., Yang, P., Li, C., Wang, L., Lian, H., Quan, Z., & Lin, J. (2008). Preparation and luminescence properties of YVO_4:Ln and Y(V, P)O_4:Ln (Ln = Eu^{3+}, Sm^{3+}, Dy^{3+}) nanofibers and microbelts by sol − gel/electrospinning process. *Chemistry of Materials*, *20*, 6686–6696.

Hu, J., & Wang, Q. (2014). New synthesis for a group of tetragonal $LnVO_4$ and their luminescent properties. *Materials Letters*, *120*, 20–22.

Huang, H., Li, D., Lin, Q., Zhang, W., Shao, Y., Chen, Y., ... Fu, X. (2009). Efficient degradation of benzene over $LaVO_4/TiO_2$ nanocrystalline heterojunction photocatalyst under visible light irradiation. *Environmental Science & Technology*, *43*, 4164–4168.

Huang, Z., Huang, S., Ou, G., & Pan, W. (2012). Systhesis, phase transformation and photoluminescence properties of $Eu:La_{1-x}Gd_xVO_4$ nanofibers by electrospinning method. *Nanoscale*, *4*, 5065–5070.

Huignard, A., Buissette, V., Franville, A.-C., Gacoin, T., & Boilot, J.-P. (2003). Emission processes in YVO_4:Eu nanoparticles. *The Journal of Physical Chemistry B*, *107*, 6754–6759.

Huignard, A., Buissette, V., Laurent, G., Gacoin, T., & Boilot, J. P. (2002). Synthesis and characterizations of YVO_4:Eu colloids. *Chemistry of Materials*, *14*, 2264–2269.

Huignard, A., Gacoin, T., & Boilot, J.-P. (2000). Synthesis and luminescence properties of colloidal YVO_4:Eu phosphors. *Chemistry of Materials*, *12*, 1090–1094.

Huong, T. T., Vinh, L. T., Phuong, H. T., Khuyen, H. T., Anh, T. K., Tu, V. D., & Minh, L. Q. (2016). Controlled fabrication of the strong emission YVO_4:Eu^{3+} nanoparticles and nanowires by microwave assisted chemical synthesis. *Journal of Luminescence*, *173*, 89–93.

Jeong, H., Lee, B.-I., & Byeon, S.-H. (2014). A new route through the layered hydroxide form for the synthesis of $GdVO_4$ dispersible in polar solvents. *New Journal of Chemistry*, *38*, 5691–5694.

Jeyaraman, J., Shukla, A., & Sivakumar, S. (2016). Targeted stealth polymer capsules encapsulating Ln^{3+}-doped $LaVO_4$ nanoparticles for bioimaging applications. *ACS Biomaterials Science & Engineering*, *2*, 1330–1340.

Jia, C.-J., Sun, L.-D., You, L.-P., Jiang, X.-C., Luo, F., Pang, Y.-C., & Yan, C.-H. (2005). Selective synthesis of monazite- and zircon-type $LaVO_4$ nanocrystals. *The Journal of Physical Chemistry B*, *109*, 3284–3290.

Jiang, G., Wei, X., Chen, Y., Duan, C., & Yin, M. (2013). Broadband downconversion in YVO_4:Tm^{3+},Yb^{3+} phosphors. *Journal of Rare Earths*, *31*, 27–31.

Jin, L.-N., Liu, Q., & Sun, W.-Y. (2013). Synthesis and photoluminescence of octahedral $GdVO_4$ microcrystals by hydrothermal conversion of $Gd_2(CO_3)_3 \cdot xH_2O$ nanospheres. *Solid State Sciences*, *19*, 45–50.

Jo, D. S., Luo, Y. Y., Senthil, K., Masaki, T., & Yoon, D. H. (2011). Synthesis of high efficient nanosized $Y(V,P)O_4$:Eu^{3+} red phosphors by a new technique. *Optical Materials*, *33*, 1190–1194.

Jovanović, D. J. (2020). 6 - Lanthanide-doped orthovanadate phosphors: Syntheses, structures, and photoluminescence properties. In S. J. Dhoble, V. B. Pawade, H. C. Swart, & V. Chopra (Eds.), *Spectroscopy of Lanthanide Doped Oxide Materials* (pp. 235–291). Woodhead Publishing.

Kalai Selvan, R., Gedanken, A., Anilkumar, P., Manikandan, G., & Karunakaran, C. (2009). Synthesis and characterization of rare earth orthovanadate (RVO_4; R = La, Ce, Nd, Sm, Eu & Gd) nanorods/nanocrystals/nanospindles by a facile sonochemical method and their catalytic properties. *Journal of Cluster Science*, *20*, 291–305.

Kalevaru, V. N., Dhachapally, N., & Martin, A. (2016). Catalytic performance of lanthanum vanadate catalysts in ammoxidation of 2-methylpyrazine. *Catalysts*, *6*, 10.

Kanna Sharma, T. S., Hwa, K.-Y., Santhan, A., & Ganguly, A. (2021). Synthesis of novel three-dimensional flower-like cerium vanadate anchored on graphitic carbon nitride as an efficient electrocatalyst for real-time monitoring of mesalazine in biological and water samples. *Sensors and Actuators B: Chemical*, *331*, 129413.

Khorasanizadeh, M. H., Monsef, R., Amiri, O., Amiri, M., & Salavati-Niasari, M. (2019). Sonochemical-assisted route for synthesis of spherical shaped holmium vanadate nanocatalyst for polluted waste water treatment. *Ultrasonics Sonochemistry*, *58*, 104686.

Kolesnikov, I. E., Golyeva, E. V., Kurochkin, M. A., Lähderanta, E., & Mikhailov, M. D. (2016). Nd^3 + -doped YVO_4 nanoparticles for luminescence nanothermometry in the first and second biological windows. *Sensors and Actuators B: Chemical*, *235*, 287–293.

Kolesnikov, I. E., Tolstikova, D. V., Kurochkin, A. V., Platonova, N. V., Pulkin, S. A., Manshina, A. A., & Mikhailov, M. D. (2015). Concentration effect on structural and luminescent properties of YVO_4:Nd^{3+} nanophosphors. *Materials Research Bulletin*, *70*, 799–803.

Kumar, D., Sharma, M., Haranath, D., & Pandey, O. P. (2017). Facile route to produce spherical and highly luminescent Tb^{3+} doped Y_2O_3 nanophosphors. *Journal of Alloys and Compounds*, *695*, 726–736.

Lai, Y.-J., & Lee, D.-J. (2021). Pollutant degradation with mediator Z-scheme heterojunction photocatalyst in water: A review. *Chemosphere*, 131059.

Li, G., Chao, K., Peng, H., & Chen, K. (2008). Hydrothermal synthesis and characterization of YVO_4 and YVO_4:Eu^{3+} nanobelts and polyhedral micron crystals. *The Journal of Physical Chemistry C*, *112*, 6228–6231.

Li, X., Yu, M., Hou, Z., Li, G., Ma, P. A., Wang, W., ... Lin, J. (2011). One-dimensional $GdVO_4$:Ln^{3+} (Ln = Eu, Dy, Sm) nanofibers: Electrospinning preparation and luminescence properties. *Journal of Solid State Chemistry*, *184*, 141–148.

Liang, Y., Noh, H. M., Xue, J., Choi, H., Park, S. H., Choi, B. C., ... Jeong, J. H. (2017). High quality colloidal GdVO4:Yb,Er upconversion nanoparticles synthesized via a protected calcination process for versatile applications. *Materials & Design*, *130*, 190–196.

Liu, F., Shao, X., Yin, Y., Zhao, L., Shao, Z., Liu, X., & Meng, X. (2011). Shape controlled synthesis and tribological properties of $CeVO_4$ nanoparticles aslubricating additive. *Journal of Rare Earths*, *29*, 688–691.

Liu, F., Shao, X., Yin, Y., Zhao, L., Sun, Q., Shao, Z., ... Meng, X. (2011). Selective synthesis and growth mechanism of $CeVO_4$ nanoparticals via hydrothermal method. *Journal of Rare Earths*, *29*, 97–100.

Liu, G., Duan, X., Li, H., & Dong, H. (2009). Hydrothermal synthesis, characterization and optical properties of novel fishbone-like $LaVO_4$:Eu^{3+} nanocrystals. *Materials Chemistry and Physics*, *115*, 165–171.

Liu, J., & Li, Y. (2007). General synthesis of colloidal rare earth orthovanadate nanocrystals. *Journal of Materials Chemistry*, *17*, 1797–1803.

Liu, J., & Li, Y. D. (2007). Synthesis and self-assembly of luminescent Ln^{3+}-doped $LaVO_4$ uniform nanocrystals. *Advanced Materials*, *19*, 1118–1122.

Liu, J., Yao, Q., & Li, Y. (2006). Effects of downconversion luminescent film in dye-sensitized solar cells. *Applied Physics Letters*, *88*, 173119.

Liu, Y., Xiong, H., Zhang, N., Leng, Z., Li, R., & Gan, S. (2015). Microwave synthesis and luminescent properties of YVO_4:Ln^{3+} (Ln = Eu, Dy and Sm) phosphors with different morphologies. *Journal of Alloys and Compounds*, *653*, 126–134.

Lyadov, A. S., & Kurilkin, V. V. (2016). Reduction specifics of rare-earth orthovanadates (REE = La, Nd, Sm, Dy, Ho, Er, Tm, Yb, and Lu). *Russian Journal of Inorganic Chemistry*, *61*, 86–92.

Ma, J., Wu, Q., & Ding, Y. (2008). Selective synthesis of monoclinic and tetragonal phase $LaVO_4$ nanorods via oxides-hydrothermal route. *Journal of Nanoparticle Research*, *10*, 775–786.

Ma, Q., & Wang, Q. (2015). Facile synthesis of lanthanide vanadates and their luminescent properties. *Displays*, *39*, 6–10.

Mahapatra, S., Nayak, S. K., Madras, G., & Guru Row, T. N. (2008). Microwave synthesis and photocatalytic activity of nano lanthanide (Ce, Pr, and Nd) orthovanadates. *Industrial & Engineering Chemistry Research*, *47*, 6509–6516.

Martínez-Huerta, M. V., Coronado, J. M., Fernández-García, M., Iglesias-Juez, A., Deo, G., Fierro, J. L. G., & Bañares, M. A. (2004). Nature of the vanadia–ceria interface in V^{5+}/CeO_2 catalysts and its relevance for the solid-state reaction toward $CeVO_4$ and catalytic properties. *Journal of Catalysis*, *225*, 240–248.

Mentasti, L., Martínez, N., Zucchi, I. A., Santiago, M., & Barreto, G. (2020). Development of a simple process to obtain luminescent YVO_4:Eu^{3+} nanoparticles for fiber optic dosimetry. *Journal of Alloys and Compounds*, *829*, 154628.

Mialon, G., Gohin, M., Gacoin, T., & Boilot, J.-P. (2008). High temperature strategy for oxide nanoparticle synthesis. *ACS Nano*, *2*, 2505–2512.

Michalska, M., Jasiński, J. B., Pavlovsky, J., Żurek-Siworska, P., Sikora, A., Gołębiewski, P., ... Seidlerova, J. (2021). Solid state-synthesized lanthanum orthovanadate ($LaVO_4$) Co-doped with Eu as efficient photoluminescent material. *Journal of Luminescence*, *233*, 117934.

Mishra, S., Priyadarshinee, M., Debnath, A. K., Muthe, K. P., Mallick, B. C., Das, N., & Parhi, P. (2020). Rapid microwave assisted hydrothermal synthesis cerium vanadate nanoparticle and its photocatalytic and antibacterial studies. *Journal of Physics and Chemistry of Solids*, *137*, 109211.

Mohd Razali, N. A., Wan Salleh, W. N., Aziz, F., Jye, L. W., Yusof, N., & Ismail, A. F. (2021). Review on tungsten trioxide as a photocatalysts for degradation of recalcitrant pollutants. *Journal of Cleaner Production*, *309*, 127438.

Monsef, R., Ghiyasiyan-Arani, M., Amiri, O., & Salavati-Niasari, M. (2020). Sonochemical synthesis, characterization and application of $PrVO_4$ nanostructures as an effective photocatalyst for discoloration of organic dye contaminants in wastewater. *Ultrasonics Sonochemistry*, *61*, 104822.

Monsef, R., Ghiyasiyan-Arani, M., & Salavati-Niasari, M. (2018). Application of ultrasound-aided method for the synthesis of $NdVO_4$ nano-photocatalyst and investigation of eliminate dye in contaminant water. *Ultrasonics Sonochemistry*, *42*, 201–211.

Monsef, R., Ghiyasiyan-Arani, M., & Salavati-Niasari, M. (2019). Utilizing of neodymium vanadate nanoparticles as an efficient catalyst to boost the photocatalytic water purification. *Journal of Environmental Management*, *230*, 266–281.

Monsef, R., Soofivand, F., Abbas Alshamsi, H., Al-Nayili, A., Ghiyasiyan-Arani, M., & Salavati-Niasari, M. (2021). Sonochemical synthesis and characterization of $PrVO_4$/CdO nanocomposite and their application as photocatalysts for removal of organic dyes in water. *Journal of Molecular Liquids*, *336*, 116339.

Moshtaghi, S., Zinatloo-Ajabshir, S., & Salavati-Niasari, M. (2016). Nanocrystalline barium stannate: facile morphology-controlled preparation, characterization and investigation of optical and photocatalytic properties. *Journal of Materials Science: Materials in Electronics*, *27*, 834–842.

Mosleh, M., & Mahinpour, A. (2016). Sonochemical synthesis and characterization of cerium vanadate nanoparticles and investigation of its photocatalyst application. *Journal of Materials Science: Materials in Electronics*, *27*, 8930–8934.

Narsimulu, D., Kakarla, A. K., & Su Yu, J. (2021). Cerium vanadate/carbon nanotube hybrid composite nanostructures as a high-performance anode material for lithium-ion batteries. *Journal of Energy Chemistry*, *58*, 25–32.

Oka, Y., Yao, T., & Yamamoto, N. (2000). Hydrothermal synthesis of lanthanum vanadates: Synthesis and crystal structures of zircon-type LaVO4 and a new compound LaV_3O_9. *Journal of Solid State Chemistry*, *152*, 486–491.

Othman, I., Hisham Zain, J., Abu Haija, M., & Banat, F. (2020). Catalytic activation of peroxymonosulfate using $CeVO_4$ for phenol degradation: An insight into the reaction pathway. *Applied Catalysis B: Environmental*, *266*, 118601.

Paunović, V., Artusi, M., Verel, R., Krumeich, F., Hauert, R., & Pérez-Ramírez, J. (2018). Lanthanum vanadate catalysts for selective and stable methane oxybromination. *Journal of Catalysis*, *363*, 69–80.

Periša, J., Antić, Ž., Ma, C.-G., Papan, J., Jovanović, D., & Dramićanin, M. D. (2020). Pesticide-induced photoluminescence quenching of ultra-small Eu^{3+}-activated phosphate and vanadate nanoparticles. *Journal of Materials Science & Technology*, *38*, 197–204.

Phuruangrat, A., Kuntalue, B., Thongtem, S., & Thongtem, T. (2016). Effect of PEG on phase, morphology and photocatalytic activity of $CeVO_4$ nanostructures. *Materials Letters*, *174*, 138–141.

Ponnaiah, S. K., & Prakash, P. (2021). A new high-performance supercapacitor electrode of strategically integrated cerium vanadium oxide and polypyrrole nanocomposite. *International Journal of Hydrogen Energy*, *46*, 19323–19337.

Qian, L., Zhu, J., Chen, Z., Gui, Y., Gong, Q., Yuan, Y., ... Qian, X. (2009). Self-assembled heavy lanthanide orthovanadate architecture with controlled dimensionality and morphology. *Chemistry – A European Journal*, *15*, 1233–1240.

Ray, S., Banerjee, A., & Pramanik, P. (2009). Shape controlled synthesis, characterization and photoluminescence properties of $YVO_4:Dy^{3+}/Eu^{3+}$ phosphors. *Materials Science and Engineering: B*, *156*, 10–17.

Razi, F., Zinatloo-Ajabshir, S., & Salavati-Niasari, M. (2017). Preparation, characterization and photocatalytic properties of Ag_2ZnI_4/AgI nanocomposites via a new simple hydrothermal approach. *Journal of Molecular Liquids*, *225*, 645–651.

Ren, Q.-F., Zhang, B., Chen, S.-H., Wang, S.-L., Zheng, Q., Ding, Y., ... Jin, Z. (2019). Amine salts assisted controllable synthesis of the $YVO_4:Eu^{3+}$ nanocrystallines and their luminescence properties. *Physica B: Condensed Matter*, *557*, 1–5.

Riwotzki, K., & Haase, M. (1998). Wet-chemical synthesis of doped colloidal nanoparticles: YVO_4:Ln (Ln = Eu, Sm, Dy). *The Journal of Physical Chemistry B*, *102*, 10129–10135.

Salavati-Niasari, M., Saleh, L., Mohandes, F., & Ghaemi, A. (2014). Sonochemical preparation of pure t-$LaVO_4$ nanoparticles with the aid of tris(acetylacetonato)lanthanum hydrate as a novel precursor. *Ultrasonics Sonochemistry*, *21*, 653–662.

Shandilya, M., Rai, R., & Singh, J. (2016). Review: Hydrothermal technology for smart materials. *Advances in Applied Ceramics*, *115*, 354–376.

Shen, D., Zhang, Y., Zhang, X., Wang, Z., Zhang, Y., Hu, S., & Yang, J. (2018). Morphology/phase controllable synthesis of monodisperse $ScVO_4$ microcrystals and tunable multicolor luminescence properties in $Sc(La)VO_4(PO4):Bi^{3+}$, Ln^{3+} phosphors. *CrystEngComm*, *20*, 5180–5190.

Shen, J., Sun, L.-D., & Yan, C.-H. (2008). Luminescent rare earth nanomaterials for bioprobe applications. *Dalton Transactions*, 5687–5697.

Stouwdam, J. W., Raudsepp, M., & van Veggel, F. C. J. M. (2005). Colloidal nanoparticles of Ln^{3+}-doped $LaVO_4$: Energy transfer to visible- and near-infrared-emitting lanthanide ions. *Langmuir: The ACS Journal of Surfaces and Colloids*, *21*, 7003–7008.

Sun, Y., Liu, H., Wang, X., Kong, X., & Zhang, H. (2006). Optical spectroscopy and visible upconversion studies of $YVO_4:Er^{3+}$ nanocrystals synthesized by a hydrothermal process. *Chemistry of Materials*, *18*, 2726–2732.

Takeshita, S., Isobe, T., Sawayama, T., & Niikura, S. (2011). Low-temperature wet chemical precipitation of $YVO_4:Bi^{3+}$, Eu^{3+} nanophosphors via citrate precursors. *Progress in Crystal Growth and Characterization of Materials*, *57*, 127–136.

Tamilmani, V., Nair, B. U., & Sreeram, K. J. (2017). Phosphate modulated luminescence in lanthanum vanadate nanorods—Catechin, polyphenolic ligand. *Journal of Solid State Chemistry*, *252*, 158–168.

Tang, S., Huang, M., Wang, J., Yu, F., Shang, G., & Wu, J. (2012). Hydrothermal synthesis and luminescence properties of $GdVO_4:Ln^{3+}$ (Ln = Eu, Sm, Dy) phosphors. *Journal of Alloys and Compounds*, *513*, 474–480.

Tegus, O., Amurisana, B., & Zhiqiang, S. (2019). Morphology-sensitive photoluminescent properties of $YVO_4:Ln^{3+}$ ($Ln^{3+} = Eu^{3+}$, Sm^{3+}, Dy^{3+}, Tm^{3+}) hierarchitectures. *Journal of Luminescence*, *215*, 116624.

Thirumalai, J., Chandramohan, R., & Vijayan, T. A. (2011). A novel 3D nanoarchitecture of $PrVO_4$ phosphor: Selective synthesis, characterization, and luminescence behavior. *Materials Chemistry and Physics*, *127*, 259–264.

Tian, L., Chen, S.-M., Liu, Q., Wu, J.-l, Zhao, R.-n, Li, S., & Chen, L.-j (2020). Effect of Eu^{3+}-doping on morphology and fluorescent properties of neodymium vanadate nanorod-arrays. *Transactions of Nonferrous Metals Society of China*, *30*, 1031–1037.

Tian, L., Li, Y., Wang, H., Chen, S., Wang, J., Guo, Z., ... Wu, F. (2018). Controlled preparation and self-assembly of $NdVO_4$ nanocrystals. *Journal of Rare Earths*, *36*, 179–183.

Tian, L., Sun, Q., Xu, X., Li, Y., Long, Y., & Zhu, G. (2013). Controlled synthesis and formation mechanism of monodispersive lanthanum vanadate nanowires with monoclinic structure. *Journal of Solid State Chemistry*, *200*, 123–127.

Toda, A., Toda, K., Ishigaki, T., Uematsu, K., & Sato, M. (2010). Luminescence properties of $Y(P,V)O_4$ synthesized by microwave heating. *Key Engineering Materials*, *421–422*, 356–359.

Uematsu, K., Ochiai, A., Toda, K., & Sato, M. (2006). Characterization of YVO_4:Eu^{3+} phosphors synthesized by microwave heating method. *Journal of Alloys and Compounds*, *408–412*, 860–863.

Vadivel, S., Paul, B., Kumaravel, M., Hariganesh, S., Rajendran, S., Prasanga Gayanath Mantilaka, M. M. M. G., ... Puviarasu, P. (2020). Facile synthesis of $YbVO_4$, and YVO_4 nanostructures through MOF route for photocatalytic applications. *Inorganic Chemistry Communications*, *115*, 107855.

Vairapperumal, T., Natarajan, D., Manikantan Syamala, K., Kalarical Janardhanan, S., & Unni, B. N. (2017). Catechin caged lanthanum orthovanadate nanorods for nuclear targeting and bioimaging applications. *Sensors and Actuators B: Chemical*, *242*, 700–709.

Wang, F., Liu, C., Zhou, Z., Jia, P., & Lin, J. (2012). Molten salt synthesis and luminescent properties of YVO_4:Eu nanocrystalline phosphors. *Journal of Rare Earths*, *30*, 202–204.

Wang, H., Meng, Y., & Yan, H. (2004). Rapid synthesis of nanocrystalline $CeVO_4$ by microwave irradiation. *Inorganic Chemistry Communications*, *7*, 553–555.

Wang, J., Xu, Y., Hojamberdiev, M., Cui, Y., Liu, H., & Zhu, G. (2009). Optical properties of porous YVO_4:Ln (Ln = Dy^{3+} and Tm^{3+}) nanoplates obtained by the chemical co-precipitation method. *Journal of Alloys and Compounds*, *479*, 772–776.

Wang, J., Xu, Y., Hojamberdiev, M., & Zhu, G. (2009). Influence of sodium dodecyl sulfonate (SDS) on the hydrothermal synthesis of YVO_4:Eu^{3+} crystals in a wide pH range. *Journal of Alloys and Compounds*, *487*, 358–362.

Wang, N., Chen, W., Zhang, Q., & Dai, Y. (2008). Synthesis, luminescent, and magnetic properties of $LaVO_4$:Eu nanorods. *Materials Letters*, *62*, 109–112.

Wang, X., Du, P., Liu, W., Huang, S., Hu, Z., Wang, Q., & Li, J.-G. (2020). Organic-free direct crystallization of t-$LaVO_4$:Eu nanocrystals with favorable luminescence for LED lighting and optical thermometry. *Journal of Materials Research and Technology*, *9*, 13264–13273.

Wang, Y., Wang, S., Wu, Z., Li, W., & Ruan, Y. (2013). Photoluminescence properties of Ce and Eu co-doped YVO_4 crystals. *Journal of Alloys and Compounds*, *551*, 262–266.

Wangkhem, R., Singh, N. S., Singh, N. P., Singh, S. D., & Singh, L. R. (2018). Facile synthesis of re-dispersible YVO_4:Ln^{3+} (Ln^{3+} = Dy^{3+}, Eu^{3+}, Sm^{3+}) nanocrystals: Luminescence studies and sensing of Cu^{2+} ions. *Journal of Luminescence*, *203*, 341–348.

Weng, X., Yang, Q., Wang, L., Xu, L., Sun, X., & Liu, J. (2013). General synthesis and self-assembly of lanthanide orthovanadate nanorod arrays. *CrystEngComm*, *15*, 10230–10237.

Xia, H. R., Li, L. X., Zhang, H. J., Meng, X. L., Zhu, L., Yang, Z. H., ... Wang, J. Y. (2000). Raman spectra and laser properties of Yb-doped yttrium orthovanadate crystals. *Journal of Applied Physics*, *87*, 269–273.

Xu, H., Wang, H., Jin, T., & Yan, H. (2004). Rapid fabrication of luminescent Eu:YVO_4 films by microwave-assisted chemical solution deposition. *Nanotechnology*, *16*, 65–69.

Xu, H., Wang, H., Meng, Y., & Yan, H. (2004). Rapid synthesis of size-controllable YVO_4 nanoparticles by microwave irradiation. *Solid State Communications*, *130*, 465–468.

Xu, H., Wang, H., & Yan, H. (2007). Preparation and photocatalytic properties of YVO_4 nanopowders. *Journal of Hazardous Materials*, *144*, 82–85.

Xu, Z., Kang, X., Li, C., Hou, Z., Zhang, C., Yang, D., ... Lin, J. (2010). Ln^{3+} (Ln = Eu, Dy, Sm, and Er) ion-doped YVO_4 nano/microcrystals with multiform morphologies: Hydrothermal synthesis, growing mechanism, and luminescent properties. *Inorganic Chemistry*, *49*, 6706–6715.

Xu, Z., Li, C., Hou, Z., Peng, C., & Lin, J. (2011). Morphological control and luminescence properties of lanthanide orthovanadate $LnVO_4$ (Ln = La to Lu) nano-/microcrystals via hydrothermal process. *CrystEngComm*, *13*, 474–482.

Yan, C.-H., Yan, Z.-G., Du, Y.-P., Shen, J., Zhang, C., & Feng, W. (2011). Chapter 251 - controlled synthesis and properties of rare earth nanomaterials. In K. A. Gschneidner, J.-C. G. Bünzli, & V. K. Pecharsky (Eds.), *Handbook on the Physics and Chemistry of Rare Earths* (pp. 275–472). Elsevier.

Yan, Y., Hojamberdiev, M., Xu, Y., Wang, J., & Luan, Z. (2013). Hydrothermally-induced morphological transformation of $GdVO_4$:Eu^{3+}. *Materials Chemistry and Physics*, *139*, 298–304.

Yan, Y., Wang, J., Lu, Z., Ren, B., Wang, L., & Xu, Y. (2013). Influence of TTAB on morphology and photoluminescence of $GdVO_4$:Eu^{3+} powders in a wide pH range. *Powder Technology*, *249*, 475–481.

Yang, L., Peng, S., Zhao, M., & Yu, L. (2019). A facile strategy to prepare YVO_4:Eu^{3+} colloid with novel nanostructure for enhanced optical performance. *Applied Surface Science*, *473*, 885–892.

Yang, X., Xu, L., Zhai, Z., Cheng, F., Yan, Z., Feng, X., ... Hou, W. (2013). Submicrometer-sized hierarchical hollow spheres of heavy lanthanide orthovanadates: sacrificial template synthesis, formation mechanism, and luminescent properties. *Langmuir: The ACS Journal of Surfaces and Colloids*, *29*, 15992–16001.

Yang, X., Zhang, Y., Zhang, P., He, N., Yang, Q., Peng, H., ... Gui, J. (2018). pH modulations of fluorescence $LaVO_4$:Eu^{3+} materials with different morphologies and structures for rapidly and sensitively detecting Fe^{3+} ions. *Sensors and Actuators B: Chemical*, *267*, 608–616.

Yu, C., Yu, M., Li, C., Zhang, C., Yang, P., & Lin, J. (2009). Spindle-like lanthanide orthovanadate nanoparticles: Facile synthesis by ultrasonic irradiation, characterization, and luminescent properties. *Crystal Growth & Design*, *9*, 783–791.

Yuan, H., Wang, K., Wang, C., Zhou, B., Yang, K., Liu, J., & Zou, B. (2015). Pressure-induced phase transformations of zircon-type $LaVO_4$ nanorods. *The Journal of Physical Chemistry C*, *119*, 8364–8372.

Zahedifar, M., Chamanzadeh, Z., & Hosseinpoor, S. M. (2013). Synthesis of $LaVO_4$: Dy^{3+} luminescent nanostructure and optimization of its performance as down-converter in dye-sensitized solar cells. *Journal of Luminescence*, *135*, 66–73.

Zhang, B., Ma, Q., & Xu, C.-Q. (2020). Orthogonally polarized dual-wavelength Nd:YVO_4/MgO:PPLN intra-cavity frequency doubling green laser. *Optics & Laser Technology*, *125*, 106005.

Zhang, F., Li, G., Zhang, W., & Yan, Y. L. (2015). Phase-dependent enhancement of the green-emitting upconversion fluorescence in $LaVO_4$:Yb^{3+}, Er^{3+}. *Inorganic Chemistry*, *54*, 7325–7334.

Zhang, H., Fu, X., Niu, S., Sun, G., & Xin, Q. (2004a). Low temperature synthesis of nanocrystalline YVO_4:Eu via polyacrylamide gel method. *Journal of Solid State Chemistry*, *177*, 2649–2654.

Zhang, H., Fu, X., Niu, S., Sun, G., & Xin, Q. (2004b). Photoluminescence of YVO4:Tm phosphor prepared by a polymerizable complex method. *Solid State Communications*, *132*, 527–531.

Zhang, H., Fu, X., Niu, S., & Xin, Q. (2008). Synthesis and luminescent properties of nanosized YVO_4:Ln (Ln = Sm, Dy). *Journal of Alloys and Compounds*, *457*, 61–65.

Zhang, J., Shi, J., Tan, J., Wang, X., & Gong, M. (2010). Morphology-controllable synthesis of tetragonal $LaVO_4$ nanostructures. *CrystEngComm*, *12*, 1079–1085.

Zhang, L., You, H., & Yang, M. (2012). Synthesis and luminescent properties of spindle-like $YVO_4:Ln^{3+}$ (Ln = Eu, Dy) self-assembled of nanoparticles. *Journal of Physics and Chemistry of Solids*, *73*, 368–373.

Zhang, S., Wang, L., Peng, H., Li, G., & Chen, K. (2010). Influence of P-doping on the morphologies and photoluminescence properties of $LaVO_4:Tb^{3+}$ nanostructures. *Materials Chemistry and Physics*, *123*, 714–718.

Zhang, S., Zhou, S., Li, H., & Li, L. (2008). Investigation of thermal expansion and compressibility of rare-earth orthovanadates using a dielectric chemical bond method. *Inorganic Chemistry*, *47*, 7863–7867.

Zhang, Z., & Liang, Y. (2020). Multicolor luminescent $GdVO_4:Ln^{3+}$ (Ln = Eu, Dy and Sm) nanophosphors prepared by a protected calcination method. *Optik*, *207*, 163825.

Zheng, Y., Sun, X., Su, H., Sun, L., & Qi, C. (2018). Monodisperse $YVO_4:Eu^{3+}$ nanospindles: Rapid converted growth and luminescence properties. *Materials Research Bulletin*, *105*, 149–153.

Zhong, J., & Zhao, W. (2015). Novel dumbbell-like $LaVO_4:Eu^{3+}$ nanocrystals and effect of Ba^{2+} codoping on luminescence properties of $LaVO_4:Eu^{3+}$ nanocrystals. *Journal of Sol-Gel Science and Technology*, *73*, 133–140.

Zhu, L., Li, J., Li, Q., Liu, X., Meng, J., & Cao, X. (2007). Sonochemical synthesis and photoluminescent property of YVO_4:Eu nanocrystals. *Nanotechnology*, *18*, 055604.

Zhu, L., Li, Q., Li, J., Liu, X., Meng, J., & Cao, X. (2007). Selective synthesis of mesoporous and nanorod $CeVO_4$ without template. *Journal of Nanoparticle Research*, *9*, 261–268.

Zhu, Y., Ni, Y., & Sheng, E. (2016). Mixed-solvothermal synthesis and applications in sensing for Cu^{2+} and Fe^{3+} ions of flowerlike LaVO4:Eu3 + nanostructures. *Materials Research Bulletin*, *83*, 41–47.

Zinatloo-Ajabshir, S., Heidari-Asil, S. A., & Salavati-Niasari, M. (2021a). Recyclable magnetic $ZnCo_2O_4$-based ceramic nanostructure materials fabricated by simple sonochemical route for effective sunlight-driven photocatalytic degradation of organic pollution. *Ceramics International*, *47*, 8959–8972.

Zinatloo-Ajabshir, S., Heidari-Asil, S. A., & Salavati-Niasari, M. (2021b). Simple and eco-friendly synthesis of recoverable zinc cobalt oxide-based ceramic nanostructure as high-performance photocatalyst for enhanced photocatalytic removal of organic contamination under solar light. *Separation and Purification Technology*, *267*, 118667.

Zinatloo-Ajabshir, S., Morassaei, M. S., Amiri, O., & Salavati-Niasari, M. (2020). Green synthesis of dysprosium stannate nanoparticles using Ficus carica extract as photocatalyst for the degradation of organic pollutants under visible irradiation. *Ceramics International*, *46*, 6095–6107.

Zinatloo-Ajabshir, S., Mortazavi-Derazkola, S., & Salavati-Niasari, M. (2017). Schiff-base hydrothermal synthesis and characterization of Nd_2O_3 nanostructures for effective photocatalytic degradation of eriochrome black T dye as water contaminant. *Journal of Materials Science: Materials in Electronics*, *28*, 17849–17859.

Zinatloo-Ajabshir, S., Salavati-Niasari, M., Sobhani, A., & Zinatloo-Ajabshir, Z. (2018). Rare earth zirconate nanostructures: Recent development on preparation and photocatalytic applications. *Journal of Alloys and Compounds*, *767*, 1164–1185.

Rare earth titanate ceramic nanomaterials

Ali Sobhani-Nasab[1,2] *and Saeid Pourmasud*[3]
[1]Autoimmune Diseases Research Center, Kashan University of Medical Sciences, Kashan, Iran, [2]Core Research Lab, Kashan University of Medical Sciences, Kashan, Iran, [3]Department of Physics, University of Kashan, Kashan, Iran

6.1 General introduction

Pyrochlore oxides having the general formula of $A_2B_2O_6$ belong to the space group of Fd3m, containing lanthanide with a lone pair of electrons. The other important element called B is a posttransition metal. These kinds of structures are usually explained as in the following: A_2O tetrahedra and B_2O_6 octahedra as two networks that merge together with oxygens in two sublattices represented as O and O^-, respectively. Pyrochlore lattice is served for the incorporation of dopants, interstitial oxygen, and electronic defects by limited doping of A and B sites. In pyrochlore lattice, both A and B positions generate a network of corner-sharing tetrahedra. The geometry of such oxides will be frustrated provided that either A and B or both are magnetic, and the closest neighbor with interactional interchange has an antiferromagnetic (AF) property.

There are many reports on pyrochlores making use of Raman and infrared spectroscopy. These techniques can shed light on varying dynamic-based procedures, such as phonons, charge carriers, and spin. These also affect both the physical and chemical behaviors of pyrochlores. One can gain detailed knowledge of the distributed ions at various symmetry positions and deviated from site symmetry induced by either defections at points or local stress in compound. In perfect pyrochlore, $Ln_2Ti_2O_7$, Raman-active modes associate with vibrations of only O atoms, but apart from O atoms IR-active modes take into account of the vibrations of A and B ones (Kumar & Gupta, 2012). Pyrochlore oxides have been focused because of their extensive range of physics- and chemistry-based characteristics, such as piezoelectric (Cann, Randall, & Shrout, 1996), catalytic activity (Kumar et al., 2010), fuel cells (Hwang et al., 2003a; Mims et al., 1995), high permittivity dielectrics (Hwang et al., 2003a) and ferroelectric properties (Subramanian, Aravamudan, & Rao, 1983), gas sensors (Sobhani-Nasab et al., 2019a) flexible energy storage device (Pan et al., 2007; Shimakawa, Kubo, & Manako, 1996), magnetoresistance (Shimakawa et al., 1996), optical, and photocatalytic (Coles, Bond, & Williams, 1994), and as a host material (Ewing, Weber, & Lian, 2004). Rare earth titanates have, additionally, been thoroughly studied as a result of their attractive magnetic features (Gardner, Gingras, & Greedan, 2010) like spin liquid, spin ice,

Advanced Rare Earth-Based Ceramic Nanomaterials. DOI: https://doi.org/10.1016/B978-0-323-89957-4.00008-6

accompanied by spin glass behaviors. Pyrochlore oxides have been prepared via a number of procedures like hydrothermal synthesis, the sol–gel method (Kawashima et al., 2015), coprecipitation synthesis (Yan et al., 2009), and the conventional solid-state method (Feng et al., 2014).

6.1.1 Lanthanum titanates

The first lanthanum-based structure which is studied here is lanthanum titanate, $La_2Ti_2O_7$. It belongs to a member of the perovskite layer structure with the family of ferroelectrics, which is occasionally assigned to as the strontium pyroniobate family (Taş, 2001; Zhang et al., 2007a). The structure of such compounds characteristically is a perovskite slab stacked along an axis and could be formed by corner-sharing BO6 octahedra as well as a number of 12 coordinated A cations. Each slab contains four octahedra accompanied by joining an adjacent slab by A cations, which are placed at the vicinity of the boundary (Ishizawa et al., 1980, 1982). For more clarification, an excess of O_2 layer alongside the perovskite (Winkler et al., 2014) direction after every four (n) distorted perovskite units (Brandon & Megaw, 1970) is a specific feature of $A_2B_2O_7$ structures.

The anisotropic behavior in electrical, mechanical, and dielectric properties could be originated from the anisotropy of the structure (Ahmadi, Rahimi-Nasrabadi, & Behpour, 2017). Moreover, $La_2Ti_2O_7$ does not construct the expected isometric pyrochlore structure type. It is stable because of the proportion of the cations radii in lanthanum cation (La^{3+}) to titanium cation (Ti^{4+}). For $La^{(3+)}/Ti^{(4+)}$, it is in the span of (1.46) to (1.78); hence, the produced compound will choose to prioritize $La_2Ti_2O_7$. If the proportion is bigger than 1.78, the layered perovskite type is prioritized, whereas if the proportion is lower than 1.46, such structure will fall into a defect fluorite structure (Subramanian et al., 1983). If $La^{(3+)}/Ti^{(4+)} = 1.92$ reached utilizing the ionic radii from Shannon (Bartel et al., 2019), the crystallization of lanthanum titanate composition reported in this attempt will be in a monoclinic, space group of P21. The respective obtained unit cell boundaries in $La_2Ti_2O_7$ (a, b, and $c = 7.8114(2)$ Å, 5.5474(1) Å, and 13.0185(1) Å), the number of molecules per unit cell ($Z = 4$), as well as the theoretical density (ρ) are $\rho = 5.78$ g/cm^3 (Nanamatsu et al., 1974) (Table 6.1).

Ishizawa et al. (1982) have stated that TiO_6 octahedra network is asymmetrically deformed that could be originated from any modification in structure from monoclinic (P21) to orthorhombic phase (CmC21) once it heats above 1053 K in which polarization with spontaneity comes down the b axis (Brandon & Megaw, 1970; Ishizawa et al., 1982) when the temperature is between 1053 and 993 K, which is explained by spinning that takes place through the Ti^{4+} ions with no deformation in the morphology of the octahedral. This, in turn, results in a little alteration in the value of the dipole moment in each respective TiO_6 octahedron (Ishizawa et al., 1982). It should be noted that for temperatures higher than 1773 K, the structure goes through further transformation into the paraelectric phase (Cmcm) (Ishizawa et al., 1982).

Table 6.1 Selected precursors and methods for the preparation of lanthanum titanate nanostructures and their applications.

Number	Type of earth titanate	Synthesis method	Precursors	Property	Reference
1	$La_2Ti_2O_7$	Sol−gel route	$Ti(OBu)_4$ and La_2O_3	Battery	Henao et al. (2019)
2	$La_2Ti_2O_7$ and $LaTiO_3$	Sol−gel	TiO_2 and La_2O_3	—	Herrera, Jiménez-Mier and Chavira (2014)
3	Ln_2TiO_5	Solid state	TiO_2 and Ln_2O_3	Ionic conductivity	Zhang et al. (2013)
4	$La_2Ti_2O_7$	Hydrothermal	TiO_2 and $La(NO_3)_3 \cdot 6H_2O$	—	Yang, Gordon, and Chan (2013)
5	$La_2Ti_2O_7$	Electrospinning combined with calcination method	Titanium tetraisopropanolate and $La(NO_3)_3 \cdot 6H_2O$	Flexible energy storage device	Cao et al. (2020)
6	$La_2Ti_2O_7$	Microemulsion method	$TiCl_4$ and $La(NO_3)_30.6H_2O$	Antioxidant	Akram et al. (2020)
7	$La_2Ti_2O_7$	Molten salt method	TiO_2 and La_2O_3	Photocatalytic	Huang et al. (2020)
8	$La_2Ti_2O_7$	Sol−gel autocombustion method	Lanthanum nitrate and tetra-*n*-butyl titanate	Photocatalytic	Talebi and Safazade (2016)
9	Ln_2TiO_5	Ball milling	La_2O_3 and TiO_2	—	Aughterson et al. (2015)
10	$La_2Ti_2O_7$	Liquid-feed flame spray pyrolysis	Lanthanum isobutyrate and triethanolamine titanate	Photocatalytic	Abe and Laine (2020)
11	$La_2Ti_2O_7$	Sol−gel method	Lanthanum nitrate and tetrabutyl titanate	Photocatalytic	Wang et al. (2016)
12	$La_2Ti_2O_7$	Sol−gel route	Lanthanum (III) nitrate hexahydrated and titanium (IV) *n*-butoxide	Photocatalytic and sonophotocatalytic	Leroy et al. (2020)
13	$La_2Ti_2O_7$	Microwave method	$La(NO_3)_3 \cdot 6H_2O$ and $TiCl_4$	Photocatalysis	Zhou et al. (2016)
14	$La_2Ti_2O_7$	Sol−gel	$C1_6H_{36}O_4Ti$ and $La(NO_3)_30.6H_2O$	Optical	Li et al. (2020b)
15	$La_2Ti_2O_7$	Sol−gel	Tetrabutyl titanate and lanthanum nitrate	Photocatalytic	Wang et al. (2019)
16	$La_2Ti_2O_7$	Ball milling	La_2O_3 and TiO_2	—	Kambale et al. (2019)
17	$La_2Ti_2O_7$	Sol−gel method	Tetrabutyl titanate and lanthanum nitrate	Photocatalytic	Zhang, W et al. (2019)

Lanthanum titanate has been previously fabricated through varying routes such as sol–gel (Ahmadi et al., 2017), coprecipitation (Shcherbakova, Mamsurova, & Sukhanova, 1979a; Takahashi & Ohtsuka, 1989), solid-state method (Taş, 2001), hydrothermal (Li et al., 2006; Song et al., 2007), urea precipitation (Suresh, Prasadarao, & Komarneni, 2001), thermal decomposition of metal–organic (Kim et al., 2001, 2003) and nitrates precursors (Shcherbakova, Mamsurova, & Sukhanova, 1979b), and mix wet method (Swami & Sreenivas, 2019). Such fabrications have produced tinier dimension in particles in comparison to the solid-state route. They also generally need long-lasting heating times as well as synthetic routes. Recently, it has additionally been prepared with the morphology of thin films utilizing molecular beam epitaxy (Fompeyrine, Seo, & Locquet, 1999; Seo, Fompeyrine, & Locquet, 1998), pulsed laser deposition (Havelia et al., 2008; Kushkov et al., 1993) along with laser-heated pedestal growth (Yamamoto & Bhalla, 1991b). $La_2Ti_2O_7$ with a layered perovskite structure has been thoroughly considered on account of its specific features, including dielectric (Lu et al., 2013), ferroelectric (Ishizawa et al., 1982), photocatalytic, and optical properties (Arney, Porter, Greve, & Maggard, 2008; Kim, 2002a).

Owing to attractive chemical and physical characteristics of such compounds, they have been used in many different fields by scholars as follows. In the last case, lanthanum titanate has been extensively used to exhibit high enough photocatalytic performance for CO_2 conversion (Arney, Porter, Greve, & Maggard, 2008), organic pollutant degradation (Hwang et al., 2003b; Kim et al., 2002a), and water splitting (Abe, Higashi, Sayama, Abe, & Sugihara, 2006; Arney, Porter, Greve, & Maggard, 2008). The specific and rational arrangements of nanostructures with distinct exposed facets have become important not only for the improvement in the physicochemical characteristics of these materials but also for the study of relationship between structure and activity. Huang et al. (2020) describe the synthesis of dandelion-like lanthanum titanate through the molten salt route. The minimum and maximum necessitated temperature for preparation is 700°C and 900°C, respectively (Fig. 6.1), at least 200°C lower than that required by other methods. The dandelion structure is composed of well-crystallized lanthanum titanate nanorods whose sizes are below of 300–500 nm in the axial and 100 nm in the radial direction. This is different from that commonly believed in two-dimensional photocatalytic activities of rhodamine B to degrade over the lanthanum titanate nanoparticles and is a unanimous agreement on the plate-like nanostructure. A surface heterojunction formed in the lanthanum titanate, thus, notably improves the photocatalytic behavior (Huang et al., 2020). As the need for sound and effective technologies concerning batteries has initiated the development of new type of materials, these are focused on the development of suitable materials for battery technology for adequate hydrogen storage under different operational conditions.

For more clarification, Henao et al. (2019) prepared lanthanum titanate nanomaterials to be applied in rechargeable batteries. Lanthanum titanate nanostructures were fabricated at 850°C for 5 hours by the commonly used sol–gel method. The lanthanum titanate nanomaterials had an asymmetrical morphology with a particle

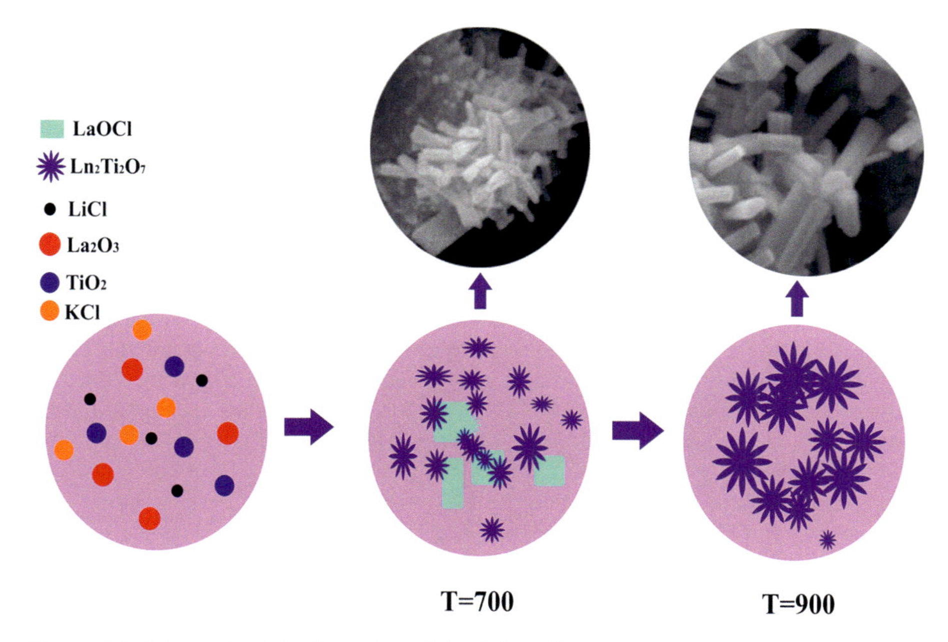

Figure 6.1 Schematic of the formation of dandelion-like lanthanum titanate nanostructures at 700°C and 900°C.

size around 100 nm, and a proclivity to make clusters. $La_2Ti_2O_7$ had a layered structure and demonstrated a highest electrochemical hydrogen uptake volume of 224 mAh/g assigned to 0.84 wt.% of hydrogen. Electrochemical reports confirmed that the lanthanum titanate structure could act as an alternative electrode material for secondary batteries because of its good cycling stability after activation. Some nanostructures are reported to be poisonous despite the fact that they are being used in medicine- and industry-based applications. Biocompatible features of these tiny materials should be evaluated before their use because of their toxic effects on the vital organs of human body as well as the possibility of deposition in our ecosystem. Any exposure to nanoparticles containing applications can increase the probability of dangerous for intake of these nanoparticles once they are able to infiltrate into the living systems through injection, inhalation, ingestion, along with dermal penetration either immediately or indirectly. There is no extensive and available report in literature regarding the biocompatibility of lanthanum titanate nanoparticles. Thus this effort has emphasized to discover the influence of lanthanum titanate nanoparticle supplementation for varying periods on blood biochemistry, behavior, together with antioxidants from main organs of albino mice in a sex-specific manner. To do so, Akram et al. (2020) synthesized and characterized biocompatible $La_2Ti_2O_7$ nanoparticles in albino mice. They made use of microemulsion method to produce lanthanum titanate nanoparticles. Albino mice with 7 weeks of age from both genders were fed 50 mg/mL saline/kg body weight of nanostructure for group

1 (15 days) and group 2 (29 days). Control groups were kept in parallel. Selected behavioral [Morris water maze, light and dark box (LDB), rotarod and open-field] experiments were performed; blood biochemical analysis was carried out; and antioxidants were evaluated in vital organs of all treatments. Males brought with remedy in lanthanum titanate nanoparticles for 15 days were remarkably subjected to more duration in light and, vice versa, less time in dark during apparatus, LDB test. While they had made considerably more platform entries and platform maximum visits in the course of acquisition phase of Morris water maze test, they were not affected in probe trail performance compared to other group.

The monocyte count, white blood cells, plus lymphocytes of these mice substantially grew, whereas their triglyceride quantity in serum decreased to a large extent. Also, the magnitude of superoxide dismutase (SOD) in heart increased while the intensity of malonaldehyde (MDA) in kidney decreased. In the meanwhile, apart from kidney, female mice treated with 15-day $Li_2Ti_2O_7$ nanoparticles had meaningfully higher standing of SOD in liver. Furthermore, males tested with lanthanum titanate nanoparticles for 29 days had better anticlockwise turning throughout open field, substantially higher standing of SOD in kidney with MDA in lung, as well as lessened extent of triglycerides in serum.

In return, females treated with nanoparticles for nearly 1 month showed higher extent of SOD in heart, apart from liver, kidney, than their control group. Oral supplementation of lanthanum titanate nanoparticles for changeable time span enriched the investigative behavior in males, and blood chemistry accompanied by antioxidants from vital organs in the process of both trial conditions spread.

6.1.2 Cerium titanates

The next titanium-based compound, which is considered in this study, is CeO_2-TiO_2, ceria–titania. The thin films containing such structures have received much attention in the past years because of their utmost importance in some applications, such as antireflective coatings (Lev et al., 1997), electrochromic layers (Ghodsi, Tepehan, & Tepehan, 1999; Setién-Fernández et al., 2013), and self-cleaning glasses (Zhao et al., 2008). Not long ago it was confirmed that CeO_2-TiO_2 films can be fabricated through process which should controlled from Ce to Ti films. Moreover, they are now pleasing as a demanding and encouraging materials thanks to their improved or new properties relating to the CeO_2-TiO_2 mixed oxides (Otsuka-Yao-Matsuo, Omata, & Yoshimura, 2004). Also they could be produced in different phases which rely on the oxidation state of cerium. In some phases, there are Ce^{3+} ions such as $Ce_4Ti_9O_{24}$, $Ce_2Ti_2O_7$, and Ce_2TiO_5, while in other phases, such as $CeTi_2O_6$ and $CeTiO_4$, Ce is in the oxidation state Ce^{4+}. The likelihood of obtaining one of these phases is dependent on the key factor of atmosphere of annealing, generally a reduction of atmosphere (Martos et al., 2008; Verma, Goyal, & Sharma, 2008) or low vacuum (Verma et al., 2008) allowing to obtain phases with Ce^{3+}, while firing in air typically results in cerium titanate phases with Ce^{4+} ions (Yoshida et al., 2007).

Kidchob et al. (2009) used a combination of far-infrared spectroscopy and X-ray diffraction (XRD) analysis to characterize cerium titanate films which has done through optimization of annealing temperature and composition. To prepare films, metal chlorides and precursors were used in a sol−gel mechanism. This provided obtaining $CeTi_2O_6$ films upon annealing in air. Annealing at 700°C and in a restricted span of ceria−titania mixed compositions are two prerequisite conditions for the formation of $CeTi_2O_6$. Once the XRD analysis is not practically useful, the far-infrared spectra could be served to monitor the construction of crystalline phases at the early stage of the current process at lower firing temperatures (Kidchob et al., 2009). On the one hand, $CeTi_2O_6$ is a titanate-based photocatalyst material that has an insubstantial bandgap value of 2.7 eV and has been shown to demonstrate practical applications in the field of semiconductor photocatalyst, abrasives together with ionic conductors. More importantly, the conduction band level of $CeTi_2O_6$ is less negative than that of other semiconductors. This can make straightforward heterojunction formation with the materials (Huynh et al., 2012; Kim et al., 2008). Since the noticeable abundance of Ce^{4+} ions as well as Ce^{3+} states, $CeTi_2O_6$ is fundamentally differing from both titanium dioxide and cerium oxide structures. This is why it can exhibit acceptable photocatalytic reaction in visible spectrum. Nevertheless, owing the property of excitons already mentioned, the photocatalytic performance of pure $CeTi_2O_6$ is markedly hampered. Hence, the fusion of any semiconductor material or metal doping with proper amount of energy is important to prepare the $CeTi_2O_6$ heterojunction with maximum amount of permanence once they are subjected to the visible spectrum and near-infrared region (Otsuka-Yao-Matsuo et al., 2004; Tu, Zhou, & Zou, 2014; Vadivel et al., 2020) (Table 6.2).

6.1.3 Praseodymium titanates

Of monoclinic lanthanide titanates, two compounds, $Ce_2Ti_2O_7$ and $Pr_2Ti_2O_2$, have extraordinary property due to inherent defects, which could be originated from the variable valence states of elements, Ce and Pr. In recent times, Atuchin et al. (2012) have studied spectroscopic and electronic behaviors of praseodymium titanate. The verification of the dielectric and optical characteristics of praseodymium titanate indicates high functionality in nonlinear optics and a dielectric constant (Wenger et al., 2008). Reports also state that the ferroelectric transition of praseodymium titanate is dependent on the nature of sample preparation. The ferroelectric transition at nearly 740°C is the behavior which has been reported in the spark plasma-sintered sample of praseodymium titanate (Gao et al., 2013).

Sun et al. (2013) have studied coupled magnetism and ferroelectric properties in nanocrystalline praseodymium titanate, where the magnetism is linked with the oxygen vacancies whereas the ferroelectricity has been associated with the structural features. They studied the frequency and temperature-dependent dielectric and proved an anomaly near the proposed ferroelectric transition temperature (570 K). Also it is claimed that Tc could come down for the crystallite size effects. The minimum temperature investigated for each monoclinic $Ln_2Ti_2O_7$ phases is measured for described transition temperature in praseodymium titanate.

Table 6.2 Selected precursors and methods for the preparation of cerium titanate nanostructures and their applications.

Number	Type of earth titanate	Synthesis method	Precursors	Property	Reference
1	$Ce_2Ti_2O_7$	Mixed oxide route	CeO_2 and TiO_2	Ferroelectric	Gao et al. (2015)
2	$CeTi_2O_6$	Sol−gel	Cerium (III) nitrate hexahydrate and titanium (IV) isopropoxide	Optical	Valeš et al. (2014)
3	$Ce_2Ti_2O_7$	Sol−gel associated with spin-coating	Titanium isopropoxide	Electrical	Bayart et al. (2016)
4	$Ce_2Ti_2O_7$	Chemical solution deposition process	Titanium isopropoxide and cerium acetylacetonate	Ferroelectric	Kim et al. (2008)
5	$Ce_2Ti_2O_7$	Deposited by thermal evaporator	Cerium oxide and titanium target	Flash memory devices	Kao et al. (2017)

So the structure has been studied by Koz'min (1997). At normal temperature, praseodymium titanate has a monoclinic structure with space group of P21. The lattice constants are ($a = 12.996$, $b = 7.704$, $c = 5.485$ Å). A survey for the second-harmonic generation in praseodymium titanate was indeterminate because its non-centrosymmetric structure has not been clarified so far. Using differential thermal analysis, a maximum point related to the phase transition was observed at 1770°C. Nevertheless, this did not suggest direct evidence that such maximum point corresponds to the ferroelectric Tc, because there was no report to prove if praseodymium titanate is ferroelectric. On the other hand, the application of such nanoparticles as photocatalysts has been reported in many literatures. For example, Long et al. (2017) synthesized praseodymium titanate by sol−gel process making use of praseodymium nitrate, citric acid, ethylene glycol, and tetrabutyl titanate as main raw materials. The influence of varying illumination time and calcination temperature on the photocatalytic property of praseodymium titanate was studied. It was understood that the single-phase praseodymium titanate may be reached by sol−gel procedure. The praseodymium titanate calcination at 1000°C was uniform. The final product had dimension of 200 nm. For ultraviolet (UV) irradiation, the decomposition of methyl orange reached 80.11% for 180 minutes of photocatalytic reaction. The praseodymium titanate samples demonstrated a maximum amount of photocatalytic behavior for the degradation of methyl orange (Long et al., 2017) (Table 6.3).

Table 6.3 Selected precursors and methods for the preparation of praseodymium titanate nanostructures and their applications.

Number	Type of earth titanate	Synthesis method	Precursors	Property	Reference
1	$Pr_2Ti_2O_7$	Sol−gel method	Tetra-*n*-butyl titanate and Pr_2O_3	Magnetoelectric	Sun et al. (2013)
2	$Ln_2Ti_2O_7$ (Ln = La, Pr, Sm, and Y)	Coprecipitation	$La(NO_3)_3 \cdot nH_2O$, Pr_6O_{11}, $Sm_2(CO_3)_3$, Y $(NO_3)_3 \cdot 6H_2O$, and $C_{16}H_{36}O_4Ti$	Catalyst	Fang et al. (2019)
3	$Pr_2Ti_2O_7$	Solid state	Pr_6O_{11} and TiO_2	Ferroelectric	Atuchin et al. (2012)
4	$Pr_2Ti_2O_7$	Solid state	Pr_6O_{11} and TiO_2	Electrical	Patwe et al. (2015)
5	$Pr_2Ti_2O_7$	Spark plasma sintering	Pr_6O_{11} and TiO_2	Piezoelectric	Gao et al. (2013)
6	$Pr_2Ti_2O_7$	Solid state	Purity Pr_6O_{11} and TiO_2	Temperature-dependent Raman spectra	Saha et al. (2011)
7	$Pr_2Ti_2O_7$	Solid state	Pr_6O_{11} and TiO_2	—	Kesari et al. (2016)

6.1.4 Neodymium titanates

Neodymium titanate with the chemical formula of $Nd_2Ti_2O_7$ is a member of small category of rare earth titanates (Ghodsi et al., 1999). It has ferroelectric properties at normal temperature; $Nd_2Ti_2O_7$ is monoclinic with space group of P21 (Scheunemann & Müller-Buschbaum, 1975). There are eight positions for titanium atoms which can occupy, and Ti–O intervals which are in the span of 182–223 pm. Also, Nd atoms are placed in eight sites, all coordinated by 12 oxygens with Nd–O spaces in the limit of 240–385 pm. Since both Nd^{3+} and Ti^{4+} can occupy eight various places and there are substantial changes in distance of Ti–O and Nd–O in the crystal lattice, neodymium titanate is able to have a highly capacity for solid solution formation with possible substitution in Ti or Nd places. Concerning physical properties, neodymium titanate is characterized by an extremely high Curie temperature $Tc > 1500°C$, spontaneous polarization $Ps = 9$ lC/cm^2, great permittivity $e = 31-47$, and high piezoelectric properties (Atuchin et al., 2008; Winfield, Azough, & Freer, 1992; Yamamoto & Bhalla, 1991a, 1991b). This is why such compound could be served for the manufacturing of ferroelectric random access memory elements (Kim et al., 2002b; Lee et al., 2005). Being thermally stable up to $T = 1400°C$ and low dielectric lost can pave the way for an extensive application of neodymium titanate as a main factor to fabricate microwave ceramics and nanocomposites (Atuchin et al., 2008; Prasadarao et al., 1991). Furthermore, outstanding refractive indices $n_a = 2.15$, $n_b = 2.27$, and $n_c = 2.23$ ($k = 526.5$ nm) and a fairy good level of nonlinear optical susceptibility were evaluated for neodymium titanate (Stefanovich, Malhasyan, & Venevtsev, 1980; Zakharov et al., 1978). Also, NTO has multiferroic feature and a magnetochiral effect has been seen in this structure, short time ago (Kidyarov and Atuchin, 2007; Shimada, Kiyama, & Tokura, 2008; Zakharov et al., 1978).

Ahmadi et al. (2017) for the first time prepared Nd_2TiO_5 nanostructures via sol–gel route with alanine, leucine, and histidine as new template. A simple approach for Nd_2TiO_5 nanoparticles was proposed by using natural template permitting the reaction to proceed usually in milder conditions. Even though the present chemistry-based strategies have successfully fabricated clearly defined Nd_2TiO_5 nanostructures, they are not generally cost-effective and involve the use of toxic and risky chemicals. Furthermore, Nd_2TiO_5 nanostructures are using as photocatalytics to destruct methyl orange in the process of UV irradiation (Ahmadi et al., 2017; Sobhani-Nasab et al., 2015b) (Table 6.4).

6.1.5 Samarium titanates

Samarium titanate ($Sm_2Ti_2O_7$) is being used as an optical and electronic material (Kawashima et al., 2015; Yan et al., 2009). Notwithstanding, it is usually fabricated by sol–gel and solid-state method. Samarium titanate has usually lack porous structure and is not believed to be an encouraging photocatalyst owing to two key factors: not having an acceptable amount of adsorption capacity and insufficient quantum efficiency. Many studies have been reported in relation to the high-pressure behavior of $A_2B_2O_7$ pyrochlores, but generally these reports have been restricted to a

Table 6.4 Selected precursors and methods for the preparation of neodymium titanate nanostructures and their applications.

Number	Type of earth titanate	Synthesis method	Precursors	Property	Reference
1	$Nd_2Ti_2O_7$	Solid state	TiO_2 and Nd_2O_3	Electronic	Atuchin et al. (2008)
2	$Nd_2Ti_2O_7$	Coprecipitation method	$K_2[TiO(C_2O_4)_2]_2H_2O$ (KTO) and $TiCl_4$	—	Suresh et al. (2001)
3	$Nd_2Ti_2O_7$	Polymerized complex method	$Ti[OCH(CH_3)_2]_4$ and Pr^{3+}	—	Milanova et al. (1996)
4	$Nd_2Ti_2O_7$	Sputtering	Nd_2O_3 and TiO_2	Biosensors	Pan et al. (2009)
5	$Nd_2Ti_2O_7$	Self-propagating high-temperature synthesis	Nd_2O_3 and TiO_2	Photoluminescence	Joseph et al. (2008)

qualitative interpretation of Raman spectra or XRD (Kumar, Shekar, & Sahu, 2008; Scott et al., 2011; Zhang et al., 2007b). Particularly, the high-pressure behavior of samarium titanate has been studied by Zhang (Zhang and Saxena, 2005; Zhang et al., 2005). In these two reviews, a single set of XRD evaluation was interpreted to demonstrate partial amorphization at 51 GPa, and from complementary Raman measurements, it was concluded that higher than 34 GPa anion ordering can happen. Hence, authors found that at 40 GPa, samarium titanate crystallizes in a distorted pyrochlore structure. Zhang, Haolun, and Ling (2018) prepared samarium titanate using sol–gel method. They evaluated photocatalytic degradation of Reactive Brilliant Red X3B (RBR X3B), to measure the influence of polyethylene glycol as template in sol–gel on photocatalytic activity of samarium titanate. The PEG4000-modified samarium titanate was analyzed using scanning electron microscopy and XRD techniques. The aim of this effort was to clarify the effects of polyethylene glycol on both physical and chemical properties of the samarium titanate and encourage the functionality of this material (Zhang et al., 2018) (Table 6.5).

6.1.6 Europium titanates

The introduction of lanthanide ions as centers of luminescence in an extensive range of host lattices could be considered as a development in the luminescence materials (Yen and Weber, 2004). However, the ideal pyrochlore crystals, for example, europium titanate have no indication on strong light emission. This could be referred to the point that Eu^{3+} ions are regularly located at the centrosymmetric positions in the cubic pyrochlore lattice of $Eu_2Ti_2O_7$. Since positions of Eu^{3+} have centrosymmetric property, all the electric dipole transitions are not allowed, whereas the magnetic dipole ones may be supposed in this pyrochlore phosphor. Furthermore, europium titanate is a stoichiometric phosphor, in which all (100%) the host constituent cations (Eu^{3+}) are replaced with the activator ions. This could lead to an appearance of noticeable concentration quenching of the Eu^{3+} emission intensity.

The photoluminescence (PL) behaviors of pyrochlore europium titanate have been reported by many authors (Faucher and Caro, 1975; Garbout, Kallel-Kchaou, & Férid, 2016). As far as we know, however, the data on detailed experimental data of its PL excitation properties and temperature dependence of the PL properties are available. The Eu_2TiO_5 crystallizes in the orthorhombic structure. Of many kinds of properties, up to the present time, only the crystallographic and thermal features of this material have been investigated (Kandan et al., 2013; Syamala et al., 2008).

Mrázek et al. inquired into the fabrication of nanocrystalline europium titanate with sol–gel method. The particle size of the formed europium titanates was to be between 20 and 100 nm (Mrázek et al., 2014) (Table 6.6).

6.1.7 Gadolinium titanates

From a technological point of view, $Gd_2Ti_2O_7$ is a challenged material that shows greatly attractive physical, structural, chemical, and properties for different applications. It could be used in high-temperature thermal barrier coatings, catalysts, and

Table 6.5 Selected precursors and methods for the preparation of samarium titanate nanostructures and their applications.

Number	Type of earth titanate	Synthesis method	Precursors	Property	Reference
1	$Sm_2Ti_2O_7$	Solid state	TiO_2 and Sm_2O_3	—	Rabanal et al. (1999)
2	$Sm_2Ti_2O_7$	Modified Pechini	$Sm(NO_3)_3 \cdot 6H_2O$ and titanium (IV) isopropoxide	Electrical conductivity	Cioatera et al. (2019)
3	Sm_2TiO_5	Spin-coating a layer of photoresist	Sm_2TiO_5	Sensor	Wu et al. (2010)
4	$Sm_2Ti_2O_7$	Optical floating zone method	TiO_2 and Sm_2O_3	—	Winkler et al. (2014)
5	Sm_2TiO_5	Deposition	Samarium and titanium metal	Biosensor	Pan et al. (2010a)
6	Sm_2TiO_5	Solid state	TiO_2 and Sm_2O_3	Heat capacity and thermal expansion	Panneerselvam et al. (2011)

Table 6.6 Selected precursors and methods for the preparation of europium titanate nanostructures and their applications.

Number	Type of earth titanate	Synthesis method	Precursors	Property	Reference
1	$Eu_2Ti_2O_7$	Sol−gel	$Eu(NO_3)_3$,$6H_2O$ $Ti(OC_4H_9)_4$	Electronic	Popov et al. (2020)
2	$Eu_2Ti_2O_7$	Sol−gel	Europium salt, and tetra-*n*-butyl titanate	Photocatalyst	Sobhani-Nasab and Behpour (2016)
3	$Eu_2Ti_2O_7$	High-temperature solid	Eu_2O_3 and TiO_2	High-temperature piezoelectric sensing	Gao et al. (2018)
4	$Eu_2Ti_2O_7$ and Eu_2TiO_5	Solid-state reaction	Eu_2O_3 and TiO_2	Photoluminescence	Orihashi, Nakamura, and Adachi (2016)
5	$Eu_2Ti_2O_7$	Sputtering	Europium and titanium	Biosensor	Pan et al. (2011)
6	$Eu_2Ti_2O_7$	Sol−gel	Eu_2O_3 and TiO_2	Optical	Mrazek et al. (2018)
7	$Eu_2Ti_2O_7$	Solid-state reaction	Eu_2O_3 and TiO_2	—	Pal et al. (2018)
8	$EuTiO_3$ and $Eu_2Ti_2O_7$	Ball milling	Eu_2O_3 and TiO_2		Schell et al. (2019)
9	$Eu_2Ti_2O_7$	Sol−gel	Titanium(IV)butoxide and europium(III) nitrate pentahydrate	Optical	Mrázek et al. (2019)
10	Eu_2TiO_5	Solid-state and wet chemical methods	Europium oxide and titanium sponge	Thermal expansion	Syamala et al. (2008)

fast ion conductors. $Gd_2Ti_2O_7$ is also a potential waste forms to immobilize actinides in a manner that all actinides can be taken in Gd sublattice in the pyrochlore structure and was therefore proposed as a model system for the current attempt (Padture, Gell, & Jordan, 2002; Shamblin et al., 2016; Wuensch et al., 2000). Gadolinium titanate has been synthesized by various procedures including solid-state reaction (Maczka et al., 2008; Perera et al., 2002), sol−gel (Pang et al., 2004; Zhang et al., 2009), single crystal growth (Petrenko et al., 2012), Pechini route (Matteucci et al., 2007), calcinations and chemical coprecipitation method (Liu, Ouyang, & Sun, 2011).

For example, Dharuman and Berchmans (2013) fabricated nanocrystalline gadolinium titanate powder via one-step molten salt method. Gadolinium titanate was prepared by X-ray photoelectron spectroscopy (XPS), Fourier-transform infrared spectroscopy (FT-IR), energy-dispersive X-ray spectroscopy (EDAX), UV, and XRD analyses (Dharuman and Berchmans, 2013). Li et al. (2020a) prepared gadolinium titanate using polyethylene glycol with a sol−gel process. The influence of polyethylene glycol on the features of $Gd_2Ti_2O_7$ photocatalysts was investigated. Polyethylene glycol could prevent crystal growth of pyrochlore-structured gadolinium titanate in the products. By adding polyethylene glycol, absorption age underwent a blue shift for porous gadolinium titanate samples, whose bandgap energies were from 3.48 to 3.75 eV. The surface area and pore volume of the porous gadolinium titanates were simply increased from 5.08 m^2/g and 0.0024 cm^3/g to 21.55 m^2/g and 0.0090 cm^3/g, respectively. Both adsorption capacity and photocatalytic performance of the gadolinium titanates are depend on the surface area of the materials. The photocatalytic property of the gadolinium titanate samples was in a manner confirming with the produced hydroxyl radicals. All the porous gadolinium titanates were highly functional on RBR X3B degradation, and the gadolinium titanate (1.5) sample had the highest photocatalytic property. The in respective constants of reaction rate for gadolinium titanate of 0, 1.5, and 4.0 were 0.00347, 0.03055, and 0.01656/minute, respectively. Also, the destruction of RBR X3B molecule was implemented via photocatalytic oxidation process which was confirmed via any reduction in absorption intensities of the organic groups (Li et al., 2020a) (Table 6.7).

6.1.8 Terbium titanates

In recent years, the Tb_2TiO_5 and $Tb_2Ti_2O_7$ have been well focused by engineers and scholars, as it is one of the primary ceramic phases considered to immobilize surplus weapon and high-level wastes (Zhang et al., 2013). Also, as absorption cross section for thermal neutrons is large enough for terbium titanate ceramics, they could be considered as a good candidate to keep in check rod materials used in the nuclear facilities (Niu et al., 2012). Aughterson, Zaluzec, and Lumpkin (2021) conducted an investigation into the critical amorphization temperature of Ln_2TiO_5 systematically. He realized that terbium titanate can be simply amorphous. The space group of this composition is associated with Pnam, No. 62 containing as many as 32 atoms per unit cell. On one hand, those Ti cations which are smaller than others are five-coordinated and formed approximately square pyramidal shape.

Table 6.7 Selected precursors and methods for the preparation of gadolinium titanate nanostructures and their applications.

Number	Type of earth titanate	Synthesis method	Precursors	Property	Reference
1	$Gd_2Ti_2O_7$	Solid state	TiO_2 and Gd_2O_3	—	Jafar et al. (2021)
2	$Gd_2Ti_2O_7$	Sol−gel	Gd_2O_3 and tetrabutyl titanate	Luminescence and optical	Liao et al. (2018)
3	$Gd_2Ti_2O_7$	Solid state	TiO_2 and Gd_2O_3		Taylor et al. (2016)
4	$Gd_2Ti_2O_7$	Sol−gel	$Gd(NO_3)_3 \cdot 6H_2O$ and tetrabutyl titanate	Photocatalytic	Zhang, Liu, and Li (2018)
5	$Gd_2Ti_2O_7$	Single-step molten salt technique	TiO_2 and Gd_2O_3		Dharuman and Berchmans (2013)
6	$Gd_2Ti_2O_7$	High-energy ball milling	TiO_2 and Gd_2O_3		Kulriya et al. (2017)
	$Gd_2Ti_2O_7$	Modified single-step autoignition combustion	TiO_2 and Gd_2O_3	Optical and dielectric	Jeyasingh, Saji, and Wariar (2017)

The adjacent polyhedral are shared at the corner of structure, making chains parallel to *c* axis. On the other hand, Tb cations with bigger dimensions have sevenfold coordination, whereas coordination number takes the value of eight for rare earth. Through conventional solid reaction methods, Zhang et al. (2019) fabricated complex oxide pellets with high purity of TiO_2 and Tb_2O_3 powders. Polycrystalline product terbium titanate pellets with orthorhombic structure were illuminated with 400 keV He + and 800 keV Kr2 + ions at normal temperature (Zhang et al., 2019) (Table 6.8).

6.1.9 Dysprosium titanates

Dysprosium titanates have some specific properties, especially, great magnetocaloric as well as spin-ice properties. It takes into account as a favorable material for sensor electronic and photonic utilization (Chen and Pan, 2012; Lau et al., 2006; Mrázek et al., 2015; Pan and Lin, 2010; Scharffe et al., 2015; Su et al., 2012). This structure has two phases, namely, Dy_2TiO_5 and $Dy_2Ti_2O_7$. The latter has the pyrochlore structure and is famous for its potential to incorporate actinides as well as for its consideral amount of refusal to accept damage once it is subjected to particles with a great amount of energy (Lang et al., 2010; Lumpkin et al., 2007; Xiao et al., 2007). It has an appealing control rod composition which could be used for thermal neutron nuclear reactors including pressurized water reactor and boiling water reactor. The major benefits of such structure are as follows: no realeasing of gas at the time of neutron exposure, high radiation resistance and neutron efficiency, almost nonswelling, a high melting point (1870°C), having no interaction with the cladding when it

Table 6.8 Selected precursors and methods for the preparation of terbium titanate nanostructures and their applications.

Number	Type of earth titanate	Synthesis method	Precursors	Property	Reference
1	$Tb_2Ti_2O_7$	Ball milling	Tb_4O_7 and TiO_2	High-temperature magnetic	Kumar and Venkateswaran (2019)
2	$Tb_2Ti_2O_7$	Firing in air with intermittent grinding	TiO_2 and Tb_2O_3	Heat capacity	Cornelius et al. (2005)
3	$Tb_2Ti_2O_7$	High-energy ball milling	Tb_4O_7 and TiO_2	—	Kumar et al. (2020)
4	$Tb_2Ti_2O_7$	Czochralski method	Tb_4O_7 and TiO_2	Magneto-optical properties	Klimm et al. (2017)
5	$Tb_2Ti_2O_7$	Solid-state reaction	Tb_4O_7 and TiO_2	—	Han, Gardner, and Booth (2004)

is heated higher than 1000°C, relatively straightforward preparation, nonradioactive waste, and effortlessness of reprocessing (Galahom, 2016).

Eremeeva et al. (2018) suggested a generic sol−gel route to be able to fabricate nanocrystalline $Dy_2Ti_2O_7$. Therefore they managed to prepare powder and highly transparent thin films containing nanocrystals with size of interest. The thermal evolution of these nanocrystals was accompanied by orthodox structural course of action. Furthermore both kinetic factors of the nanocrystal nucleation process from amorphous xerogel and crystallization mechanism were clarified (Table 6.9).

6.1.10 Holmium titanate

Holmium titanate is a structure with the potential to split unsafe contaminating through photocatalytic procedure (Williams, 1926). Its dielectric permittivity may range from 30 to 100 (Ahn and Forbes, 2007). They have been considered to have a composition which is proposed to be able to be tailored in sensoric applications. So that it can evaluate the level of pH along with the detection of biomaterials (Pan et al., 2010b). Another ternary holmium titanate−based structure is a preferable candidate to be used in (Macalik et al., 2009; Panitz et al., 2000) spectroscopy, in particular, Raman spectroscopy. Holmium titanate has also demonstrated multiferroic feature, meaning that its both ferromagnetic and ferroelectric aspects should be taken into account (Dong et al., 2009). Additionally, from magnetic material point of view, some have paid attention to holmium titanates, in which at enough low temperatures, the magnetic moments of ions continue to exist disordered. So, it is reasonable that they may be contemplated as an array of nanoscale magnets (Bramwell, 2006), which would be constant at normal temperature and show magnetic moments that are large to be sufficient for immediate monitoring. The holmium titanate has thus been called as "spin ice" in several studies (Shastry, 2003), in addition to its similar compound, $Dy_2Ti_2O_7$. Holmium titanium oxides have been scarcely discussed, and it is different from methodological point of view. The diversity, from methodological point of view, of this compound has caused to be scarcely discussed, and little information would be available in the literature. It is synthesized in the shape of single crystals by a coating zone method (Li et al., 2013), and as a polycrystalline material through solid-state reactions (Dong et al., 2009). Such compounds have been prepared, in advanced, employing sol−gel mechanism from tetra-*n*-butyl titanate along with holmium metal salt (Shi, Zheng, & Wu, 2009). Also, one may find an information about its synthesis by atomic layer deposition applying holmium beta-diketonate in the role of a precursor (Ahn and Forbes, 2007). Besides, through sputtering from Ho and Ti targets assisted by oxygen, holmium titanates could be deposited in the shape of thin films (Pan et al., 2010b) (Table 6.10).

6.1.11 Erbium titanates

Erbium titanates, with chemical formula of $Er_2Ti_2O_7$ and Er_2TiO_5, have been interest of study since 10 years ago (Gardner et al., 2010). $Er_2Ti_2O_7$ with pyrochlore

Table 6.9 Selected precursors and methods for the preparation of dysprosium titanate nanostructures and their applications.

Number	Type of earth titanate	Synthesis method	Precursors	Property	Reference
1	$Dy_2Ti_2O_7$	Combustion	Dy_2O_3 and TiO_2	Dielectric	Jeyasingh et al. (2018)
2	Dy_2TiO_5	Cosputtering	Dy and Ti from a pure	—	Pan and Lu (2011)
3	Dy_2TiO_5	Mechanochemical synthesis	Dysprosium oxide and anatase	—	Eremeeva et al. (2018)
4	Dy_2TiO_5	Polymer carrier chemical	Titanium (IV) isopropoxide and dysprosium nitrate pentahydrate	—	Jung, Kim, and Lee (2006)
5	Dy_2TiO_5	Milling balls	Dy_2O_3 and TiO_2	—	Aughterson et al. (2021)
6	$Dy_2Ti_2O_7$	Sol−gel	Tetra-*n*-butyl titanate and dysprosium nitrate	Photocatalytic	Rahimi-Nasrabadi, Mahdavi, and Adib (2017)
7	$Dy_2Ti_2O_7$	Sol−gel	Titanium (IV) butoxide and dysprosium (III) nitrate pentahydrate	Photonic	Mrázek et al. (2015)
8	Dy_2TiO_5	Molten salt	Dy_2O_3 and TiO_2	Neutron adsorption	Guo et al. (2017)
9	$Dy_2Ti_2O_7$	Sol−gel autocombustion	Dysprosium nitrate and tetra-*n*-butyl titanate	Photocatalytic	Khademolhoseini (2016)
10	$Dy_2Ti_2O_7$	Solid-state reaction	Dy_2O_3 and TiO_2	Dielectric	Zhang et al. (2016)

Table 6.10 Selected precursors and methods for the preparation of holmium titanate nanostructures and their applications.

Number	Type of earth titanate	Synthesis method	Precursors	Property	Reference
1	$Ho_2Ti_2O_7$	Solid state	Ho_2O_3 and TiO_2	Magnetic and optical	Yadav et al. (2020)
2	$Ho_2Ti_2O_7$	Mixing	Ho_2O_3 and TiO_2	Magnetic	Ghasemi et al. (2018)
3	$Ho_2Ti_2O_7$	Floating zone	—	Optical absorption and fluorescence	Macalik et al. (2009)
4	$Ho_2Ti_2O_7$	Solid-phase method	Ho_2O_3 and TiO_2	Magnetic properties	Nemytova et al. (2020)

lattice composed of tetrahedral linked with their vertices. Therefore exchange interaction could be prevented in some determinate positions. These compounds show a great notable variety in low-temperature behaviors by the virtue of varying crystal-field properties, which is independent of having Kramers or non-Kramers based behavior, to the relative significance of the dipolar interaction with reference to exchange. Of these characteristics, the most famous is the "spin-ice" ground state taking place at $Ho_2Ti_2O_7$ and $Dy_2Ti_2O_7$ (Bramwell et al., 2001; Ramirez et al., 1999). In recent years, with respect to the recognition of the anisotropy in the nearest adjacent exchange interaction there has been a much development to fully comprehend the low-temperature behaviors of the Tb, Er, and Yb members of the series (Cao et al., 2009; Onoda and Tanaka, 2011; Thompson et al., 2011). Researchers tend towards the magnetization at very low temperatures in erbium titanates, in the AF phase, and we will demonstrate that the magnetization graphs in different field alignments, along with the neutron diffuse scattering in the paramagnetic graph, can be explained through unique anisotropic exchange tensor. Hence, erbium titanates have an unorthodox magnetic order.

The second-order transition of such material takes place once temperature changes from 1.2 K into a noncoplanar 0 K AF structure (Champion et al., 2003; Poole, Wills, & Lelievre-Berna, 2007). Thermal and quantum fluctuations can make stable abnormality for magnetic order of this compound. This phenomenon is called order by disorder (Savary et al., 2012; Zhitomirsky et al., 2012). Regardless of many theory- and experiment-based studies which have been done for this structure and other closely related pyrochlore materials (Petrenko, Lees, & Balakrishnan, 2011; Ross et al., 2011), there are some concerns, and many hypothetical enquiries about erbium titanates have remained unanswered. Also recently Kristina Vlášková et al. (2020b) reported a new straightforward preparation procedure for an erbium titanate single crystal using the optical coating zone technique and a nonprereacted mixture of initial oxides. The existed route removes all the inconveniences that are related to the fabrication of polycrystalline precursor and eliminates the risk of contaminated single crystal. Prepared ingot is verified via X-ray, electron, and neutron diffraction methods, as well as by specific heat and magnetization measurements, concerning to the crystal quality. All evaluated bulk characteristics are well matched with preceding outcomes. The introduced preparation route can be also used for other pyrochlore titanates along with other 227 oxides (Vlášková et al., 2020b) (Table 6.11).

6.1.12 Ytterbium titanates

Recently, the material $Yb_2Ti_2O_7$ has been extensively studied (Hodges et al., 2001; Ross et al., 2009, 2011). Despite the fact that the detailed form of the ground state of this compound once it is in low span temperatures yet to be more thoroughly discussed, according to the nature of the magnetic interactions $Yb_2Ti_2O_7$ could lie in the quantum spin ice. The exchange interaction parameters at low temperature have been clarified through high-field inelastic neutron scattering. This suggests that in the compound, ferromagnetic ising type exchange interaction is more common than the other terms in the hamiltonin, which ends in considerable amount of quantum fluctuations and dynamics (Ross et al., 2011). When it comes to the verification of crystal-field levels, the low-energy effective spin sector for the structure could be decreased in spin-1/2 moment. Several absorbing effects have been suggested for the compound, including having quantum spin liquid property, whose elementary excitations hold fractional quantum numbers. This state supplies an emerging quantum electrodynamics with a photon mode at low-energy range (Benton, Sikora, & Shannon, 2012; Savary and Balents, 2012). It has recently been suggested that in weak applied <001> magnetic field (or when spontaneous magnetization exists),

Table 6.11 Selected precursors and methods for the preparation of erbium titanate nanostructures and their applications.

Number	Type of earth titanate	Synthesis method	Precursors	Property	Reference
1	$Er_2Ti_2O_7$	Floating zone technique	Er_2O_3 and TiO_2	—	Vlášková et al. (2020a)
2	Er_2TiO_5	Cosputtering	Er, and Ti	Biosensor	Pan et al. (2013)
3	Er_2TiO_5	Ball milling	Er_2O_3 and TiO_2	—	Aughterson et al. (2018)
4	$Er_2Ti_2O_7$	Mechanochemical synthesis	Er_2O_3 and TiO_2	—	Chung et al. (2019)

the elementary excitations of materials for example $Yb_2Ti_2O_7$ appears in extended "quantum strings" composed of fluctuating multiple flipped spins linking monopole pairs (Wan and Tchernyshyov, 2012). With inherent quantum dynamics, these novel excitations are extended objects rather than point particles, an interesting feature of quantum spin ice. The time-domain terahertz spectroscopy (TDTS) analysis was used to investigate the inherent features of the excitations in quantum spin ice, measuring $Yb_2Ti_2O_7$ single crystals under magnetic field applied along the <001> direction. Spectroscopic features matched with the picture of quantum strings are seen in the low-field range. In the strong-field range, a crossover towards field-induced order is monitored, where the excitations appear in magnons and two-magnon excitations. So ytterbium titanate has the potential to be served as a photocatalyst. Enayat (2018) prepared ytterbium titanate nanocomposite with glucose, lactose, as well as starch as gelling agent via sol–gel method. Then, the influence of these gelling agents and solvent on the dimension of final products was verified, and as fabricated ytterbium titanate nanocomposite was carried out to degrade rhodamine B when it is subjected to the UV irradiation. Ytterbium titanate nanocomposite was highly functional in photocatalytic property and removed approximately 96% of rhodamine B within 100-minute irradiation (Enayat, 2018) (Table 6.12).

6.1.13 Lutetium titanates

The last structure which is studied here is $Lu_2Ti_2O_7$. It has the cubic pyrochlore structure (space group: $Fd\bar{3}m$; $a = 1.0019$ nm) and is an encouraging transparent ceramic. This structure is an ordered defect fluorite one having Lu^{3+} and Ti^{4+} cations, which constitutes a cubic close-packed cation array, with 7/8 of the tetrahedral interstices taken up by oxygen ions (Subramanian et al., 1983). Being chemically stable as well as having the potential to accept foreign elements in solid solution, it has been tailored to be used in sensors (Pan et al., 2013), ionic conductors (Shlyakhtina, Ukshe, & Shcherbakova, 2005), and as a host material for nuclear waste immobilization (Wen et al., 2016). Moreover, lutetium titanate is a favorable host material to be used in optical imaging systems as a consequence of its high refractive index (Wang et al., 2018; Weber, 2002). Because lutetium titanate powder is not commercially available, it would be prepared by many methods such as

Table 6.12 Selected precursors and methods for the preparation of ytterbium titanate nanostructures and their applications.

Number	Type of earth titanate	Synthesis method	Precursors	Property	Reference
1	$Yb_2Ti_2O_7$	Coprecipitation	Yb^{3+} and Ti^{4+}	Electrical conductivity	Shlyakhtina et al. (2005)
2	$Yb_2Ti_2O_7$	Solid state	Yb_2O_3 and TiO_2	—	Uno et al. (2018)

coprecipitation (Shlyakhtina et al., 2006), sol–gel technique (Begg et al., 2001), solid-state reaction (Brixner, 1964), molten salt method (Li et al., 2011), as well as Pechini process (Abe, Higashi, Zou, Sayama, & Abe, 2004). Among the powder preparation techniques, the solid-state reaction method is the most uncomplicated and convenient one. In temperature range between 800°C and 1100°C and various salt-to-oxide mole ratios of 1:1, 3:1, and 5:1, Li et al. (2011) prepared lutetium titanate ceramic powder with pyrochlore structure in molten salt with NaCl as salt flux. It was shown that the synthesizing temperature and proportion of salt to oxide could strongly influence the morphology of lutetium titanate powder. Moreover, it was clarified that, by the observation of microstructure, crystal shape of lutetium titanate powder can convert from spheroid into octahedron, and eventually alters into irregular after further increase in temperature. At temperatures above 1000°C, the formation of a duplex microstructure could be seen. Further studies disclosed that even though the grain size of the synthesized powder was not found to considerably improve by an increment of the molten salt content, the proportion already mentioned can influence the crystalline of lutetium titanate from the molten salt flux (Li et al., 2011) (Table 6.13).

6.2 Fabrication of lanthanide titanate nanostructures

The conventional routes for the fabrication of structures are also the foundation of great number of preparation procedures for nanomaterials. Nonetheless, by taking advantage of new technologies some have been able to prepare such tiny structures. To advance the most up-to-date technologies, the fusion of a number of various procedures is still straightforward. Different methods including ultrasonic irradiation, coprecipitation, sol–gel, hydrothermal, microwave, and thermal decomposition have been used for the fabrication of nanostructures (Feng et al., 2014; Hosseinpour-Mashkani and Sobhani-Nasab, 2017; Kawashima et al., 2015; Salavati-Niasari et al., 2016; Sobhani-Nasab and Behpour, 2016; Sobhani-Nasab et al., 2019a, 2019b; Suresh et al., 2001; Yan et al., 2009). Fig. 6.2 displays various procedures for the synthesis of nanosized lanthanide titanate.

Of several numbers of being used to prepare nanosized materials, sol–gel is a widespread and, at the same time, of utmost important preparation mechanism to be able to produce various lanthanide titanates. This route supplies an inventive tool for the synthesis of both inorganic polymer and organic/inorganic hybrid materials. From historical point of view, starting from the mid-1800s scientists used the sol–gel method to prepare nanomaterials, as Schott Glass Company (Jena, Germany) performed its experiments with 100 years later.

Because this method can be implemented under extraordinary mild conditions, there is no limitation on shapes, sizes, morphology, and formats of final products, for example, films, as well as monoliths and monosized particles (Gandomi et al., 2020; Kooshki et al., 2019; Rahimi-Nasrabadi et al., 2019; Salavati-Niasari et al., 2016; Sobhani-Nasab et al., 2015a).

Table 6.13 Selected precursors and methods for the preparation of lutetium titanate nanostructures and their applications.

Number	Type of earth titanate	Synthesis method	Precursors	Property	Reference
1	$Lu_2Ti_2O_7$	Ball milling	Lu_2O_3 and TiO_2	—	An, Zhang, Fan, Goto, & Wang (2019)
2	$Lu_2Ti_2O_7$	Coprecipitation	Lu^{3+} and Ti^{4+}	Impedance of electrochemical cells	Shlyakhtina, Ukshe et al. (2005)
3	$Lu_2Ti_2O_7$	Traditional ceramic processing	Lu_2O_3 and TiO_2	—	Xie et al. (2015)
4	$Lu_2Ti_2O_7$	Traditional ceramic processing	Lu_2O_3 and TiO_2	—	Qiu-Rong et al. (2015)
5	$Lu_2Ti_2O_7$	Reactive spark plasma sintering	Lu_2O_3 and TiO_2	—	An, Ito, and Goto (2011)
6	$Lu_2Ti_2O_7$	Molten salt technique	Lu_2O_3 and TiO_2	—	Zhang, Li, and Liu (2013)

Figure 6.2 Schematic representation of various synthesis methods for the preparation of lanthanide titanate nanostructures.

The following steps are necessary, in order, to perform sol−gel preparation method: (1) subject to hydrolysis, (2) condensation, and (3) thermal decomposition of metal precursors or metal alkoxides in solution. During this process, at first, a solution with stability containing all essential reagents, called sol, is constructed through metal alkoxides or precursors. Then, to make a networked structure (gel), it should undergo hydrolysis and afterwards be condensed. This can lead to a noticeable improvement in the viscosity. The control process for the kinetics of the reactions is done by water, alcohol, acid or base, and to achieve the favorable particle dimension we can modify temperature of the precursor, concentration, together with pH values. Once the gel is formed, an aging step is necessitated to enable the formation of a solid mass. The aging step is associated with the expulsion of solvent, Ostwald ripening, along with phase transformation which step may take up too long (Ghaemifar et al., 2020).

6.3 Conclusion and outlook

Different chemical procedures for the preparation of lanthanide titanate are introduced in this chapter.

The solid-state and sol–gel procedure could be regarded as the common procedures for the production of lanthanide titanates.

Pyrochlore oxides having the general formula of $A_2B_2O_6$ belong to the space group of Fd3m, which contains lanthanide or an element with lone pair of electrons.

These pyrochlores are usually explained as in the following: two interpenetrating networks of A_2O tetrahedra and B_2O_6 octahedra, with oxygens in two sublattices represented as O and O^-, respectively.

Apart from having a very broad range of structural properties, lanthanide titanates can demonstrate both chemical and physical properties.

Lanthanide titanates have various properties, such as piezoelectric, catalytic activity, fuel cells, high permittivity dielectrics, ferroelectric properties, dielectric, photoluminescence, photocatalytic, battery electrocatalytic, sensor as well as optical behaviors.

Lanthanide titanate nanostructures could be used in various applications, and they have pervoskite structures.

References

Abe, R., Higashi, M., Sayama, K., Abe, Y., & Sugihara, H. (2006). Photocatalytic activity of R3MO7 and R2Ti2O7 (R = Y, Gd, La; M = Nb, Ta) for water splitting into H2 and O2. *The Journal of Physical Chemistry B*, *110*(5), 2219–2226.

Abe, R., Higashi, M., Zou, Z., Sayama, K., & Abe, Y. (2004). Photocatalytic water splitting into H2 and O2 over R2Ti2O7 (R = Y, rare earth) with pyrochlore structure. *Chemistry Letters*, *33*(8), 954–955.

Abe, Y., & Laine, R. M. (2020). Photocatalytic plate-like La2Ti2O7 nanoparticles synthesized via liquid-feed flame spray pyrolysis (LF-FSP) of metallo-organic precursors. *Journal of the American Ceramic Society*, *103*(9), 4832–4839.

Ahmadi, F., Rahimi-Nasrabadi, M., & Behpour, M. (2017). Synthesis Nd 2 TiO 5 nanoparticles with different morphologies by novel approach and its photocatalyst application. *Journal of Materials Science: Materials in Electronics*, *28*(2), 1531–1536.

Ahn, K., & Forbes, L. (2007). *ALD of amorphous lanthanide doped TiOx films*, Patent Application United States 187772, p. A1.

Akram, I. N., Akhtar, S., Khadija, G., Awais, M. M., Latif, M., Noreen, A., et al. (2020). Synthesis, characterization, and biocompatibility of lanthanum titanate nanoparticles in albino mice in a sex-specific manner. *Naunyn-Schmiedeberg's Archives of Pharmacology*, 1–13.

An, L., Ito, A., & Goto, T. (2011). Effects of sintering and annealing temperature on fabrication of transparent Lu2Ti2O7 by spark plasma sintering. *Journal of the American Ceramic Society*, *94*(11), 3851–3855.

An, L., Zhang, J., Fan, R., Goto, T., & Wang, S. (2019). Impedance study of spark–plasma–sintered lutetium titanate ceramics: Effect of post–annealing. *Ceramics International*, *45*(13), 16317–16322.

Arney, D., Porter, B., Greve, B., & Maggard, P. A. (2008). New molten-salt synthesis and photocatalytic properties of La2Ti2O7 particles. *Journal of Photochemistry and Photobiology A: Chemistry*, *199*(2–3), 230–235.

Atuchin, V., Gavrilova, T., Grivel, J.-C., & Kesler, V. (2008). Electronic structure of layered titanate Nd2Ti2O7. *Surface Science*, *602*(19), 3095–3099.

Atuchin, V., Gavrilova, T., Grivel, J.-C., Kesler, V., & Troitskaia, I. (2012). Electronic structure of layered ferroelectric high-k titanate Pr2Ti2O7. *Journal of Solid State Chemistry*, *195*, 125–131.

Aughterson, R. D., Lumpkin, G. R., Thorogood, G. J., Zhang, Z., Gault, B., & Cairne, J. M. (2015). Crystal chemistry of the orthorhombic Ln2TiO5 compounds with Ln = La, Pr, Nd, Sm, Gd, Tb and Dy. *Journal of Solid State Chemistry*, *227*, 60–67.

Aughterson, R. D., Lumpkin, G. R., Smith, K. L., de los Reyes, M., Davis, J., Avdeev, M., et al. (2018). The ion-irradiation tolerance of the pyrochlore to fluorite Ho (x) Yb (2-x) TiO5 and Er2TiO5 compounds: A TEM comparative study using both in-situ and bulk ex-situ irradiation approaches. *Journal of Nuclear Materials*, *507*, 316–326.

Aughterson, R. D., Zaluzec, N. J., & Lumpkin, G. R. (2021). Synthesis and ion-irradiation tolerance of the Dy2TiO5 polymorphs. *Acta Materialia*, *204*, 116518.

Bartel, C. J., Sutton, C., Goldsmith, B. R., Ouyang, R., Musgrave, C. B., Ghiringhelli, L. M., et al. (2019). New tolerance factor to predict the stability of perovskite oxides and halides. *Science Advances*, *5*(2), eaav0693.

Bayart, A., Shao, Z., Ferri, A., Roussel, P., Desfeux, R., & Saitzek, S. (2016). Epitaxial growth and nanoscale electrical properties of Ce2Ti2O7 thin films. *RSC Advances*, *6* (39), 32994–33002.

Begg, B. D., Hess, N. J., Weber, W. J., Devanathan, R., Icenhower, J. P., Thevuthasan, S., et al. (2001). Heavy-ion irradiation effects on structures and acid dissolution of pyrochlores. *Journal of Nuclear Materials*, *288*(2–3), 208–216.

Benton, O., Sikora, O., & Shannon, N. (2012). Seeing the light: Experimental signatures of emergent electromagnetism in a quantum spin ice. *Physical Review B*, *86*(7), 075154.

Bramwell, S., Harris, M., Den Hertog, B., Gingras, M., Gardner, J., McMorrow, D., et al. (2001). Spin correlations in Ho 2 Ti 2 O 7: A dipolar spin ice system. *Physical Review Letters*, *87*(4), 047205.

Bramwell, S. T. (2006). Great moments in disorder. *Nature*, *439*(7074), 273–274.

Brandon, J., & Megaw, H. D. (1970). On the crystal structure and properties of Ca2Nb2O7,"calcium pyroniobate". *Philosophical Magazine*, *21*(169), 189–194.

Brixner, L. (1964). Preparation and properties of the Ln2Ti2O7-type rare earth titanate. *Inorganic Chemistry*, *3*(7), 1065–1067.

Cann, D. P., Randall, C. A., & Shrout, T. R. (1996). Investigation of the dielectric properties of bismuth pyrochlores. *Solid State Communications*, *100*(7), 529–534.

Cao, H., Gukasov, A., Mirebeau, I., Bonville, P., Decorse, C., & Dhalenne, G. (2009). Ising vs X Y Anisotropy in Frustrated R 2 Ti 2 O 7 Compounds as "Seen" by Polarized Neutrons. *Physical Review Letters*, *103*(5), 056402.

Cao, Y., Tang, P., Han, Y., & Qiu, W. (2020). Synthesis of La2Ti2O7 flexible self-supporting film and its application in flexible energy storage device. *Journal of Alloys and Compounds*, *842*, 155581.

Champion, J., Harris, M., Holdsworth, P., Wills, A., Balakrishnan, G., Bramwell, S., et al. (2003). Er2Ti2O7: Evidence of quantum order by disorder in a frustrated antiferromagnet. *Physical Review B*, *68*(2), 020401.

Chen, F.-H., & Pan, T.-M. (2012). Physical and Electrical Properties of Dy 2 O 3 and Dy 2 TiO 5 Metal Oxide–High-κ Oxide–Silicon-Type Nonvolatile Memory Devices. *Journal of Electronic Materials*, *41*(8), 2197–2203.

Chung, C.-K., O'Quinn, E. C., Neuefeind, J. C., Fuentes, A. F., Xu, H., Lang, M., et al. (2019). Thermodynamic and structural evolution of mechanically milled and swift heavy ion irradiated Er2Ti2O7 pyrochlore. *Acta Materialia*, *181*, 309–317.

Cioatera, N., Voinea, E.-A., Osiceanu, P., Papa, F., Duă, A., Resceanu, I., et al. (2019). Vanadium-substituted Sm2Ti2O7 pyrochlore. Insight into the structure and electrical conductivity under oxidizing and highly reducing atmosphere. *Solid State Ionics*, *339*, 114995.

Coles, G. S., Bond, S. E., & Williams, G. (1994). Metal stannates and their role as potential gas-sensing elements. *Journal of Materials Chemistry*, *4*(1), 23–27.

Cornelius, A., Light, B., Kumar, R. S., Eichenfield, M., Dutton, T., Pepin, R., et al. (2005). Disturbing the spin liquid state in Tb2Ti2O7: Heat capacity measurements on rare earth titanates. *Physica B: Condensed Matter*, *359*, 1243–1245.

Dharuman, N., & Berchmans, L. J. (2013). Low temperature synthesis of nano-crystalline gadolinium titanate by molten salt route. *Ceramics International*, *39*(8), 8767–8771.

Dong, X., Wang, K., Luo, S., Wan, J., & Liu, J. M. (2009). Coexistence of magnetic and ferroelectric behaviors of pyrochlore Ho 2 Ti 2 O 7. *Journal of Applied Physics*, *106*(10), 104101.

Enayat, M. J. (2018). Photocatalytic studies of Yb 2 TiO 5/Yb 2 Ti 2 O 7 nanocomposite synthesized by new technique. *Journal of Materials Science: Materials in Electronics*, *29*(5), 3829–3835.

Eremeeva, J., Vorotilo, S., Kovalev, D. Y., Gofman, A., & Lopatin, V. (2018). Mechanochemical synthesis of Dy2 TiO 5 single-phase crystalline nanopowders and investigation of their properties. *Inorganic Materials: Applied Research*, *9*(2), 291–296.

Ewing, R. C., Weber, W. J., & Lian, J. (2004). Nuclear waste disposal—pyrochlore (A 2 B 2 O 7): Nuclear waste form for the immobilization of plutonium and "minor" actinides. *Journal of Applied Physics*, *95*(11), 5949–5971.

Fang, X., Xu, L., Zhang, X., Zhang, K., Dai, H., Liu, W., et al. (2019). Effect of rare earth element (Ln = La, Pr, Sm, and Y) on physicochemical properties of the Ni/Ln2Ti2O7 catalysts for the steam reforming of methane. *Molecular Catalysis*, *468*, 130–138.

Faucher, M., & Caro, P. (1975). Ordre et desordre dans certains composes du type pyrochlore. *Journal of Solid State Chemistry*, *12*(1–2), 1–11.

Feng, J., Luo, W., Fang, T., Lv, H., Wang, Z., Gao, J., et al. (2014). Highly photo-responsive LaTiO2N photoanodes by improvement of charge carrier transport among film particles. *Advanced Functional Materials*, *24*(23), 3535–3542.

Fompeyrine, J., Seo, J. W., & Locquet, J.-P. (1999). Growth and characterization of ferroelectric LaTiO3·5 thin films. *Journal of the European Ceramic Society*, *19*(6–7), 1493–1496.

Galahom, A. A. (2016). Investigation of different burnable absorbers effects on the neutronic characteristics of PWR assembly. *Annals of Nuclear Energy*, *94*, 22–31.

Gandomi, F., Sobhani-Nasab, A., Pourmasoud, S., Eghbali-Arani, M., & Rahimi-Nasrabadi, N. (2020). Synthesis of novel Fe 3 O 4@ SiO 2@ Er 2 TiO 5 superparamagnetic core–-shell and evaluation of their photocatalytic capacity. *Journal of Materials Science: Materials in Electronics*, *31*, 10553–10563.

Gao, Z., Liu, L., Han, X., Meng, X., Cao, L., Ma, G., et al. (2015). Cerium titanate (Ce2Ti2O7): A ferroelectric ceramic with perovskite-like layered structure (PLS). *Journal of the American Ceramic Society*, *98*(12), 3930–3934.

Gao, Z., Yan, H., Ning, H., & Reece, M. (2013). Ferroelectricity of Pr2Ti2O7 ceramics with super high Curie point. *Advances in Applied Ceramics*, *112*(2), 69–74.

Gao, Z., Liu, Y., Lu, C., Xia, Y., Fang, L., Ma, Y., et al. (2018). Phase transition of Eu2Ti2O7 under high pressure and a new ferroelectric phase with perovskite-like layered structure. *Journal of the American Ceramic Society*, *101*(6), 2571–2577.

Garbout, A., Kallel-Kchaou, N., & Férid, M. (2016). Relationship between the structural characteristics and photoluminescent properties of LnEuTi2O7 (Ln = Gd and Y) pyrochlores. *Journal of Luminescence*, *169*, 359–366.

Gardner, J. S., Gingras, M. J., & Greedan, J. E. (2010). Magnetic pyrochlore oxides. *Reviews of Modern Physics*, *82*(1), 53.

Ghaemifar, S., Rahimi-Nasrabadi, M., Pourmasud, S., Eghbali-Arani, M., Behpour, M., & Sobhani-Nasab, A. (2020). Preparation and characterization of MnTiO 3, FeTiO 3, and CoTiO 3 nanoparticles and investigation various applications: A review. *Journal of Materials Science: Materials in Electronics*, *31*(9), 6511–6524.

Ghasemi, A., Scheie, A., Kindervater, J., & Koohpayeh, S. M. (2018). The pyrochlore Ho2Ti2O7: Synthesis, crystal growth, and stoichiometry. *Journal of Crystal Growth*, *500*, 38–43.

Ghodsi, F., Tepehan, F., & Tepehan, G. (1999). Optical and electrochromic properties of sol–gel made CeO2–TiO2 thin films. *Electrochimica Acta*, *44*(18), 3127–3136.

Guo, X., Feng, Y., Zhao, J., Ma, L., Li, Q., Zhang, Z., et al. (2017). Neutron adsorption performance of Dy2TiO5 materials obtained from powders synthesized by the molten salt method. *Ceramics International*, *43*(2), 1975–1979.

Han, S.-W., Gardner, J. S., & Booth, C. H. (2004). Structural properties of the geometrically frustrated pyrochlore Tb 2 Ti 2 O 7. *Physical Review B*, *69*(2), 024416.

Havelia, S., Balasubramaniam, K., Spurgeon, S., Cormack, F., & Salvador, P. (2008). Growth of La2Ti2O7 and LaTiO3 thin films using pulsed laser deposition. *Journal of Crystal Growth*, *310*(7–9), 1985–1990.

Henao, J., Pacheco, Y., Sotelo, O., Casales, M., & Martinez-Gómez, L. (2019). Lanthanum titanate nanometric powder potentially for rechargeable Ni-batteries: Synthesis and electrochemical hydrogen storage. *Journal of Materials Research and Technology*, *8*(1), 759–765.

Herrera, G., Jiménez-Mier, J., & Chavira, E. (2014). Layered-structural monoclinic–orthorhombic perovskite La2Ti2O7 to orthorhombic LaTiO3 phase transition and their microstructure characterization. *Materials Characterization*, *89*, 13–22.

Hodges, J., Bonville, P., Forget, A., Rams, M., Królas, K., & Dhalenne, G. (2001). The crystal field and exchange interactions in Yb2Ti2O7. *Journal of Physics: Condensed Matter*, *13*(41), 9301.

Hosseinpour-Mashkani, S. M., & Sobhani-Nasab, A. (2017). Green synthesis and characterization of NaEuTi 2 O 6 nanoparticles and its photocatalyst application. *Journal of Materials Science: Materials in Electronics*, *28*(5), 4345–4350.

Huang, Z., Liu, J., Huang, L., Tian, L., Wang, S., Zhang, G., et al. (2020). One-step synthesis of dandelion-like lanthanum titanate nanostructures for enhanced photocatalytic performance. *NPG Asia Materials*, *12*(1), 1–12.

Huynh, L. T., Eger, S. B., Walker, J. D., Hayes, J. R., Gaultois, M. W., & Grosvenor, A. P. (2012). How temperature influences the stoichiometry of CeTi2O6. *Solid State Sciences*, *14*(6), 761–767.

Hwang, D. W., Cha, K. Y., Kim, J., Kim, H. G., Bae, S. W., & Lee, J. S. (2003a). Electronic band structure and photocatalytic activity of Ln2Ti2O7 (Ln = La, Pr, Nd). *The Journal of Physical Chemistry B*, *107*(21), 4963–4970.

Hwang, D. W., Lee, J. S., Li, W., & Oh, S. H. (2003b). Photocatalytic degradation of CH3Cl over a nickel-loaded layered perovskite. *Industrial & Engineering Chemistry Research*, *42*(6), 1184–1189.

Ishizawa, N., Marumo, F., Iwai, S., Kimura, M., & Kawamura, T. (1980). Compounds with perovskite-type slabs. III. The structure of a monoclinic modification of Ca2Nb2O7. *Acta Crystallographica Section B: Structural Crystallography and Crystal Chemistry*, *36*(4), 763–766.

Ishizawa, N., Marumo, F., Iwai, S., Kimura, M., & Kawamura, T. (1982). Compounds with perovskite-type slabs. V. A high-temperature modification of La2Ti2O7. *Acta Crystallographica Section B: Structural Crystallography and Crystal Chemistry*, *38*(2), 368–372.

Jafar, M., Phapale, S., Nigam, S., Achary, S., Mishra, R., Majumder, C., et al. (2021). Implication of aliovalent cation substitution on structural and thermodynamic stability of Gd2Ti2O7: Experimental and theoretical investigations. *Journal of Alloys and Compounds*, *859*, 157781.

Jeyasingh, T., Saji, S., & Wariar, P. (2017). *Synthesis of nanocrystalline Gd2Ti2O7 by combustion process and its structural, optical and dielectric properties. AIP Conference Proceedings*. AIP Publishing LLC.

Jeyasingh, T., Saji, S., Kavitha, V., & Wariar, P. (2018). *Frequency dependent dielectric properties of combustion synthesized Dy2Ti2O7 pyrochlore oxide. AIP Conference Proceedings*. AIP Publishing LLC.

Joseph, L. K., Dayas, K., Damodar, S., Krishnan, B., Krishnankutty, K., Nampoori, V., et al. (2008). Photoluminescence studies on rare earth titanates prepared by self-propagating high temperature synthesis method. *Spectrochimica Acta Part A: Molecular and Biomolecular Spectroscopy*, *71*(4), 1281–1285.

Jung, C.-H., Kim, C.-J., & Lee, S.-J. (2006). Synthesis and sintering studies on Dy2TiO5 prepared by polymer carrier chemical process. *Journal of Nuclear Materials*, *354*(1–3), 137–142.

Kambale, K. R., Kulkarni, A. R., Narayanan Venkataramani, A., & Vairagade, S. B. (2019). Synthesis of high Curie temperature $La_2Ti_2O_7$ piezoceramic by mechanochemical activation: A preliminary investigation. *Advances in Ceramics for Environmental, Functional, Structural, and Energy Applications II*, *266*, 59.

Kandan, R., Reddy, B. P., Panneerselvam, G., & Nagarajan, K. (2013). Enthalpy increment measurements on europium titanate. *Journal of Thermal Analysis and Calorimetry*, *112*(1), 59–61.

Kao, C. H., Chen, S. Z., Luo, Y., Chiu, W. T., Chiu, S. W., Chen, I. C., et al. (2017). The influence of Ti doping and annealing on Ce2Ti2O7 flash memory devices. *Applied Surface Science*, *396*, 1673–1677.

Kawashima, K., Hojamberdiev, M., Wagata, H., Yubuta, K., Vequizo, J. J. M., Yamakata, A., et al. (2015). NH3-assisted flux-mediated direct growth of LaTiO2N crystallites for visible-light-induced water splitting. *The Journal of Physical Chemistry C*, *119*(28), 15896–15904.

Kesari, S., Atuchin, N. P., Patwe, S. J., Achary, S. N., Sinha, A. K., Sastry, P. U., et al. (2016). Structural stability and anharmonicity of Pr2Ti2O7: Raman spectroscopic and XRD studies. *Inorganic Chemistry*, *55*(22), 11791–11800.

Khademolhoseini, S. (2016). Sol–gel auto-combustion synthesis of dysprosium titanate nanoparticles using tyrosine as a novel fuel. *Journal of Materials Science: Materials in Electronics*, *27*(10), 10759–10763.

Kidchob, T., Malfatti, L., Marongiu, D., Enzo, S., & Innocenzi, P. (2009). Formation of cerium titanate, CeTi 2 O 6, in sol–gel films studied by XRD and FAR infrared spectroscopy. *Journal of Sol-Gel Science and Technology*, *52*(3), 356–361.

Kidyarov, B., & Atuchin, V. (2007). Universal crystal classification system "point symmetry–physical property". *Ferroelectrics*, *360*(1), 96–99.

Kim, H. G., Hwang, D. W., Bae, S. W., Jung, J. H., & Lee, J. S. (2003). Photocatalytic water splitting over La 2 Ti 2 O 7 synthesized by the polymerizable complex method. *Catalysis Letters*, *91*(3), 193–198.

Kim, J., Hwang, D. W., Bae, S. W., Kim, Y. G., & Lee, J. S. (2001). Effect of precursors on the morphology and the photocatalytic water-splitting activity of layered perovskite La 2 Ti 2 O 7. *Korean Journal of Chemical Engineering*, *18*(6), 941–947.

Kim, J., Hwang, D. W., Kim, H.-G., Bae, S. W., Ji, S. M., & Lee, J. S. (2002a). Nickel-loaded La 2 Ti 2 O 7 as a bifunctional photocatalyst. *Chemical Communications*, *21*, 2488–2489.

Kim, W. S., Ha, S.-M., Yun, S., & Park, H.-H. (2002b). Microstructure and electrical properties of Ln2Ti2O7 (Ln = La, Nd). *Thin Solid Films*, *420*, 575–578.

Kim, W. S., Yang, J.-K., Lee, C.-K., Lee, H.-S., & Park, H.-H. (2008). Synthesis and characterization of ferroelectric properties of Ce2Ti2O7 thin films with Ce3 + by chemical solution deposition. *Thin Solid Films*, *517*(2), 506–509.

Klimm, D., Guguschev, C., Kok, D., Naumann, M., Ackermann, L., Rytz, D., et al. (2017). Crystal growth and characterization of the pyrochlore Tb 2 Ti 2 O 7. *CrystEngComm*, *19*(28), 3908–3914.

Kooshki, H., Sobhani-Nasab, A., Eghbali-Arani, M., Ahmadi, F., Ameri, V., & Rahimi-Nasrabadi, M. (2019). Eco-friendly synthesis of PbTiO3 nanoparticles and PbTiO3/carbon quantum dots binary nano-hybrids for enhanced photocatalytic performance under visible light. *Separation and Purification Technology*, *211*, 873–881.

Koźmin, P. A., Zakharov, N. A., Surazhskaya, M.D. (1997) Crystal structure of $Pr_2Ti_2O_7$. *Inorganic Materials, 33*(8), 850–852.

Kulriya, P., Yao, T., Scott, S. M., Nanda, S., & Lian, J. (2017). Influence of grain growth on the structural properties of the nanocrystalline Gd2Ti2O7. *Journal of Nuclear Materials*, *487*, 373–379.

Kumar, B. S., & Venkateswaran, C. (2019). *First observed metal to insulator transition in the vacant 3d orbital quantum spin liquid Tb $ _2 $ Ti $ _2 $ O $ _7$*. arXiv preprint arXiv:1904.12478.

Kumar, B. S., Kumar, Y. N., Kamalarasan, V., & Venkateswaran, C. (2020). Non-adiabatic small polaron hopping transport above metal-like to insulator transition in the vacant 3d-orbital Tb 2 Ti 2 O 7 pyrochlore oxide. *Journal of Materials Science: Materials in Electronics*, *31*(24), 22312–22322.

Kumar, N., Tripathi, R., Dogra, A., Awana, V., & Kishan, H. (2010). Effect of Pr doping in La–Sn–Mn–O system. *Journal of Alloys and Compounds*, *492*(1–2), L28–L32.

Kumar, N. S., Shekar, N. C., & Sahu, P. C. (2008). Pressure induced structural transformation of pyrochlore Gd2Zr2O7. *Solid State Communications*, *147*(9–10), 357–359.

Kumar, S., & Gupta, H. (2012). First principles study of zone centre phonons in rare-earth pyrochlore titanates, RE2Ti2O7 (RE = Gd, Dy, Ho, Er, Lu; Y). *Vibrational Spectroscopy*, *62*, 180–187.

Kushkov, V., Zverlin, A., Zaslavskii, A., Slivinskaya, A., & Melnikov, A. (1993). Structure of the Ln 2 Ti 2 O 7 thin films prepared by pulsed-laser evaporation. *Journal of Materials Science*, *28*(2), 361–363.

Lang, M., Zhang, F., Zhang, J., Wang, J., Lian, J., Weber, W. J., et al. (2010). Review of A2B2O7 pyrochlore response to irradiation and pressure. *Nuclear Instruments and Methods in Physics Research Section B: Beam Interactions with Materials and Atoms*, *268*(19), 2951–2959.

Lau, G., Muegge, B., McQueen, T., Duncan, E., & Cava, R. (2006). Stuffed rare earth pyrochlore solid solutions. *Journal of Solid State Chemistry*, *179*(10), 3126–3135.

Lee, C. K., Kim, W. S., Park, H.-H., Jeon, H., & Pae, Y. H. (2005). Thermal-stress stability of yttrium oxide as a buffer layer of metal-ferroelectric-insulator-semiconductor field effect transistor. *Thin Solid Films*, *473*(2), 335–339.

Leroy, S., Blach, J.-F., Huvé, M., Léger, B., Kania, N., Henninot, J.-F., et al. (2020). Photocatalytic and sonophotocatalytic degradation of rhodamine B by nano-sized La2Ti2O7 oxides synthesized with sol-gel method. *Journal of Photochemistry and Photobiology A: Chemistry*, *401*, 112767.

Lev, O., Wu, Z., Bharathi, S., Glezer, V., Modestov, A., Gun, J., et al. (1997). Sol – gel materials in electrochemistry. *Chemistry of Materials*, *9*(11), 2354–2375.

Li, B.-r., Chang, H.-b., Liu, D.-y., & Yuan, X.-n. (2011). Synthesis of Lu2Ti2O7 powders by molten salt method. *Materials Chemistry and Physics*, *130*(1–2), 755–759.

Li, K., Wang, Y., Wang, H., Zhu, M., & Yan, H. (2006). Hydrothermal synthesis and photocatalytic properties of layered La2Ti2O7 nanosheets. *Nanotechnology*, *17*(19), 4863.

Li, Q., Xu, L., Fan, C., Zhang, F., Lv, Y., Ni, B., et al. (2013). Single crystal growth of the pyrochlores R2Ti2O7 (R = rare earth) by the optical floating-zone method. *Journal of Crystal Growth*, *377*, 96–100.

Li, X., Yang, J., Zhang, Y., & Zhang, W. (2020a). Polyethylene glycol in sol-gel precursor to prepare porous Gd2Ti2O7: Enhanced photocatalytic activity on Reactive Brilliant Red X-3B degradation. *Materials Science in Semiconductor Processing*, *117*, 105181.

Li, Y., Jiang, L., Chen, Q., & Zhu, J. (2020b). Regulate the microstructure and band gap of La 2 Ti 2 O 7. *Journal of Materials Science: Materials in Electronics*, *31*(1), 52–59.

Liao, J., Wang, Q., Kong, L., Ming, Z., Wang, Y., Li, Y., et al. (2018). Effect of Yb3 + concentration on tunable upconversion luminescence and optically temperature sensing behavior in Gd2TiO5: Yb3 + /Er3 + phosphors. *Optical Materials*, *75*, 841–849.

Liu, Z. G., Ouyang, J. H., & Sun, K. N. (2011). Electrical conductivity improvement of Nd2Ce2O7 ceramic co-doped with Gd2O3 and ZrO2. *Fuel Cells*, *11*(2), 153–157.

Long, C. Y., Peng, F. F., Jin, M. M., Tang, P. S., & Chen, H. F. (2017). Preparation and characterization of Pr2Ti2O7 by sol-gel process. *Key Engineering Materials*, *748*, 413–417.

Lu, Y., Le Paven, C., Nguyen, H. V., Benzerga, R., Le Gendre, L., Rioual, S., et al. (2013). Reactive sputtering deposition of perovskite oxide and oxynitride lanthanum titanium films: Structural and dielectric characterization. *Crystal Growth & Design*, *13*(11), 4852–4858.

Lumpkin, G. R., Pruneda, M., Rios, S., Smith, K. L., Trachenko, K., Whittle, K. R., et al. (2007). Nature of the chemical bond and prediction of radiation tolerance in pyrochlore and defect fluorite compounds. *Journal of Solid State Chemistry*, *180*(4), 1512–1518.

Macalik, L., Mączka, M., Solarz, P., Fuentes, A., Matsuhira, K., & Hiroi, Z. (2009). Optical spectroscopy of the geometrically frustrated pyrochlore Ho2Ti2O7. *Optical Materials*, *31*(6), 790–794.

Maczka, M., Hanuza, J., Hermanowicz, K., Fuentes, A., Matsuhira, K., & Hiroi, Z. (2008). Temperature-dependent Raman scattering studies of the geometrically frustrated pyrochlores Dy2Ti2O7, Gd2Ti2O7 and Er2Ti2O7. *Journal of Raman Spectroscopy*, *39*(4), 537–544.

Martos, M., Julián-López, B., Folgado, J. V., Cordoncillo, E., & Escribano, P. (2008). *Sol–gel synthesis of tunable cerium titanate materials*. Wiley Online Library.

Matteucci, F., Cruciani, G., Dondi, M., Baldi, G., & Barzanti, A. (2007). Crystal structural and optical properties of Cr-doped Y2Ti2O7 and Y2Sn2O7 pyrochlores. *Acta Materialia*, *55*(7), 2229–2238.

Milanova, M. M., Kakihana, M., Arima, M., Yashima, M., & Yoshimura, M. (1996). A simple solution route to the synthesis of pure La2Ti2O7 and Nd2Ti2O7 at 700–800° C by polymerized complex method. *Journal of Alloys and Compounds*, *242*(1–2), 6–10.

Mims, C., Jacobson, A., Hall, R., & Lewandowski, J. (1995). Methane oxidative coupling over nonstoichiometric bismuth-tin pyrochlore catalysts. *Journal of Catalysis*, *153*(2), 197–207.

Mrazek, J., Aubrecht, J., Todorov, F., Buršík, J., Puchý, V., Džunda, R., et al. (2014). Synthesis and crystallization mechanism of europium-titanate Eu2Ti2O7. *Journal of Crystal Growth*, *391*, 25–32.

Mrázek, J., Potel, M., Buršík, J., Mráček, A., Kallistová, A., Jonášová, Š., et al. (2015). Sol–gel synthesis and crystallization kinetics of dysprosium-titanate Dy2Ti2O7 for photonic applications. *Materials Chemistry and Physics*, *168*, 159–167.

Mrázek, J., Surýnek, M., Bakardjieva, S., Buršík, J., & Kašík, I. (2018). CO2 laser-assisted preparation of transparent Eu2Ti2O7 thin films. *Ceramics International*, *44*(8), 9479–9483.

Mrázek, J., Vytykáčová, S., Buršík, J., Puchý, V., Girman, V., Peterka, P., et al. (2019). Sol-gel route to nanocrystalline Eu2Ti2O7 films with tailored structural and optical properties. *Journal of the American Ceramic Society*, *102*(11), 6713–6723.

Nanamatsu, S., Kimura, M., Doi, K., Matsushita, S., & Yamada, N. (1974). A new ferroelectric: La2Ti2O7. *Ferroelectrics*, *8*(1), 511–513.

Nemytova, O., Piir, I., Koroleva, M., Perov, D., & Rinkevich, A. (2020). Magnetic properties of nanocomposite and bulk rare earth titanates Ho2Ti2O7 and Yb2Ti2O7. *Journal of Magnetism and Magnetic Materials*, *494*, 165800.

Niu, H., Gou, H., Ewing, R. C., & Lian, J. (2012). First principles investigation of structural, electronic, elastic and thermal properties of rare-earth-doped titanate Ln2TiO5. *AIP Advances*, *2*(3), 032114.

Onoda, S., & Tanaka, Y. (2011). Quantum fluctuations in the effective pseudospin-1 2 model for magnetic pyrochlore oxides. *Physical Review B*, *83*(9), 094411.

Orihashi, T., Nakamura, T., & Adachi, S. (2016). Synthesis and unique photoluminescence properties of Eu2Ti2O7 and Eu2TiO5. *Journal of the American Ceramic Society*, *99*(9), 3039–3046.

Otsuka-Yao-Matsuo, S., Omata, T., & Yoshimura, M. (2004). Photocatalytic behavior of cerium titanates, CeTiO4 and CeTi2O6 and their composite powders with SrTiO3. *Journal of Alloys and Compounds*, *376*(1–2), 262–267.

Padture, N. P., Gell, M., & Jordan, E. H. (2002). Thermal barrier coatings for gas-turbine engine applications. *Science (New York, N.Y.)*, *296*(5566), 280–284.

Pal, A., Singh, A., Ghosh, A., & Chatterjee, S. (2018). High temperature spin-freezing transition in pyrochlore Eu2Ti2O7: A new observation from ac-susceptibility. *Journal of Magnetism and Magnetic Materials*, *462*, 1–7.

Pan, T.-M., Lin, J.-C., Wu, M.-H., & Lai, C.-S. (2009). Structural properties and sensing performance of high-k Nd2TiO5 thin layer-based electrolyte–insulator–semiconductor for pH detection and urea biosensing. *Biosensors and Bioelectronics*, *24*(9), 2864–2870.

Pan, T.-M., Huang, M.-D., Lin, C.-W., & Wu, M.-H. (2010a). A urea biosensor based on pH-sensitive Sm2TiO5 electrolyte–insulator–semiconductor. *Analytica Chimica Acta*, *669*(1–2), 68–74.

Pan, T.-M., Huang, M.-D., Lin, W.-Y., & Wu, M.-H. (2010b). Development of high-κ HoTiO3 sensing membrane for pH detection and glucose biosensing. *Sensors and Actuators B: Chemical*, *144*(1), 139–145.

Pan, T.-M., & Lin, C.-W. (2010). Structural and sensing characteristics of Dy2O3 and Dy2TiO5 electrolyte – insulator – semiconductor pH sensors. *The Journal of Physical Chemistry C, 114*(41), 17914–17919.

Pan, T.-M., Chang, K.-Y., Wu, M.-H., Tsai, S.-W., Ko, F.-H., & Chi, L. (2011). Development of high-κ Eu 2 Ti 2O 7 EIS devices using poly-N-isopropylacrylamide as an enzyme encapsulation material for pH detection and uric acid biosensing. In *5th European Conference of the International Federation for Medical and Biological Engineering*. Springer.

Pan, T.-M., & Lu, C.-H. (2011). Effect of postdeposition annealing on the structural and electrical properties of thin Dy2TiO5 dielectrics. *Thin Solid Films, 519*(22), 8149–8153.

Pan, T.-M., Liao, P.-Y., Chang, K.-Y., & Chi, L. (2013). Structural and sensing characteristics of Gd2Ti2O7, Er2TiO5 and Lu2Ti2O7 sensing membrane electrolyte–insulator–semiconductor for bio-sensing applications. *Electrochimica Acta, 89*, 798–806.

Pan, W., Wan, C., Xu, Q., Wang, J., & Qu, Z. (2007). Thermal diffusivity of samarium–gadolinium zirconate solid solutions. *Thermochimica Acta, 455*(1–2), 16–20.

Pang, M., Lin, J., Fu, J., & Cheng, Z. (2004). Luminescent properties of Gd2Ti2O7: Eu3 + phosphor films prepared by sol–gel process. *Materials Research Bulletin, 39*(11), 1607–1614.

Panitz, J.-C., Mayor, J.-C., Grob, B., & Durisch, W. (2000). A Raman spectroscopic study of rare earth mixed oxides. *Journal of Alloys and Compounds, 303*, 340–344.

Panneerselvam, G., Krishnan, R. V., Nagarajan, K., & Antony, M. (2011). Heat capacity and thermal expansion of samarium titanate. *Materials Letters, 65*(12), 1778–1780.

Patwe, S. J., Katari, V., Salke, N. P., Deshpande, S. K., Rao, R., Gupta, M. K., et al. (2015). Structural and electrical properties of layered perovskite type Pr2Ti2O7: Experimental and theoretical investigations. *Journal of Materials Chemistry C, 3*(17), 4570–4584.

Perera, D. S., Stewart, M. W., Li, H., Day, R. A., & Vance, E. R. (2002). Tentative Phase Relationships in the System CaHfTi2O7-Gd2Ti2O7 with up to 15 mol% Additions of Al2TiO5 and MgTi2O5. *Journal of the American Ceramic Society, 85*(12), 2919–2924.

Petrenko, O., Lees, M., & Balakrishnan, G. (2011). Titanium pyrochlore magnets: How much can be learned from magnetization measurements? *Journal of Physics: Condensed Matter, 23*(16), 164218.

Petrenko, O., Lees, M., Balakrishnan, G., Glazkov, V., & Sosin, S. (2012). *Novel magnetic phases in a Gd2Ti2O7 pyrochlore for a field applied along the [100] axis*. arXiv:1203.6326.

Poole, A., Wills, A., & Lelievre-Berna, E. (2007). Magnetic ordering in the XY pyrochlore antiferromagnet Er2Ti2O7: A spherical neutron polarimetry study. *Journal of Physics: Condensed Matter, 19*(45), 452201.

Popov, V., Menushenkov, A., Molokova, A. Y., Ivanov, A., Rudakov, S., Boyko, N., et al. (2020). Rearrangement in the local, electronic and crystal structure of europium titanates under reduction and oxidation. *Journal of Alloys and Compounds, 831*, 154752.

Prasadarao, A., Selvaraj, U., Komarneni, S., & Bhalla, A. S. (1991). Grain orientation in sol-gel derived Ln2Ti2O7 ceramics (Ln = La, Nd). *Materials Letters, 12*(5), 306–310.

Qiu-Rong, X., Jian, Z., Dong-Min, Y., Qi-Xun, G., & Ning, L. (2015). Krypton ion irradiation-induced amorphization and nano-crystal formation in pyrochlore Lu2Ti2O7 at room temperature. *Chinese Physics B, 24*(12), 126103.

Rabanal, M., Várez, A., Amador, U., y Dompablo, E. A., & García-Alvarado, F. (1999). Structure and reaction with lithium of tetragonal pyrochlore-like compound Sm2Ti2O7. *Journal of Materials Processing Technology, 92*, 529–533.

Rahimi-Nasrabadi, M., Mahdavi, S., & Adib, K. (2017). Controlled synthesis and characterization of Dy 2 Ti 2 O 7 nanoparticles through a facile approach. *Journal of Materials Science: Materials in Electronics*, *28*(21), 16133–16140.

Rahimi-Nasrabadi, M., Ghaderi, A., Banafshe, H. R., Eghbali-Arani, M., Akbari, M., Ahmadi, F., ... Sobhani-Nasab, A. (2019). Preparation of Co 2 TiO 4/CoTiO 3/ Polyaniline ternary nano-hybrids for enhanced destruction of agriculture poison and organic dyes under visible-light irradiation. *Journal of Materials Science: Materials in Electronics*, *30*(17), 15854–15868.

Ramirez, A. P., Hayashi, A., Cava, R. J., Siddharthan, R., & Shastry, B. (1999). Zero-point entropy in 'spin ice'. *Nature*, *399*(6734), 333–335.

Ross, K. A., Savary, L., Gaulin, B. D., & Balents, L. (2009). Two-dimensional Kagome correlations and field induced order in the ferromagnetic X Y pyrochlore Yb 2 Ti 2 O 7. *Physical Review Letters*, *103*(22), 227202.

Ross, K., Ruff, J., Adams, C., Gardner, J., Dabkowska, H., Qiu, Y., et al. (2011). Quantum excitations in quantum spin ice. *Physical Review X*, *1*(2), 021002.

Saha, S., Prusty, S., Singh, S., Suryanarayanan, R., Revcolevschi, A., & Sood, A. (2011). Pyrochlore "dynamic spin-ice" Pr2Sn2O7 and monoclinic Pr2Ti2O7: A comparative temperature-dependent Raman study. *Journal of Solid State Chemistry*, *184*(8), 2204–2208.

Salavati-Niasari, M., Soofivand, F., Sobhani-Nasab, A., Shakouri-Arani, M., Faal, A. Y., & Bagherid, S. (2016). Synthesis, characterization, and morphological control of ZnTiO3 nanoparticles through sol-gel processes and its photocatalyst application. *Advanced Powder Technology*, *27*(5), 2066–2075.

Savary, L., & Balents, L. (2012). Coulombic quantum liquids in spin-1/2 pyrochlores. *Physical Review Letters*, *108*(3), 037202.

Savary, L., Ross, K. A., Gaulin, B. D., Ruff, J. P., & Balents, L. (2012). Order by quantum disorder in Er 2 Ti 2 O 7. *Physical Review Letters*, *109*(16), 167201.

Scharffe, S., Kolland, G., Valldor, M., Cho, V., Welter, J., & Lorenz, T. (2015). Heat transport of the spin-ice materials Ho2Ti2O7 and Dy2Ti2O7. *Journal of Magnetism and Magnetic Materials*, *383*, 83–87.

Schell, J., Kamba, S., Kachlik, M., Maca, K., Drahokoupil, J., Rano, B. R., et al. (2019). Thermal annealing effects in polycrystalline EuTiO3 and Eu2Ti2O7. *AIP Advances*, *9* (12), 125125.

Scheunemann, K., & Müller-Buschbaum, H. (1975). Zur kristallstruktur von Nd2Ti2O7. *Journal of Inorganic and Nuclear Chemistry*, *37*(11), 2261–2263.

Scott, P. R., Midgley, A., Musaev, O., Muthu, D., Singh, S., Suryanarayanan, R., et al. (2011). High-pressure synchrotron X-ray diffraction study of the pyrochlores: Ho2Ti2O7, Y2Ti2O7 and Tb2Ti2O7. *High Pressure Research*, *31*(1), 219–227.

Seo, J., Fompeyrine, J., & Locquet, J.-P. (1998). *Microstructural investigation of La2Ti2O7 thin films grown by MBE. Superconducting and related oxides: physics and nanoengineering III*. International Society for Optics and Photonics.

Setién-Fernández, I., Echániz, T., & González-Fernández, L. (2013). First spectral emissivity study of a solar selective coating in the 150–600 ºC temperature, p. 390–395.

Shamblin, J., Feygenson, M., Neuefeind, J., Tracy, C. L., Zhang, F., Finkeldei, S., et al. (2016). Probing disorder in isometric pyrochlore and related complex oxides. *Nature Materials*, *15*(5), 507–511.

Shastry, B. S. (2003). Spin ice and other frustrated magnets on the pyrochlore lattice. *Physica B: Condensed Matter*, *329*, 1024–1027.

Shcherbakova, L., Mamsurova, L., & Sukhanova, G. (1979a). Rare earth titanates. *Uspekhi Khimii*, *48*(3), 423–447.

Shcherbakova, L. G. e., Mamsurova, L. G., & Sukhanova, G. (1979b). Lanthanide titanates. *Russian Chemical Reviews*, *48*(3), 228.

Shi, J.-W., Zheng, J.-t., & Wu, P. (2009). Preparation, characterization and photocatalytic activities of holmium-doped titanium dioxide nanoparticles. *Journal of Hazardous Materials*, *161*(1), 416–422.

Shimada, Y., Kiyama, H., & Tokura, Y. (2008). Nonreciprocal directional dichroism in ferroelectric Nd2Ti2O7. *Journal of the Physical Society of Japan*, *77*(3), 033706.

Shimakawa, Y., Kubo, Y., & Manako, T. (1996). Giant magnetoresistance in Ti 2 Mn 2 O 7 with the pyrochlore structure. *Nature*, *379*(6560), 53–55.

Shlyakhtina, A., Ukshe, A., & Shcherbakova, L. (2005). Ionic conduction of a high-temperature modification of Lu 2 Ti 2 O 7. *Russian Journal of Electrochemistry*, *41*(3), 265–269.

Shlyakhtina, A., Abrantes, J., Larina, L., & Shcherbakova, L. (2005). Synthesis and conductivity of Yb2Ti2O7 nanoceramics. *Solid State Ionics*, *176*(17–18), 1653–1656.

Shlyakhtina, A., Abrantes, J. C. C., Levchenko, A. V., Knot'ko, A. V., Karyagina, O. K., & Shcherbakova, L. G. (2006). Synthesis and electrical transport properties of Lu2 + xTi2 − xO7 − x/2 oxide-ion conductors. *Solid State Ionics*, *177*(13–14), 1149–1155.

Sobhani-Nasab, A., Hosseinpour-Mashkani, S. M., Salavati-Niasari, M., Taqriri, H., Bagheri, S., & Saberyan, K. (2015a). Controlled synthesis of CoTiO 3 nanostructures via two-step sol–gel method in the presence of 1, 3, 5-benzenetricarboxylic acid. *Journal of Cluster Science*, *26*(4), 1305–1318.

Sobhani-Nasab, A., Hosseinpour-Mashkani, S. M., Salavati-Niasari, M., Taqriri, H., Bagheri, S., & Saberyan, K. (2015b). Synthesis, characterization, and photovoltaic application of NiTiO 3 nanostructures via two-step sol–gel method. *Journal of Materials Science: Materials in Electronics*, *26*(8), 5735–5742.

Sobhani-Nasab, A., & Behpour, M. (2016). Synthesis, characterization, and morphological control of Eu 2 Ti 2 O 7 nanoparticles through green method and its photocatalyst application. *Journal of Materials Science: Materials in Electronics*, *27*(11), 11946–11951.

Sobhani-Nasab, A., Behpour, M., Rahimi-Nasrabadi, M., Ahmadi, F., & Pourmasoud, S. (2019a). New method for synthesis of BaFe 12 O 19/Sm 2 Ti 2 O 7 and BaFe 12 O 19/Sm 2 Ti 2 O 7/Ag nano-hybrid and investigation of optical and photocatalytic properties. *Journal of Materials Science: Materials in Electronics*, *30*(6), 5854–5865.

Sobhani-Nasab, A., Behpour, M., Rahimi-Nasrabadi, M., Ahmadi, F., Pourmasoud, S., & Sedighi, F. (2019b). Preparation, characterization and investigation of sonophotocatalytic activity of thulium titanate/polyaniline nanocomposites in degradation of dyes. *Ultrasonics Sonochemistry*, *50*, 46–58.

Song, H., Peng, T., Cai, P., Yi, H., & Yan, C. (2007). Hydrothermal synthesis of flaky crystallized La 2 Ti 2 O 7 for producing hydrogen from photocatalytic water splitting. *Catalysis letters*, *113*(1), 54–58.

Stefanovich, S. Y., Malhasyan, S., & Venevtsev, Y. N. (1980). Photovoltaic properties and photoconductivity of A2B2O7 ferroelectrics. *Ferroelectrics*, *29*(1), 59–62.

Su, Y., Sui, Y., Wang, X., Cheng, J., Wang, Y., Liu, W., et al. (2012). Large magnetocaloric properties in single-crystal dysprosium titanate. *Materials Letters*, *72*, 15–17.

Subramanian, M., Aravamudan, G., & Rao, G. S. (1983). Oxide pyrochlores—a review. *Progress in Solid State Chemistry*, *15*(2), 55–143.

Sun, L., Ju, L., Qin, H., Zhao, M., Su, W., & Hu, J. (2013). Room-temperature magnetoelectric coupling in nanocrystalline nanocrystalline Pr2Ti2O7. *Physica B: Condensed Matter*, *431*, 49–53.

Suresh, M., Prasadarao, A., & Komarneni, S. (2001). Mixed hydroxide precursors for La2Ti2O7 and Nd2Ti2O7 by homogeneous precipitation. *Journal of electroceramics*, *6*(2), 147–151.

Swami, R., & Sreenivas, K. (2019). *Dielectric properties of La2Ti2O7 ceramics. AIP Conference Proceedings*. AIP Publishing LLC.

Syamala, K., Panneerselvam, G., Subramanian, G., & Antony, M. (2008). Synthesis, characterization and thermal expansion studies on europium titanate (Eu2TiO5). *Thermochimica Acta*, *475*(1–2), 76–79.

Takahashi, J., & Ohtsuka, T. (1989). Vibrational spectroscopic study of structural evolution in the coprecipitated precursors to La2Sn2O7 and La2Ti2O7. *Journal of the American Ceramic Society*, *72*(3), 426–431.

Talebi, R., & Safazade, S. (2016). Auto-combustion preparation and characterization of lanthanum titanate nanoparticles by using tyrosine as fuel and its photocatalyst application. *Journal of Materials Science: Materials in Electronics*, *27*(8), 8294–8298.

Taş, A. C. (2001). Molten salt synthesis of calcium hydroxyapatite whiskers. *Journal of the American Ceramic Society*, *84*(2), 295–300.

Taylor, C. A., Patel, M. K., Aguiar, J. A., Zhang, Y., Crespillo, M. L., Wen, J., et al. (2016). Combined effects of radiation damage and He accumulation on bubble nucleation in Gd2Ti2O7. *Journal of Nuclear Materials*, *479*, 542–547.

Thompson, J. D., McClarty, P. A., Rønnow, H. M., Regnault, L. P., Sorge, A., & Gingras, M. J. (2011). Rods of neutron scattering intensity in Yb 2 Ti 2 O 7: Compelling evidence for significant anisotropic exchange in a magnetic pyrochlore oxide. *Physical Review Letters*, *106*(18), 187202.

Tu, W., Zhou, Y., & Zou, Z. (2014). Photocatalytic conversion of CO2 into renewable hydrocarbon fuels: State-of-the-art accomplishment, challenges, and prospects. *Advanced Materials*, *26*(27), 4607–4626.

Uno, W., Fujii, K., Niwa, E., Torii, S., Miao, P., Kamiyama, T., et al. (2018). Experimental visualization of oxide-ion diffusion paths in pyrochlore-type Yb2Ti2O7. *Journal of the Ceramic Society of Japan*, *126*(5), 341–345.

Vadivel, S., Hariganesh, S., Paul, B., Mamba, G., & Puviarasu, P. (2020). Highly active novel CeTi2O6/g-C3N5 photocatalyst with extended spectral response towards removal of endocrine disruptor 2, 4-dichlorophenol in aqueous medium. *Colloids and Surfaces A: Physicochemical and Engineering Aspects*, *592*, 124583.

Valeš, V., Matějová, L., Matěj, Z., Brunátová, T., & Holý, V. (2014). Crystallization kinetics study of cerium titanate CeTi2O6. *Journal of Physics and Chemistry of Solids*, *75*(2), 265–270.

Verma, A., Goyal, A., & Sharma, R. (2008). Microstructural, photocatalysis and electrochemical investigations on CeTi2O6 thin films. *Thin Solid Films*, *516*(15), 4925–4933.

Vlášková, K., Proschek, P., Pospíšil, J., & Klicpera, M. (2020a). High temperature study on Er 2 Ti 2 O 7 single crystal. *Acta Physica Polonica*, *137*, 753–755.

Vlášková, K., Vondráčková, B., Daniš, S., & Klicpera, M. (2020b). Low-temperature study of an Er2Ti2O7 single crystal synthesized by floating zone technique and simplified feed rod preparation route. *Journal of Crystal Growth*, *546*, 125783.

Wan, Y., & Tchernyshyov, O. (2012). Quantum strings in quantum spin ice. *Physical Review Letters*, *108*(24), 247210.

Wang, H., Du, L., Yang, L., Zhang, W., & He, H. (2016). Sol-gel synthesis of La2Ti2O7 modified with PEG4000 for the enhanced photocatalytic activity. *Journal of Advanced Oxidation Technologies*, *19*(2), 366–371.

Wang, H., Zhang, Y., Ma, Z., & Zhang, W. (2019). Role of PEG2000 on sol-gel preparation of porous La2Ti2O7 for enhanced photocatalytic activity on ofloxacin degradation. *Materials Science in Semiconductor Processing*, *91*, 151–158.

Wang, Z., Zhou, G., Jiang, D., & Wang, S. (2018). Recent development of A 2 B 2 O 7 system transparent ceramics. *Journal of Advanced Ceramics*, *7*(4), 289–306.

Weber, M. J. (2002). *Handbook of optical materials* (Vol. 19). CRC Press.

Wen, J., Sun, C., Dholabhai, P., Xia, Y., Tang, M., Chen, D., et al. (2016). Temperature dependence of the radiation tolerance of nanocrystalline pyrochlores A2Ti2O7 (A = Gd, Ho and Lu). *Acta Materialia*, *110*, 175–184.

Wenger, C., Lupina, G., Lukosius, M., Seifarth, O., Müssig, H.-J., Pasko, S., et al. (2008). Microscopic model for the nonlinear behavior of high-k metal-insulator-metal capacitors. *Journal of Applied Physics*, *103*(10), 104103.

Williams, E. (1926). Note on the magnetic properties of rare earth oxides. *Physical Review*, *27*(4), 484.

Winfield, G., Azough, F., & Freer, R. (1992). DiP224: Neodymium titanate (Nd2Ti2O7) ceramics. *Ferroelectrics*, *133*(1), 181–186.

Winkler, B., Friedrich, A., Morgenroth, W., Haussühl, E., Milman, V., Stanek, C. R., et al. (2014). Compression behavior of Sm 2 Ti 2 O 7-pyrochlore up to 50 GPa: Single-crystal X-ray diffraction and density functional theory calculations. *Chinese Science Bulletin*, *59*(36), 5278–5282.

Wu, M.-H., Lin, T.-W., Huang, M.-D., Wang, H.-Y., & Pan, T.-M. (2010). Label-free detection of serum uric acid using novel high-k Sm2TiO5 membrane-based electrolyte–insulator–semiconductor. *Sensors and Actuators B: Chemical*, *146*(1), 342–348.

Wuensch, B. J., Eberman, K. W., Heremans, C., Ku, E. M., Onnerud, P., Yeo, E. M., et al. (2000). Connection between oxygen-ion conductivity of pyrochlore fuel-cell materials and structural change with composition and temperature. *Solid State Ionics*, *129*(1–4), 111–133.

Xiao, H., Wang, L., Zu, X., Lian, J., & Ewing, R. C. (2007). Theoretical investigation of structural, energetic and electronic properties of titanate pyrochlores. *Journal of Physics: Condensed Matter*, *19*(34), 346203.

Xie, Q., Zhang, J., Dong, X., Guo, Q., & Li, N. (2015). Heavy ion irradiation-induced microstructural evolution in pyrochlore Lu2Ti2O7 at room temperature and 723 K. *Journal of Solid State Chemistry*, *231*, 159–162.

Yadav, P. K., Singh, P., Shukla, M., Banik, S., & Upadhyay, C. (2020). Effect of B-site substitution on structural, magnetic and optical properties of Ho2Ti2O7 pyrochlore oxide. *Journal of Physics and Chemistry of Solids*, *138*, 109267.

Yamamoto, J., & Bhalla, A. (1991a). Microwave dielectric properties of layered perovskite A2B2O7 single-crystal fibers. *Materials Letters*, *10*(11–12), 497–500.

Yamamoto, J. K., & Bhalla, A. S. (1991b). Piezoelectric properties of layered perovskite A 2Ti2O7 (A = La and Nd) single-crystal fibers. *Journal of Applied Physics*, *70*(8), 4469–4471.

Yan, H., Ning, H., Kan, Y., Wang, P., & Reece, M. J. (2009). Piezoelectric ceramics with super-high curie points. *Journal of the American Ceramic Society*, *92*(10), 2270–2275.

Yang, T., Gordon, Z. D., & Chan, C. K. (2013). Synthesis of hyperbranched perovskite nanostructures. *Crystal Growth & Design*, *13*(9), 3901–3907.

Yen, W. M., & Weber, M. J. (2004). *Inorganic phosphors: Compositions, preparation and optical properties*. CRC Press.

Yoshida, M., Koyama, N., Ashizawa, T., Sakata, Y., & Imamura, H. (2007). A new cerium-based ternary oxide slurry, CeTi2O6, for chemical-mechanical polishing. *Japanese Journal of Applied Physics*, *46*(3R), 977.

Zakharov, N., Krikorov, V., Kustov, E., & Stefanovich, S. Y. (1978). New non-linear crystals in the A2B2O7 series. *Physica Status Solidi (A)*, *50*(1), K13–K16.

Zhang, F., & Saxena, S. (2005). Structural changes and pressure-induced amorphization in rare earth titanates RE2Ti2O7 (RE: Gd, Sm) with pyrochlore structure. *Chemical Physics Letters*, *413*(1–3), 248–251.

Zhang, F., Manoun, B., Saxena, S., & Zha, C. (2005). Structure change of pyrochlore Sm 2 Ti 2 O 7 at high pressures. *Applied Physics Letters*, *86*(18), 181906.

Zhang, F., Lian, J., Becker, U., Ewing, R., Wang, L., Hu, J., et al. (2007a). Structural change of layered perovskite La2Ti2O7 at high pressures. *Journal of Solid State Chemistry*, *180*(2), 571–576.

Zhang, F., Lian, J., Becker, U., Wang, L., Hu, J., Saxena, S., et al. (2007b). Structural distortions and phase transformations in Sm2Zr2O7 pyrochlore at high pressures. *Chemical Physics Letters*, *441*(4–6), 216–220.

Zhang, J., Zhang, F., Lang, M., Lu, F., Lian, J., & Ewing, R. C. (2013). Ion-irradiation-induced structural transitions in orthorhombic Ln2TiO5. *Acta Materialia*, *61*(11), 4191–4199.

Zhang, J., Xie, Q., Dong, X., Jiao, X., & Li, N. (2019). Light He and heavy Kr ions irradiation effects in orthorhombic Tb2TiO5 ceramics. *Nuclear Instruments and Methods in Physics Research Section B: Beam Interactions with Materials and Atoms*, *441*, 88–92.

Zhang, N., Wang, H., Li, Y., Li, Q., Huang, S., Yu, Y., et al. (2016). Incipient ferroelectricity and conductivity relaxations in Dy2Ti2O7. *Journal of Alloys and Compounds*, *683*, 387–392.

Zhang, S. Y., Li, B. R., & Liu, R. H. (2013). *The influence of oxide-salt ratio on crystallization of Lu2Ti2O7 phase from molten salt. Applied Mechanics and Materials*. Trans Tech Publications.

Zhang, W., Haolun, L., & Ling, D. (2018). A novel porous Sm2Ti2O7 material synthesized in sol-gel process using polyethylene glycol 4000 as template. *Materials Science*, *24*(4), 417–420.

Zhang, W., Liu, Y., & Li, C. (2018). Photocatalytic degradation of ofloxacin on Gd2Ti2O7 supported on quartz spheres. *Journal of Physics and Chemistry of Solids*, *118*, 144–149.

Zhang, W., Li, H., Ma, Z., Li, H., & Wang, H. (2019). Photocatalytic degradation of azophloxine on porous La2Ti2O7 prepared by sol-gel method. *Solid State Sciences*, *87*, 58–63.

Zhang, Y., Linghong, D., Xinling, P., & Zhang, W. (2009). Influence of annealing temperature on luminescent properties of Eu3 + /V5 + co-doped nanocrystalline Gd2Ti2O7 powders. *Journal of Rare Earths*, *27*(6), 900–904.

Zhao, X., Zhao, Q., Yu, J., & Liu, B. (2008). Development of multifunctional photoactive self-cleaning glasses. *Journal of Non-Crystalline Solids*, *354*(12–13), 1424–1430.

Zhitomirsky, M., Gvozdikova, M., Holdsworth, P., & Moessner, R. (2012). Quantum order by disorder and accidental soft mode in Er 2 Ti 2 O 7. *Physical Review Letters*, *109*(7), 077204.

Zhou, Y., Mei, Z. Y., Ye, H. S., Su, W. W., Zhao, X., & Tang, P. S. (2016). *Preparation and characterization of nanocrystalline La2Ti2O7 by microwave assisted process. Materials Science Forum*. Trans Tech Publications.

Rare-earth-based tungstates ceramic nanomaterials: recent advancements and technologies

7

Ali Salehabadi
Environmental Technology Division, School of Industrial Technology, Universiti Sains Malaysia, Penang, Malaysia

7.1 General introduction

Solid-state materials including nanomaterials are commonly categorized under three major classes: insulators, semiconductors, and conductors (Atkins, Overton, Rourke, Weller, & Armstrong, 2006). These materials can be in the form of metal or in combination with other materials to form alloys, oxides, sulfides, etc. Metal oxides (MOs) and mixed metal oxides (MMOs) are two large groups of solid-state materials with several properties such as porosity, magnetic properties, morphology, and reaction/interaction profiles. These materials can be used in several novel applications in modern industries such as solar cells (Morassaei, Salehabadi, Akbari, & Salavati-Niasari, 2019), fuel cells (Fernandes, Woudstra, van Wijk, Verhoef, & Aravind, 2016), hydrogen storage systems (Salehabadi, Ahmad, Morad, Salavati-Niasari, & Enhessari, 2019), sensors (Enhessari & Salehabadi, 2018), biosensors (Salehabadi & Enhessari, 2019), etc.

In storage techniques, the energy is delivered utilizing the surpassing energy, which can store and recover from the stored site. The final stage comprises in an electrical vitality generation by utilizing either a conventional inside combustion motor or a fuel cell (Amirante, Cassone, Distaso, & Tamburrano, 2016). Physical storage such as compressed gas and solid-state materials-based storage like physisorption or chemisorption are two major categories of energy storage technologies (Esaka, 2004; Jurczyk, Nowak, Szajek, & Jezierski, 2012; Kombarakkaran, Noveron, Helgesen, Shen, & Pietra, 2009; Li, Deng, Zhang, Liu, & Yu, 2012; Zeng et al., 2012). The conventional high-pressure gas/liquid is not capable to fill the gap in "future storage goals." Therefore joint chemical/physical storage systems are urgently required with potential over traditional vitality capacity strategies (Sakintuna, Lamaridarkrim, & Hirscher, 2007). In renewable energy class, hydrogen storage is a fundamental fixing for tackling hydrogen (Yuan, Li, & Wu, 2017). The main solid-state compounds in this field are metal hydrides and alloys, MMOs, MOs, metal organic frameworks, and carbon materials.

In common, negative oxytungstate ions are called tungstates (Williams, 2004). In the mineral kingdom, the simple tetraoxotungstate(VI), or tungstate ion, WO_4^{2-},

Advanced Rare Earth-Based Ceramic Nanomaterials. DOI: https://doi.org/10.1016/B978-0-323-89957-4.00003-7

is present. Like molybdate (Mo), tungstate ions (W^{6+}) in acidic conditions can polymerize to form (hetero)polytungstates. The minerals containing (hetero)polytungstates are rare, and still requires more investigations (Joaquín-Morales et al., 2018). Commercially, very rare numbers of minerals with essential tungstate are exhibited. However, since several performances of these materials are reported, they have great commercial significance. Tungsten (W) in its highest oxidation state, that is, +6, with the standard potential −1.26 V (at room temperature), plays and important roles in the mineralogy of the elements.

High-specific gravity of tungsten is reported; therefore, it is utilized in many academic and industrial applications of applications (Neikov & Yefimov, 2019). For instant, a related electronics application is the vibration mechanism of mobile phones. Tungstates are an important class of visible light responsive photocatalysts for the photocatalytic degradation of pollutants from wastewater. In addition, tungsten itself (tungsten powder) is used as a plastic filler that often ends up in enclosures and housings. Therefore tungsten has many different application areas in electronic enclosures, housings, and packages, yet many applications are often overlooked when a conflict minerals audit is performed (Kaczmarek, Ndagsi, Van Driessche, Van Hecke, & Van Deun, 2015).

Dual phase tungstate complex is a renowned combination of tungsten with other elements, with enhanced properties for the production of advanced materials. For example, tungsten carbide (WC), tungstate alloys like Ni-W, tungstate based MOs like Bi_2WO_6, tungsten disulfide (WS_2) have very high hardness and thus great wear resistance an excellent choice in applications like drill bits and other cutting tool surfaces (Sundaresan, Yamuna, & Chen, 2020). Tungstenite, WS_2, is an extremely rare species, isomorphous with its common congener molybdenite, MoS_2. In addition, mixed metal-based tungstate is also exhibited high melting temperature. Tungsten's very high melting temperature is utilized in electronics applications such as filaments, heating elements, and interconnecting material between the transistors and silicon dioxide (Ningombam & Singh, 2019).

The lanthanoids (Ln) are sometimes referred to as the "rare-earth elements;" however, rare-earth is inappropriate because they are not particularly rare, except for promethium (Pr), which has no stable isotope. The reduction potentials of the lanthanoids are all similar, with values ranging from −1.99 V for Eu^{3+}/Eu to −2.38 V for La^{3+}/La (a honorary member of the f block). Lanthanoids are categorized into two groups: light rare-earth metals (LREMs), and heavy rare-earth metals (HREMs). Lanthanum (La), cerium (Ce), praseodymium (Pr), neodymium (Nd), samarium (Sm) are LREMs, while europium (Eu), gadolinium (Gd), terbium (Tb), dysprosium (Dy), holmium (Ho), erbium (Er), thulium (Tm), ytterbium (Yb), lutetium (Lu), yttrium (Y) are HREMs (Wall, 2021).

Solid-state MMOs-based tungstates with lanthanoids are reported for several applications starting from the 60's and 70's by scientists. The chemical formula of Ln and W are limited into three structures: $Ln_2(WO_4)_3$, $Ln_2W_2O_9$, and $Gd_{14}W_4O_{33}$, where Ln is the charge carrier in these phases. Several categories of these MOs and their respective applications are listed in Table 7.1. In these structures, lanthanoids are used generally their 3 oxidation states (Ln^{3+}) for combination with tungsten. In our upcoming sections, we will discuss the details of the MMOs-based Ln^{3+} and W^{6+}. The details of the important materials from this table will be discussed in Section 7.2. The +3 - oxidation state is common at the left of the period and is the only oxidation state normally encountered for scandium. In this section, various complex oxides based rare-earth elements in combination with tungstate will be discussed in terms of "synthesis" and "structure."

In this chapter, primarily, some important compositions of Ln−W−O (from Table 7.1) will be discussed in terms of their bulk structures and general applications (Section 7.2). In Section 7.3, the crystal structures of the Ln−W−O materials will be explained, both theoretically and experimentally. These discussions will be completed with their synthesis techniques (Section 7.4), their properties (Section 7.5), and applications (Section 7.6).

7.2 Characteristics of common Ln−W−O compounds

Inorganic Ln−W−O materials consist lanthanoids (Ln in its mostly trivalent states), and tungsten (W in its hexavalent states). These materials (as summarized in Table 7.1) have been achieved great interests due to their abilities to perform in various applications ranging from optics to energies and biomedicals. The Ln−W−O inorganic systems are known to be more favorable as compared to other inorganic and organic systems, owing to their superior properties. In this section, some common Ln−W−O materials will be discussed in detail.

7.2.1 *Scandium tungstates*

Scandium (Sc) adopts its group oxidation state when exposed to air and moisture, so these metal-metal bonded compounds are prepared out of contact with air and moisture. Ion transport in $Ln_2(MO_4)_3$ with the scandium (Sc) tungstate (W) structure, where A^{3+} represents a trivalent Sc^{3+}, and M^{6+} is W^{6+}. It is known that trivalent cations could be the mobile charge carriers in these compounds. Indeed, a charge carrier is a particle or quasiparticle that is free to move, carrying an electric charge such as electrons, ions, and holes. In addition, trivalent ions are renowned as the highly interactive carriers which can easily interact/react with adjoining anions. This interaction causes "immobilization." The mobility of A^{3+} in solids are confirmed experimentally such as:

- DC polarization experiments with blocking electrodes
- Cross-sectional electron probe microanalysis

Almost a low activation energy by reductive oxygen loss leads to predominantly electronic conductivity. Low energy defects and mobility are two factors in $Sc_2(WO_4)_3$, which can form by aliovalent doping with Sc^{3+} cations.

Ion transport studies showed that $Sc_2(WO_4)_3$ is an ionic conductor with an ionic transference number 1.00 (Zhou, Neiman, & Adams, 2011). Based on the structure

Table 7.1 Common combinations of Ln^{3+} and W^{6+} in oxides.

Popular RE ions	MMO	Important properties	Application	References
Sc^{3+}	$Sc_2(WO_4)_3$	– Negative thermal expansion – Large structural voids – Reducible – Ionic Conductor	– Lithium-ion batteries – Electronic devices	(Andersen et al., 2018; Zhou, Rao, & Adams, 2011)
Y^{3+}	$Y_2W_3O_{12}$ Y_2WO_6 Y_6WO_{12}	– Negative to zero thermal expansion – Thermal Stability – Mechanical stability – Optical properties	– LED – Composites – Electronic devices	(Cao et al., 2020; Marzano, Pontón, & Marinkovic, 2019)
La^{3+}	$La_2W_2O_9$ $La_2W_3O_{12}$ $La_{14}W_8O_{45}$ $La_6W_2O_{15}$	– High quantum efficiency – Phase stability – Ionic conductor	– Electronic devices	(Cheng, Ren, Lin, Tong, & Miao, 2019)
Ce^{3+}	$Ce_2(WO_4)_3Ce_{10}W_{22}O_{81}$	– Ionic Conductor – Optical properties	– Catalysts – Luminescence	(Anjana, Joseph, & Sebastian, 2010)
Pr^{3+}	$Pr_2(WO_4)_3$	– Magnetic properties	– Catalysts – Drug	(Logvinovich, Arakcheeva, Pattison, Eliseeva, & Tomě, 2010)

		– Optical properties – Electrochemical – Electrocatalytic		
Nd^{3+}	$Nd_2(WO_4)_3$	N.R	N.R	(Weil, Stöger, & Aleksandrov, 2009)
Pm^{3+}	N.R			
Sm^{3+}	$Sm_2(WO_4)_3$	– Photocatalytic properties – Electrochemical activity	– Catalysts – LED – Antiandrogen drug – Sensor	(Sundaresan et al., 2020)
Eu^{3+}	$EuWO_4$	Not reported	Not reported	(López-Moreno, Rodríguez-Hernández, Muñoz, Romero, & Errandonea, 2011)
Gd^{3+}	$Gd_2W_2O_9$ $Gd_{14}W_4O_{33}$ $Gd_2(WO_4)_3$	– Optical properties	– Biological application – Luminescence – FED devices – LED – Sensor	(Inomata, Kishida, Maruyama, & Watanabe, 2015; Nilsson, Grins, Käll, & Svensson, 1996; Wang, Li, Zhang, & Liu, 2019)
Tb^{3+}	$Tb_2(WO_4)_3$	– Optical properties	– Catalyst – Sensor	(Azizi, Darroudi, Soleymani, & Shadjou, 2021)
Dy^{3+}	Dy_2WO_6 $Dy_2(WO_4)_3$	– Optical properties – Electrocatalysis	– Catalyst – Sensor	(Thirumalai, Shanthi, & Swaminathan, 2017; Zhang, Huang, Xu, & Novel, 2020)
Ho^{3+}	Ho_2WO_6	– Optical properties – Electrocatalysis	– Catalyst – Solar cell – Fuel cell	(Thirumalai, Shanthi, & Swaminathan, 2019)
Er^{3+}	$Er_2(WO_4)_3$	– Photovoltaic properties – Magnetic properties	– Catalyst – Solar cell – Electronic devices	(Makhlouf, Abdulkarim, Adam, & Qiao, 2020)

(Continued)

Table 7.1 (Continued)

Popular RE ions	MMO	Important properties	Application	References
Tm^{3+}	N.R			
Yb^{3+}	Yb_6WO_{12}	– Thermal conductor	Not reported	(Zheng et al., 2011)
Lu^{3+}	$Lu_2(WO_4)_3$	– Ionic conductor – Negative thermal expansion	– Luminescence – Composite	(Liu, Secco, Imanaka, & Adachi, 2002)

Note: RE, MMO, and N.R represent rare-earth, mixed metal oxides, and not reported, respectively.

of this MMO, the presence of Sc^{3+}, WO_4^{2+} and O^{2-} govern the mobile charge carrier performance. The conductivity σ of the materials is directly dependent on the carrier concentration (q) and the mobility of the charge carriers (μ) as Eq. 7.1,

$$\sigma = q \times \mu \tag{7.1}$$

Charge carriers in materials particularly MMOs are generated by ionization of the atoms in the lattice structures. In general, all atoms in a conductor structure (here MMOs) are ionized at specific temperatures, where the number of presented electrons becomes constant with temperature (Pavlidis, Savidis, & Friedman, 2017). Upon increasing the temperature, resistance is also increased. This phenomenon is due to the higher rate of collision with the lattice, impurities, and grain boundaries of the conductor material, which can further cause reduction in electrons mobility (Kafizas, Godin, & Durrant, 2017).

7.2.2 Yttrium tungstates

As mentioned in Section 7.1, the tungstate (W) is renowned as a self-activated phosphor. This element can emit under UV excitation (Cao et al., 2020). Metal tungstates can be coordinated regularly to form WO_4^{2-} tetrahedra or sixfold coordinated to form WO_6^{6-} octahedra. Yttrium (Y) is a well-known silvery-metallic metal under the rare-earth elements in the periodic table, which is used widely in LEDs, superconductors, and lasers. Yttrium is lithophile, that is, preferentially partitioned into silicate minerals, and refractory, that is, high melting temperature and mechanical properties potential for high-temperature nuclear systems in its trivalent state (Y^{3+}). The properties of Y are something between the smaller lanthanide such as dysprosium (Dy) and holmium (Ho). The Y^{3+} can be joined W^{6+} (both tetrahedra octahedra) in order to form Y_6WO_{12} (Bharat, Il Jeon, & Yu, 2016), $Y_2(WO_4)_3$ (Marzano et al., 2019) or Y_2WO_6 (Xing, Shen, & Li, 2020).

In WO_6^{6-} (with the charge transfer band at around 300 nm), the local energy can transfer to Y^{3+} ($WO_6^{6-} \rightarrow Y^{3+}$), which can further lead to an intense luminescence property of Y–W–O. For instance, theoretically, the crystal structure of Y_2WO_6 is consisted of two sites: (1) the frameworks of WO_6, and (2) the interstitial Y ions. The Y ions are localized in three different subsites (Y1, Y2, and Y3). Out of three Y, two Y ions form YO_8 decanedron, whereas the third Y ion is seven-coordinated with O^{2-} ions (Cao et al., 2020).

Various compositions of Y–W–O have been utilized for advanced applications such as laser (optical), wireless communications, and electronics. Integrated optics are the technologies of fabrication of integrated (1) optical devices, (2) photonic circuits, or (3) planar light-wave circuits is called integrated optics. The so-called contains combined optical components. Lasers, optical filters, amplifiers, and photodetectors are some examples of integrated optics (Ding et al., 2015). In addition, wireless communications in general definition, it is transmitting on one of the industrial, scientific, and medical (ISM) bands with potential for interference and thus use technologies to protect themselves from interference. Ceramic and piezoelectric materials are used recently for advanced wireless communication systems such as 5 G.

7.2.3 Lanthanum tungstates

In periodic table, the lanthanide series start with lanthanum (La). It is usually in its trivalent oxidation states (La^{3+}) with 1.16 Å ionic radius. As compared to other elements in lanthanides series, this element is more common in various industrial applications, particularly in combination with other elements. The major application of La is in catalysis, and also coordination chemistry such as in cyclopentadienyl lanthanum complexes (Mo & Hou, 2020).

Tungstate ions (W^{6+}) in combination with other materials in the form of oxides have been investigated in various bulk structures and morphologies such as scheelite and wolframite structures. Several applications of tungstate-based MOs have been reported ranging from electrolytes in fuel cells to laser and supercapacitors electrodes. Tungsten (W) can impart several attractive performances in the final properties of the tungsten-based MOs such as photoluminescence, conductive, and magnetic properties.

Lanthanum tungstates (La−W−O) have been widely studied in various compositions such as $La_2W_2O_9$, $La_2W_3O_{12}$, $La_{14}W_8O_{45}$, and $La_6W_2O_{15}$ owing to their unique properties. A monoclinic structure of lanthanum tungstates is reported with a $C2/c\text{-}C_{2h}{}^6$ space group, having Raman-active WO^{4-} groups. Among them, $La_2W_3O_{12}$ has been achieved a great interest due to easy crystal growth, thermal, and mechanical properties (Ran, Deibert, Ivanova, Meulenberg, & Mayer, 2019). In addition, semiconducting photocatalytic performance of $La_2W_3O_{12}$ in environmental studies has been recently found a center of attention. Both wet and dry chemical techniques are used for synthesis of La−W−O materials. However, dry chemical requires serving high amount of energy and time. In addition, the starting materials, that is, WO_3 and La_2O_3 in dry chemical method may decompose at such a high temperature.

7.2.4 Cerium tungstates

Cerium (Ce) is a rare-earth element, most abundant member of the lanthanide in earth's crust, and discovered primarily by Jons J, and developed by Berzelius and Hisinger. Cerium is usually in its two oxidation states: tri- and tetra- valent states (i.e., Ce^{3+} and Ce^{4+}), with important applications in ceramics and catalysts. Cerium is the only material known to have a solid-state critical point. Cerium is used in metallurgy as a stabilizer in alloys and in welding electrodes.

Cerium tungstates are well-known rare-earth tungstates materials with several applications. These combinations lead to fabrication of two major compositions of Ce and W ions such as $Ce_2(WO_4)_3$, $Ce_{10}W_{22}O_{81}$. These materials are synthesized mostly *via* wet chemical techniques and exploited as adsorbent material. The current application is important for the decontamination study of cobalt and europium ions from radioactive waste solutions (Kaczmarek et al., 2015).

In the proposed composition M-WO_4 (where M is a cation), the crystal structure consists of a tetragonal crystal structure (space group $I4_{1/a}$). Here, the tungsten (W) atom is coordinated with four oxygen (O) atoms to form WO_4, and M cation with 8 times coordinated. In $Ce_2(WO_4)_3$ structure, which is the stable form of Ce−W−O, Ce is a trivalent (Ce^{3+}) lacunar scheelite cation, with missed 1/3 of the Ce^{3+}, which linked with distorted WO_4. It is known that in general there are two crystal systems: (1) tetragonal scheelite crystal system, (2) monoclinic distorted scheeite system (Damascena dos Passos, Pereira de Souza, Bernard-Nicod, Leroux, & Arab, 2020).

Cerium tungstates are utilized as supercapacitors. Supercapacitors are devices for energy storage with rapid charge/discharge capability, long cycle lives and stability in various wearable and portable electronic and hybrid vehicles. Cerium tungstates have large specific capacitance, with stable and reversible cyclic performance. These class of materials are thermodynamically favorable for energy storage. In addition, the presence of Ce^{3+} in the structure of $Ce_2(WO_4)_3$ oxide can impart catalytic performances particularly for oxidative coupling of methane (OCM).

7.2.5 Gadolinium tungstates

Gadolinium (Gd: [Xe] $4f^7$ $5d^1$ $6s^2$) is a rare-earth element with atomic number 64. It is a silvery-white metal, slightly malleable and ductile. Gd when expose to oxygen and moisture can oxidized to form a black coating. The number of unpaired electrons in Gd govern the reactivity of this element. One of the main applications of Gd is in magnetic resonance imaging as an image contrast agent (Zeng et al., 2020). Rare-earth elements like Gd are commonly surface active, therefore are important in the metallurgy such as alloying, purification, and metamorphosis of inclusions. In addition, several features of these elements, particularly when mixed with other elements, are fundamental in advanced studies such as energy (batteries, superconductors, solar cells, etc.), optics (laser), and luminescence.

The presence of tungstate ions in the structure of rare-earth elements impart many unusual effects on the mentioned properties of these elements. Various compositions of Gd and W have been reported such as $Gd_2W_2O_9$, $Gd_{14}W_4O_{33}$, and $Gd_2(WO_4)_3$. $Gd_2(WO_4)_3$ is a type of salt that adopts an orthorhombic space group crystallographic structure, which usually shows a ferroelectric−paraelectric phase transition at about 150°C−190°C in an isomorph crystal. The ferroelectric property of $Gd_2(WO_4)_3$ make it suitable for catalytic applications, laser hosts, phosphors. These properties are almost similar in other Gd−W−O salts. In the lattice structure of $Gd_2(WO_4)_3$, tungsten atoms are located into two sites; in one site, W atom is surrounded by four oxygen atoms (i.e., WO_4) with a close tetrahedral conformation, while the other site is rather distorted (Rahimi-Nasrabadi et al., 2017).

The catalytic activities of Gd−W−O nanoparticles have been widely reported by several researchers. Degradation of methylene blue (MB) under UV irradiation shows that the MB can be degraded in the presence of Gd−W−O nanoparticles, superior to the famous TiO_2 nanoparticles. A maximum conversion of 98% is reported for the Gd−W−O nanoparticles as compared to TiO_2 nanoparticles with around 50% of conversion (Rahimi-Nasrabadi et al., 2017).

In the same vein, Periyasamy et al. (2019) synthesized two-dimensional Gd2 (WO4)3 nanoflakes via coprecipitation method, and examined its catalytic activities against degradation of the postharvest fungicide carbendazim (CBZ). Primarily, they found a great electrochemical response for CZB of the $Gd_2(WO_4)_3$ nanoflakes with an interesting sensitivity of 0.39 $\mu A/\mu M/cm^2$. In addition, the $Gd_2(WO_4)_3$ degraded CBZ at 98% efficiency in the presence of visible light irradiation.

7.2.6 Dysprosium tungstates

Dysprosium (Dy) is another rare-earth element, metallic silver, with atomic number 66. Dy concentration is not available in the form of elemental in nature but existed in combination with other lanthanides such as fergusonite, euxenite, etc. Yttrium (Y) ores that contain Dy are the most abundant heavy lanthanides with 7%−8% of Dy. Naturally occurring dysprosium from other elements is composed of the stable isotopes such as ^{156}Dy, ^{158}Dy, ^{160}Dy, ^{161}Dy, ^{162}Dy, ^{163}Dy, and ^{164}Dy. In between, ^{164}Dy and ^{162}Dy are the two most abundant isotopes of Dy with around 54% abundancy as compared to other isotopes.

Mixtures of Dy with other elements are widely investigated for optics, nuclear, and energy storage systems. Particularly, the combinations of Dy with tungsten (W) have been recently highlighted as host for many advanced applications. It is known that the $Dy_2(WO_4)_3$ is formed in a monoclinic lattice, with space group C-2/c, Z = 4. Upon increasing the temperature, the phase transitions from monoclinic to orthorhombic occur. Dy_2WO_6 and $Dy_2(WO_4)_3$ are two well-known composition of Dy−W−O. These materials are used in terms of "catalytic activity" and "self-cleaning" properties. The catalytic activities of Dy−W−O materials can be improved when doped with other MOs like ZnO. The doped Dy_2WO_6-ZnO showed a critical effect on the photocatalytic activity, with high reusability upto four runs. It is also reported a promising potential of Dy_2WO_6-ZnO in methanol fuel cells (Thirumalai et al., 2017). In the same vein, the graphene-Dy_2WO_6-ZnO nanocomposites are utilized for electrochemical supercapacitor applications with capacitance performance upto 752 F/g at the current density of 1 A/g (Raja, Selvakumar, Swaminathan, & Kang, 2021).

7.3 Crystal structures

Lattice structures are commonly used to connect various loads within a volume of space. Lattice structure in solid-state materials is defined as an abstract structure which can obtain from the mathematical theories and abstract algebra. In theory, each element is partially ordered, where every two elements have two fingerprints: a supremum and an infimum. X-ray analyses like X-Ray diffraction analysis are developed for studying the crystal structure of crystalline materials, where the

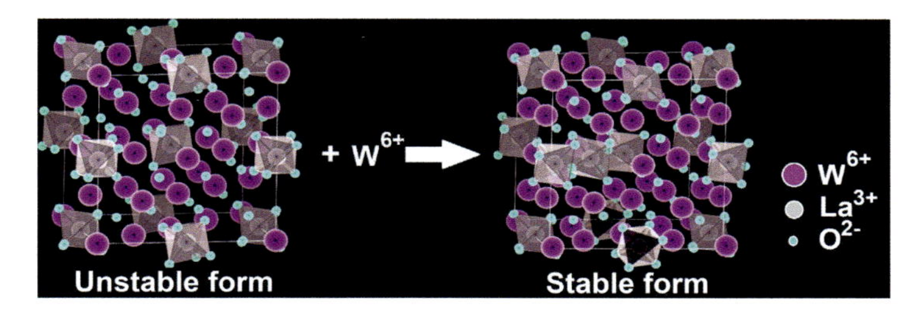

Figure 7.1 The atomic arrangements of two "ideal" compositions from DFT calculations of $La_{28-x}W_{4+x}O_{54+\delta}v_{2-\delta}$ for $x = 0$ and 1.
Source: Reproduced from Magrasó, A. & Haugsrud, R., (2014) Effects of the La/W ratio and doping on the structure, defect structure, stability, and functional properties of proton-conducting lanthanum tungstate $La_{28-x}W_{4+x}O_{54+\delta}$, A review, *Journal of Material Chemistry A. 2*, 12630–12641. https://doi.org/10.1039/c4ta00546e.

scattering of X-rays by periodic array of atoms cause definite diffraction patterns. This pattern is a quantitative image of atomic arrangements within the crystal lattice. Here, all presented materials in Table 7.1 are listed based on their crystal structures and parameters.

Lanthanoids and tungsten are generally formed in polyhedron form of $Ln_2(WO_4)_3$ or $Ln_2W_3O_{12}$ consisting LnO_6 octahedra and WO_4 tetrahedra. It is reported that the heat treatment plays a crucial role in determining the structure of this class of materials. For example, in the thermal expansion profile of $Ho_2W_3O_{12}$, a monoclinic structure is obtained when sample is cooled down slowly in furnace, while an orthorhombic structure is observed when sample is quenched in water (fast cooling) (Xiao et al., 2008).

The chemical formula of rare-earth tungstate primarily emanated from their respective MOs, Ln_2O_3 and WO_3, in order to form single phase Ln_6WO_{12}. Further studies have led to introduce several compositions of tungstate with large rare-earth cations, having a range of the Ln^{3+}/W^{6+} ratio. In these complementary studies, it was found that Ln_6WO_{12} is not thermodynamically stable below 1740°C, but more stable rather in more tungsten rich phase. For example, in the case of La^{3+} and W^{6+}, the proposed compositions like La_6WO_{12} or $La_{10}W_2O_{21}$ are not single phase. Therefore, the primary assumption (mentioned above) depends on the temperature of calcination. These two compositions, that is, La_6WO_{12} and $La_{10}W_2O_{21}$ refer to the same material of a more complex formula such as $La_{28-x}W_{4+x}O_{54+3x/2}$. The JCPDS card for these two compositions is filled as 30–0687 and 30–0686. Based on the reflections and peak intensities, it can be deduced that these two compositions are almost identical, with only a small difference in the lattice parameters (Magrasó & Haugsrud, 2014).

The lattice structure of the tungstates with other lanthanide cations have not been reported in detail. Computational studies support experimental observations, to fulfill the shortcomings in experimental results. For example, the density of La_6WO_{12} powders is measured experimentally to be around $\sim 6.4\ g/cm^3$. In the same vein, the computational studies are reported to be around 5.75 g/cm^3, in a crystal with the stoichiometry amounts of the ions, and the cell parameter ~ 11.19 Å. The lower density implies that, (1) the unit cell contains more than 24 lanthanum ions, (2) more than 4 tungstens, and more than 48 oxygens. It is theoretically confirmed that the La occupies two Wyckoff positions. The first La resides fully in its position, while the second La close to full occupancy. An almost high thermal expansion factors of second La indicated disorder in the respective sublattices (Magrasó, Frontera, Marrero-López, & Núñez, 2009). Experimentally, high resolution X-ray synchrotron also confirmed the obtaining results from theoretical-calculations, that is, a cubic cell, with the atomic positions of the structural model reported in previous reference (Magrasó et al., 2009). Fig. 7.1 clearly shows the bulk lattice structure of the unstable (stochiometric) and stable (nonstochiometric) forms of La-W-O, where the stable form has nearly one more tungsten in the lanthanum sites.

In summary, rare-earth tungstates with the general chemical formula $Ln_2(WO_4)_3$ are interesting materials due to their negative thermal expansion behavior. This structure of this class of materials are considered as an ordered defect variant with a threefold scheelite supercell. In this structure, one rare-earth site is unoccupied, thus leading to above formula and distortions of the coordination polyhedra consequently. The Ln^{3+} cations are coordinated by eight oxygen atoms in the form of a distorted bi-capped trigonal prism. The W atoms are tetrahedrally surrounded by oxygen atoms.

7.4 Synthesis techniques

Several approaches have been developed to find appropriate methods for the synthesis of Ln−W−O based ceramics. As we will discuss later (Section 7.5), the rare-earth tungstate materials have negative to zero thermal expansion coefficients which can affect the final properties of the products. In addition, the Ln−W−O materials are potential in terms of "energy storage" and "energy conversion" such as sensing devices, solar cells, and fuel cells. In the previous sections, we have collected and listed the most abundant MMOs-based Ln−W−O (nano)structures and their respective applications (see Table 7.1). In this section, primarily, the common synthesis methods (wet and dry chemical methods) for production of MMOs will be discussed. At the end of this section, the common preparation methods for synthesis of $Ln_xW_yO_z$ materials will be presented.

7.4.1 Wet chemical methods

These are several popular techniques for the preparation of a wide range of nanomaterials at lower temperature as compared to the dry chemical methods. In wet chemical methods, a series of chemical reaction of the precursor species take place in order to produce nucleation followed by formation of nanomaterials. In general,

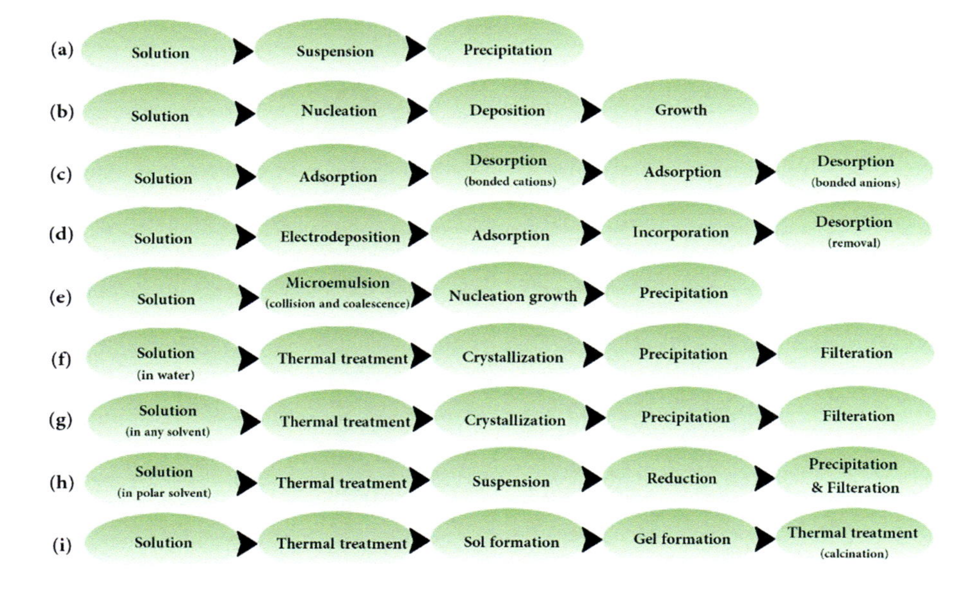

Figure 7.2 Wet-chemical synthesis methods for production of mixed metal oxide NPs including $Ln_2(WO_4)_3$; (A) Chemical precipitation, (B) Chemical bath deposition (CBD), (C) Successive ionic layer adsorption and reaction (SILAR), (D) Electrochemical deposition, (E) Microemulsion, (F) Hydrothermal, (G) Solvothermal, (H) Polyol, and (I) Sol-gel.

the obtaining particles from these methods have broad particle size distribution with complex morphology. This deficiency can be prevented by changing the precursors, synthesis parameters and dynamics of chemical reaction. In addition, provisional sufficient supersaturation is another factor which govern the morphology of the final product. This method performed almost at low temperatures, ambient conditions using simple equipment, which make these methods economical and more popular (Majid & Bibi, 2018).

Wet chemical methods offer good control of stoichiometry to mix the precursors at atomic scale. The materials of good stoichiometry and size ranging from 1 nm to a few microns can be prepared and easily be stabilized using capping agents. Other crucial points in these methods can be categorized as:

- Reactant concentration
- Precursor pH
- Mixing temperature
- Reaction time
- Calcination temperature
- Calcination time

The production has implemented in several pass ways such as:

1. Chemical precipitation: It is the process of conversion solution → solid by converting the substance into insoluble form. The overall reaction profile is like Fig. 7.2A.

2. Chemical bath deposition (CBD): It involves several techniques that produce films of solid inorganic, nonmetallic compounds on substrates by immersing the substrate in a precursor solution (De Guire, Bauermann, Parikh, & Bill, 2013) (Fig. 7.2B).
3. Successive ionic layer adsorption and reaction (SILAR): It is based on sequential immersion of the substrate into separately placed cations and anions for making uniform and large area thin films (Fig. 7.2C). The method is inexpensive, simple and convenient for large area (Pathan & Lokhande, 2004).
4. Electrochemical deposition: It is a technique that has been widely used for the synthesis of metal nanoparticles. Electrochemical deposition occurs at the interface of an electrolyte solution containing the metal to be deposited and an electrically conductive metal substrate (Arulmani, Anandan, & Ashokkumar, 2018) (Fig. 7.2D).
5. Microemulsion: This method is used for the formation of various solid-state nanoparticles (NPs) at which dual phases immiscible liquids dispersed in another (for example oil and water) (Singh, Madhu, & Awasthi, 2013). This reaction occurs in the presence of atleast one surfactant. Since this method is thermodynamically favorable, NPs can be continuously grown within the micron size water droplets (Fig. 7.2E).
6. Hydrothermal: Here a chemical reaction occurs in water, and in a sealed pressure stainless steel vessel (called reactor), under controllable temperature and pressure, for almost a long time (Thirumalai et al., 2017). During crystallization processes, growing crystals/crystallites tend to reject impurities (Fig. 7.2F).
7. Solvothermal: It is a chemical reaction in a solvent at temperatures slightly higher than the solvent boiling point and pressures above atmospheric pressure. The reaction profile (medium) can be any solvents ranging from organic to inorganic solvent (Shaikh, Ubaidullah, Mane, & Al-Enizi, 2020). In case of using water as a solvent, the method is called "hydrothermal." Both hydrothermal and solvothermal occur in a closed reactor and under controlled temperature, pressure, and time (Fig. 7.2G).
8. Polyol: It is a nonaqueous synthesis method of metal NPs, polyol-based processes that involve the thermal reduction of a precursor (generally a metal salt), with a polymeric stabilizer dissolved in an organic solvent (Rao & Cölfen, 2017). A well-known reaction profile is the reduction of metal salts in a liquid α-diol (a highly polar solvent), which allows the conversion of metal salts into metal NPs at almost high boiling point (Ott et al., 2015). The particles can be finally collected by centrifugation/decantation (Fig. 7.2H).
9. Sol-gel: It involves the formation of inorganic colloidal suspension or "sol" followed by gelation of the sol. The continuous liquid phase after sol formation is called "gel." Gel is in fact a 3D network structure. The gel requires further heating in order to form highly crystallin NPs (Yilmaz & Soylak, 2019) (Fig. 7.2I).

In summary, the wet chemical techniques are widely used for production of various structures of crystalline $Ln_xW_yO_z$, with several applications in science and technology. Through these techniques, selective surface structures with controlled shapes and sizes can be obtained, which can further lead to a set of desired properties.

7.4.2 Dry-chemical methods

Dry chemical method or solid-state chemical method is the simplest and traditional method for the production of single crystal and polycrystalline materials. In this technique, the overall reaction profiles occur in a single step at high temperature. Though the number of acting parameters in this technique is few, however this

Table 7.2 Synthesis routes for preparation of crystalline $Ln_xW_yO_z$.

Materials	Method	Starting materials	References
$Ho_2W_3O_{12}$	– Solid state	– Ho_2O_3 & WO_3	(Xiao et al., 2008)
$Er_2(WO_4)_3$	– Ultrasonication	– $Er(NO_3)_3 0.5H_2O$ & $Na_2WO_4 0.2H_2O$	(Makhlouf et al., 2020)
$Dy_2(WO_4)_3$	– Hydrothermal	– $(Dy_2O_3 + HNO_3)$ & Na_2WO_4	(Zhang et al., 2020)
$Gd_2(WO_4)_3$	– Sol-gel – Coprecipitation	– $Gd(NO_3)_3$ & $H_{26}N_6O_{41}W_{12}0.18H_2O$ – $GdCl_3 0.6H_2O$ & $Na_2WO_4 0.2H_2O$	(Liu et al., 2013; Wang et al., 2012)
$Ce_2(WO_4)_3$	– Hydrothermal	– $Ce(OAc)_3$ & $(Na_2WO_4 + H_2O)$	(Kaczmarek et al., 2015)
$Y_2W_3O_{12}$	– Solid state – Coprecipitation	– Y_2O_3 & WO_3 – $Y(NO_3)_3 \cdot \times H_2O$ & $Na_2WO_4 0.2H_2O$	(Cao et al., 2020; Marzano et al., 2019)

technique is relatively slow, difficult to control, and needs high amount of energy ($500 < \text{temperature} < 2000°C$) (Ben Smida et al., 2020). The growth of crystals in dry-chemical method happens in five major steps:

- Reorganization of the atoms: the reorganization is defined as the distortion of the atoms from their relaxed configurations. This reorganization needs energy, which is known as "reorganization energy."
- Adsorption: this takes place on the surface of the solid, and also known as physisorption.
- Diffusion: in this step, the individual molecules or atoms move from an area of higher concentration to a lower concentration.
- Fixation: this step is sometimes called "termination," where any ongoing reactions/interactions stop, and the atoms locate on their final sites.
- Aggregation: in aggregation, a significant increase in the average crystal size takes place. The current happens slightly at temperatures larger than the aggregation point.
- The conventional solid-state synthesis routs have been used for production of Ln-W-O crystals. For example, Xiao and his coworkers (Xiao et al., 2008) synthesized $Ho_2W_3O_{12}$ and $Tm_2W_3O_{12}$ *via* solid-state reaction using the analytical grade of Ho_2O_3, Tm_2O_3, and WO_3 (purity 99.5%). They ground all starting material for about 1 hour and treated thermally primarily at 800°C and then at 900°C for 24 hours.

$Ln_2(WO_4)_3$ (Ln is lanthanoid (III) such as La, Sm, Eu, Gd, etc.) is utilized to study their electrical properties (Pestereva, Guseva, Vyatkin, & Lopatin, 2017). In this study, the samples are prepared *via* solid-state method from their oxide powders (i.e., Ln_2O_3 and WO_3). The results illustrated the formation of the extra-pure grade products in air. It is expressed that the purity of the samples is directly related to the calcination temperature (in the range of 700°□-1000°□) in five steps and annealing time (from 10 to 30 hours).

In summary, though the solid-sate method is a simple method for production of crystalline MMOs such as $Ln_2(WO_4)_3$, however, the method is not kinetically and

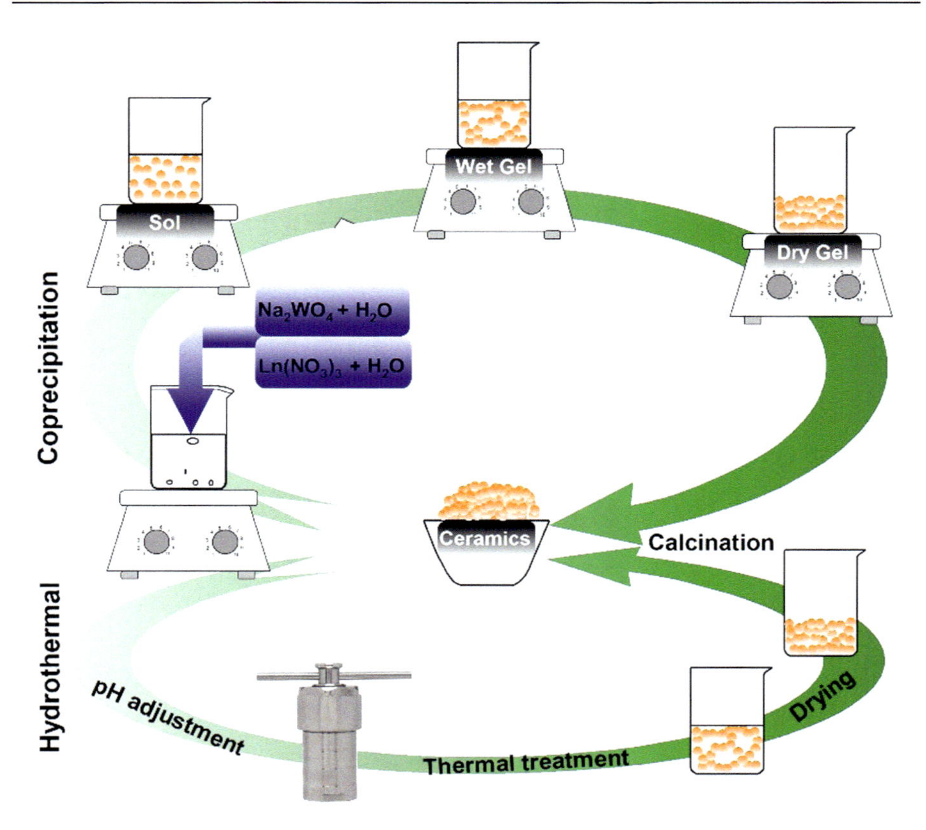

Figure 7.3 Preparation of rare-earth tungetates (Ln–W–O) via coprecipitation and hydrothermal method.

thermodynamically favorable. In addition, the thermal treatment at high temperature can decompose the desired products.

7.4.3 Preparation of rare-earth-based tungstates (Ln–W–O)

Both wet and dry chemical methods are arranged for synthesis of Ln–W–O materials. Wet chemical synthesis method is a widely used technique for growing $Ln_xW_yO_z$ crystals such as hydrothermal, solvothermal, sol–gel method, ultrasonic, and coprecipitation since these synthesis routes offer a high degree of controllability and reproducibility for 2D nanomaterial fabrication. In addition, several templated and nontemplated methods have been examined for controlling the size and shape of nanocrystals, for example, the use of surfactants and polymers. Table 7.2 shows some examples of synthesis methods (both wet and dry chemical techniques) for production of $Ln_xW_yO_z$.

Hydrothermal synthesis is one of the widespread techniques for production of highly dense and pure $Ln_xW_yO_z$ crystals, and reported as a successful technique for synthesis of Gd_2WO_6 (Karuppaiah et al., 2019), Dy_2WO_6 (Zhang et al., 2020), Y_2WO_6 (Kaczmarek, Van Hecke, & Van Deun, 2014), Ho_2WO_6 (Thirumalai et al., 2019), $Ce_2(WO_4)_3$ (Kaczmarek et al., 2015), etc. In this technique, water soluble starting materials used as the starting materials such as hydrated sodium tungstate ($Na_2WO_4.xH_2O$) and lanthanoids acetates ($Ln(CH_3CO_2)_3.xH_2O$) or nitrates ($Ln(NO_3)_3.xH_2O$). The starting materials, after pH adjustments, are primarily dissolved in water and then treated hydrothermally in a Teflon lined stainless steel autoclave under controllable temperature and time. The obtaining precipitates from autoclave are dried in air and finally calcined in order to obtain a highly crystalline and pure sample.

In the same vein, coprecipitation method is used for synthesis of $Ln_xW_yO_z$ (nano)materials. In this easy, low cost, and reproducible method, stoichiometric amount of water-soluble metal salts (like those in hydrothermal method) is mixed and then treated thermally. The thermal treatment in this method is important in terms of "particle formation." The overall admixture during heating slowly from sol, wet gel, dry gel. The final step is formation of dense particles after calcination at an appropriate temperature. Fig. 7.3 illustrates the hydrothermal and coprecipitation methods. This method is utilized for production of $La_2(WO_4)_3$ (Kuriakose, H, Jose, John, & Varghese, 2020), $Gd_2(WO_4)_3$ (Mosleh, 2017), $Y_2W_3O_{12}$ (Marzano et al., 2019), etc.

Sonochemical method is also recently used for production of Ln–W–O materials. Samarium tungstate ($Sm_2W_3O_{12}$) and praseodymium tungstate ($Pr_2W_3O_{12}$) nanoparticles are synthesized via sonochemical assisted precipitation method. The former is synthesized using stoichiometric amount of $Sm(NO_3)_3$ and Na_2WO_4 solutions. This admixture is then sonicated with a power of 60 W for 1 hour. The fine yellowish precipitates obtained from this step is then centrifuged, collected, washed and dried at 50°C overnight. Finally, the obtaining solid materials are annealed at 800°C for 4 hours (Sundaresan et al., 2020). The later ($Pr_2W_3O_{12}$) is also synthesized in a similar way, where 50 mL solution of 0.2 M $Pr(NO_3)_3$ is mixed with a solution of 0.3 M of Na_2WO_4 to obtain greenish precipitate. The reaction container is then sonicated for 30 minutes with the passed energy of 67.3 kJ. The reaction profile is completed after stirring at 700 rpm for 2 hours. The product from this step is then centrifuged, washed, dried (at 50°C), and calcined at 600°C for 3 hours (Yamuna, Sundaresan, Chen, & Shih, 2020).

Traditional chemical decomposition or solid-state reaction is one of the facile methods for production of $Ln_xW_yO_z$ materials. In this method, a mixture of solid reagents, mostly MOs, is treated thermally at almost high temperature ($>1000°C$) to synthesize new solid-state MMOs. For instant, yttrium tungstate (Y_2WO_6) is synthesized *via* a solid reaction process using Y_2O_3 and WO_3, where these MOs are directly mixed in a ball mill for 4 hours. Final step is calcination at 1200°C for 4 hours. After calcination, the sample is crashed and sieved to collect homogenous product with uniform morphology. Though this method is a simple method, however, this technique is slow, difficult to control, not suitable for mass production and needs high amount of energy.

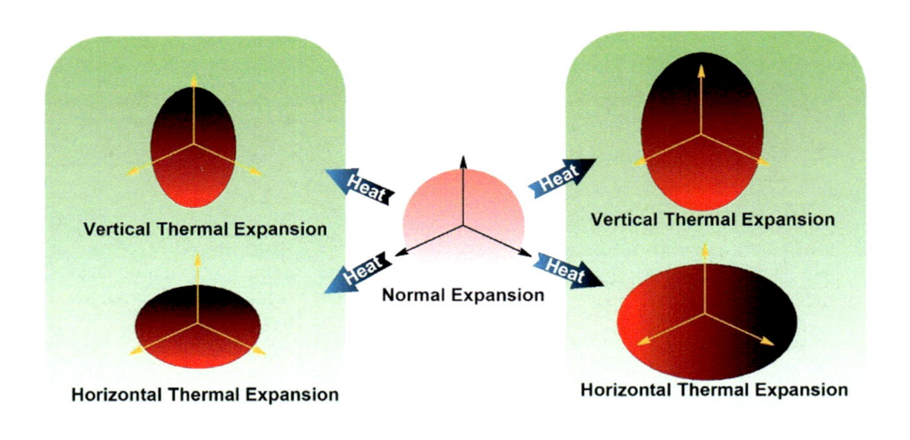

Figure 7.4 Negative and positive thermal expansions.

7.5 Common properties

As mentioned before, MMOs-based $Ln_2(WO_4)_3$ are renowned as prime candidates for utilizing in various application, owing to their superior properties. This section introduces common (major) properties of these materials by providing some introductory background on ionic conduction and thermal expansion.

7.5.1 Ionic conduction

The ionic conduction in solid-state materials is defined as the tendency of materials toward ions movements from one site to another. This happens particularly within the defects of the crystal lattice. The term ionic bonding is an important factor of many compounds which allows ionic diffusion. In other words, the ionic diffusion in an electrical field is called ionic conduction. The ionic conduction is almost a new term in solid-state materials and has been a subject of interest when the "ceramic" materials have been commercialized for several industrial applications. The term "fast ion conductor" has been defined when mobile ions are complexed with the host ceramics. The ionic conduction is important in energy storage systems such as battery, fuel cell, and sensor technologies. Solids exhibit levels of ionic conductivity comparable to those of liquids.

The ionic conduction properties of Ln−W−O materials have been investigated for energy applications. Zhou et al. (2011) reported the ionic conductivity of Sc−W−O by studying the structural defects. They found that the intrinsic formation of polyatomic defects in $Sc_2(WO_4)_3$ structures govern the ionic conductivity. In this structure, there are two types of defects; Frenkel and Schottky, and the defects of the WO_4^{2-} tetrahedron are thermodynamically favorable as compared to isolated W and O defects.

7.5.2 Thermal expansion

In material sciences, the propensity of materials to alter in volume in response to temperature (but not phase transition) is called thermal expansion. It is logical when the materials are heated, its particles move. This continuous motion leads to higher average separation. There are two categories of thermal expansions: Negative and positive (Fig. 7.4). The empty internal cages cause negative thermal expansion (NTE), while the existence of additional atoms or molecules in internal voids is called positive thermal expansion (PTE). The small concentration of intercalated ions has a strong influence through steric hindrance of ion vibrations, which directly controls the thermal expansion (Chen et al., 2017).

Rare-earth tungstate materials (Ln−W−O) are mostly NTE, without volume producing phase alterations. These characteristics along with almost high thermal conductivity are characteristics that greatly increase its productivity. Several applications of NTE crystal materials have been proposed such as composites. Low thermal expansion in composite industries can improve the controllability of bulk thermal expansion. Zero−TE, or close to zero−TE, is highly recommended for optics, and electronics. Among all rare-earth tungstate materials, $Sc_2(WO_4)_3$ has been achieved a great interest due to its isotropic, and NTE over a wide range of temperature from −263°C to 927°C (Liu, Cheng, & Yang, 2012).

7.6 Common applications

Rare-earth-based tungstates (nano)materials are promising family of MMOs synthesized in two compositions: stoichiometric or nonstoichiometric, typically in $Ln_2(WO_4)_3$ (Ln^{3+}: summarized in Table 7.1) structure. These solid-state materials have been recently achieved great research interest in academia for various industrial applications. In this section, major applications of Ln−W−O materials will be discussed including composite, solar cell, wastewater, and fuel cell.

7.6.1 Composite technology

(Nano) composite materials have been commercialized some decades ago for various industrial applications. There are several advantages of using composite materials as compared to conventional one phase matrices such as easy processability, and design flexibility. In addition, the composite materials are thermodynamically and kinetically favorable, owing to their reinforcement effects of the fibers and/or particles.

As discussed before, negative thermal expansion (NTE) is the fingerprint of almost all Ln−W−O materials. Among them, $Sc_2(WO_4)_3$ illustrates a stronger NTE as compared to other NTE materials. The combination of $Sc_2(WO_4)_3$ with positive thermal expansion

(PTE) can impart several features into the final properties of the composites. In a study by Lio and his coworkers (Liu et al., 2012), the development of low thermal expansion $Sc_2(WO_4)_3/ZrO_2$ and $Sc_2(WO_4)_3/Cu$ are investigated. They reported that the thermal expansion coefficient of above composites can be controlled to be positive (PTE), negative (NTE) or zero. They found that the thermal expansion in these composites is directly depended on volume fraction of the $Sc_2(WO_4)_3$. The zero thermal expansion is important in ceramic technologies, for example, in the ceramic substrate of optical fiber gratings and also microelectronics. The zero expansion can eliminate the thermal stresses due to the thermal gradient or transient temperature change.

7.6.2 Solar cell

Energy storage and energy conversion technologies are two important terms recently attracted by the scientists. Dye-sensitized solar cells (DSSCs) are one of the technological advancements of the solar systems which are widely developed owing to their almost low fabrication costs, environment concerns, and high efficiency. In DSSCs, the counter electrode plays an important role on efficiency of the system. Platinum is a well-known counter electrode with enhanced electrocatalytic efficiency to electrolytes. However, the high cost, and also low abundance have prevented the wide range applications of this electrode in both academia and industries. Several alternatives have been investigated such as MMOs, polymers, composites, nanocarbon, etc. The admixture of two or more nanomaterials are utilized in modern solar cell technologies to enhance the photocatalyst performances. These admixture shows a strong interaction between the materials in the interfaces in order to depress the recombination rate. Ln−W−O is one of the recent nanomaterials in this field which can reduce the recombination rate of charge carriers with efficient light harvesting. In this class of materials, the electrocatalytic activity of the Ln−W−O counter electrodes are enhanced through the reduction process in the presence of H_2 or N_2 gases.

Erbium tungstate, $Er_2(WO_4)_3$, is used as a counter electrode in a DSSCs (Makhlouf et al., 2020). The oxygen vacancies and defects in this structure is assumed to be a potential candidate for photovoltaic applications. The optical band gap is found to be around 2.05 eV. This value tends to decrease upon addition of urea. Upon addition of 5 wt% urea into the ceramic texture, that is, $Er_2(WO_4)_3$, under constant flow rate of nitrogen gas, the power of DSSC conversion efficiency is enhanced from $\sim$ 0.12% to 4.18%.

7.6.3 Catalytic activity

Semiconductors like MOs and MMOs are widely used in photocatalytic applications. However, the environmental concerns and toxicity of some MOs and MMOs have been limited to their applications. Therefore new materials with improved properties are urgently required in order to fill the gaps. For example, zinc oxide (ZnO) is one of the nontoxic materials with biochemical stability and photocatalytic applications. However, ZnO required a coupling agent to enhance its photocatalytic activity. Lanthanum or rare-earth-based tungstate materials have been utilized as photocatalysts in wastewater treatments.

The visible emission of these materials is due to their 4f shell electron transitions. For instance, Sm^{3+} and Dy^{3+} doped metastable Y_2WO_6 microspheres is known for their luminescence properties (Kaczmarek et al., 2014). In the same vein, Ho_2WO_6 doped ZnO nanoparticles are synthesized by hydrothermal method and used for photodegradation and electrooxidation (Thirumalai et al., 2019). Based on the photocatalytic properties of Ho_2WO_6 doped ZnO, it has assumed that this admixture of materials can be potentially utilized as an energy effective and eco-friendly degradation of unwanted pollutants present in the industry wastewater. It is found that the Ho_2WO_6/ZnO has a high-specific surface area of about 16.44 m^2/g which is an important factor in photocatalytic performances.

As mentioned before, destruction of organic pollutants using semiconductor nanoscales MOs and MMOs have been widely studied owing to their photocatalytic activities against UV and solar lights. Here, ZnO nanoparticles (ZnONPs) are the potential heterogenous candidates in this field. ZnONPs are renowned along their wide band gap, and large excitation binding energy. However, the large band gap of ZnONPs barricade their applications, and requires further improvements such as nanostructuring in order to form nanocomposites or modification of the bulk structure by changing their surrounding elements (to form a MMOs). The semiconductor photocatalytic properties of ZnO/Dy_2WO_6 nanoparticles have been recently utilized as an ecofriendly nanocomposite for production of a potential heterogenous catalyst. This nanocomposite is synthesized via a template-free hydrothermal process. The photocatalytic activity of ZnO/Dy_2WO_6 nanocomposite examined for two organic pollutants: Rhodamine-B (Rh-B) and Trypan Blue (TB). Interestingly, irradiation of solar light has confirmed a great improvement of the photocatalytic performances of ZnO when combined with Dy_2WO_6. In addition, Dy_2WO_6/ZnO

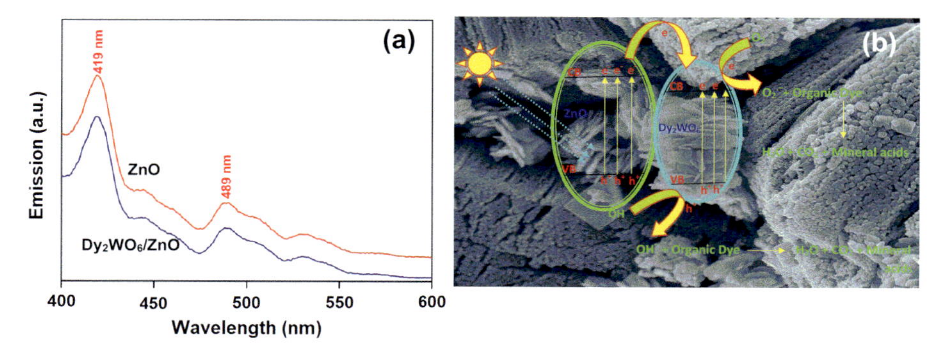

Figure 7.5 (A) Photoluminescence spectrum of Dy_2WO_6-ZnO (B) prepared ZnO and (B) Dy_2WO_6 doped ZnO, and (B) degradation mechanism.
Source: Reprinted from Thirumalai, K., Shanthi, M., & Swaminathan, M., (2017) Hydrothermal fabrication of natural sun light active Dy_2WO_6 doped ZnO and its enhanced photo-electrocatalytic activity and self-cleaning properties, *RSC Advances* 7 7509–7518. https://doi.org/10.1039/C6RA24843H.

exhibits superior electrocatalytic activity as compared to pristine ZnO for methanol electrooxidation as the anode catalyst in direct methanol fuel cells. The mechanism of photocatalytic activities includes five major steps: (1) activation by light, (2) generation of electrons and holes, (3) recombination of generated electrons and holes, (4) energy release, and (5) fluorescence emission. Fluorescence emissions happen in dual recombination rates; lower and higher. Lower emission implies lower electron–hole recombination rate (EHRR), while higher emission implies higher EHRR. Recombination of electron-hole pairs govern the amount of fluorescence emission. It is known that there is a direct relationship between photoluminescence (PL) and photocatalytic activity.

Thirumalai and his coworkers (Thirumalai et al., 2017) compared the photoluminescence (PL) profiles of ZnO, and its respective nanocomposites; Dy_2WO_6/ZnO. They reported an excitation wavelength at around 380 nm and two successive emissions at 419 (electron–hole recombination) and 489 nm. They found that the position of the PL bands of both ZnO and Dy_2WO_6/ZnO nanocomposites are almost same, while their PL intensities are varied. The PL intensity of ZnO is higher than nanocomposites (Fig. 7.5a). It seems Dy_2WO_6 decreases the electron-hole recombination, which can further cause decrement in intensity. On the other word, the Dy_2WO_6 causes an increment in electron and hole availability, that is, generation of more radicals for degradation.

In the catalytic profiles of the MMOs, having lanthanide ions (Ln^{m+}), the degradation mechanism happens in f-shells, where the electrons created through the f-f transition are trapped by the photocatalyst in visible light. In Dy_2WO_6, the Dy (atomic number 66) utilizes its 4f electrons ([Xe]$0.4f^{10}0.6s^2$). The overall mechanism of photocatalytic degradation can be proposed as:

Transition from $4f^{10}$ electrons → implementation → optical adsorption → electron–hole pairs separation → photodegradation.

Theoretically, in Dy_2WO_6 doped ZnO, Dy exists in its dual oxidation states, Dy (III) and Dy (IV). Here, Dy(III) donates electron to oxygen (O_2) molecules in order to form radical ($\cdot O_2$)by transforming into Dy(IV). This phenomenon causes several successive charged migrations, and therefore incrementation of rate of photodegradation as compared to pristine ZnO (Fig. 7.5b).

7.6.4 Fuel cell

A fuel cell is an electrochemical device which uses hydrogen or another fuel (chemical energy) to create power, clean and efficient. When hydrogen is fuel, then electricity, water, and heat are the products. Fuel cells can provide power for systems as large as a utility power station and as small as a laptop computer. Proton-conducting solid oxide fuel cells (p-SOFCs) is a type of electrochemical device which shows almost a high energy conversion efficiency. High proton conduction is the key enabling an efficient p-SOFC.

Various materials have been utilized and reported as an efficient material for fuel cells. Lanthanum tungsten oxide (Ln–W–O) has been developed as an electrolyte material for p-SOFCs. The presence of Ln^{3+} imparts high proton conductivity.

In this structure, the W^{6+} occupies the Ln^{3+} sites, that is, W^{6+} acts as donor dopants. In this structure, there are oxygen vacancies which governed by the Ln/W ratio. It is known that the acceptor-doped materials are hydrated *via* the interaction between oxygen vacancies and water vapor (Mielewczyk-Gryń et al., 2019). The ratio of Ln/W is directly related to the amount of oxygen vacancies. Upon increase in oxygen vacancies, the proton conductivity will be increased. It is renowned that the amount of oxygen vacancies can be increased via (A) doping acceptor dopants, and (B) increasing the La/W ratio. Several atoms can dope in Ln−W−O structure, particularly at the W sites. Aluminum (Al), Molybdate (Mo), and zirconium (Zr) have been used as dopants at the W sites, and Ca in Ln sites, but there is not any effect on proton conductivity (Porras-Vázquez et al., 2016).

Molybdenum substituted lanthanum tungstate, $La_{28-y}(W_{1-x}Mo_x)_{4+y}O_{54+\delta}$ ($x = 0-1$, $y = 0.923$) is investigated for enhancement of the n-type electronic conductivity in a typical fuel cell (Amsif et al., 2012). In this study, Amsif and his coworkers reported that these materials can be potentially utilized as electron−proton conductors under wet reducing conditions for $x \leq 0.4$, but less stable at $x \geq 0.6$. The total conductivities in humidified hydrogen are calculated to be around 0.016 S/cm for $x = 0.2$ and 0.043 S/cm for $x = 0.4$ at 900°C.

7.7 Conclusion and outlook

In this chapter, an attempt has been made to investigate the developments of the rare-earth tungstate (or lanthanum tungstate) materials in terms of "synthesis," "structural features," "properties" and "applications." The following features of the Ln−W−O materials are discussed:

- Ln−W−O materials can be synthesized via both wet and dry chemical techniques.
- Wet chemical technique is more favorable due to its lower formation temperatures, ambient conditions using simple equipment, higher yields, and controllable morphology as compared to dry techniques.
- Lanthanoids and tungsten are generally formed in polyhedron form of $Ln_2(WO_4)_3$ or $Ln_2W_3O_{12}$ consisting LnO_6 octahedra and WO_4 tetrahedra.
- Ionic conduction and thermal expansion are two important properties of Ln−W−O materials.
- Rare-earth tungstate materials (Ln−W−O) are mostly NTE, without volume producing phase alterations.
- Owing to their unique properties, Ln−W−O can be potentially used in energy conversion and storage systems.
- Though rare-earth tungstate ceramic materials have been studied, however, based on the available literatures, more attention is required in order to improve the properties of this class of materials for commercialization. Except those toxic rare-earth elements, the rest can be used in combination with tungsten for various applications. To the best of my knowledge, these materials can be potentially used either individually or in combination with conductive polymers for "energy" applications.

References

Amirante, R., Cassone, E., Distaso, E., & Tamburrano, P. (2016). Overview on recent developments in energy storage: Mechanical, electrochemical and hydrogen technologies. *Energy Conversion and Management*, *132*, 372–387. Available from https://doi.org/10.1016/j.enconman.2016.110.046.

Amsif, M., Magrasó, A., Marrero-López, D., Ruiz-Morales, J. C., Canales-Vázquez, J., & Núñez, P. (2012). Mo-Substituted lanthanum tungstate La 28- yW 4 + yO 54 + δ: A competitive mixed electron-proton conductor for gas separation membrane applications. *Chemistry of Materials: A Publication of the American Chemical Society*, *24*, 3868–3877. Available from https://doi.org/10.1021/cm301723a.

Andersen, H. L., Al Bahri, O. K., Tsarev, S., Johannessen, B., Schulz, B., Liu, J., ... Sharma, N. (2018). Structural evolution and stability of Sc2(WO4)3 after discharge in a sodium-based electrochemical cell. *Dalton Transactions*, *47*, 1251–1260. Available from https://doi.org/10.1039/c7dt04374k.

Anjana, P. S., Joseph, T., & Sebastian, M. T. (2010). Low temperature sintering and microwave dielectric properties of Ce 2(WO4)3 ceramics. *Ceramics International*, *36*, 1535–1540. Available from https://doi.org/10.1016/j.ceramint.2010.020.026.

Arulmani, S., Anandan, S., & Ashokkumar, M. (2018). *Introduction to advanced nanomaterials. Nanomaterials for green energy* (pp. 1–53). Elsevier. Available from https://doi.org/10.1016/B978-0-12-813731-4.00001-1.

Atkins, P., Overton, T., Rourke, J., Weller, M., Armstrong, F., Inorganic Chemistry, Fourth, USA, 2006.

Azizi, S., Darroudi, M., Soleymani, J., & Shadjou, N. (2021). Tb2(WO4)3@N-GQDs-FA as an efficient nanocatalyst for the efficient synthesis of β-aminoalcohols in aqueous solution. *J. Mol. Liq.*, *329*, 115555. Available from https://doi.org/10.1016/j.molliq.2021.115555.

Ben Smida, Y., Marzouki, R., Kaya, S., Erkan, S., Faouzi Zid, M., & Hichem Hamzaoui, A. (2020). *Synthesis methods in solid-state chemistry. Synthesis methods and crystallization* (pp. 1–16). IntechOpen. Available from https://doi.org/10.5772/intechopen.93337.

Bharat, L. K., Il Jeon, Y., & Yu, J. S. (2016). RE3 + (RE3 + = Tm3 + , Tb3 + and Sm3 +) ions activated Y6WO12 phosphors: Synthesis, photoluminescence, cathodoluminescence and thermal stability. *Journal of Alloys and Compounds*, *685*, 559–565. Available from https://doi.org/10.1016/j.jallcom.2016.050.321.

Cao, C., Wei, S., Zhu, Y., Liu, T., Xie, A., Noh, H. M., & Jeong, J. H. (2020). Synthesis, optical properties, and packaging of Dy3 + doped Y2WO6, Y2W3O12, and Y6WO12 phosphors. *Materials Research Bulletin*, *126*, 110846. Available from https://doi.org/10.1016/j.materresbull.2020.110846.

Chen, J., Gao, Q., Sanson, A., Jiang, X., Huang, Q., Carnera, A., ... Xing, X. (2017). Tunable thermal expansion in framework materials through redox intercalation. *Nature Communications*, *8*, 1–7. Available from https://doi.org/10.1038/ncomms14441.

Cheng, Q., Ren, F., Lin, Q., Tong, H., & Miao, X. (2019). High quantum efficiency red emitting α-phase La2W2O9:Eu3 + phosphor. *Journal of Alloys and Compounds*, *772*, 905–911. Available from https://doi.org/10.1016/j.jallcom.2018.080.320.

Damascena dos Passos, R. H., Pereira de Souza, C., Bernard-Nicod, C., Leroux, C., & Arab, M. (2020). Structural and electrical properties of cerium tungstate: Application to methane conversion. *Ceramics International*, *46*, 8021–8030. Available from https://doi.org/10.1016/j.ceramint.2019.120.026.

De Guire, M. R., Bauermann, L. P., Parikh, H., & Bill, J. (2013). *Chemical bath deposition. Chemical solution deposition of functional oxide thin films* (pp. 319–339). Wien: Springer-Verlag. Available from https://doi.org/10.1007/978-3-211-99311-8_14.

Ding, B., Han, C., Zheng, L., Zhang, J., Wang, R., & Tang, Z. (2015). Tuning oxygen vacancy photoluminescence in monoclinic Y2WO6 by selectively occupying yttrium sites using lanthanum. *Scientific Reports*, *5*, 1–10. Available from https://doi.org/10.1038/srep09443.

Enhessari, M., & Salehabadi, A. (2018). Perovskites-based nanomaterials for chemical sensors. *Progresses in Chemical Sensor*, 13.

Esaka, T. (2004). Hydrogen storage in proton-conductive perovskite-type oxides and their application to nickel hydrogen batteries. *Solid State Ionics*, *166*, 351–357. Available from https://doi.org/10.1016/j.ssi.2003.110.023.

Fernandes, A., Woudstra, T., van Wijk, A., Verhoef, L., & Aravind, P. V. (2016). Fuel cell electric vehicle as a power plant and SOFC as a natural gas reformer: An exergy analysis of different system designs. *Applied Energy*, *173*, 13–28. Available from https://doi.org/10.1016/j.apenergy.2016.030.107.

Inomata, M., Kishida, K., Maruyama, Y., & Watanabe, T. (2015). Synthesis of a new scheelite-type Eu3 + -doped Gd2W2O9 red light emitting phosphor by the polymerized complex method. *Solid State Sciences*, *48*, 251–255. Available from https://doi.org/10.1016/j.solidstatesciences.2015.080.003.

Joaquín-Morales, M. G., Fuentes, A. F., Montemayor, S. M., Meléndez-Zaragoza, M. J., Gutiérrez, J. M. S., Ortiz, A. L., & Collins-Martínez, V. (2018). Synthesis conditions effect on the of photocatalytic properties of MnWO4 for hydrogen production by water splitting. *International Journal of Hydrogen Energy*. Available from https://doi.org/10.1016/J.IJHYDENE.2018.100.075.

Jurczyk, M., Nowak, M., Szajek, A., & Jezierski, A. (2012). Hydrogen storage by Mg-based nanocomposites. *International Journal of Hydrogen Energy*, *37*, 3652–3658. Available from https://doi.org/10.1016/j.ijhydene.2011.040.012.

Kaczmarek, A. M., Van Hecke, K., & Van Deun, R. (2014). Enhanced luminescence in Ln3 + -doped Y2WO6 (Sm, Eu, Dy) 3D microstructures through Gd3 + codoping. *Inorganic chemistry*, *53*, 9498–9508. Available from https://doi.org/10.1021/ic5005837.

Kaczmarek, A. M., Ndagsi, D., Van Driessche, I., Van Hecke, K., & Van Deun, R. (2015). Green and blue emitting 3D structured Tb:Ce2(WO4)3 and Tb:Ce10W22O81 micromaterials. *Dalton Transactions*, *44*, 10237–10244. Available from https://doi.org/10.1039/c5dt00764j.

Kafizas, A., Godin, R., & Durrant, J. R. (2017). *Charge carrier dynamics in metal oxide photoelectrodes for water oxidation. Semiconductors and semimetals* (pp. 3–46). Academic Press Inc. Available from https://doi.org/10.1016/bs.semsem.2017.02.002.

Karuppaiah, S., Annamalai, R., Muthuraj, A., Kesavan, S., Palani, R., Ponnusamy, S., … Meenakshisundaram, S. (2019). Efficient photocatalytic degradation of ciprofloxacin and bisphenol A under visible light using Gd 2 WO 6 loaded ZnO/bentonite nanocomposite. *Applied Surface Science*, *481*, 1109–1119. Available from https://doi.org/10.1016/j.apsusc.2019.03.178.

Kombarakkaran, J., Noveron, J. C., Helgesen, M., Shen, K., & Pietra, T. (2009). Hydrogen storage in dinuclear Pt(II) metallacycles. *International Journal of Hydrogen Energy*, *34*, 5704–5709. Available from https://doi.org/10.1016/j.ijhydene.2009.050.018.

Kuriakose, S., H, H., Jose, A., John, M., & Varghese, T. (2020). Structural and optical characterization of lanthanum tungstate nanoparticles synthesized by chemical precipitation route and their photocatalytic activity. *Optical Materials. (Amst).*, *99*, 109571. Available from https://doi.org/10.1016/j.optmat.2019.109571.

Li, P., Deng, S. H., Zhang, L., Liu, G. H., & Yu, J. Y. (2012). Hydrogen storage in lithium-decorated benzene complexes. *International Journal of Hydrogen Energy*, *37*, 17153–17157. Available from https://doi.org/10.1016/j.ijhydene.2012.090.007.

Liu, H., Secco, R. A., Imanaka, N., & Adachi, G. (2002). X-ray diffraction study of pressure-induced amorphization in Lu2(WO4)3. *Solid State Communications*, *121*, 177–180. Available from https://doi.org/10.1016/S0038-1098(01)00458-6.

Liu, Q. Q., Cheng, X. N., & Yang, J. (2012). Development of low thermal expansion Sc 2 (WO 4) 3 containing composites. *Materials Technology*, *27*, 388–392. Available from https://doi.org/10.1179/1066785712Z.00000000094.

Liu, W., Sun, J., Li, X., Zhang, J., Tian, Y., Fu, S., ... Chen, B. (2013). Laser induced thermal effect on upconversion luminescence and temperature-dependent upconversion mechanism in Ho3 + /Yb 3 + -codoped Gd2(WO4)3 phosphor. *Optical Materials. (Amst).*, *35*, 1487–1492. Available from https://doi.org/10.1016/j.optmat.2013.03.008.

Logvinovich, D., Arakcheeva, A., Pattison, P., Eliseeva, S., Tomĕs, P., Marozau, I., ... Chapuis, G. (2010). Crystal structure and optical and magnetic properties of Pr 2 (MoO4)3. *Inorganic Chemistry*, *49*, 1587–1594. Available from https://doi.org/10.1021/ic9019876.

López-Moreno, S., Rodríguez-Hernández, P., Muñoz, A., Romero, A. H., & Errandonea, D. (2011). First-principles calculations of electronic, vibrational, and structural properties of scheelite EuWO4 under pressure. *Physical Review B — —Covering Condensed Matter and Material Physics*, *84*, 064108. Available from https://doi.org/10.1103/PhysRevB.84.064108.

Magrasó, A., & Haugsrud, R. (2014). Effects of the La/W ratio and doping on the structure, defect structure, stability and functional properties of proton-conducting lanthanum tungstate La28-xW4 + xO54 + δ. A review. *Journal of Materials Chemistry A.*, *2*, 12630–12641. Available from https://doi.org/10.1039/c4ta00546e.

Magrasó, A., Frontera, C., Marrero-López, D., & Núñez, P. (2009). New crystal structure and characterization of lanthanum tungstate "La6WO12" prepared by freeze-drying synthesis. *Journal of the Chemical Society Dalton Transactions*, 10273–10283. Available from https://doi.org/10.1039/b916981b.

Majid, A., & Bibi, M. (2018). *Wet chemical synthesis methods. Cadmium based II-VI semiconducting nanomaterials* (pp. 43–101). Cham: Springer. Available from https://doi.org/10.1007/978-3-319-68753-7_3.

Makhlouf, M. M., Abdulkarim, S., Adam, M. S. S., & Qiao, Q. (2020). Unraveling urea pretreatment correlated to activate Er2(WO4)3 as an efficient and stable counter electrode for dye-sensitized solar cells. *Electrochimica acta*, *333*, 135540. Available from https://doi.org/10.1016/j.electacta.2019.135540.

Marzano, M., Pontón, P. I., & Marinkovic, B. A. (2019). Co-precipitation of low-agglomerated Y2W3O12 nanoparticles: The effects of aging time, calcination temperature and surfactant addition. *Ceramics International*, *45*, 20189–20196. Available from https://doi.org/10.1016/j.ceramint.2019.060.288.

Mielewczyk-Gryń, A., Wachowski, S., Prześniak-Welenc, M., Dzierzgowski, K., Regoutz, A., Payne, D. J., & Gazda, M. (2019). Water uptake analysis of acceptor-doped lanthanum orthoniobates. *Journal of Thermal Analysis. Calorimetry*, *138*, 225–232. Available from https://doi.org/10.1007/s10973-019-08208-6.

Mo, Z., & Hou, Z. (2020). *Lanthanum. Reference module in chemistry, molecular. sciences and chemical engineering*. Elsevier. Available from https://doi.org/10.1016/B978-0-08-102688-5.00004-0.

Morassaei, M. S., Salehabadi, A., Akbari, A., & Salavati-Niasari, M. (2019). A potential photovoltaic material for dye sensitized solar cells based BaCe2(MoO4)4 doped Er3 + / Yb3 + nanostructures. *Journal of Cleaner Production*, *209*, 762–768. Available from https://doi.org/10.1016/J.JCLEPRO.2018.100.296.

Mosleh, M. (2017). Controllable synthesis of gadolinium tungstate nanoparticles with different surfactants by precipitation route. *Journal of Materials Science: Materials in Electronics*, *28*, 8494–8499. Available from https://doi.org/10.1007/s10854-017-6571-9.

Neikov, O. D., & Yefimov, N. A. (2019). *Nanopowders. Handbook of non-ferrous metal powders* (pp. 271–311). Elsevier. Available from https://doi.org/10.1016/b978-0-08-100543-9.00009-9.

Nilsson, M., Grins, J., Käll, P. O., & Svensson, G. (1996). Synthesis, structural characterisation and magnetic properties of Gd14W4O33 - XNy($0 \leq x \leq 17 \pm 2{,}0 \leq y \leq 9 \pm 2$), a new fluorite-related oxynitride. *Journal of Alloys and Compounds*, *240*, 60–69. Available from https://doi.org/10.1016/0925-8388(96)02325-0.

Ningombam, G. S., & Singh, N. R. (2019). *Lanthanide-doped orthometallate phosphors. Spectroscopy of lanthanide doped oxide materials* (pp. 113–234). Elsevier Inc. Available from https://doi.org/10.1016/B978-0-08-102935-0.00005-8.

Ott, F., Panagiotopoulos, I., Anagnostopoulou, E., Pousthomis, M., Lacroix, L. M., Viau, G., & Piquemal, J. Y. (2015). *Soft chemistry nanowires for permanent magnet fabrication. Magnetic nano and microwires, design, synthesis, properties and applications* (pp. 629–651). Elsevier. Available from https://doi.org/10.1016/B978-0-08-100164-6.00021-7.

Pathan, H. M., & Lokhande, C. D. (2004). Deposition of metal chalcogenide thin films by successive ionic layer adsorption and reaction (SILAR) method. *Bulletin of Material Science*, *27*, 85–111. Available from https://doi.org/10.1007/BF02708491.

Pavlidis, V. F., Savidis, I., & Friedman, E. G. (2017). *Electrical properties of through silicon vias. Three-dimensional integrated circuit design* (pp. 67–117). Elsevier. Available from https://doi.org/10.1016/b978-0-12-410501-0.00004-6.

Periyasamy, S., Vinoth Kumar, J., Chen, S. M., Annamalai, Y., Karthik, R., & Erumaipatty Rajagounder, N. (2019). Structural Insights on 2D gadolinium tungstate nanoflake: A promising electrocatalyst for sensor and photocatalyst for the degradation of postharvest fungicide (Carbendazim). *ACS Applied Material Interfaces.*, *11*, 37172–37183. Available from https://doi.org/10.1021/acsami.9b07336.

Pestereva, N., Guseva., Vyatkin, I., & Lopatin, D. (2017). Electrotransport in tungstates Ln2 (WO4)3 (Ln = La, Sm, Eu, Gd). *Solid State Ionics*, *301*, 72–77. Available from https://doi.org/10.1016/j.ssi.2017.01.009.

Porras-Vázquez, J. M., Dos Santos-Gómez, L., Marrero-López, D., Slater, P. R., Masó, N., Magrasó, A., & Losilla, E. R. (2016). Effect of tri- and tetravalent metal doping on the electrochemical properties of lanthanum tungstate proton conductors. *Dalton Transactions*, *45*, 3130–3138. Available from https://doi.org/10.1039/c5dt03833b.

Rahimi-Nasrabadi, M., Pourmortazavi, S. M., Aghazadeh, M., Ganjali, M. R., Karimi, M. S., & Novrouzi, P. (2017). Optimizing the procedure for the synthesis of nanoscale gadolinium(III) tungstate as efficient photocatalyst. *Journal of Materials Science: Materials in Electronics*, *28*, 3780–3788. Available from https://doi.org/10.1007/s10854-016-5988-x.

Raja, A., Selvakumar, K., Swaminathan, M., & Kang, M. (2021). Redox additive based rGO-Dy2WO6-ZnO nanocomposite for enhanced electrochemical supercapacitor applications. *Synthetic metals*, *276*, 116753. Available from https://doi.org/10.1016/j.synthmet.2021.116753.

Ran, K., Deibert, W., Ivanova, M. E., Meulenberg, W. A., & Mayer, J. (2019). Crystal structure investigation of La 5.4 W 1 − y Mo y O 12 − δ for gas separation by high-resolution transmission electron microscopy. *Scientific Reports*, *9*, 1−9. Available from https://doi.org/10.1038/s41598-019-39758-2.

Rao, A., & Cölfen, H. (2017). *Comprehensive supramolecular chemistry ii: facet control in nanocrystal growth. Comprehensive supramolecular chemistry II* (pp. 129−156). Elsevier Inc. Available from https://doi.org/10.1016/B978-0-12-409547-2.12638-1.

Sakintuna, B., Lamaridarkrim, F., & Hirscher, M. (2007). Metal hydride materials for solid hydrogen storage: A review. *International Journal of Hydrogen Energy*, *32*, 1121−1140. Available from https://doi.org/10.1016/j.ijhydene.2006.110.022.

Salehabadi, A., & Enhessari, M. (2019). *Application of (mixed) metal oxides-based nanocomposites for biosensors*. Available from https://doi.org/10.1016/B978-0-08-102814-8.00013-5.

Salehabadi, A., Ahmad, M. I., Morad, N., Salavati-Niasari, M., & Enhessari, M. (2019). Electrochemical hydrogen storage properties of Ce<inf>0.75</inf>Zr<inf>0.25</inf>O<inf>2</inf> nanopowders synthesized by sol-gel method. *Journal of Alloys and Compounds*. Available from https://doi.org/10.1016/j.jallcom.2019.03.160.

Shaikh, S. F., Ubaidullah, M., Mane, R. S., & Al-Enizi, A. M. (2020). *Types, Synthesis methods and applications of ferrites. Spinel ferrite nanostructures for energy storage devices* (pp. 51−82). Elsevier. Available from https://doi.org/10.1016/b978-0-12-819237-5.00004-3.

Singh, R. N., Madhu., & Awasthi, R. (2013). *Alcohol fuel cells. New and future developments in catalysis batteries, hydrogen storage fuel cells* (pp. 453−478). Elsevier B.V. Available from https://doi.org/10.1016/B978-0-444-53880-2.00021-1.

Sundaresan, P., Yamuna, A., & Chen, S. M. (2020). Sonochemical synthesis of samarium tungstate nanoparticles for the electrochemical detection of nilutamide. *Ultrasonics Sonochemistry*, *67*, 105146. Available from https://doi.org/10.1016/j.ultsonch.2020.105146.

Thirumalai, K., Shanthi, M., & Swaminathan, M. (2017). Hydrothermal fabrication of natural sun light active Dy2WO6 doped ZnO and its enhanced photo-electrocatalytic activity and self-cleaning properties. *RSC Advances*, *7*, 7509−7518. Available from https://doi.org/10.1039/c6ra24843h.

Thirumalai, K., Shanthi, M., & Swaminathan, M. (2019). Ho2WO6/ZnO nanoflakes for photo−electrochemical and self cleaning applications. *Materials Science in Semiconductor Processing*, *90*, 78−86. Available from https://doi.org/10.1016/j.mssp.2018.100.007.

Wall, F. (2021). *Rare earth elements. Encyclopedia of geology* (pp. 680−693). Elsevier. Available from https://doi.org/10.1016/b978-0-08-102908-4.00101-6.

Wang, D., Yang, P., Cheng, Z., Wang, W., Hou, Z., Dai, Y., . . . Lin, J. (2012). Patterning of Gd2(WO4)3:Ln3 + (Ln = Eu, Tb) luminescent films by microcontact printing route. *Journal of Colloid and Interface Science*, *365*, 320−325. Available from https://doi.org/10.1016/j.jcis.2011.09.008.

Wang, W., Li, J., Zhang, Z., & Liu, Z. (2019). The synthesis and luminescent properties of Dy/Re (Re = Tb or Eu) co-doped Gd2(WO4)3 phosphor with tunable color via energy transfer. *Journal of Luminescence.*, *207*, 114−122. Available from https://doi.org/10.1016/j.jlumin.2018.100.122.

Weil, M., Stöger, B., & Aleksandrov, L. (2009). Nd2 (WO4)3. *Acta Crystallographica Section E Structure Reports Online.*, *65*, i45. Available from https://doi.org/10.1107/S1600536809018108, −i45.

Williams, P. A. (2004). *Minerals: Tungstates. Encyclopedia of geology* (pp. 586−588). Elsevier Inc. Available from https://doi.org/10.1016/B0-12-369396-9/00286-0.

Xiao, X. L., Cheng, Y. Z., Peng, J., Wu, M. M., Chen, D. F., Hu, Z. B., ... Jorgensen, J. (2008). Thermal expansion properties of A2(MO4)3 (A = Ho and Tm; M = W and Mo). *Solid State Science*, *10*, 321−325. Available from https://doi.org/10.1016/j.solidstatesciences.2007.090.001.

Xing, Z., Shen, C., & Li, C. (2020). Synthesis and microwave dielectric properties of an electronic ceramic Y2WO6 for wireless communications. *Physics Letters*, *384*, 126811. Available from https://doi.org/10.1016/j.physleta.2020.126811.

Yamuna, A., Sundaresan, P., Chen, S. M., & Shih, W. L. (2020). Ultrasound assisted synthesis of praseodymium tungstate nanoparticles for the electrochemical detection of cardioselective β-blocker drug. *Microchemical Journal, Devoted to the Application of Microtechniques in all Branches of Science*, *159*, 105420. Available from https://doi.org/10.1016/j.microc.2020.105420.

Yilmaz, E., & Soylak, M. (2019). *Functionalized nanomaterials for sample preparation methods. Handbook of nanomaterials in analytical chemistry, modern trends in analysis* (pp. 375−413). Elsevier. Available from https://doi.org/10.1016/B978-0-12-816699-4.00015-3.

Yuan, J., Li, W., & Wu, Y. (2017). Hydrogen storage and low-temperature electrochemical performances of A2B7 type La-Mg-Ni-Co-Al-Mo alloys. *Progress in Natural Science: Materials International*, *27*, 169−176. Available from https://doi.org/10.1016/j.pnsc.2017.030.010.

Zeng, L., Shimoda, K., Zhang, Y., Miyaoka, H., Ichikawa, T., & Kojima, Y. (2012). Lithium hydrazide as a potential compound for hydrogen storage. *International Journal of Hydrogen Energy*, *37*, 5750−5753. Available from https://doi.org/10.1016/j.ijhydene.2011.120.144.

Zeng, Y., Li, H., Li, Z., Luo, Q., Zhu, H., Gu, Z., ... Luo, K. (2020). Engineered gadolinium-based nanomaterials as cancer imaging agents. *Applied Materials Today.*, *20*, 100686. Available from https://doi.org/10.1016/j.apmt.2020.100686.

Zhang, Y. M., Huang, H. P., Xu, L., & Novel, A. (2020). Electrochemical sensor based on Au-Dy2(WO4)3 nanocomposites for simultaneous determination of uric acid and nitrite. *Chinese Journal of Analytical Chemistry*, *48*, e20032−e20037. Available from https://doi.org/10.1016/S1872-2040(20)60005-6.

Zheng, Y., Kurosaki, K., Tokushima, K., Ohishi, Y., Muta, H., & Yamanaka, S. (2011). Thermal conductivity of Y6WO12 and Yb 6WO12 ceramics. *Journal of Nuclear Materials, North-Holland*, 357−360. Available from https://doi.org/10.1016/j.jnucmat.2011.060.028.

Zhou, Y., Rao, R. P., & Adams, S. (2011). *Intrinsic polyatomic defects in Sc2(WO4)3. Solid state ionics* (pp. 34−37). Elsevier. Available from https://doi.org/10.1016/j.ssi.2010.06.024.

Zhou, Y., Neiman, A., & Adams, S. (2011). Novel polyanion conduction in Sc2(WO4)3 type negative thermal expansion oxides. *Physica Status Solidi*, *248*, 130−135. Available from https://doi.org/10.1002/pssb.201083969.

Rare earth-based ceramic nanomaterials—manganites, ferrites, cobaltites, and nickelates

Razieh Razavi[1] *and Mahnaz Amiri*[2,3]
[1]Department of Chemistry, Faculty of Science, University of Jiroft, Jiroft, Iran, [2]Neuroscience Research Center, Institute of Neuropharmacology, Kerman University of Medical Science, Kerman, Iran, [3]Department of Hematology and Medical Laboratory Sciences, Faculty of Allied Medicine, Kerman University of Medical Sciences, Kerman, Iran

8.1 General introduction

The past five decades have witnessed considerable research interests by the scientists over the great applicability potential of nanophase ferrites in the fields of high frequency and power equipment particularly in respect of electromagnetic interference suppression (Valenzuela, 2012). The numerous flexible electrical and magnetic characteristics of spinel ferrites thanks to its viability to produce a great amount of differing cations of metal as well as vast compositional variability turns them into invaluable material of future. As a rigid spinel, cobalt ferrite is known for its powerful saturation magnetization, mechanical strength, excellent magneto crystalline anisotropy, and coercive magnetic field (Ben Tahar et al., 2008). Ferrite, representing the crystalline arrangement of the spinel, constitutes a FCC using its (O^{2-}) anions interlinked with 2 sublattices called as octahedral (B) and tetrahedral (A). The trivalent and divalent cations distribution on the A and B sites shares certain features with the spinel construct. As partially inverse spinel exhibiting ferromagnetic (FM) effect, $CoFe_2O_4$ has low eddy current losses and high specific resistance when used in high frequency uses (Elkestawy, Abdel Kader, & Amer, 2010).

Fig. 8.1 indicates the spinel structure of cubic ferrite. Such extraordinary physico-chemical properties are obtained by selecting the synthesis, cations doping inside the host crystal construct as well as the sizing and particle morphology. Thanks to its unique properties said above, nanoscaled cobalt ferrites are increasingly used in biomedicine, ferro fluids, advanced microelectronics, microwave absorbing, high frequency data storage and materials (including isolators, phase shifters, circulators, etc.) (Bate, 1991; Goh et al., 2010; Rao, Mahesh, & Kumar, 2005; Sugimoto, 1999). Rare-earth oxides also can effectively be used for insulation due to their excellent electrical resistance. A proper selection of rare-earth cation would provide the possibility to change the electromagnetic properties of spinel ferrites; this simple modification would entail decisive effect on the system magnetic

Advanced Rare Earth-Based Ceramic Nanomaterials. DOI: https://doi.org/10.1016/B978-0-323-89957-4.00012-8

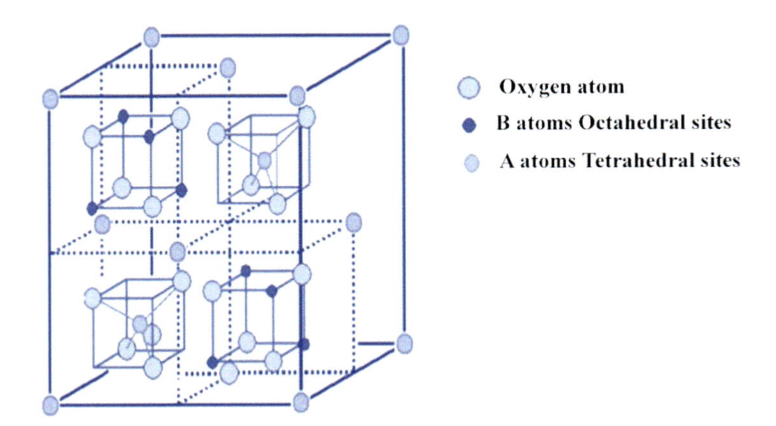

Figure 8.1 Spinel structure of cubic ferrite.

anisotropy and gives the spinel ferrite the adequate competency to be used in replacement of hexaferrites or garnets. The intentional introduction of rare-earth ions (Y^{3+}, Gd^{3+}, Ho^{3+}, Sm^{3+}, Nd^{3+}) into spinel ferrite (doping) will lead to disorderliness of its structure and causes lattice strain, which in turn would increase its electromagnetic parameters (Jacobo, Duhalde, & Bertorello, 2004). The influence of rare-earth ions replacement upon the Curie temperature was investigated by Cheng et al. (1999). In another study Kumar (Vijaya Kumar & Ravinder, 2002) examined the thermoelectric characteristics of spinel ferrites when doped by Gd^{3+}. Gd^{3+} doped ferrites magnetic and electrical properties were investigated by Bharathia (Kamala Bharathi, Arout Chelvane, & Markandeyulu, 2009). Rana (Rana, Thakur, & Kumar, 2011) showed the effect of Gd replacement on cobalt ferrite dielectric properties. To realize the localized charge carrier's behaviors, the electrical conduction and the polarity effect, the knowledge of cubic spinel ferrite's dielectric parameters are so important. Scrutinizing the impedance and electric modulus are other important methods for thorough analysis of the factors contributing in the resistance properties of nanoscale materials and further assessment of the relaxation behavior. According to the literature (Ravinder, Vijaya Kumar, & Balaya, 2001) the rare-earth concentration in doping of ferrites is very important in terms of describing their electromagnetic characteristics. The rare-earth-doped systems has such unique properties thanks to the 4fn electronic states (Maqsood, 2009). In the first chapters of the book, a comprehensive and complete description of the rare-earth elements is given. Fig. 8.2 displays the properties and applications of these elements in summery.

The aim of this extent research was to study the rare-earth-based ceramic nanomaterials—manganites, ferrites, cobaltites, and nickelates (recent advances related to preparation methods and applications) in order to give valuable information for readers and researchers for better understanding of advantages or disadvantages of various synthesis methods and applications of rare-earth ferrites.

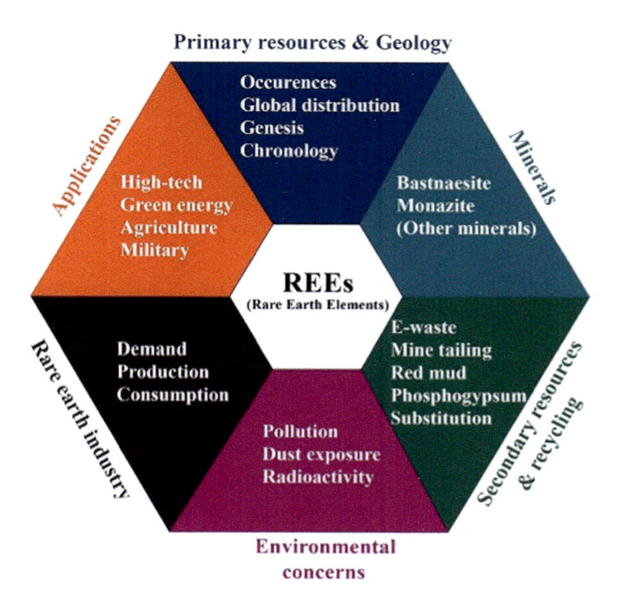

Figure 8.2 Schematic of properties and applications of rare-earth elements in summary. *Source*: Reproduced with permissions from Dushyantha, N., Batapola, N., Ilankoon, I. M. S. K, Rohitha, S., Premasiri, R., Abeysinghe, B., . . . Dissanayake, K., (2020). The story of rare-earth elements (REEs): Occurrences, global distribution, genesis, geology, mineralogy and global production, *Ore Geology Reviews*, *122* 103521.

8.2 Rare-earth ferrites

8.2.1 Short introduction of rare-earth ferrites

Rare-earth ferrites with its generic formula of MFe_2O_4 (M = Y or Dy to Lu) are crystallized in rhombohedral medium with the group space of R − 3m (Ikeda et al., 2005). Having a layered structure, rare-earth ferrites comprise irregular loading of the triangular lattices 2D of the rare-earth, oxygen and iron ions to the c direction. Ions of Fe in both mixture of valent states (Fe^{2+} and Fe^{3+}) have identical amounts. The mentioned ions electrostatical interactions as well as the antiferromagnetic (AFM) super exchange interactions among the closest neighboring spins residing on the triangular net are believed to be major cause for their extraordinary characteristics. Both the iron ion's charges and the spins may form perfectly ordered structures in the low-temperature phase. The rare-earth ferrites, almost all, exhibit ferrimagnetic order at approx. 250 k owing to the powerful magnetic interaction among the locally situated Fe moments (Matsumoto, Môri, Iida, Tanaka, & Siratori, 1992; Yoshii, Ikeda, & Nakamura, 2006; Yoshii, Ikeda, Matsuo, Horibe, & Mori, 2007; Yoshii, Ikeda, & Mori, 2007). While an electron hopping among the layer's Fe^{2+} and the Fe^{3+} ions is expected, this hopping of electrons brings about large electric conductivity and competition among the charges of the closest and following closest neighboring ions present on the triangular lattice, resulting in ordering of charge (Ikeda et al., 2004; Ikeda, Mori, & Kohn, 2005; Nagano & Ishihara, 2007). The above-mentioned Fe^{2+} and Fe^{3+} ions ordering will cause a large number of ferrites to demonstrate phase transition (Tanaka et al., 1982).

Researches undertaken on $LuFe_2O_4$ neutron diffraction have reported the material experiences successive several phase transition starting from a disorganized status to a 2D charge density wave status toward a 3D wave of charge density status and afterwards to a 3D ones (Ikeda et al., 2000; Yamada, Kitsuda, Nohdo, & Ikeda, 2000). $LuFe_2O_4$ microscopy experiments of in situ cooling transmission electron revealed that the ground point charges were adequately CSP (crystallized in a charge-strip phase) while the charge focus in the mentioned CSP could be explained as a no sinusoidal charge density wave, giving an increase in the electrical polarity (Zhang et al., 2019). A number of the ferrites show ferroelectricity as it can be inferred from the low-frequency dielectric dispersion (Ikeda et al., 2004). Fe^{2+} and Fe^{3+} ions develop a superstructure that stimulates electrical polarization comprising the distributed polar symmetry electrons because the two ions centers are not coinciding (Ikeda, Mori, & Yoshii, 2007; Nagano, Naka, Nasu, & Ishihara, 2007). The MFe_2O_4 dielectric dispersion may be realized qualitatively based on the Fe^{+n} ions' charge frustration in the 2D triangular lattice, at which the electronshop among the two ions (Ikeda, Kohn, Kito, Akimitsu, & Siratori, 1995; Siratori, Akimitsu, & Kohn, 1994) and the ferroelectricity origin fundamentally pertains to the charge-ordering transition emanating from the robust electron (Yamada & Nohdo, 1997). Some studies have recently reported magnetodielectric response in $LuFe_2O_4$ in the ambient temperature representing a coupling among the spins on the one hand and electric dipoles on the other hand (Subramanian et al., 2006). The charge fluctuations resulting from interconversion in between two types of charge order (hindered by the magnetic field application) is believed to be the cause of magnetocapacitance effect in $LuFe_2O_4$ at ambient temperature (Xiang & Whangbo, 2007). Fig. 8.3 indicates the rare-earth iron garnets crystal structure, by Fe and R ions from their neighboring oxygen ions formed from local polyhedra environment (Nakamoto, Xu, Xu, Xu, & Bellaiche, 2017).

The nanoferrites properties are affected by the synthesis conditions, composition of chemicals, temperature of sintering, additives as doping, and the preparation method, therefore in the next part, various synthesis methods are discussed.

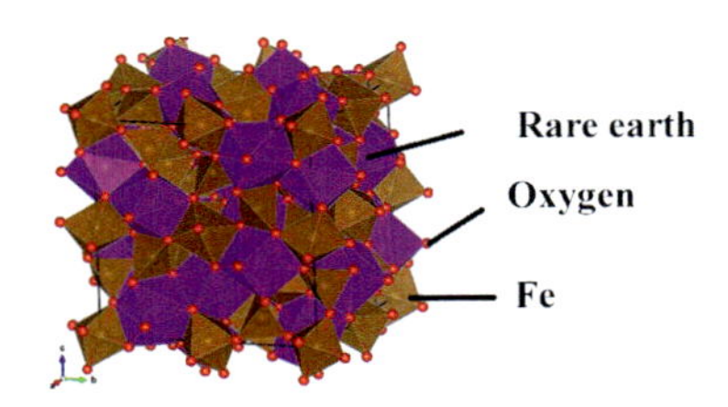

Figure 8.3 The crystal structure of rare-earth of iron garnets.

8.2.2 Synthesis methods of rare-earth ferrites

Nanoferrites exhibit excellent magnetic and dielectric materials characteristics. Such characteristics are determined based on the cations choice and their manner of distribution among the octahedral and tetrahedral points of the spinel lattice. The nanoferrites' characters were influenced by the provisioning circumstances, chemical compounds involved, sintering heat, doping material and the preparation method (Gul & Maqsood, 2008). The spinel ferrite has a cubic closed pack (FCC) crystal structure, while its anions (O^{2-}) are linked with two sublattices called as octahedral (B) and tetrahedral (A). The distribution manner of trivalent and divalent cations across A and B points contributes certain properties to the structure of the spinel. As a partially inverse spinel by FM effect, $CoFe_2O_4$ has elevated particular resistance and lower eddy current losses when used at increased rate of recurrence uses (Elkestawy et al., 2010). Such extraordinary physico-electric characteristics may be accessible through selecting the synthesis method, cation doping in the crystal structure of the host as well as the morphology and size of the particles. The synthesis of cobalt ferrites is effected through various methods such as sol-gel (Kim, Yi, Park, Namgung, & Lee, 1999), hydrothermal (Peng et al., 2011), citrate precursor (Vijaya Kumar & Ravinder, 2002), chemical coprecipitation (Il Kim, Kim, & Lee, 2003), solvothermal (Cai & Wan, 2007), reverse micelle (Sileo & Jacobo, 2004), solid state reaction (Rao, S, AM, MC, & RK, 2009), microemulsion (Pillai & Shah, 1996), and the ferrofluid, advanced microelectronics, technology biomedicine, high microwave absorbing materials (circulators, phase shifters, isolators), frequency data storage, are among the fields. The application of nanoscaled cobalt ferrites are warmly welcomed thanks to the above sais properties (Bate, 1991; Goh et al., 2010; Sankaranarayanan & Sreekumar, 2003; Sugimoto, 1999). Suitable electric insulation as well as considerable electric conductivity is demonstrated by the rare-earth oxides. Suitable selection of rare-earth cation in spinel ferrites can change its electromagnetic characteristics and decisively impact the system magnetic anisotropy so that the spinel ferrite turns out to become a unique substitute for garnets or hexaferrites. Parent spinel ferrite doped with rare-earth ions (Y^{3+}, Gd^{3+}, Ho^{3+}, Sm^{3+}, Nd^{3+}) will lead to disorderliness of the structure and strain development in the lattice which in turn leads to the increase in the electrical and magnetic parameters values (Jacobo et al., 2004). The rare-earth ion replacement impact on the Curie temperature was investigated by Cheng et al. (1999) Schmool, Keller, Guyot, Krishnan, and Tessier (1999) and Kumar (Vijaya Kumar & Ravinder, 2002) determined the thermoelectric characteristics of Gd^{3+} doped spinel ferrites. In a research conducted by Bharathia (Meng et al., 2009) the magnetoelectric characteristics of Gd^{3+} doped ferrites was studied. In another study Rana (Rana et al., 2011) investigated the Gd replacement effect on the dielectric characteristics of cobalt ferrites. The cubic spinel ferrites dielectric parameters are crucial in realizing the electrical conductance, localized charge carrier's behavior as well as the polarization issue. Investigating the impedance and electric modulus are other important methods for thorough analysis of the factors involved in the resistance properties of nanoscale materials and further assessment of the relaxation effect. According to the literature (Ravinder et al., 2001; Vijaya Kumar & Ravinder, 2002) the rare-earth concentration in doping of ferrites is very important in terms of describing their electromagnetic characteristics. Such excellent properties of doped rare-earth systems results from the 4fn electronic states (Maqsood, 2009). Sol–gel process is a synthesis method attracting considerable attention to itself owing to its immense advantages like lower required process heat and homogenous distribution of reactant. Using sol-gel process, the products obtained demonstrate good crystal quality, narrower distribution of size and shape uniformity (Jacob, Thankachan, Xavier, & Mohammed, 2011). It has been shown that the introduction of rare-earth ions to the spinel ferrite structure will lead to structural disorder that in turn would induce strain in the system and thereby modifying its magnetic and electrical properties (Sileo & Jacobo, 2004). In a study conducted by Vijaya et al. (Vijaya Kumar and Ravinder, 2002), the changes of the magnetic characteristics of samarium replaced $CoFe_2O_4$ that underwent synthetization using citrate precursor technique were demonstrated; they reported that with the addition of Sm^{3+} ions, the saturation magnetization and coercivity were reduced. In a study undertaken by Peng et al. (2011) the growth in the size of crystallite of nanoparticles of cobalt ferrite via doping of gadolinium was reported. Guo, Shen, Meng, and Feng (2010) in their study concluded that the replacement of Sm^{3+} into $NiFe_2O_4$ would increase the lattice parameters and reduce the size of crystal of materials. Currently, with the speedy improvement of electronic devices to diminishment, the feature sizes of the microelectronic devices based on rare-earth-doped composites are economized into nanoscale sizes. Therefore after investigation of synthesis methods, in the next part various application of rare-earth ferrites are argued as well.

8.2.3 Application of rare-earth ferrites

Mechanical hardness, thermal stability, high magnetostriction coefficient, large coercive field and anisotropy constant are the properties attracting intense interest in undertaking basic research to provide for novel applications. All such properties raise all usage of them in a extensive spectrum of uses from medicine (for example, magnetically activated drug delivery (Amiri & Shokrollahi, 2013a, 2013b), DNA isolation (Rittich et al., 2006), MRI contrast agents (Liang, Ravi, Sampath, & Gambino, 2007) to electronics (like optoelectronics, storage media (Mukherjee et al., 2013), magnetostrictive and microwave frequency devices (Choueikani et al., 2009). Considering wholly such a great potential for new applications, various research groups have conducted investigations on the effect of rare-earth on the $CoFe_2O_4$ chattels in bulk (Dascalu et al., 2013), thin film (Zhou, Zhang, Liao, Yan, & Chen, 2004) or nanoparticle scales (Amiri & Shokrollahi, 2013a, 2013b; Mohaideen & Joy, 2012). Cobalt ferrite on the basis of its huge Faraday effect in 400–500 nm and 700–800 ranges, was viewed like as a potential alternative for magneto-optical (MO) equipment that were involved both in the field of magnetic recording (Himcinschi et al., 2013) and light modulation and deflectors (Rai, Mishra, Nguyen, & Liu, 2013). The application of $CoFe_2O_4$ as MO recording of media is impeded under the effect of high Curie temperature that would challenge the thermal-magnetic writing. Rare-earth ions among other impacts have the ability to reduce the Curie temperature and heighten the MO response when they exist in ferrite constructs (Nikumbh et al., 2014; Rai et al., 2013). In a research conducted by Cheng et al. (1999) a rise in polar Kerr rotation from 0.6° to 1° was observed pertaining to Er and Tm doped $CoFe_2O_4$ films that underwent 800°C annealing for 1 hour. This was while very slight than no changes of MO response was recorded relative to the Ho, Yb, and Lu ion replacement. Also the RE replacement would reduce the particle size that represents a significant reason in lower noise media (Cheng et al., 1999). $CoFe_2O_4$ has another important property, that is, the infrared emissivity which is further enhanced by effecting change in the cations distribution and nature of the spinel ferrite (Wu, Yu, Dong, & Geng, 2014). In one study, Zang and et al. found a growth of approx. 7% in infrared emissivity pertaining to the 8–14 mm wave bands following the addition of rare-earth elements into cobalt-based ferrites (Zhang & Wen, 2010; Zhang & Wen, 2012). All the above declared achievements emphasize the significance of investigations on the effect of RE count to the cobalt ferrite properties. The great importance in elements of rare-earth and their impact on the magnetic and microstructure characteristics of the replaced ferrite lies in the occupancy of the 4f electron shell [from 0 (La) to 14 (Lu)] and magnetic momentums [from 0 (La) to 10.6 mB (Dy)]. The lanthanide metals because of their relatively mild constants and the large orbital component present in their momentums, exhibit the biggest known magnetostrictions (Sinning et al., 2010). The RE elements provide huge ionic radii that upon replacing the cations with smaller ionic radii in structures of different type, would specify the change in the cell symmetry and thus develops internal stress. Consequently, beside the changes of the structural characteristics of the material such as greater cell parameter, lower average dimensions of crystallite and grain, etc., the magnetic and magnetostrictive properties of replacement materials and dielectric properties are also changed (Pervaiz & Gul, 2013; Zhao et al., 2006). While the RE elements' presence in bulk material lead to the formation of residual phase and reduced magnetic reaction, in nanocrystals and thin films state, however, only the spinel lattice can be identified (Kumar & Kar, 2014; Meng et al., 2009; Panda, Shih, & Chin, 2003) and in the meantime a higher magnetization for the rare-earth elements with higher magnetic momentum than Fe (Dascalu et al., 2013) are observed. Nanocrystal of rare-earth orthoferrites $LnFeO_3$ (Ln) lanthanide elements had attracted abundant interest owing to their excellent physico-chemical characteristics applicable to many applied fields. Such compounds having perovskite construct are very promising if used as catalyst (Shen & Weng, 1998), gas separator, (Kharton et al., 1999; Kharton, Yaremchenko, & Naumovich, 1999) solid oxide fuel cells (SOFCs) with cathodes, (Bahlakeh, Ramezanzadeh, Dehghani, & Ramezanzadeh, 2019) sensor and magneto-optic materials, (Schmool et al., 1999; Traversa et al., 1995) and spin valves (Sakakima, Satomi, Hirota, & Adachi, 1999).

8.3 Rare-earth manganites

8.3.1 Short introduction of rare-earth manganites

The materials made of manganese are called manganites and many compounds (manganese oxides, etc.) contain manganese. Recent decade has witnessed abundant interest by the researchers as to investigating the RE manganese oxides existing inside the perovskites-like construction. Such manganese oxides possess such characteristics like colossal magnetoresistance (CMR) (Abdel-Latif et al., 2008) and multiferroic special effects (Uusi-Esko, Malm, Imamura, Yamauchi, & Karppinen, 2008). The mentioned combinations may be utilizing magnetic storage of media as well as magnetic sensors (Abdel-Latif & El-Sherbini, 2010; Abdel-Latif et al., 2006). The researchers' interests were further intensified after the discovery of the potential of rare-earth manganese oxides for the spintronics (Fiebig, Lottermoser, Fröhlich, Goltsev, & Pisarev, 2002; Mendoza-Huizar & Rios-Reyes, 2011) and ferroelectromagnets applications (Sharan et al., 2004) and this was additional to their importance in the particular physics of simultaneous existence of ferromagnetism and ferroelectricity (Petrovic et al., 2012). With the common formulation of ABO_3, the A and B cations of Perovskites are the metallic cations while the O plays the role of a nonmetallic anion (Coey, Viret, & Von Molnár, 1999). A is a big cation, identical in size to O^{2-} whereas the B is a small cationlike Mn^{3+} or Mn^{4+}, which are coordinated octahedrally by oxygen. In our case, the A represents the rare-earth element (Nd, Eu, Sm...), B is the 3d transition metal element (Fe, Co, Mn...) and O represents the oxygen. It can be said that perovskite structure is known to have a cubic close-packed array shaped of O^{-2} anions and A^{+3} cations, having small B^{+3} cations in octahedral interstitial sites. Extensive research has been devoted to partial substitution of rare-earth element with divalent element (e.g., Ca, Sr, Ba, etc.) in this compound (Coey et al., 1999). The construction so distorted is often orthorhombic. Accordingly, it can be said that the crystalline structure of such materials in addition to having cubic structure, has orthorhombic, rhomohedral, and hexagonal forms.

Ferroelectrics are the kind of materials having spontaneous electric polarization property capable of reversal under the effect of external electric field. Ferroelectric constituents are essentially like FM constituents possessing magnetic field exchanging magnetization property, although being less famous because methodical research was not in place before the year 1920 when the American scientist recognized the ferroelectricity in Rochelle salt. Both FM and ferroelectric constituents had extensive applicability in different energy transformation equipment, integrated circuits, information storage devices, and transducers (Auciello, Scott, & Ramesh, 1998; Scott, Dawber, & Rabe, 2005; Setter et al., 2006). Simultaneously, much of the remarkable results in improper multiferroics and ferroelectrics are definitely interrelated to the manganite and ferrite families, consequently it is needless to say that the manganites and ferrites rediscovery greatly initiated the renaissance of multiferroics. Because there has been limited number of reviews on $BiFeO_3$ (Catalan & Scott, 2009; Silva, Reyes, Esparza, Camacho, & Fuentes, 2011) and the type-II multiferroics of perovskite-type rare-earth manganites, (Martins et al., 2017; Prellier, Singh, & Murugavel, 2005), the present review focuses on the hexagonal RE ferrites and manganites that contain h-$RFeO_3$, h-$RMnO_3$, and RFe_2O_4 (R = rare-earth ion). Hexagonal RE ferrites and manganites are of utmost importance not only because of including two kinds of typical improper ferroelectrics by geometric ferroelectricity, by charge-order induced ferroelectricity, but also for the reason that such systems entail abundant fascinating physical phenomena.

8.3.2 Synthesis methods of rare-earth manganites

A synthesis of the under investigation rare-earth manganites was made by using various approaches. Reaction of solid states were applied to provide Samarium (Bashkirov, Parfenov, Abdel-Latif, & Zaripova, 2005; Bouziane, Yousif, Abdel-Latif, & Hricovini, 2005), Europium (Abdel-Latif, 2011; Farag & Mostafa, 2007; Hendi & Abdel-Latif, 2011) and a part of manganites of ytterbium (Abdel-Latif, 2010) that chemical reactions were employed to supply nanocrystal scale of ytterbium manganites (Abdel-Latif, 2011; RöBler et al., 2009). Samarium ferrimanganites $SmFe_xMn_{1-x}O_3$ were produced from pure oxides; Sm_2O_3, Fe_2O_3, Fe_2O_3, and Mn_2O_3 with appropriate ratio through solid state method. Fe_2O_3 is employed for enabling the Mossbauer spectra, given in the reference Bashkirov et al. (2005). Based on the reference (Abdel-Latif et al., 2008), strontium doping gives samarium manganite during which the synthetization of $Sm_{0.6}Sr_{0.4}MnO_3$ is carried out from the initial pure oxides Sm_2O_3, Mn_2O_3, and carbonate $SrCO_3$. We mixed the above-mentioned oxides and carbonate with suitable proportions, then the mixture was milled and pressed to take the form of a disk. The produced disk was set to fire for 12 hours in the free air at 950°C. Afterwards the milling and pressing processes were applied and the sample was then fired once more at 1350°C for 72 hours. In reference Abdel-Latif (2011), solid state reaction method was used to produce $Eu_{0.65}$ $Sr_{0.35}$ $Mn_{1-x}FexO_3$ from the primary pure oxides Eu_2O_3, Fe_2O_3, Mn_2O_3, and SrO. The above pure oxides were thoroughly mixed at suitable ratios and then milled with each other by agate mortar and pressed to take a disk form under the pressure of 15ton/cm^2. The disks so prepared were set to fire for 12 hours in the open air at 1200°C. The specimens so presintered were powdered once more and were forced under similar amount of pressure to take the form of the disk with 12 mm diameter. All the specimens were set to fire once more at 1350°C for 72 hours while an intermediate grinding was also performed to make sure of homogenization; afterwards a natural furnace cooling was followed. The Yb nanocomposites specimens were prepared from pure oxides through standard solid solution method (SrO, Yb_2O_3, Fe_2O_3, Fe_2O_3, and Mn_2O_3) (Abdel-Latif, 2010). The initial oxides were 99.9% purified. A mixture of the mentioned oxides was made, milled, and pressed and finally calcined at 1050°C for 25 hours. Afterwards the compound was milled and put under pressure and eventually was fired at 1250°C for 12 hours. In the end, the previous process was performed once again with the difference that the sintering temperature was set to 1350°C that was lasted for 40 hours in the open air.

8.3.3 Application of rare-earth manganites

Currently, with the increasingly fast improvement of electronic means to integrated systems and device diminishment, the trait sizes of such devices are decreasing into nanoscale dimensions using rare-earth-doped perovskite manganite. Differing limited size impacts in RE-doped perovskite manganite oxide at nanoscale structures would lead to newer more attractive characteristics of this system. Studies conducted during recent years have reported much advancement concerning RE-doped perovskite manganite oxide nanostructures following remarkable empirical and theoretical attempts. This article provides the reader with an impression of the latest advances on the production, fundamental features, physico-mechanical properties as well as the functional applicability of the rare-earth-doped perovskite manganite oxide nanomaterials. Based on their composition, they show different magnetoelectric features like charging, FM, AFM, and diverse orbital ordering. Hence, such characteristics have the capacity to be used for sensors and spintronic applications. The primary research on the perovskite manganites dates back to Jonker & Santen (1950). The application of oxides of perovskite as catalyst actors for combustion systems in high temperature is fascinating in particular owing to their thermochemical durability in oxygen and steam having atmosphere at high degree temperatures (Uemura S, Mitsudo, & Haruta, 1998). The catalytic system but should be properly formed to attain mechanical/chemical adequate stability. In one study (Isupova et al., 1995) the researchers tried to develop rings with diverse perovskites with no addition of binders; the attempt was in vain because the rings showed weak mechanical strength. One option could be the deposition of a catalytic compound on a support, complying with some criteria. It's worth noting that a major advantage of catalytic combustion is reducing the NO at the source; the reason which lies in the considerable decrease in the number of high temperature spots that are common to noncatalytic combustion systems. Principally, the availability of a reducing agent speeds up the wash-up of the catalyst surface. Nonetheless it can be said that the conclusion reached could be reasonable in case the very mechanism can work in the reducing agent's presence if such an agent is CO or hydrocarbons (Isupova et al., 1995). Lanthanum manganite ($LaMnO_3$) among the $LnMnO_3$ materials demonstrates large thermal stability as well as oxygen transition. Several volatile organic compounds (VOC) were catalytically oxidated in quest for extensive study of the above compound. In a recent work, the oxidation of acetone, isopropanol, and benzene in the presence of $LaMnO_3$ perovskite was investigated (Spinicci, Faticanti, Marini, De Rossi, & Porta, 2003). Decrease of activity starts from acetone and continues to benzene. Lanthanum manganite has been employed for the purpose of destructing the chlorinated VOC methanol and is actively present in the reduction of the nitrogen oxides (Chirilă, Papp, Suprun, & Balasanian, 2007; Yonghua, Futai, & Hui, 1988). The development of some by-products including C-C coupling and higher chlorinated or cracking compounds requires deactivation of catalyst. As for the hydrocarbon VOC abatement field, the $LaMnO_3$ perovskite-type catalysts have seldom been evaluated.

8.4 Rare-earth cobaltites

8.4.1 Short introduction of rare-earth cobaltites

During the past decades, many researches have been undertaken about ABO_3-type perovskite oxides with universal formula of $Ln_{1-x}Ak_xBO_3$ (Ln = trivalent lanthanide, Ak = divalent alkaline earth, B = transition metal) (Rao, Kundu, Seikh, & Sudheendra, 2004). Especially the discovery of perovskite cobaltites and the magnetic ordering were first reported in 1950s and 1960s respectively (Jonker & Santen, 1953) and thereafter abundant works could be found in the literature reporting fascinating phenomena. A number of the advantageous characteristics of the perovskite cobaltites are already known, particularly the rock crystal structure alteration, the paramagnetic (PM) toward FM change at temperature of Curie (TC) and the relevant insulator-metal transition, etc. Finding CMR in manganites-doped has revitalized scientists' attention in perovskite oxides from the beginning of the 90's (Rao et al., 2004). In the year 1997, researchers (Martin, Maignan, Pelloquin, Nguyen, & Raveau, 1997) come across huge amounts of magnetoresistance pertaining to the freshly found ordered cobaltite, (Ln = Eu, Gd): $LnBaCo_2O_{5.4}$, called as 112-phases layered. This results in even further interest owing to their capacity for the improvement of storage of magnetic data. Additionally, the perovskite cobaltites have concerned interest owing to their probable applicability as catalysts for oxidation, sensor of gas, solid oxides fuel cells besides membranes for oxygen separation and electron conduction (Teraoka, Nobunaga, Okamoto, Miura, & Yamazoe, 1991). As a result, researchers in the year 2003 found superconductivity in hydrated sodium cobaltite phase, and cobaltites have attracted much more attention ever since (Takada et al., 2003). In addition to the applicability potential, the abilities of the mentioned materials demonstrate rich phase diagram covering an extensive spectrum of magnetic characteristics and such spectacles as charge/orbital ordering, spin/cluster-glass performance, phase separation via electronic, etc. (Fauth & Suard, 2001; Kundu & Ramesha, 2004; La et al., 1994; Mahendiran, Raychaudhuri, Chainani, Sarma, & Roy, 1995; Nakajima, Ichihara, & Ueda, 2005; Wu et al., 2005). Such phenomena are indicative of a shared interaction among the spin/lattice/charge and the orbital degrees of freedom, providing improved realization of some powerfully correlated electronic behaviors. Such interactions exhibit themselves in single crystal, polycrystalline samples and in thin films as well. External factors and chemical means can be used to regulate the perovskite cobaltites capabilities. The RE cobaltites show instantaneous occurrence of ferromagnetism and metallicity.

8.4.2 Synthesis methods of rare-earth cobaltites

One of the materials extensively investigated owing to its attractive physical properties is the 'layered oxygen deficient "112" cobalt pervoskites' with the general formulation $LnBaCo_2O_{5+d}$ which exhibits such characteristics as, metal-insulator transition, ferromagnetism, and high magnetoresistance (Raveau & Seikh, 2012). The oxygen content in this structural family, together with the organization of the Ln^{3+} and Ba^{2+} cations and of the anion cavities have an important role in the magnetic characteristics, for example, their capability to show metal-insulator transition, (Maignan, Martin, Pelloquin, Nguyen, & Raveau, 1999; Martin et al., 1997; Seikh et al., 2008), giant magnetoresistance, (Liu et al., 2010; Ma et al., 2012; Suard & Fauth, 2000; Troyanchuk, Kasper, & Khalyavin, 1998; Yuan et al., 2007), charge-ordering, (Vogt et al., 2000), and lastly spin-state transition (Moritomo, Akimoto, Takeo, & Machida, 2000; Respaud et al., 2001). Nonetheless such systems are quite noticeable by the particular difficulty they are encountered with to perform synthetization of the well-ordered stoichiometric perovskite, conforming to d = 1. The stoichiometric perovskite (ordered) $LnBaCo_2O_6$ consisting LnO and BaO layers 1:1 ordered stacking, may only undergo synthetization until now to produce Ln = La and Nd. In other words, the synthetization of the "112" layered $LaBaCo_2O_6$ perovskite needs particular synthesis conditions, reduction of cobaltite $LaBaCo_2O_5$ then annealing it in oxygen at 350°C, to evade the formation of $La_{0.5}Ba_{0.5}CoO_3$ (Asish et al., 2007; Rautama et al., 2008). The direct solid state reaction method could not be used for obtaining ordered layered "112" perovskite $NdBaCo_2O_6$, and soft chemistry synthetization would be needed (Pralong, Caignaert, Hebert, & Maignan, 2006). Literature shows that the recent research on "112" layered cobaltites $PrBaCo_2O_{5+d}$ (Carlos Frontera, García-Muñoz, Carrillo, Aranda, & Margiolaki, 2006; Ganorkar, Priolkar, Sarode, & Banerjee, 2011; Garcıa-Mu ~ noz et al., 2004) have demonstrated intense competition among the AFM and FM interactions in the mentioned system. It is fascinating that the authors have illustrated that the dependency of the magnetic characteristics on oxygen content in this family of materials can be traced in the distortion of the cobalt oxygen polyhedral but unable to reach d values larger than 0.9, because of their experimental lab situations.

8.4.3 Application of rare-earth cobaltites

Oxides materials were among the materials not deemed as proper materials, though Terasaki et al. in 1997 (Terasaki, Sasago, & Uchinokura, 1997) claimed discovering productive thermoelectric characteristics in $NaCo_2O_4$. Cobaltites similar to $Ca_3Co_4O_9$ demonstrate two monoclinic subsystems. The $[Ca_2Co_3O_6]$ and $[CoO_2]$ in such compounds respectively are representing the rock salt (RS)-type block and the CdI2-type layers. These structures therefore are formed by two monoclinic subsystems, in other words, the RS- block layer and CoO_2 layer. Considering $Ca_3Co_4O_9$ and $Bi_2Sr_2Co_2O_9$, a lattice misfit is found among the hexagonal CoO_2 sheet and the square Ca_2CoO_3 and $Bi_2Sr_2O_4$ substructures, confirming their misfit layered structure (He, Liu, & Funahashi, 2011; Lambert, Leligny, & Grebille, 2001). Cobaltites because of their improved thermoelectric characteristics are one of the best candidates for the thermoelectric applications (Pelloquin, Hebert, Maignan, & Raveau, 2004) and great thermopower (Luo, Jing, Chen, & Chen, 2007) are among the most notable properties of these materials. Benefits such as nontoxicity, temperature stability, and great oxidation resistance prepares good context for research activities (Rowe, 2006). Additionally, it's worth noting that cobaltites are prospective materials applicable to numerous technologies, especially hydrogen power engineering. This would raise further interest due to their capacity for wider applications in further enhancement of magnetic data storage. Furthermore, the perovskite cobaltites are attracting research interests due to their potential applicability in various uses especially in high oxygen diffusivity and electron conduction. As for the applicability, $LaCoO_3$ is known as prospective contact component for the cathode part in the SOFCs (Singhal 2002; Sun, Hui, & Roller, 2010). There have been reports also stating that Sr-replaced $LaCoO_3$ apparently may act as a quick ion conductance and as a membrane for high-temperature oxygen separation (Kharton et al., 1999; Kovalevsky et al., 1998).

8.5 Rare-earth nickelates

8.5.1 Short introduction of rare-earth nickelates

Correlated transition metal oxides (TMOs) from among the functional materials have allocated particular attention to themselves because of their notable magnetic, electronic, structural, and optical characteristics. This behavioral spectrum spreads the possibility for developing new electronic architecture. A peculiar characteristic of TMOs complex is the presence of some rivalry among the electronic phases near boundaries phase that are usually very sensitive to delicate structural variations and hence a fine control technique is proposed. In respect of the ever-increasing attention to the realization and execution of the TMOs, outstanding progresses have been achieved in recent years in relation to qualified epitaxial TMO heterostructures. This embraces the production of digital heterostructures oxide, making probable the modification of electronic states of TMOs by benefitting from the interface persuaded phenomena and decrease dimensions. These epitaxial heterostructures have paved the way for the introduction of novel, and realization of current physical one so that it is nowadays sensible to predict oxide electronic devices with the capacity to undergo competition with semiconductor architecture (Bibes, Villegas, & Barthélèmy, 2011; Hwang et al., 2012; Mannhart & Schlom, 2010; Zhang, 2011) like perovskite nickelates, with the generic $RNiO_3$ formula, where R is a trivalent RE, R = La, Pr, Nd, Sm, . . ., Lu. $RNiO_3$ have many useful structural/physical characteristics. The phase diagram of the nickelates family is a typical instance of the phase transitions physical science in correlated oxides in which the electron-lattice coupling significantly donates. An outstanding characteristic of the nickelates, exclusive of R = La, is a rapid metal/insulator transition (MIT), principally strong-minded by the Ni-O-Ni bond angle. Consequently, the perovskite nickelates physical properties can be compromised by numerous parameters ranged from temperature, pressure and R-size, to epitaxial strain in heterostructures, as well as the stoichiometry or electrostatic doping. The best part of the recent attention allocated to nickelates has been concentrated on thin films and their incorporation into superlattices with the general purpose of monitoring their physical properties via strain control, confinement and interface influences. Furthermore, owing to the shortage of large nickelate single crystals, epitaxial thin films also represent the superior system worth of studying the wonderful physics of such compounds.

8.5.2 Synthesis methods of rare-earth nickelates

Moritomo et al. was the first group that successfully synthesized the bulk $RNiO_3$s. (Moritomo et al., 2000), while their physical characteristics were widely identified only in the early 90s in the original works undertaken by García-Muñoz, Rodríguez-

Carvajal, Lacorre, and Torrance (1992). Finding of a specific MIT existed in all the family members, eliminating the case of R = La, and scrutinizing a strange AFM structure representing the characteristics of the insulating phase, ignited unprecedented attention in the complexes. The experimental investigation of nickelates however was restricted to polycrystalline ceramic specimens, the single crystal size of which being restricted to the micron scale. Additionally, the synthesis materials involved problematic situations, dictating enormously high pressure of oxygen as well as high temperature applicable for the chemically steadying of the perovskite phase. According to the literature, the great quality $LaNiO_3$ single crystals has been lately synthesized; this is while obtaining high quality thin film nickelate specimens is currently crucial for the empirical investigations of such compounds. Previous works on nickelates have been essentially restricted to polycrystalline ceramic specimens produced under severe oxygen pressure/high temperature, equally vital for the perovskite stabilization phase with Ni being in its nominal 3 + valence state. The lack of large crystals has prevented a more precision examination on and a profounder realization of fundamental components and combination mechanism. In this respect, the synthetization constitutes a major advance in terms of further scrutinizing the rich potential of the structural and physical characteristics of nickelates. Also the synthetization of high quality $LaNiO_3$ single crystals has lately been described (fang Li et al., 2017; Zhang, Zheng, Ren, & Mitchell, 2017); removing a long-run barricade hampering the bulk $RNiO_3$ investigations and opening new windows on further recognition of such materials. Nowadays most researchers of the field are concentrating on $LaNiO_3$, $NdNiO_3$, $SmNiO_3$ and solid solutions thereto. But $RNiO_3$ with smaller RE needs future studies to be started. Progress in thin film synthetization methods are currently providing for the synthesis of high quality oxide heterostructures which are increasingly analogous to that of semiconductors. Normally, heterostructures introduce great tools for the electronic status manipulation of transitory metal oxides through benefitting from the strain impacts, interface-driven phenomena or reduction of dimensions. Latest research on $RNiO_3$-based superlattices have created a diversity of routs to even more controlling their applications and even realizing new collective phases. Using modern methods and techniques is essential for realization and characterization of the mentioned potentials. Until now, most of the studies have focused on $LaNiO_3$-based superlattices and some more recent research would integrate other $RNiO_3$ compounds; comparatively, all-nickelate multilayers (not counting $LaNiO_3$) have seldom been examined. Moreover, the primary prediction made by Chaloupka & Khaliullin (2008) of a superconducting phase that is stabilized in $RNiO_3$ films via confinement technique and heterostructuring (Chaloupka & Khaliullin, 2008) has yet to be established.

8.5.3 Application of rare-earth nickelates

The $LaNiO_3$ in some applications is exploited more directly as an active and functional part of the equipment, for examples, oxygen reduction in the catalysis area (Cao, Lin, Sun, Yang, & Zhang, 2015; Catalano et al., 2018; Singh et al., 1994; Wang et al., 2013), in electrolysis by water (Singh et al., 1994), in fuel cells, in gas sensors (Wang et al., 2013) and in super-capacitors (Cao et al., 2015). Eventually, and for the discussion completeness, it's worth noting that $LaNiO_3$ is considered a conductive oxide buffer layer on surfaced nickel tapes as well, applicable for high-Tc-coated superconductors (HTS) (Sun et al., 2005); it provides both epitaxial template and electrically coupling of the HTS layer to the metallic tape substrate, for example, for the RABiTS technology. Although the $LaNiO_3$ applications are based on its great and relatively steady conductivity over a wide range of temperature, the extra rare-earth nickelates family members interest attention for their sudden and rapid MIT and its relevant properties. Utmost of the studies in the literature have focused either on $SmNiO_3$ or $NdNiO_3$, mostly on the previous, which has an MIT overhead room temperature (130°C) permitting its possible integration onto the oxide-electronics based on CMOS (complimentary metal oxide semiconductor) (Ha, Aydogdu, & Ramanathan, 2011). The acceptable adjustment of the MIT transition temperatures on thin films of nickelate by numerous parameters is vital for the development of their capacity for different uses. One of the typical uses recently argued in the literature on nickelates concerns with their memoristive performance that is completely linked to the MIT. Nickelates aside, nanosize resistive changing devices that are occasionally called as memristors, are currently attracting considerable attention for their usage in logic, memory, and neuromorphic uses (Jo et al., 2010). As for the $SmNiO_3/LaAlO_3$ structures, there are reports that could open the door to practical use based on such resistive switching. Later, in another study they introduced a $SmNiO_3$-based transistor that could mimic a biologic synapse through $SmNiO_3$ doping using an ionic solution (Shi, Zhou, & Ramanathan, 2014). Lastly the colossal resistance switching detected from electron doping could be one of the routes to larger conductance changes, spreading a superior perspective for upcoming uses (Shi, Ha, Zhou, Schoofs, & Ramanathan, 2013).

8.6 Conclusion and outlook

High-quality nanocrystalline rare-earth materials have indicated great attention due to their distinctive chemical/physical possessions for numerous applications. These perovskite structured compounds are talented as gas separators, catalysts, cathodes actor in solid oxide in fuel cells, sensor, magnetooptic materials, spin valves, and so forth. The rare-earth-based nanocomposites were usually synthesized by different methods. Most of excellent reviews of past decades on rare-earth by means of nanoparticles focused on synthesis and properties of these materials. The present chapter investigated not only the synthesis method but also their applications in various fields. Although there are numerous attractive properties of rare-earth-based ceramic nanomaterials (ferrites), they are made from the relations among the charge/spin/orbital and freedom of lattice degrees, whereas there is quite an extended way to go for gaining a full consideration of the interaction mechanisms among those. It is estimated that in the following years, additional progress will be attained in the experimental/theoretical investigations on rare earth-based ceramic nanomaterials.

References

Abdel-Latif, I. A. (2010). Study on microstructure and electrical properties of europium manganites. *Arab. J. Nucl. Sci. Appl.*, *43*.

Abdel-Latif, I. A. (2011). Study on the effect of nano size of strontium-ytterbium manganites on some physical properties. *AIP Conf. Proc.*, *1370*, 108–115. Available from https://doi.org/10.1063/1.3638090.

Abdel-Latif, I. A., & El-Sherbini, M. (2010). Low temperature study of resistivity of Sm0.6Sr0.4MnO3. *Mater. Sci. An Indian J.*, *6*, 109–111. Available from https://www.tsijournals.com/articles/low-temperature-study-of-resistivity-of-sm06sr04mno3.pdf.

Abdel-Latif, I. A., Hassen, A., Zybill, C., Abdel-Hafiez, M., Allam, S., & El-Sherbini, T. (2008). The influence of tilt angle on the CMR in Sm0.6Sr0.4MnO3. *J. Alloys Compd.*, *452*, 245–248. Available from https://doi.org/10.1016/j.jallcom.2007.07.022.

Abdel-Latif, V. A. T. I. A., Khramov, A. S., Smirnov, V. V. P. A. P., Bashkirov, S. S., Tserkovnaya, Z. I. E. A., & Gumarov, G. G. (2006). Electrical and magnetic properties - structure correlation on Nd0.65Sr0.35Mn1-XFeXO3. *Egypt. J. Solids*, *29*, 341–350. Available from https://doi.org/10.21608/ejs.2006.149284.

Amiri, S., & Shokrollahi, H. (2013a). Magnetic and structural properties of RE doped Co-ferrite (RE = Nd, Eu, and Gd) nano-particles synthesized by co-precipitation. *Journal of Magnetism and Magnetic Materials*, *345*, 18–23. Available from https://doi.org/10.1016/j.jmmm.2013.05.030.

Amiri, S., & Shokrollahi, H. (2013b). The role of cobalt ferrite magnetic nanoparticles in medical science. *Mater. Sci. Eng. C.*, *33*, 1–8. Available from https://doi.org/10.1016/j.msec.2012.09.003.

Asish, B. R., Kundu, K., Rautama, E.-L., Boullay, P., Caignaert, V., & Pralong, V. (2007). Spin-locking effect in the nanoscale ordered perovskite cobaltite La Ba Co_2O_6. *Phys. Rev. B.*, *76*.

Auciello, O., Scott, J. F., & Ramesh, R. (1998). The physics of ferroelectric memories. *Physics Today*, *51*, 22–27. Available from https://doi.org/10.1063/1.882324.

Bahlakeh, G., Ramezanzadeh, B., Dehghani, A., & Ramezanzadeh, M. (2019). Novel cost-effective and high-performance green inhibitor based on aqueous Peganum harmala seed extract for mild steel corrosion in HCl solution: Detailed experimental and electronic/atomic level computational explorations. *J. Mol. Liq.*, *283*, 174–195. Available from https://doi.org/10.1016/j.molliq.2019.03.086.

Bashkirov, S. S., Parfenov, V. V., Abdel-Latif, I. A., & Zaripova, L. D. (2005). Mössbauer effect and electrical properties studies of SmFe xMn1-xO3 (x = 0.7, 0.8 and 0.9). *J. Alloys Compd*, *387*, 70–73. Available from https://doi.org/10.1016/j.jallcom.2004.06.070.

Bate, G. (1991). Magnetic recording materials since 1975. *Journal of Magnetism and Magnetic Materials*, *100*, 413–424. Available from https://doi.org/10.1016/0304-8853(91)90831-T.

Ben Tahar, L., Artus, M., Ammar, S., Smiri, L. S., Herbst, F., Vaulay, M. J., … Fiévet, F. (2008). Magnetic properties of CoFe1.9RE0.1O4 nanoparticles (RE = La, Ce, Nd, Sm, Eu, Gd, Tb, Ho) prepared in polyol. *Journal of Magnetism and Magnetic Materials*, *320*, 3242–3250. Available from https://doi.org/10.1016/j.jmmm.2008.060.031.

Bibes, M., Villegas, J. E., & Barthélèmy, A. (2011). Ultrathin oxide films and interfaces for electronics and spintronics. *Adv. Phys.*, *60*, 5–84. Available from https://doi.org/10.1080/00018732.2010.534865.

Bouziane, C. R. K., Yousif, A., Abdel-Latif, I. A., & Hricovini, K. (2005). Electronic and magnetic properties of SmFe1 − xMnxO3 orthoferrites (x = 0.1, 0.2, and 0.3). *Journal of Applied Physics*, *97*.

Cai, W., & Wan, J. (2007). Facile synthesis of superparamagnetic magnetite nanoparticles in liquid polyols. *Journal of Colloid and Interface Science*, *305*, 366−370. Available from https://doi.org/10.1016/j.jcis.2006.10.023.

Cao, Y., Lin, B. P., Sun, Y., Yang, H., & Zhang, X. Q. (2015). *Electrochimica Acta*, *174*, 41.

Carlos Frontera, A. C., García-Muñoz, J. L., Carrillo, A. E., Aranda, M. A. G., & Margiolaki, I. (2006). Spin state of Co 3 + and magnetic transitions in R BaCo 2 O 5.50 (R = Pr, Gd) : Dependence on rare-earth size. *Phys. Rev. B.*, *74*.

Catalan, G., & Scott, J. F. (2009). Physics and applications of bismuth ferrite. *Adv. Mater.*, *21*, 2463−2485. Available from https://doi.org/10.1002/adma.200802849.

Catalano, S., Gibert, M., Fowlie, J., Iñiguez, J., Triscone, J. M., & Kreisel, J. (2018). Rare-earth nickelates RNiO3: Thin films and heterostructures. *Reports Prog. Phys.*, *81*. Available from https://doi.org/10.1088/1361-6633/aaa37a.

Chaloupka, J., & Khaliullin, G. (2008). Orbital order and possible superconductivity in $LaNiO_3$/$LaMO_3$ superlattices. *Physical Review Letters*, *100*. Available from https://doi.org/10.1103/PhysRevLett.100.016404.

Cheng, F. X., Jia, J. T., Xu, Z. G., Zhou, B., Liao, C. S., Yan, C. H., ... Bin Zhao, H. (1999). Microstructure, magnetic, and magneto-optical properties of chemical synthesized Co-RE (RE-Ho, Er, Tm, Yb, Lu) ferrite nanocrystalline films. *Journal of Applied Physics*, *86*, 2727−2732. Available from https://doi.org/10.1063/1.371117.

Chirilă, L. M., Papp, H., Suprun, W., & Balasanian, I. (2007). Synthesis, characterization and catalytic reduction of NOx emissions over $LaMnO_3$ perovskite. *Environ. Eng. Manag. J.*, *6*, 549−553. Available from https://doi.org/10.30638/eemj.2007.070.

Choueikani, F., Royer, F., Jamon, D., Siblini, A., Rousseau, J. J., Neveu, S., & Charara, J. (2009). Magneto-optical waveguides made of cobalt ferrite nanoparticles embedded in silica/zirconia organic-inorganic matrix. *Applied Physics Letters*, *94*. Available from https://doi.org/10.1063/1.3079094.

Coey, J. M. D., Viret, M., & Von Molnár, S. (1999). Mixed-valence manganites. *Adv. Phys.*, *48*, 167−293. Available from https://doi.org/10.1080/000187399243455.

Dascalu, G., Pompilian, G., Chazallon, B., Nica, V., Caltun, O. F., Gurlui, S., & Focsa, C. (2013). Rare earth doped cobalt ferrite thin films deposited by PLD. *Appl. Phys. A Mater. Sci. Process.*, *110*, 915−922. Available from https://doi.org/10.1007/s00339-012-7196-8.

Elkestawy, M. A., Abdel kader, S., & Amer, M. A. (2010). AC conductivity and dielectric properties of Ti-doped CoCr1.2Fe0.8O4 spinel ferrite. *Phys. B Condens. 8uMatter*, *405*, 619−624. Available from https://doi.org/10.1016/j.physb.2009.09.076.

Fang Li, M., Guo Liu, Y., Ming Zeng, G., Bo Liu, S., Jiang Hu, X., Shu, D., ... Li Yan, Z. (2017). Tetracycline absorbed onto nitrilotriacetic acid-functionalized magnetic graphene oxide: Influencing factors and uptake mechanism. *Journal of Colloid and Interface Science*, *485*, 269−279. Available from https://doi.org/10.1016/j.jcis.2016.09.037.

Farag, I. A. A.-L. I., & Mostafa, A. (2007). Preparation and Structural Characterization of Eu0.65Sr0.35Mn1-xFexO3. *Egypt. J. Solids.*, *30*, 149−155. Available from https://doi.org/10.21608/ejs.2007.149070.

Fauth, V. C. F., & Suard, E. (2001). Intermediate spin state of Co 3 + and Co 4 + ions in La 0.5 Ba 0.5 CoO_3 evidenced by Jahn-Teller distortions. *Phys. Rev. B.*, *65*.

Fiebig, M., Lottermoser, T., Fröhlich, D., Goltsev, A. V., & Pisarev, R. V. (2002). Observation of coupled magnetic and electric domains. *Nature*, *419*, 818–820. Available from https://doi.org/10.1038/nature01077.

Ganorkar, S., Priolkar, K. R., Sarode, P. R., & Banerjee, A. (2011). Effect of oxygen content on magnetic properties of layered cobaltites PrBaCo2O5 + δ. *Journal of Applied Physics*, *110*. Available from https://doi.org/10.1063/1.3633521.

García-Muñoz, J. L., Rodríguez-Carvajal, J., Lacorre, P., & Torrance, J. B. (1992). Neutron-diffraction study of RNiO3 (R = La,Pr,Nd,Sm): Electronically induced structural changes across the metal-insulator transition. *Phys. Rev. B.*, *46*, 4414–4425. Available from https://doi.org/10.1103/PhysRevB.46.4414.

Garcıa-Muñoz, E. D. J. L., Frontera, C., Llobet, A., Carrillo, A. E., Caneiro, A., Aranda, M. A. G., . . . Ritter, C. (2004). Magnetic and electronic properties of Eu1 − xSrxMnO3. pdf. *Journal of Magnetism and Magnetic Materials*, 272–276.

Goh, S. C., Chia, C. H., Zakaria, S., Yusoff, M., Haw, C. Y., Ahmadi, S., . . . Lim, H. N. (2010). Hydrothermal preparation of high saturation magnetization and coercivity cobalt ferrite nanocrystals without subsequent calcination. *Mater. Chem. Phys.*, *120*, 31–35. Available from https://doi.org/10.1016/j.matchemphys.2009.100.016.

Gul, I. H., & Maqsood, A. (2008). Structural, magnetic and electrical properties of cobalt ferrites prepared by the sol-gel route. *J. Alloys Compd.*, *465*, 227–231. Available from https://doi.org/10.1016/j.jallcom.2007.11.006.

Guo, L., Shen, X., Meng, X., & Feng, Y. (2010). Effect of Sm3 + ions doping on structure and magnetic properties of nanocrystalline $NiFe_2O_4$ fibers. *J. Alloys Compd.*, *490*, 301–306. Available from https://doi.org/10.1016/j.jallcom.2009.09.182.

Ha, S. D., Aydogdu, G. H., & Ramanathan, S. (2011). Metal-insulator transition and electrically driven memristive characteristics of SmNiO3 thin films. *Applied Physics Letters*, *98*. Available from https://doi.org/10.1063/1.3536486.

He, J., Liu, Y., & Funahashi, R. (2011). Oxide thermoelectrics: The challenges, progress, and outlook. *Journal of Materials Research*, *26*, 1762–1772. Available from https://doi.org/10.1557/jmr.2011.108.

Hendi, S. A. S. A., & Abdel-Latif, I. A. (2011). Structure, electrical and dielectric properties of strontium europium ferrimanganites. *J. Am. Sci.*, *7*, 749.

Himcinschi, C., Vrejoiu, I., Salvan, G., Fronk, M., Talkenberger, A., Zahn, D. R. T., . . . Kortus, J. (2013). Optical and magneto-optical study of nickel and cobalt ferrite epitaxial thin films and submicron structures. *Journal of Applied Physics*, *113*. Available from https://doi.org/10.1063/1.4792749.

Hwang, H. Y., Iwasa, Y., Kawasaki, M., Keimer, B., Nagaosa, N., & Tokura, Y. (2012). Emergent phenomena at oxide interfaces. *Nature Materials*, *11*, 103–113. Available from https://doi.org/10.1038/nmat3223.

Ikeda, N., Kohn, K., Kito, H., Akimitsu, J., & Siratori, K. (1995). Anisotropy of dielectric dispersion in $ErFe_2O_4$ single crystal. *J. Phys. Soc. Japan.*, *64*, 1371–1377. Available from https://doi.org/10.1143/JPSJ.64.1371.

Ikeda, N., Kohn, K., Myouga, N., Takahashi, E., Kitôh, H., & Takekawa, S. (2000). Charge frustration and dielectric dispersion in $LuFe_2O_4$. *J. Phys. Soc. Japan.*, *69*, 1526–1532. Available from https://doi.org/10.1143/JPSJ.69.1526.

Ikeda, N., Ohsumi, H., Mizumaki, M., Mori, S., Horibe, Y., & Kishimoto, K. (2004). Frustration and ordering of iron ions on triangular iron mixed valence system RFe_2O_4. *Journal of Magnetism and Magnetic Materials*, 272–276. Available from https://doi.org/10.1016/j.jmmm.2003.12.155.

Ikeda, N., Ohsumi, H., Ohwada, K., Ishii, K., Inami, T., Kakurai, K., ... Kitô, H. (2005). Ferroelectricity from iron valence ordering in the charge-frustrated system $LuFe_2O_4$. *Nature*, *436*, 1136–1138. Available from https://doi.org/10.1038/nature04039.

Ikeda, N., Mori, S., & Kohn, K. (2005). Charge ordering and dielectric dispersion in mixed valence oxides RFe_2O_4. *Ferroelectrics*, *314*, 41–56. Available from https://doi.org/10.1080/00150190590926085.

Ikeda, N., Mori, S., & Yoshii, K. (2007). Ferroelectricity from valence ordering in RFe2O4. *Ferroelectrics*, *348*, 38–47. Available from https://doi.org/10.1080/00150190701196112.

Il Kim, Y., Kim, D., & Lee, C. S. (2003). Synthesis and characterization of $CoFe_2O_4$ magnetic nanoparticles prepared by temperature-controlled coprecipitation method. *Phys. B Condens. Matter.*, *337*, 42–51. Available from https://doi.org/10.1016/S0921-4526(03)00322-3.

Isupova, L. A., Sadykov, V. A., Solovyova, L. P., Andrianova, M. P., Ivanov, V. P., Kryukova, G. N., ... Tretyakov, V. F. (1995). Monolith perovskite catalysts of honeycomb structure for fuel combustion. *Studies in Surface Science and Catalysis*, *91*, 637–645. Available from https://doi.org/10.1016/S0167-2991(06)81803-3.

Jacob, B. P., Thankachan, S., Xavier, S., & Mohammed, E. M. (2011). Effect of Gd3 + doping on the structural and magnetic properties of nanocrystalline Ni-Cd mixed ferrite. *Phys. Scr.*, *84*. Available from https://doi.org/10.1088/0031-8949/84/04/045702.

Jacobo, S. E., Duhalde, S., & Bertorello, H. R. (2004). Rare earth influence on the structural and magnetic properties of NiZn ferrites. *J. Magn. Magn. Mater. 272–*, *276*, 2253–2254. Available from https://doi.org/10.1016/j.jmmm.2003.120.564.

Jo, S. H., Chang, T., Ebong, I., Bhadviya, B. B., Mazumder, P., & Lu, W. (2010). Nanoscale memristor device as synapse in neuromorphic systems. *Nano Letters*, *10*, 1297–1301. Available from https://doi.org/10.1021/nl904092h.

Jonker, G. H., & Santen, J. H. V. (1950). Ferromagnetic compounds of manganese with perovskite structure. *Physica*, *16*, 337–349. Available from https://doi.org/10.1016/0031-8914(50)90033-4.

Jonker, G. H., & Santen, J. H. V. (1953). Magnetic compounds wtth perovskite structure III. ferromagnetic compounds of cobalt. *Physica.*, *19*, 120–130.

Kamala Bharathi, K., Arout Chelvane, J., & Markandeyulu, G. (2009). Magnetoelectric properties of Gd and Nd-doped nickel ferrite. *Journal of Magnetism and Magnetic Materials*, *321*, 3677–3680. Available from https://doi.org/10.1016/j.jmmm.2009.070.011.

Kharton, V. V., Yaremchenko, A. A., Kovalevsky, A. V., Viskup, A. P., Naumovich, E. N., & Kerko, P. F. (1999). Perovskite-type oxides for high-temperature oxygen separation membranes. *Journal of Membrane Science*, *163*, 307–317. Available from https://doi.org/10.1016/S0376-7388(99)00172-6.

Kharton, V. V., Yaremchenko, A. A., & Naumovich, E. N. (1999). Research on the electrochemistry of oxygen ion conductors in the former Soviet Union. II. Perovskite-related oxides. *J. Solid State Electrochem*, *3*, 303–326. Available from https://doi.org/10.1007/s100080050161.

Kim, C. S., yi, Y. S., Park, K. T., Namgung, H., & Lee, J. G. (1999). Growth of ultrafine Co–Mn ferrite and magnetic properties by a sol–gel method. *Journal of Applied Physics*, *85*, 5223–5225. Available from https://doi.org/10.1063/1.369950.

Kovalevsky, A. V., Kharton, V. V., Tikhonovich, V. N., Naumovich, E. N., Tonoyan, A. A., Reut, O. P., & Boginsky, L. S. (1998). Oxygen permeation through Sr(Ln)CoO3-δ (Ln = La, Nd, Sm, Gd) ceramic membranes. *Mater. Sci. Eng. B.*, *52*(B52), 105–116. Available from https://doi.org/10.1016/s0921-5107(97)00292-4.

Kumar, L., & Kar, M. (2014). Effect of Ho3 + substitution on the cation distribution, crystal structure and magnetocrystalline anisotropy of nanocrystalline cobalt ferrite. *J. Exp. Nanosci.*, *9*, 362–374. Available from https://doi.org/10.1080/17458080.2012.661474.

Kundu, R. S. A. K., & Ramesha, K. (2004). Magnetic and electron transport properties of the rare earth cobaltates, La0. 7 – xLnxCa0. 3CoO3 (Ln = Pr, Nd, Gd and Dy): a case of phase separation. *Journal of Physics*, *16*.

La, L., Itoh, M., Inaguma, Y., Jung, W., Chen, L., & Nakamura, T. (1994). High lithium ion conductivity in the perovskite-type compounds Ln1/2Li1/2TiO3 (Ln = La, Pr, Nd, Sm). *Solid State Ionics*, *70/71*, 203–207.

Lambert, S., Leligny, H., & Grebille, D. (2001). Three forms of the misfit layered cobaltite [Ca2CoO3] [CoO2]1.62 · A 4D structural investigation. *J. Solid State Chem*, *160*, 322–331. Available from https://doi.org/10.1006/jssc.2001.9235.

Liang, S., Ravi, B. G., Sampath, S., & Gambino, R. J. (2007). Atmospheric plasma sprayed cobalt ferrite coatings for magnetostrictive sensor applications. *IEEE Transactions on Magnetics*, *43*, 2391–2393. Available from https://doi.org/10.1109/TMAG.2007.893840.

Liu, M., Liu, J., Collins, G., Ma, C. R., Chen, C. L., He, J., ... Zhang, Q. Y. (2010). Magnetic and transport properties of epitaxial (LaBa) $Co_2O_5.5 + \delta$ thin films on (001) SrTiO3. *Applied Physics Letters*, *96*. Available from https://doi.org/10.1063/1.3378877.

Luo, X. G., Jing, Y. C., Chen, H., & Chen, X. H. (2007). Intergrowth and thermoelectric properties in the Bi-Ca-Co-O system. *Journal of Crystal Growth*, *308*, 309–313. Available from https://doi.org/10.1016/j.jcrysgro.2007.07.037.

Ma, C., Liu, M., Collins, G., Liu, J., Zhang, Y., Chen, C., ... Meletis, E. I. (2012). Thickness effects on the magnetic and electrical transport properties of highly epitaxial LaBaCo 2O 5.5 + δ thin films on MgO substrates. *Applied Physics Letters*, *101*. Available from https://doi.org/10.1063/1.4734386.

Mahendiran, R., Raychaudhuri, A. K., Chainani, A., Sarma, D. D., & Roy, S. B. (1995). Large magnetoresistance in La1-xSrxMnO_3 and its dependence on magnetization. *Applied Physics Letters*, 233. Available from https://doi.org/10.1063/1.113556.

Maignan, A., Martin, C., Pelloquin, D., Nguyen, N., & Raveau, B. (1999). Structural and magnetic studies of ordered oxygen-deficient perovskites LnBaCo2O5 + δ, closely related to the "112" structure. *J. Solid State Chem*, *142*, 247–260. Available from https://doi.org/10.1006/jssc.1998.7934.

Mannhart, J., & Schlom, D. G. (2010). *Science (New York, N.Y.)*, *327*(5973), 1607.

Maqsood, A. (2009). Phase transformations in $Ho_2Si_2O_7$ ceramics. *J. Alloys Compd.*, *471*, 432–434. Available from https://doi.org/10.1016/j.jallcom.2008.030.108.

Martin, C., Maignan, A., Pelloquin, D., Nguyen, N., & Raveau, B. (1997). Magnetoresistance in the oxygen deficient LnBaCo2O5.4(Ln = Eu, Gd) phases. *Applied Physics Letters*, *71*, 1421–1423. Available from https://doi.org/10.1063/1.119912.

Martins, H. P., Mossanek, R. J. O., Martí, X., Sánchez, F., Fontcuberta, J., & Abbate, M. (2017). Mn 3d bands and Y-O hybridization of hexagonal and orthorhombic $YMnO_3$ thin films. *Journal of Physics. Condensed Matter: an Institute of Physics Journal*, *29*. Available from https://doi.org/10.1088/1361-648X/aa75e3.

Matsumoto, T., Môri, N., Iida, J., Tanaka, M., & Siratori, K. (1992). Magnetic properties of the two dimensional antiferromagnets RFe2O4 (R = Y, Er) at high pressure. *J. Phys. Soc. Japan.*, *61*, 2916–2920. Available from https://doi.org/10.1143/JPSJ.61.2916.

Mendoza-Huizar, L. H., & Rios-Reyes, C. H. (2011). Chemical reactivity of Atrazine employing the Fukui function. *J. Mex. Chem. Soc.*, *55*, 142–147.

Meng, X., Li, H., Chen, J., Mei, L., Wang, K., & Li, X. (2009). Mössbauer study of cobalt ferrite nanocrystals substituted with rare-earth Y3 + ions. *Journal of Magnetism and Magnetic Materials, 321*, 1155–1158. Available from https://doi.org/10.1016/j.jmmm.2008.10.041.

Mohaideen, K. K., & Joy, P. A. (2012). Enhancement in the magnetostriction of sintered cobalt ferrite by making self-composites from nanocrystalline and bulk powders. *ACS Appl. Mater. Interfaces., 4*, 6421–6425. Available from https://doi.org/10.1021/am302053q.

Moritomo, Y., Akimoto, T., Takeo, M., & Machida, A. (2000). Metal-insulator transition induced by a spin-state transition. *Phys. Rev. B - Condens. Matter Mater. Phys., 61*, R13325–R13328. Available from https://doi.org/10.1103/PhysRevB.61.R13325.

Mukherjee, D., Hordagoda, M., Hyde, R., Bingham, N., Srikanth, H., Witanachchi, S., & Mukherjee, P. (2013). Nanocolumnar interfaces and enhanced magnetic coercivity in preferentially oriented cobalt ferrite thin films grown using oblique-angle pulsed laser deposition. *ACS Appl. Mater. Interfaces., 5*, 7450–7457. Available from https://doi.org/10.1021/am401771z.

Nagano, A., & Ishihara, S. (2007). Spin-charge-orbital structures and frustration in multiferroic RFe 2O4. *Journal of Physics. Condensed Matter: an Institute of Physics Journal, 19*. Available from https://doi.org/10.1088/0953-8984/19/14/145263.

Nagano, A., Naka, M., Nasu, J., & Ishihara, S. (2007). Electric polarization, magnetoelectric effect, and orbital state of a layered iron oxide with frustrated geometry. *Physical Review Letters, 99*. Available from https://doi.org/10.1103/PhysRevLett.99.217202.

Nakajima, T., Ichihara, M., & Ueda, Y. (2005). New A-site ordered perovskite cobaltite $LaBaCo_2O_6$: Synthesis, structure, physical property and cation order-disorder effect. *J. Phys. Soc. Japan., 74*, 1572–1577. Available from https://doi.org/10.1143/JPSJ.74.1572.

Nakamoto, R., Xu, B., Xu, C., Xu, H., & Bellaiche, L. (2017). Properties of rare-earth iron garnets from first principles. *Physical Review Letters, B95*, 024434. Available from https://doi.org/10.1103/PhysRevB.95.024434.

Nikumbh, A. K., Pawar, R. A., Nighot, D. V., Gugale, G. S., Sangale, M. D., Khanvilkar, M. B., & Nagawade, A. V. (2014). Structural, electrical, magnetic and dielectric properties of rare-earth substituted cobalt ferrites nanoparticles synthesized by the co-precipitation method. *Journal of Magnetism and Magnetic Materials, 355*, 201–209. Available from https://doi.org/10.1016/j.jmmm.2013.11.052.

Panda, R. N., Shih, J. C., & Chin, T. S. (2003). Magnetic properties of nano-crystalline Gd- or Pr-substituted CoFe2O4 synthesized by the citrate precursor technique. *Journal of Magnetism and Magnetic Materials, 257*, 79–86. Available from https://doi.org/10.1016/S0304-8853(02)01036-3.

Pelloquin, D., Hebert, S., Maignan, A., & Raveau, B. (2004). A New Thermoelectric Misfit Cobaltite: [Sr2CoO3][CoO2]1.8. *ChemInform, 35*. Available from https://doi.org/10.1002/chin.200443014.

Peng, J., Hojamberdiev, M., Xu, Y., Cao, B., Wang, J., & Wu, H. (2011). Hydrothermal synthesis and magnetic properties of gadolinium-doped CoFe2O4 nanoparticles. *Journal of Magnetism and Magnetic Materials, 323*, 133–137. Available from https://doi.org/10.1016/j.jmmm.2010.08.048.

Pervaiz, E., & Gul, I. H. (2013). Influence of rare earth (Gd3 +) on structural, gigahertz dielectric and magnetic studies of cobalt ferrite. *J. Phys. Conf. Ser., 439*. Available from https://doi.org/10.1088/1742-6596/439/1/012015.

Petrovic, V., Marcincak, S., Popelka, P., Simkova, J., Martonova, M., Buleca, J., ... Kovac, G. (2012). The effect of supplementation of clove and agrimony or clove and lemon balm on growth performance, antioxidant status and selected indices of lipid profile of broiler chickens. *Journal of Animal Physiology and Animal Nutrition*, *96*, 970–977. Available from https://doi.org/10.1111/j.1439-0396.2011.01207.x.

Pillai, V., & Shah, D. O. (1996). Synthesis of high-coercivity cobalt ferrite particles using water-in-oil microemulsions. *Journal of Magnetism and Magnetic Materials*, *163*, 243–248. Available from https://doi.org/10.1016/S0304-8853(96)00280-6.

Pralong, B. R. V., Caignaert, V., Hebert, S., & Maignan, A. (2006). The ordered double perovskite PrBaCo2O6: Synthesis. *SolidState Ionics.*, *177*.

Prellier, W., Singh, M. P., & Murugavel, P. (2005). The single-phase multiferroic oxides: From bulk to thin film. *Journal of Physics. Condensed Matter: an Institute of Physics Journal*, *17*. Available from https://doi.org/10.1088/0953-8984/17/30/R01.

Rai, B. K., Mishra, S. R., Nguyen, V. V., & Liu, J. P. (2013). Synthesis and characterization of high coercivity rare-earth ion doped Sr0.9RE0.1Fe10Al2O19 (RE: Y, La, Ce, Pr, Nd, Sm, and Gd). *J. Alloys Compd.*, *550*, 198–203. Available from https://doi.org/10.1016/j.jallcom.2012.09.021.

Rana, A., Thakur, O. P., & Kumar, V. (2011). Effect of Gd3 + substitution on dielectric properties of nano cobalt ferrite. *Mater. Lett.*, *65*, 3191–3192. Available from https://doi.org/10.1016/j.matlet.2011.060.076.

Rao, C. N. R., Kundu, A. K., Seikh, M. M., & Sudheendra, L. (2004). Electronic phase separation in transition metal oxide systems. *Dalt. Trans.*, 3003–3011. Available from https://doi.org/10.1039/b406785a.

Rao, K. J., Mahesh, K., & Kumar, S. (2005). A strategic approach for preparation of oxide nanomaterials. *Bull. Mater. Sci.*, *28*, 19–24. Available from https://doi.org/10.1007/BF02711166.

Rao, R. K. H., S, K., AM, K., MC, V., & RK, C. (2009). Cation distribution of Ti doped cobalt ferrites. *J. Alloy. Compd.*, *488*.

Rautama, E., Boullay, P., Kundu, A. K., Caignaert, V., Pralong, V., Karppinen, M., & Raveau, B. (2008). Cationic ordering and microstructural effects in the ferromagnetic properties. *Chemistry of Materials: a Publication of the American Chemical Society*, 2742–2750.

Raveau, B., & Seikh, M. M. (2012). *Cobalt Oxides: From Crystal Chemistry to Physics*. Available from https://doi.org/10.1002/9783527645527.

Ravinder, D., Vijaya Kumar, K., & Balaya, P. (2001). High-frequency dielectric behaviour of gadolinium substituted Ni-Zn ferrites. *Mater. Lett.*, *48*, 210–214. Available from https://doi.org/10.1016/S0167-577X(00)00305-0.

Respaud, M., Frontera, C., García-Muñoz, J. L., Aranda, M. Á. G., Raquet, B., Broto, J. M., ... Rodríguez-Carvajal, J. (2001). Magnetic and magnetotransport properties of GdBaCo2O5 + δ: A high magnetic-field study. *Phys. Rev. B - Condens. Matter Mater. Phys.*, *64*, 2144011–2144017.

Rittich, B., Španová, A., Horák, D., Beneš, M. J., Klesnilová, L., Petrová, K., & Rybnikář, A. (2006). Isolation of microbial DNA by newly designed magnetic particles. *Colloids Surfaces B Biointerfaces*, *52*, 143–148. Available from https://doi.org/10.1016/j.colsurfb.2006.04.012.

Rowe, D.M. (ed.). (2006). Thermoelectrics handbook: Macro to nano (1st ed.). (p. 1022). CRC Press, ISBN 9780849322648. Published December 9, 2005, 14 Color & 622 B/W Illustrations.

RöBler, S., Jesudasan, J., Bajaj, K., Raychaudhuri, P., Steglich, F., & Wirth, S. (2009). Influence of microstructure on local conductivities in La 0.7Ce0.3MnO3 thin film. *J. Phys. Conf. Ser.*, *150*. Available from https://doi.org/10.1088/1742-6596/150/4/042164.

Sakakima, H., Satomi, M., Hirota, E., & Adachi, H. (1999). Spin-valves using perovskite antiferromagnets as the pinning layers. *IEEE Transactions on Magnetics*, *35*, 2958–2960. Available from https://doi.org/10.1109/20.801046.

Sankaranarayanan, V. K., & Sreekumar, C. (2003). Precursor synthesis and microwave processing of nickel ferrite nanoparticles. *Curr. Appl. Phys.*, *3*, 205–208. Available from https://doi.org/10.1016/S1567-1739(02)00202-X.

Schmool, D. S., Keller, N., Guyot, M., Krishnan, R., & Tessier, M. (1999). Magnetic and magneto-optic properties of orthoferrite thin films grown by pulsed-laser deposition. *Journal of Applied Physics*, *86*, 5712–5717. Available from https://doi.org/10.1063/1.371583.

Scott, J. F., Dawber, M., & Rabe, K. M. (2005). Physics of thin-film ferroelectric oxides. *Rev. Mod. Phys.*, *77*, 1048–1083.

Seikh, M. M., Simon, C., Caignaert, V., Pralong, V., Lepetit, M. B., Boudin, S., & Raveau, B. (2008). New magnetic transitions in the ordered oxygen-deficient perovskite LnBaCo2O5.50 + δ. *Chemistry of Materials: A Publication of the American Chemical Society*, *20*, 231–238. Available from https://doi.org/10.1021/cm7026652.

Setter, N., Damjanovic, D., Eng, L., Fox, G., Gevorgian, S., Hong, S., ... Streiffer, S. (2006). Ferroelectric thin films: Review of materials, properties, and applications. *Journal of Applied Physics*, *100*. Available from https://doi.org/10.1063/1.2336999.

Sharan, A., Lettieri, J., Jia, Y., Tian, W., Pan, X., Schlom, D. G., & Gopalan, V. (2004). Bismuth manganite: A multiferroic with a large nonlinear optical response. *Phys. Rev. B - Condens. Matter Mater. Phys.*, *69*. Available from https://doi.org/10.1103/PhysRevB.69.214109.

Shen, S. T., & Weng, H. S. (1998). Comparative study of catalytic reduction of nitric oxide with carbon monoxide over the La1-xSrxBO3 (B = Mn, Fe, Co, Ni) catalysts. *Industrial & Engineering Chemistry Research*, *37*, 2654–2661. Available from https://doi.org/10.1021/ie970691g.

Shi, J., Ha, S. D., Zhou, Y., Schoofs, F., & Ramanathan, S. (2013). A correlated nickelate synaptic transistor. *Nat. Commun.*, *4*. Available from https://doi.org/10.1038/ncomms3676.

Shi, J., Zhou, Y., & Ramanathan, S. (2014). Colossal resistance switching and band gap modulation in a perovskite nickelate by electron doping. *Nat. Commun.*, *5*. Available from https://doi.org/10.1038/ncomms5860.

Sileo, E. E., & Jacobo, S. E. (2004). Gadolinium-nickel ferrites prepared from metal citrates precursors. *Phys. B Condens. Matter.*, *354*, 241–245. Available from https://doi.org/10.1016/j.physb.2004.10.002.

Silva, J., Reyes, A., Esparza, H., Camacho, H., & Fuentes, L. (2011). BiFeO3: A review on synthesis, doping and crystal structure. *Integr. Ferroelectr.*, *126*, 47–59. Available from https://doi.org/10.1080/10584587.2011.574986.

Singh, R. N., Bahadur, L., Pandey, J. P., Singh, S. P., Chartier, P., & Poillerat, G. (1994). Preparation and characterization of thin films of LaNiO3 for anode application in alkaline water electrolysis. *J. Appl. Electrochem.*, *24*, 149–156. Available from https://doi.org/10.1007/BF00247787.

Singhal, K. K. S. C. (2002). High-temperature solid oxide fuel cells: Fundamentals, design and applications. *Mater. Today*, *5*, 55. Available from https://doi.org/10.1016/s1369-7021(02)01241-5.

Sinning, S., Musgaard, M., Jensen, M., Severinsen, K., Celik, L., Koldsø, H., ... Wiborg, O. (2010). Binding and orientation of tricyclic antidepressants within the central substrate site of the human serotonin transporter. *The Journal of Biological Chemistry*, *285*, 8363–8374. Available from https://doi.org/10.1074/jbc.M109.045401.

Siratori, K., Akimitsu, J., & Kohn, K. (1994). Possibility of magnetoelectric effect in antiferromagnetic rfe2o4. *Ferroelectrics*, *161*, 111–115. Available from https://doi.org/10.1080/00150199408213359.

Spinicci, R., Faticanti, M., Marini, P., De Rossi, S., & Porta, P. (2003). Catalytic activity of LaMnO3 and LaCoO3 perovskites towards VOCs combustion. *J. Mol. Catal. A Chem.*, *197*, 147–155. Available from https://doi.org/10.1016/S1381-1169(02)00621-0.

Suard, E., & Fauth, F. (2000). Charge ordering in the layered Co-based perovskite. *Phys. Rev. B - Condens. Matter Mater. Phys.*, *61*, R11871–R11874. Available from https://doi.org/10.1103/PhysRevB.61.R11871.

Subramanian, M. A., He, T., Chen, J., Rogado, N. S., Calvarese, T. G., & Sleight, A. W. (2006). Giant room-temperature magnetodielectric response in the electronic ferroelectric LuFe2O4. *Adv. Mater.*, *18*, 1737–1739. Available from https://doi.org/10.1002/adma.200600071.

Sugimoto, M. (1999). The past, present, and future of ferrites. *J. Am. Ceram. Soc.*, *2*.

Sun, C., Hui, R., & Roller, J. (2010). Cathode materials for solid oxide fuel cells: A review. *J. Solid State Electrochem*, *14*, 1125–1144. Available from https://doi.org/10.1007/s10008-009-0932-0.

Sun, Z., Wu, P. H., Feng, Y., Ji, Z. M., Yang, S. Z., Wang, M., ... Kang, L. (2005). *Thin Solid Films*, *47*(1-2), 248.

Takada, K., Sakurai, H., Takayama-Muromachi, E., Izumi, F., Dilanian, R. A., & Sasaki, T. (2003). Superconductivity in two-dimensional CoO2 layers. *ChemInform*, *34*. Available from https://doi.org/10.1002/chin.200324014.

Tanaka, M., Akimitsu, J., Inada, Y., Kimizuka, N., Shindo, I., & Siratori, K. (1982). Conductivity and specific heat anomalies at the low temperature transition in the stoichiometric YFe2O4. *Solid State Commun*, *44*, 687–690. Available from https://doi.org/10.1016/0038-1098(82)90583-X.

Teraoka, Y., Nobunaga, T., Okamoto, K., Miura, N., & Yamazoe, N. (1991). Influence of constituent metal cations in substituted LaCoO3 on mixed conductivity and oxygen permeability. *Solid State Ionics*, *48*, 207–212. Available from https://doi.org/10.1016/0167-2738(91)90034-9.

Terasaki, I., Sasago, Y., & Uchinokura, K. (1997). Large thermoelectric power in single crystals. *Phys. Rev. B - Condens. Matter Mater. Phys.*, *56*, R12685–R12687. Available from https://doi.org/10.1103/PhysRevB.56.R12685.

Traversa, Y., Matsushima, E., Okada, S., Sadaoka, G., Sakai, Y., & Watanabe, W. (1995). NO2 sensitive LaFeO3 thin films prepared by rf sputtering. *Sensors Actuators B*, *25*.

Troyanchuk, I. O., Kasper, N. V., & Khalyavin, D. D. (1998). *Phase Transitions in the Gd 0. 5 Ba 0. 5 CoO 3 Perovskite*, 13–16.

Uemura, S. I. T., Mitsudo, T., & Haruta, M., (1998). Frontiers and tasks of catalysis towards the next century, Proc. Int. Symp. Honour Profr. Tomoyuki Inui. Utr.

Uusi-Esko, K., Malm, J., Imamura, N., Yamauchi, H., & Karppinen, M. (2008). Characterization of RMnO3 (R = Sc, Y, Dy-Lu): High-pressure synthesized metastable perovskites and their hexagonal precursor phases. *Mater. Chem. Phys.*, *112*, 1029–1034. Available from https://doi.org/10.1016/j.matchemphys.2008.07.009.

Valenzuela, R. (2012). Novel applications of ferrites. *Phys. Res. Int.* Available from https://doi.org/10.1155/2012/591839.

Vijaya Kumar, K., & Ravinder, D. (2002). Electrical conductivity of Ni-Zn-Gd ferrites. *Mater. Lett.*, *52*, 166–168. Available from https://doi.org/10.1016/S0167-577X(01)00385-8.

Vogt, T., Woodward, P. M., Karen, P., Hunter, B. A., Henning, P., & Moodenbaugh, A. R. (2000). Low to high spin-state transition induced by charge ordering in antiferromagnetic YBaCo 2O 5. *Physical Review Letters*, *84*. Available from https://doi.org/10.1103/PhysRevLett.84.2969.

Wang, B., Gu, S., Ding, Y., Chu, Y., Zhang, Z., Ba, X., ... Li, X. (2013). A novel route to prepare LaNiO3 perovskite-type oxide nanofibers by electrospinning for glucose and hydrogen peroxide sensing. *Analyst*, *138*, 362–367. Available from https://doi.org/10.1039/c2an35989h.

Wu, J., Lynn, J. W., Glinka, C. J., Burley, J., Zheng, H., Mitchell, J. F., & Leighton, C. (2005). Intergranular giant magnetoresistance in a spontaneously phase separated perovskite oxide. *Physical Review Letters*, *94*. Available from https://doi.org/10.1103/PhysRevLett.94.037201.

Wu, X., Yu, H., Dong, H., & Geng, L. (2014). Enhanced infrared radiation properties of CoFe2O4 by single Ce3þ-doping with energy-efficient preparation. *Ceram. Int.*, *40*, 5905–5911. Available from https://doi.org/10.1016/j.ceramint.2013.11.035.

Xiang, H. J., & Whangbo, M. H. (2007). Charge order and the origin of giant magnetocapacitance in LuFe2O4. *Physical Review Letters*, *98*. Available from https://doi.org/10.1103/PhysRevLett.98.246403.

Yamada, N. I. Y., & Nohdo, S. (1997). Incommensurate charge ordering in charge-frustrated LuFe2O4 system.pdf. *J. Phys. Soc. Jpn.*, *66*. Available from https://doi.org/10.1143/JPSJ.66.3733.

Yamada, Y., Kitsuda, K., Nohdo, S., & Ikeda, N. (2000). Charge and spin ordering process in the mixed-valence system LuFe2O4: Charge ordering. *Phys. Rev. B - Condens. Matter Mater. Phys.*, *62*, 12167–12174. Available from https://doi.org/10.1103/PhysRevB.62.12167.

Yonghua, C., Futai, M., & Hui, L. (1988). Catalytic properties of rare earth manganites and related compounds. *React. Kinet. Catal. Lett.*, *37*, 37–42. Available from https://doi.org/10.1007/BF02061707.

Yoshii, K., Ikeda, N., & Nakamura, A. (2006). Magnetic and dielectric properties of frustrated ferrimagnet TmFe2O4. *Phys. B Condens. Matter.*, *378–380*, 585–586. Available from https://doi.org/10.1016/j.physb.2006.01.155.

Yoshii, K., Ikeda, N., & Mori, S. (2007). Magnetic and dielectric behavior of and. *Journal of Magnetism and Magnetic Materials*, *310*, 1154–1156. Available from https://doi.org/10.1016/j.jmmm.2006.10.327.

Yoshii, K., Ikeda, N., Matsuo, Y., Horibe, Y., & Mori, S. (2007). Magnetic and dielectric properties of R Fe2 O4, RFeMO4, and R GaCu O4 (R = Yb and Lu, M = Co and Cu). *Phys. Rev. B - Condens. Matter Mater. Phys.*, *76*. Available from https://doi.org/10.1103/PhysRevB.76.024423.

Yuan, Z., Liu, J., Chen, C. L., Wang, C. H., Luo, X. G., Chen, X. H., ... Donner, W. (2007). Epitaxial behavior and transport properties of PrBa Co2O5 thin films on (001) SrTi O3. *Applied Physics Letters*, *90*. Available from https://doi.org/10.1063/1.2741407.

Zhang, J., Zheng, H., Ren, Y., & Mitchell, J. F. (2017). High-pressure floating-zone growth of perovskite nickelate LaNiO3 single crystals. *Cryst. Growth Des.*, *17*, 2730–2735. Available from https://doi.org/10.1021/acs.cgd.7b00205.

Zhang, J. Z. (2011). Metal oxide nanomaterials for solar hydrogen generation from photoelectrochemical water splitting. *MRS Bulletin/Materials Research Society*, *36*, 48–55. Available from https://doi.org/10.1557/mrs.2010.9.

Zhang, Y., & Wen, D. (2010). Effect of RE/Ni (RE = Sm, Gd, Eu) addition on the infrared emission properties of Co-Zn ferrites with high emissivity. *Mater. Sci. Eng. B Solid-State Mater. Adv. Technol.*, *172*, 331–335. Available from https://doi.org/10.1016/j.mseb.2010.06.011.

Zhang, Y., & Wen, D. (2012). Infrared emission properties of RE (RE = La, Ce, Pr, Nd, Sm, Eu, Gd, Tb, and Dy) and Mn co-doped Co 0.6Zn 0.4Fe 2O 4 ferrites. *Mater. Chem. Phys.*, *131*, 575–580. Available from https://doi.org/10.1016/j.matchemphys.2011.09.003.

Zhang, Y., Wang, Y., Zhou, W., Fan, Y., Zhao, J., Zhu, L., … Liu, H. (2019). A combined drug discovery strategy based on machine learning and molecular docking. *Chemical Biology & Drug Design*, *93*, 685–699. Available from https://doi.org/10.1111/cbdd.13494.

Zhao, L., Yang, H., Zhao, X., Yu, L., Cui, Y., & Feng, S. (2006). Magnetic properties of CoFe2O4 ferrite doped with rare earth ion. *Mater. Lett.*, *60*, 1–6. Available from https://doi.org/10.1016/j.matlet.2005.07.017.

Zhou, S.-Y. W. B., Zhang, Y.-W., Liao, C.-S., Yan, C.-H., & Chen, L.-Y. (2004). Rare-earth-mediated magnetism and magneto-optical Kerr effects in nanocrystalline CoFeMn0.9RE0.1O4 thin films. *J.Magn.Magn.Mater.*, *280*.

Rare earth–doped SnO_2 nanostructures and rare earth stannate ($Re_2Sn_2O_7$) ceramic nanomaterials

Hossein Safardoust-Hojaghan
Young Researchers and Elite Club, Marand Branch, Islamic Azad University, Marand, Iran

9.1 General introduction

At the nanoscale, nanostructures play a critical role in the development of science and engineering technologies (Ben Saber, Mezni, Alrooqi, & Altalhi, 2020; Jadhav, Shinde, Mane, & O'Dwyer, 2019; Nasrollahzadeh, Issaabadi, Sajjadi, Sajadi, & Atarod, 2019; Singh et al., 2020). Nanostructures have recently attracted a lot of attention since their excellent features affect physical, thermal, stability, electrical, chemical, biological, and optoelectrical properties (Nunes et al., 2019; Pascariu et al., 2019). The nanostructures can include a variety of structures, such as bimetallic, carbon-based nanostructures, nonmetallic, semiconductors, nanocomposites, and quantum dots (Maiti, Kim, & Lee, 2021; Oliveira et al., 2021; Pan, Yang, Bian, Hu, & Zhang, 2019; Rafique, Tahir, Rafique, Safdar, & Tahir, 2020). In recent years, rare earth (Sc, Y, and La–Lu)-based nanomaterials are found great attention. The rare earth, Re, is currently a set of 17 elements with positive nuclei charges in the units + e of the protonic charge, which are either 21 (scandium), 39 (yttrium), or one of the integers Z in the range 57–71. In the nanoscale area, rare earth compounds continue to demonstrate new properties and applications (Binetti, Longo, & Carotenuto, 2015; Bouzigues, Gacoin, & Alexandrou, 2011; Yu, Eich, & Cruz, 2020). Rare earth elements have many unique properties that have piqued the curiosity of researchers due to their possible uses in advanced materials. In the field of luminescence and ceramic-based fields, several significant advances have already been made. The rare earth–based nanomaterials, unlike quantum dots, exhibit size-insensitive luminescence characteristics. Their quasi-line emissions are primarily based on intra-4f electron transitions and have significantly longer decay lifetimes (Jain et al., 2018; Yan et al., 2011; Zhao, He, & Tan, 2016). Rare earth stannate is the important pyrochlore-type oxide compound with general formula $Re_2Sn_2O_7$ (Re = lanthanum, niobium, samarium, gadolinium, erbium, ytterbium) and have been widely applied as catalysts, high-temperature pigments, magnets, optical emission materials, electronic materials, conductors, and hosts for radioactive wastes (Feng, Xiao, Qu, Zhou, & Pan, 2011; Park, Hwang, & Moon, 2003; Prabhakaran,

Advanced Rare Earth-Based Ceramic Nanomaterials. DOI: https://doi.org/10.1016/B978-0-323-89957-4.00007-4

Wang, & Boothroyd, 2017). Rare earth stannate can be used as a high-temperature catalyst due to their high fusing point ($> 2000°C$), and the application of uniqueness is arranged aspect control in the processing of vehicle exhaust. When $Re_2Sn_2O_7$ materials are prepared as nanostructures, the smaller particle size and increased specific surface area can result in different phase transition temperatures, increased catalytic activity, and better process ability (Escudero et al., 2017; Yan et al., 2011).

It has been found that the r_{RE}/r_{Sn} radius ratio is responsible for the stable crystalline phase structure of the $Re_2Sn_2O_7$ compound. Feng et al. found that for all $Re_2Sn_2O_7$ structures, lattice constants that is theoretically calculated are lower than those found in experiments. It is reported that the covalent bonds in other rare earth stannates are stronger than lanthanum stannate, and by reducing the atomic weight of Re, the covalent bonds are somewhat weakend (Kennedy, Hunter, & Howard, 1997).

9.1.1 Rare earth–doped SnO_2 nanostructures

The electronic and optical properties of pure and doped SnO_2 have been extensively studied as one of the most important members of the n-type direct gap semiconducting metal oxide (Divya, Pramothkumar, Joshua Gnanamuthu, Bernice Victoria, & Jobe Prabakar, 2020; Feng et al., 2020; Gao, Hou, & Liu, 2020; Matussin, Harunsani, Tan, Cho, & Khan, 2020). The bandgap (E_g) of bulk crystalline SnO_2 with a tetragonal rutile structure is 3.6 eV (344 nm). The bandgap engineering has been utilized to improve the band structure and subsequently optoelectronic properties of SnO_2 nanomaterials (Ganose & Scanlon, 2016; Karmaoui et al., 2018; Mounkachi et al., 2016). In this regard, the preparation route plays a key role since the properties of SnO_2-based nanomaterials depend heavily on the preparation methods. Till now, a wide range of methods have been applied for the synthesis of SnO_2-based ceramic nanomaterials including coprecipitation, electrospinning, microwave-assisted, photochemical growth at the air–water interface, hydrothermal, solvothermal, sol–gel, and ultrasonic-assisted methods (Alam, Asiri, & Rahman, 2021; Asaithambi et al., 2021; Chen, Xu, & Zhao, 2021; Soussi et al., 2020). SnO_2-based nanomaterials have been applied in ceramic glazes, lithium-ion batteries, supercapacitor, gas sensor, nanophotonics, and catalyst so far (Deng et al., 2021; Dung, Giang, Binh, Hieu, & Van, 2021; Luque et al., 2021). It is well known that dopant as an impurity can improve the physical and chemical properties of nanomaterials. In this regard, doping rare earth ions into the SnO_2-based nanostructures modifies the characteristics of ceramic nanomaterials remarkably (Hu, Li, Xu, & Yang, 2021; Sivakumar et al., 2021).

It is anticipated that choosing a luminescence method that is more high-responsive to the local structure would reveal more about the dopant position and distribution in SnO_2. The trivalent metal Eu, and Ln with their distinctive orange-red emission, is expected to be extremely responsive to their surrounding conditions. It is helpful to understand the reason for most studies on Eu as luminescent-stimulated in rare earth stannates and rare earth–doped SnO_2. The information gleaned from Eu luminescence can be used to determine whether dopants are set apart on the surface of SnO_2 nanostructures or go in for the lattice of oxide through substitutional or interstitial routes. For the different ionic radii of tin and europium ions (0.076 nm as opposed to 0.095 nm), the replacement of tin ions by bulkier europium ions causes both electric and strain impact. In the following, some reported studies in this field are discussed (Cojocaru et al., 2017; Kaur, Bhatti, & Singh, 2019).

Aragon et al. prepared Er-doped SnO_2 (SnO_2:Er) ceramic nanomaterials via polymer precursor method successfully. In this work, the content of Er tuned from 1% to 10%. The crystalline properties of samples were studied via X-ray diffraction pattern (XRD) analysis. The crystalline size calculated for prepared samples decreased from 9.5 to 3.7 nm by increasing the content of Er. This trend was also observed for mean particle size. The Magnetization (M)–magnetic field (H) curves confirm the ferromagnetic and paramagnetic properties for $SnO_2:Er^{3+}$. The improvement of the ferromagnetic part intensively relies on the erbium amount and diminishes in prepared products via erbium amount higher than 5.0%. Since the concentration of magnetic cations is less than $xp \approx 0.25$, the formation of long-range magnetic order-dependent interactions of double-exchange or super-exchange cannot be the cause of the ferromagnetic feature created in Er-doped tin oxide nanostructures. Furthermore, the interactions of super-exchange are primarily short-ranged and antiferromagnetic. Erbium ions appear to play a crucial role in this long-range ferromagnetic order, as they trigger further oxygen vacancies in the tin oxide phase and make localized spins that are paired through the spin-split impurity band ferromagnetically (Aragón, Chitta, Coaquira, Hidalgo, & Brito, 2013).

Suryavanshi et al. prepared pure SnO_2 (S0), Sm_2O_3-doped SnO_2, 2 mol.% (S1), 4 mol.% (S2), 6 mol.% (S3), and 8 mol.%. Transmission electron microscopy (TEM) images of S0, S1, S2, S3, and S4 are shown in Fig. 9.1A–E. High-resolution transmission electron microscopy (HRTEM) representations of S0 and S3 are shown in Fig. 9.1F and G. Samples S0, S1, S2, S3, and S4 display nanoparticles with irregular shapes with decreasing in size in TEM images. The TEM image of S3 shows that the 6 mol.% Sm_2O_3-doped SnO_2 sample has particular more modest particles (5–8 nm). Investigation of optical properties showed that the bandgap of SnO_2 depends strongly on the content of Sm_2O_3. The bandgap of S0 is 3.61 eV, while the bandgap of 2.0%, 4.0%, 6.0%, and 8.0% (molar) Sm_2O_3 dopant are 3.74, 3.86, 4.07, and 4.24 eV, respectively. This demonstrates that the optical energy bandgap increases as Sm_2O_3 concentration rises. This blue shift can be attributed to the quantum confinement effect on the nanoparticles (Shaikh, Chikhale, Patil, Mulla, & Suryavanshi, 2017).

Yu et al. prepared SnO_2 with nanosheet morphology (S1), Eu-doped SnO_2 with nanosheet morphology (S2), rod-like morphology of SnO_2 (S3), Eu-doped SnO_2 nanorods (S4), SnO_2 with sphere shape (S5), and Eu-doped SnO_2 nanospheres (S6) via hydrothermal method. They investigated the morphological and optical properties of samples comprehensively. Scanning electron microscope (SEM) analysis was applied for morphological studying of SnO_2 and Eu:SnO_2 (Fig. 9.2). On the surface, S1 and S2 have very close morphologies, as shown in Fig. 9.2A and B. S1

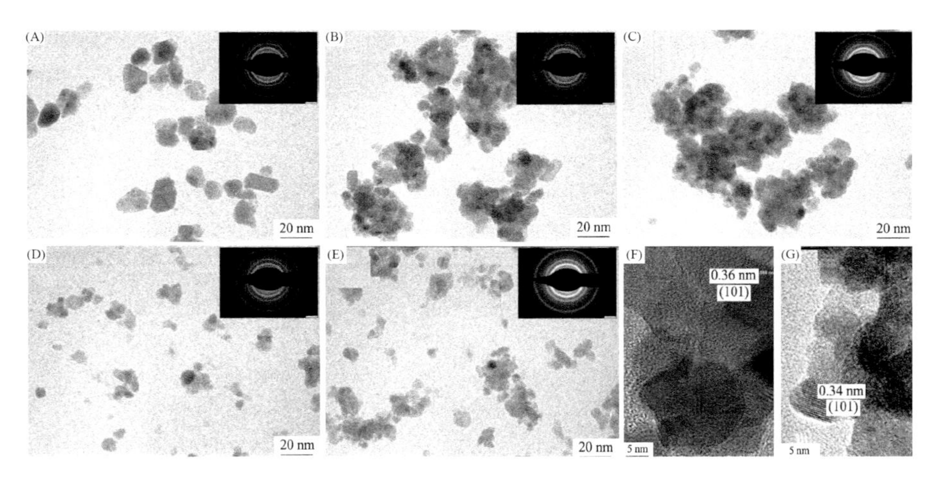

Figure 9.1 Transmission electron microscopy micrographs and SAED patterns (inset) of samples (A) S0, (B) S1, (C) S2, (D) S3, (E) S4 and high-resolution transmission electron microscopy micrographs of samples S1 (F) and S3 (G). *SAED*, selected area electron diffraction.
Source: Reproduced with permission from Shaikh, F. I., Chikhale, L. P., Patil, J. Y., Mulla, I. S., & Suryavanshi, S. S. (2017). Enhanced acetone sensing performance of nanostructured Sm_2O_3 doped SnO_2 thick films. *Journal of Rare Earths*, *35*, 813–823.

panels, on the other hand, are laid out haphazardly and exhibit different characteristics than S2. In Fig. 9.2B, nanostructures stacked together and agglomerations can be seen. The morphology of S3 and S4 is shown in Fig. 9.2C and D, respectively. In both S3 and S4, the rod-like structure is visible. The rods are approximately 1 μm in length and have a diameter of around 0.1 m. Rods in S3 distribute nonhomogeneously around the cores, while rods in S4 organize in regular form around the cores to form flower-like shapes. Fig. 9.2E and F depict SnO_2 nanospheres. Since S5 and S6 are made up of spheres of various sizes (0.1–1.5 μm), in comparison with S5, the size of particles in S6 is larger, and a more agglomerated structure is observed. It can be concluded that the morphology of the SnO_2 and europium-doped samples differ significantly. In addition, the ultraviolet–visible (UV–Vis) analysis and bandgaps of SnO_2 and europium-doped samples were studied comprehensively. It is found that SnO_2 and Eu-doped samples have a wide absorption range of 250–400 nm. Eu-doped SnO_2-related absorption edges nominated as S2, S4, and S6 are shifted to a higher wavelength. The bandgaps for tin oxide and europium-doped tin oxide nanomaterials obtained were 3.05 (S1), 3.04 eV (S2), 3.88 (S3), 3.55 eV (S4), 3.70 (S5), and 3.50 eV (S6), respectively. It can be concluded that the bandgap changes with the doping of Eu ions. Interestingly, the rate of bandgap change varies due to the presence of Eu ions in different morphologies (Li et al., 2017).

After 10 minutes, the europium-doped tin oxide was prepared hollow microspheres morphology through a microwave-solvothermal path via Lu et al. The

Figure 9.2 Morphological evolution of pure and Eu-doped nano-SnO_2: (A) pure SnO_2 nanosheets; (B) Eu-doped SnO_2 nanosheets; (C) pure SnO_2 nanorods; (D) Eu-doped SnO_2 nanorods; (E) pure SnO_2 nanospheres; (F) Eu-doped SnO_2 nanospheres.
Source: Reproduced with permission from Li, S., Yu, L., Man, X., Zhong, J., Liao, X., & Sun, W. (2017). The synthesis and band gap changes induced by the doping with rare-earth ions in nano-SnO_2. *Materials Science in Semiconductor Processing*, *71*, 128–132.

synthesis process was improved by controlling microwave power, which speed up the nucleation process and increased crystal growth. Since the tin oxide and europium emissions, the photoluminescence (PL) of Eu^{3+}-doped SnO_2 displayed a wide band in 400–575 nm. The emission spectra of $Sn_{1-x}O_2{:}xEu^{3+}$ ($x = 0.25–1.0$) were monitored at excitation wavelength of 250 nm. The emission spectra showed typical orange-red emission at 587 and 613 nm, respectively, produced by the transitions $^5D_0 \rightarrow {}^7F_J$ ($J = 1, 2$) of europium. The broad emission of the host lattice in the 400–575 nm range accompanied the typical emission of europium. The electric dipole transition $^5D_0 \rightarrow {}^7F_2$ with $J = 2$ is hypersensitive, and depending on the local setting, the intensity will differ by the order of magnitude. The magnetic dipole (MD) transitions ($^5D_0 \rightarrow {}^7F_1$), on the other hand, are unaffected by site symmetry since they are parity-allowed. As a result, the ($^5D_0 \rightarrow {}^7F_1$)/($^5D_0 \rightarrow {}^7F_2$) emission ratio can be applied to determine Eu^{3+} site symmetry (Fig. 9.3). The energy transfer from the tin oxide to europium ions was revealed to be minor since after adding europium ions into tin oxide, the blue emission did not alter. The emission strength of $SnO_2{:}Eu^{3+}$ increased steadily as the europium amount rose to 0.75 mol.%, then decreased as $Eu_2Sn_2O_7$ phase formed (Das, Som, Yang, & Lu, 2018).

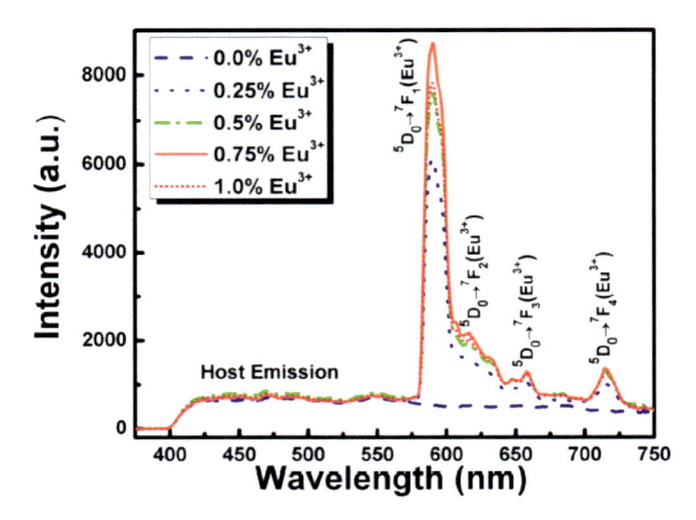

Figure 9.3 PL emission spectra of $Sn_{1-x}O_2$:$x$$Eu^{3+}$ ($x = 0-1.0$) prepared via the microwave-solvothermal route with the microwave power of 400 W and postannealed at 1000°C. *Source:* Reproduced with permission from Das, S., Som, S., Yang, C.-Y., & Lu, C.-H. (2018). Optical temperature sensing properties of SnO_2: Eu^{3+} microspheres prepared via the microwave-assisted solvothermal process. *Materials Research Bulletin*, *97*, 101–108.

9.2 Preparation methods of rare earth–doped SnO_2 nanostructures and rare earth stannate ($Re_2Sn_2O_7$) ceramic nanomaterials

Nanostructure synthesis refers to the process of preparing nanoparticles. Nanoparticles may be made from larger molecules or created from the bottom-up approach, for example, by nucleating and growing particles from fine molecular distributions in the liquid or vapor phase. Nanomaterial synthesis can be done in a variety of ways, including biological, chemical, and physical methods. Chemical and physical methods are mainly thought to be the best for producing uniformly sized nanoparticles with high stability. The synthesis of nanostructures has been faced with different challenges that limit the industrial preparation of nanostructures. The ability to synthesize nanomaterials of various shapes, monodispersity, chemical composition, and sizes is critical for their application in different fields (Ahmed, 2020; Chatterjee, Kwatra, & Abraham, 2020; Singh, Yadav, Pandey, Gupta, & Singh, 2019). Over the past decade, numerous endeavors have been centered on the control of preparation conditions of rare earth stannates and rare earth–doped SnO_2 nanomaterials to obtain desired structural and morphological properties (Yan et al., 2011).

Solid-state reactions were used to synthesis $Sm_xSn_{1-x}O_2$ (with $x = 0$, 0.005, 0.02, and 0.04) via Sundararaj et al. Using a mortar and pestle, stoichiometric amounts of high-purity Sm_2O_3 and SnO_2 were homogeneously stirred for 4 hours.

In a programmable muffle furnace, the fine powders were calcined for 10 hours at 850°C. The calcined powders were further crushed for 4 hours with a mortar and pestle to achieve fine and uniform grain size. The XRD analysis confirmed $Sm_xSn_{1-x}O_2$ formed in polycrystalline tetragonal rutile structure (JCPDS21-1250). The morphological properties were investigated via field emission scanning electron microscopy (FESEM) analysis. Results showed that the grain sizes differed linearly with the Sm dopant concentration. 0%, 0.005%, 0.02%, and 0.04% Sm content of $Sm_xSn_{1-x}O_2$ thin films had average grain sizes of 12, 24, 43, and 52 nm, respectively (Bakiya Lakshmi, Vimala Juliet, Sundararaj, Abhinav, & Chandrasekaran, 2018).

In another work, $Sm_2Sn_2O_7$ was prepared via the hydrothermal-microwave method ($Sm_2Sn_2O_7$–H). The samarium nitrate hexahydrate and tin(IV) chloride pentahydrate were dissolved in water at 1 M concentration. The 1 M NaOH was applied for fixing pH in 9. The prepared slurry was stirred for 24 hours. Then, the slurry was transferred to a reactor and applied hydrothermal condition in a microwave furnace for 4 hours at 200°C. The as-prepared samples were separated, then dried at 40°C for 8 hours. They suggested the following reactions for the preparation of $Sm_2Sn_2O_7$ ceramic:

$$
\begin{gathered}
Sm(NO_3)_3.6H_2O + H_2O(ex) \rightarrow Sm^{3+}_{aq} \\
SnCl_4.5H_2O + H_2O(ex) \rightarrow Sn^{4+}_{aq} \\
Sm^{3+}_{aq} + Sn^{4+}_{aq} + OH^{-}_{aq} \rightarrow SmSn(OH)_{7(solid)} \\
SmSn(OH)_7(solid) \rightarrow Sm_2Sn_2O_{7(solid)}
\end{gathered}
$$

$Sm_2Sn_2O_7$ was also prepared via a conventional ceramic method in which SnO_2 and Sm_2O_3 were mixed and ground in an agate mortar for 30 minutes. Then, the mixture was heated for 48 hours at 1000°C under an air atmosphere ($Sm_2Sn_2O_7$–1000). Then, the prepared solid was treated for 96 hours at 1400°C at the same condition ($Sm_2Sn_2O_7$–1400) (Trujillano, Martín, & Rives, 2016). The particle size distribution of both materials was investigated. Particle size data for the sample $Sm_2Sn_2O_7$–1400 are smaller, as expected. It is due to the sintering process that this sample obtains as a result of the high temperature used for its preparation. For preparing hydrothermal-assisted samarium stannate, the size distribution was not symmetrical, and the grain size range is larger than the sample prepared at 1400°C (Trujillano et al., 2016).

Ahmadnia-Feyzabad et al. prepared tin oxide and different amounts (2.0, 5.0, and 10.0 wt.%) of Sm_2O_3-doped tin oxide nanoparticles via combustion synthesis method according to Fig. 9.4. SEM images of tin oxide, 2.0, 5.0, and 10.0 wt.% samarium oxide–doped tin oxide samples were recorded for morphological investigation. The SEM analysis was applied for the determination of shape and size. They found that the size of secondary particles grows in proportion to the amount of Sm_2O_3 in the samples. Secondary particle sizes in pure SnO_2 were less than 100 nm, while in 2.0, 5.0, and 10.0 wt.% samarium oxide–doped tin oxide, they were about 150 nm, 250 nm, and higher than 300 nm, respectively (Ahmadnia-Feyzabad, Mortazavi, Khodadadi, & Hemmati, 2013).

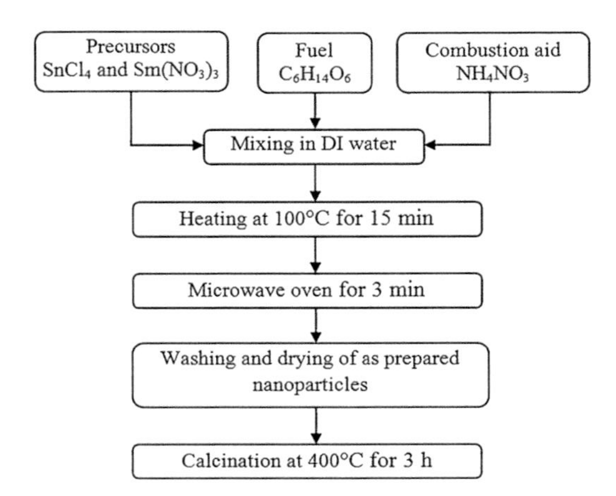

Figure 9.4 Key steps of Sm_2O_3-doped SnO_2 sample preparation by CSCS method. *CSCS*, chloride solution combustion synthesis.
Source: Reproduced with permission from Ahmadnia-Feyzabad, S., Mortazavi, Y., Khodadadi, A. A., & Hemmati, S. (2013). Sm_2O_3 doped-SnO_2 nanoparticles, very selective and sensitive to volatile organic compounds. *Sensors and Actuators B: Chemical*, *181*, 910–918.

Wang et al. prepared pure SnO_2, Eu-doped SnO_2 ceramic nanomaterials via electrospinning method: a certain amount of $SnCl_2 \cdot 2H_2O$ was dissolved in the dimethylformamide (DMF)/ethanol mixture as a solvent under stirring. Then, the sufficient content of polyvinylpyrrolidone (PVP) and $Eu(NO_3)_3 \cdot 6H_2O$ were added to the prepared solution under vigorous stirring. After that, the as-obtained viscous solution was poured into a high-voltage supply connected to a glass syringe and the electrospinning method was carried out in an ambient condition. At a distance of 20 cm, a 15 kV DC voltage was applied between the collector and syringe needle. The nanofibers were finally calcined for 5 hours at 600°C with a given heating program that was used to extract the organic portion of PVP and forming SnO_2 in crystalline form. The pure tin oxide nanomaterials were prepared via the same route. FESEM and TEM analyses were applied for in-depth morphological investigation of europium-doped tin oxide nanofibers with various molar ratios (0%, 1%, 2%, and 3%) of europium. The mean diameters of tin oxide, 1 mol.% Eu-doped tin oxide, 2 mol.% Eu-doped tin oxide, and 3 mol.% Eu-doped tin oxide were determined to be 113, 108, 103, and 100 nm, respectively (Fig. 9.5). The diameter of europium-doped tin oxide nanofibers steadily decreases with increasing europium content as compared to pure SnO_2 nanofibers. It can be related to the increasing electrostatic shielding with the total ionization of $Eu(NO_3)_3$ (Jiang et al., 2016).

Chang and Jo prepared Eu-doped SnO_2 as per the following procedures: SnO_2 and Eu_2O_3 as starting materials with $(1-x)SnO_2{:}xEu_2O_3$ composition (x: 0–0.04) were ball milled for 1 day, then dried, and put into pellets and sintered for 2 hours at 1350°C in nitrogen gas atmosphere of a tube furnace and cooled down in N_2 atmosphere. The XRD patterns of pristine tin oxide and europium oxide–doped tin

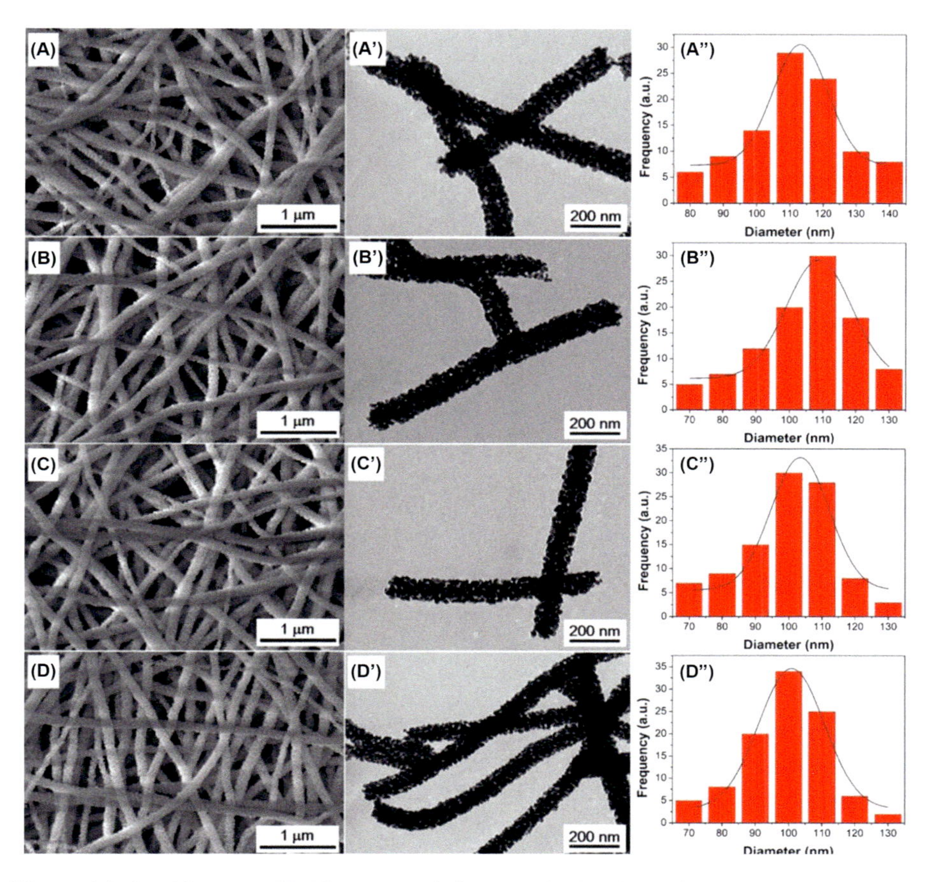

Figure 9.5 SEM images, TEM images, and diameter distributions of pure SnO_2 nanofibers (A, A′, A″), 1 mol.% Eu-doped SnO_2 nanofibers (B, B′, B″), 2 mol.% Eu-doped SnO_2 nanofibers (C, C′, C″), and 3 mol.% Eu-doped SnO_2 nanofibers (D, D′, D″).
Source: Reproduced with permission from Jiang, Z., Zhao, R., Sun, B., Nie, G., Ji, H., Lei, J., & Wang, C. (2016). Highly sensitive acetone sensor based on Eu-doped SnO_2 electrospun nanofibers. *Ceramics International*, *42*, 15881–15888.

oxide nanoparticles were studied. It was reported that the XRD pattern of the prepared sample is in good agreement with JCPDS (21-1250). No impurities were formed in sintered SnO_2. With the introduction of europium oxide, diffraction peak intensities are reduced and diffraction angles shift to high angles, resulting in a minor decrease of lattice constants and a decrease in unit cell volume. Furthermore, when the amount of Eu_2O_3 increases, the diffraction peaks related to it become more prominent.

The Raman spectra of bulk SnO_2 and the Eu_2O_3/SnO_2 are shown in Fig. 9.6. The first-order Raman active modes of rutile SnO_2 are A_{1g}, E_g, and B_{1g}, which correspond to the space group D_4h. The unique Raman peaks at 781/cm, 634/cm, and

475/cm in bulk tin oxide relate to the B_{2g}, A_{1g}, and E_g, respectively. There are also two weak Raman peaks at 692 and 501/cm. The A_{2u} and $E_{u(2)}$ modes can be ascribed to these two peaks, respectively. The introduction of Eu_2O_3 produces some intriguing Raman spectra. With the addition of Eu_2O_3, the primary vibrational peak positions related to tin oxide remain unchanged. However, as the amount of Eu added increases, the intensity of these vibrational modes decreases. In company with tin oxide vibrational modes, the Raman spectra of europium oxide/tin oxide show another new three peaks at 502, 400, and 304/cm. These newly discovered vibrational modes become stronger in intensity as additional Eu_2O_3 is added to SnO_2. The Raman peaks of Eu_2O_3 have been reported to be 315, 412, 579, and 610/cm. As a result, the detected vibrational modes have nothing to do with Eu_2O_3. The disorder is usually discovered by the spectral alterations in bulk crystal spectra. Its also worth noting that the strength of these additional bands increases when the dominant vibration mode relating to SnO_2 decreases, indicating an increase in disorder (Chang & Jo, 2007).

Liao et al. prepared $Gd_2Sn_2O_7:Eu^{3+}$ (GSO:Eu) (Eu: 7 mol. %) via simple microwave-hydrothermal method, which is described as follows:

1. Certain amounts of citric acid, europium nitrate hexahydrate, and gadolinium nitrate hexahydrate were dissolved in water through stirring.

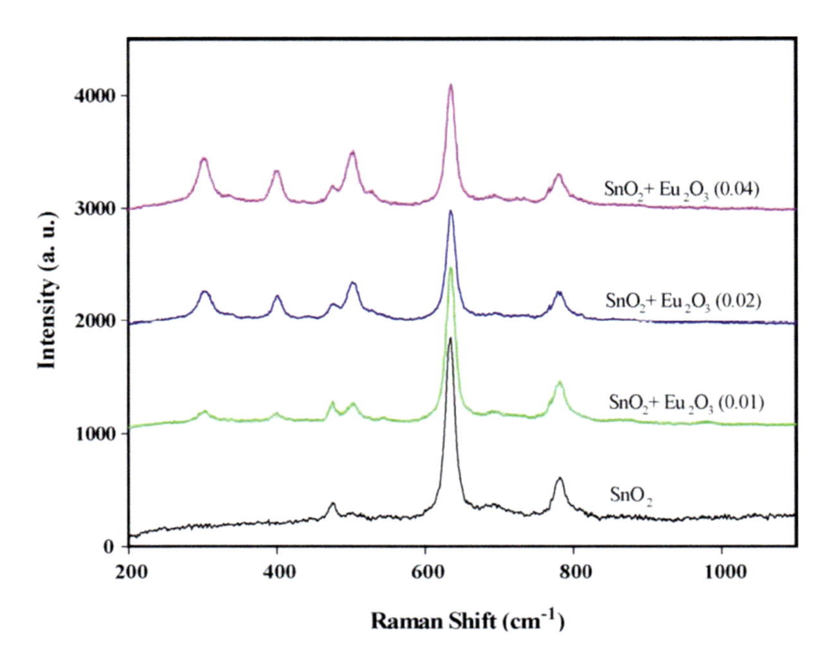

Figure 9.6 Room temperature Raman spectra of bulk SnO_2 and SnO_2 with Eu_2O_3.
Source: Reproduced with permission from Chang, S.-S., & Jo, M. S. (2007). Luminescence properties of Eu-doped SnO_2. *Ceramics International*, *33*, 511–514.

2. Sn-containing solution was provided via dissolving $SnCl_40.5H_2O$ in water and adding into the solution (I) under stirring. HNO_3 and NaOH were applied for adjusting pH. The obtained white precipitate moved into the microwave and irradiated for 1 hour at 200°C. After washing and drying of as-obtained products, the solids were annealed at 800°C for 5 hours in the air. Fig. 9.7 shows SEM images of GSO:Eu nanocrystals at different magnifications. As well as shown, the morphology of the prepared samples is rice-like, and the average size of the particles is about 50 nm, according to the SEM image. The GSO:Eu sample appears to be made up of aggregated nanoparticles based on TEM-based morphological observations (Fig. 9.7C). The energy dispersive X-ray (EDS) analysis was applied for the chemical analysis of as-prepared samples. The EDS result is in good agreement with 7% GSO:Eu nanocrystals, having gadolinium:europium molar ratio of approximately 93:7 (Fig. 9.7D). HRTEM analysis revealed that the pyrochlore structure with the *d* spacing of 0.304 nm attributes to the (222) planes in a GSO crystal. The multicrystalline character of the GSO:Eu sample was further investigated by the SAED pattern (Fig. 9.7F). The obtained results imply that the microwave-hydrothermal route may successfully produce multicrystalline nature nanocrystals (Liao et al., 2017).

The PL investigation of $Gd_{2-2x}Sn_2O_7{:}2xEu^{3+}$ (x: 0.01, 0.05, 0.07, 0.1, 0.12, 0.15, and 0.2) under the 391-nm excitation as a function of europium ion doping amounts revealed the emission spectra of GSO:Eu phosphor is dominated by narrow peaks in the region between 550 and 750 nm. This can be attributed to transitions from $^5D_0$ to the ^7F_J ($J = 0, 1, 2, 3, 4$) of the Eu^{3+} $4f^6$. At 587 and 598 nm, the MD transition of europium ion ($^5D_0 \rightarrow {}^7F_1$) produces a series of reddish-orange emissions. The strong emission peak at 587 nm is found to be exceedingly narrow, with an full width at half-maximum (FWHM) under 2 nm. That means the phosphor particles have a high degree of crystallinity and fewer defects. The electric dipole transition of the $^5D_0 \rightarrow {}^7F_2$ line is responsible for the decreased emission of europium ion at 613 nm. The $^5D_0 \rightarrow {}^7F_2$ electric dipole transition becomes greatest when europium ions are located at a region with no inversion symmetry. The $^5D_{0\rightarrow}{}^7F^1$ MD transition is prominent when they are at a place with inversion symmetry. The orange emission intensity of the $^5D_{0\rightarrow}{}^7F^1$ transition is substantially greater than the red emission intensity of the $^5D_0 \rightarrow {}^7F_2$, indicating that europium ions are mostly involved in the lattice site gadolinium ions with inversion symmetric. The solitary emission peak at 577 nm ($^5D_0 \rightarrow {}^7F_0$ transition) can be observed, indicating that Eu^{3+} only occupied one lattice position. In addition, it is found that the PL intensities are dependent on Eu^{3+} content linearly, as well as the Eu^{3+} concentration (x) in GSO: at $x = 0.07$, Eu reaches a saturation threshold. Since the concentration quenching effect, the PL intensity tends to quench with rising Eu^{3+} concentration (Liao et al., 2017).

Tian et al. prepared $La_2Sn_2O_7$ via sol–gel, coprecipitation, and hydrothermal methods. The hydrothermal process was done according to the following procedures ($La_2Sn_2O_7$-HT): sodium stannate tetrahydrate and lanthanum (III) nitrate in 1:1 molar ratio were dissolved in water. Then, the pH of the solution was kept at 9 via NH_3 and HNO_3 solution under stirring. The hydrothermal route was applied for 24 hours at 200°C. The final hydrothermal-assisted ($La_2Sn_2O_7$-HT) was provided via drying the solid at 110°C for 24 hours. $La_2Sn_2O_7$ was also synthesized via coprecipitation ($La_2Sn_2O_7$-CP): ammonia solution was added to lanthanum(III)

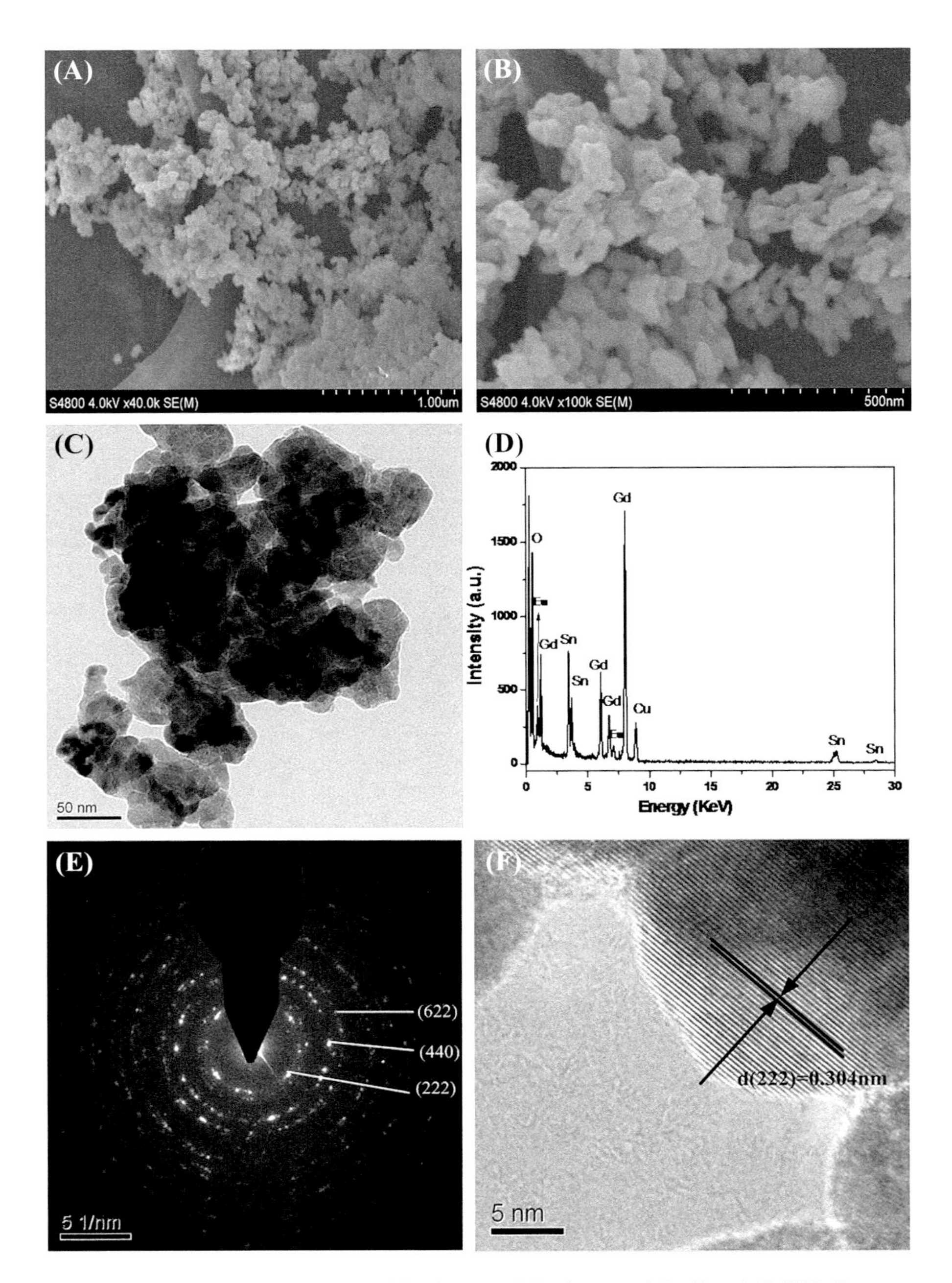

Figure 9.7 Low- (A) and high-magnification (B) SEM image of the 7 mol.% GSO:Eu sample; (C) TEM image of the nanocrystals aggregate; (D) EDS data taken from a nanocrystal aggregate; (E) HRTEM image of a nanocrystal aggregate; and (F) SAED spectrum of a nanocrystal.
Source: Reproduced with permission from Liao, J., Nie, L., Wang, Q., Liu, S., Fu, J., & Wen, H.-R. (2017). Microwave hydrothermal method and photoluminescence properties of $Gd_2Sn_2O_7$: Eu^{3+} reddish-orange phosphors. *Journal of Luminescence*, *183*, 377–382.

nitrate hexahydrate and tin(IV) chloride pentahydrate (1:1 molar ratio) mixture dropwise under stirring. The La/Sn ratio and pH were kept 1:1 and 9, respectively. Then, the prepared solid was filtered and dried at 110°C overnight. Finally, the as-obtained product was calcined for 4 hours at 800°C under an air atmosphere. The sol–gel route was done as per the following procedures ($La_2Sn_2O_7$-SG): first, a certain concentration of tin(IV) chloride solution was heated to 80°C. For the ease in which NH_3 evaporates at an elevated temperature, a sodium hydroxide solution (2 M) was applied and mixed vigorously into the tin(IV) chloride solution to prepare a tin hydroxide slurry. Lanthanum hydroxide was obtained via the same route as $Sn(OH)_4$. The tin hydroxide and lanthanum hydroxide were mixed and the pH was adjusted to 9 via NaOH. Then, cetyltrimethylammonium bromide solution was added to Sn, La-containing solution. The mixture was stirred for 5 hours at 80°C before being heated for 24 hours at 80°C and centrifuged. To obtain the finalized $La_2Sn_2O_7$-SG support, the obtained products were dried and then calcined for 4 hours at 800°C. The morphological and structural properties of samples revealed that the applied synthesis route affects shape and size intensively. They found that the hydrothermal method leads to smaller particle sizes (Tian et al., 2015).

Yang et al. prepared pyrochlore structure of europium-doped lanthanum stannate with homogeneous octahedron shape at 180°C for 36 hours via a hydrothermal route. The lanthanum (III) nitrate hexahydrate and europium (III) nitrate hexahydrate were applied as starting materials. The molar ratio of Eu^{3+} to La^{3+} was kept to 5:95 in this work. The basicity of the solution was controlled via 4 mol/L NaOH. SEM and TEM were utilized to characterize the as-prepared $La_2Sn_2O_7:Eu^{3+}$ crystals for the in-depth investigation about the impacts of experimental condition on the sizes and shape of the samples. It is found that the pH of the solution is crucial in the synthesis of the octahedral morphology of $La_2Sn_2O_7:Eu^{3+}$. They presented a possible mechanism for the octahedral morphology of products. $La_2Sn_2O_7$ nuclei were used to produce three-dimensional clusters with crucial dimensions, which eventually developed into crystallites. Since the higher atomic densities of the crystallite planes, the lone electron pairs in O atoms may significantly interact with the surface of $La_2Sn_2O_7$ nuclei under a basic condition (pH = 11) and preferred stick to particular crystalline faces (Fig. 9.8) (Yang, Su, & Liu, 2011).

In Table 9.1 some rare earth stannate ceramic nanomaterials and their synthesis methods are shown.

9.3 Applications of rare earth–doped SnO_2 nanostructures and rare earth stannate ($Re_2Sn_2O_7$) ceramic nanomaterials

9.3.1 Nanosensor

A nanosensor is a device with the ability to transmit information and data on the properties and features of nanostructures from macroscopic size to nanoscale. Chemical,

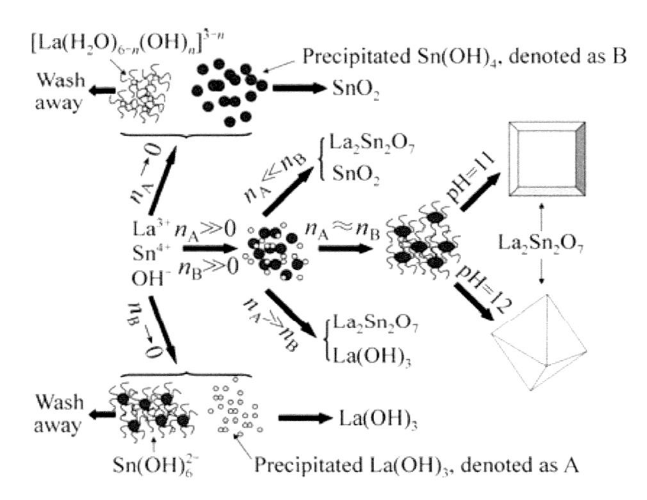

Figure 9.8 Schematic illustration of formation mechanism of samples with different phases and shapes (n denotes the number of precipitates).
Source: Reproduced with permission from Yang, J.-y., Su, Y.-c., & Liu, X.-y. (2011). Hydrothermal synthesis, characterization and optical properties of $La_2Sn_2O_7$:Eu^{3+} micro-octahedra. *Transactions of Nonferrous Metals Society of China, 21*, 535–543.

Table 9.1 Several rare earth stannate ceramic nanomaterials and their synthesis methods.

Sample	Synthesis method	Reference
$Sm_2Sn_2O_7$	Solid state	Irtyugo, Denisova, Kargin, Beletskii, and Denisov (2016)
$La_2Sn_2O_7$	Coprecipitation	Kaliyaperumal, Jayabalan, Sankarakumar, and Paramasivam (2020)
$La_2Sn_2O_7$	Solid state	Ryumin et al. (2020)
$La_2Sn_2O_7$	Coprecipitation	Ma et al. (2014)
$La_{2-x}Nd_xSn_2O_7$	Sol–gel	Quader et al. (2020)
$La_2Sn_2O_7$	Coprecipitation	Wang, Zhou, Lu, Zhou, and Yang (2006)
Dy^{3+}-doped $La_2Sn_2O_7$	Combustion	Wang et al. (2007)
$La_2Sn_2O_7$:Eu^3 +	Hydrothermal	Yang & Su (2010)
Ce-/codoped ($La_2Sn_2O_7$)	Coprecipitation	Yang, Su, Li, Liu, and Chen (2011)
$La_{2-x}RE_xSn_2O_7$ (RE = Eu and Dy)	Coprecipitation	Wang et al. (2006)
$Nd_2Sn_2O_7$	Improved Pechini process	Morassaei, Zinatloo-Ajabshir, and Salavati-Niasari (2017)
$Er_2Sn_2O_7$	Combustion	Tong, Zhao, Wang, and Lu (2009)
La^{3+}-doped $Yb_2Sn_2O_7$	Sol–gel	Wang et al. (2015)

mechanical, and optical features of nanoparticles make them suitable for application in biosensors, nanosensors, and other sensors. The various nanostructures can be applied for the detection of changes in mechanical stress, concentration, electrical forces, volume, gravitational, and magnetic. Until now, a vast number of nanomaterials have been applied in the nanosensor fields (Boomashri et al., 2021; Dong Hyun et al., 1996; Gherardi, Zonta, Astolfi, & Malagù, 2021; Jia et al., 2021; Jiang, Tian, & Yu, 2021; Li et al., 2021; Singh, Kumar, & Sharma, 2020; Sun, Guo, Qi, Liu, & Yu, 2020). In this section, the application of several rare earth stannates and rare earth element–doped SnO_2 ceramic nanomaterials in the field of nanosensor is investigated.

Liu et al. prepared SnO_2 cerium-doped SnO_2 with the morphology of nanobelts (NBs) via thermal evaporation route and subsequently investigated their application in the detection of ethanol. The response is found to reflect the change in resistance induced by the charge-transfer process between adsorbed ethanol and the designed sensor. The depletion layer is generated when oxygen adsorption and desorption reach a state of equilibrium. The reaction occurs between absorbed O^- and gas molecules when exposed to reducing gases. This can result in electrons being released into the conduction band. Finally, this process increases the conductivity. The cerium-doped tin oxide sensor displays a greater sensitivity against ethanol when compared to other gases, which might be owing to the presence of more electrons when contacting ethanol. The following are reactions that could be taken (Qin, Liu, Chen, Wu, & Li, 2016):

$$6O^- + C_2H_5OH \rightarrow 3H_2O + 2CO_2 + 6e^-$$
$$CH_3COCH_3 + O^- \rightarrow CH_3C^+O + CH_3O^- + e^-$$
$$HCHO + 2O^- \rightarrow H_2O + CO_2 + 2e^-$$
$$(CH_2OH)_2 + 5O^- \rightarrow 2CO_2 + 3H_2O + 5e^-$$
$$CO + O^- \rightarrow CO_2 + e^-$$

It is reported that the cerium-doped tin oxide–based sensor response is elevated via the introduction of cerium since (1) the addition of Ce to the tin oxide surface can cause the pH to increase, resulting in improved sensor performance. (2) When ethanol is present, the doped cerium ions accelerate the ethanol's oxidation, releasing trapped electrons and transferring them to the SnO_2 NB, resulting in an increase in electron concentration. (3) After doping, the sensor surface collects more oxygen molecules. More electrons are captured by absorbed oxygen as the contact between the gas and sensor surface gets greater in the presence of ethanol, resulting in better contact between the target gas and sensor surface (Qin et al., 2016). The sensitivity of the sensor designed toward ethanol amount in the range of 20–100 ppm is shown in Fig. 9.9A, which more exemplifies the linear connection between ethanol amount and sensitivity. Fig. 9.9B shows the response/recovery parameter of the cerium-doped tin oxide–based sensor at 230°C. The different response/recovery times were investigated to 30, 50, 100, and 200 ppm of ethanol. As shown in Fig. 9.9C, the sensitivity of the cerium-doped tin oxide–based sensor to various ethanol concentrations and humidity levels in the 30%–70% range was investigated. When the humidity changes, the sensor value practically stays the same,

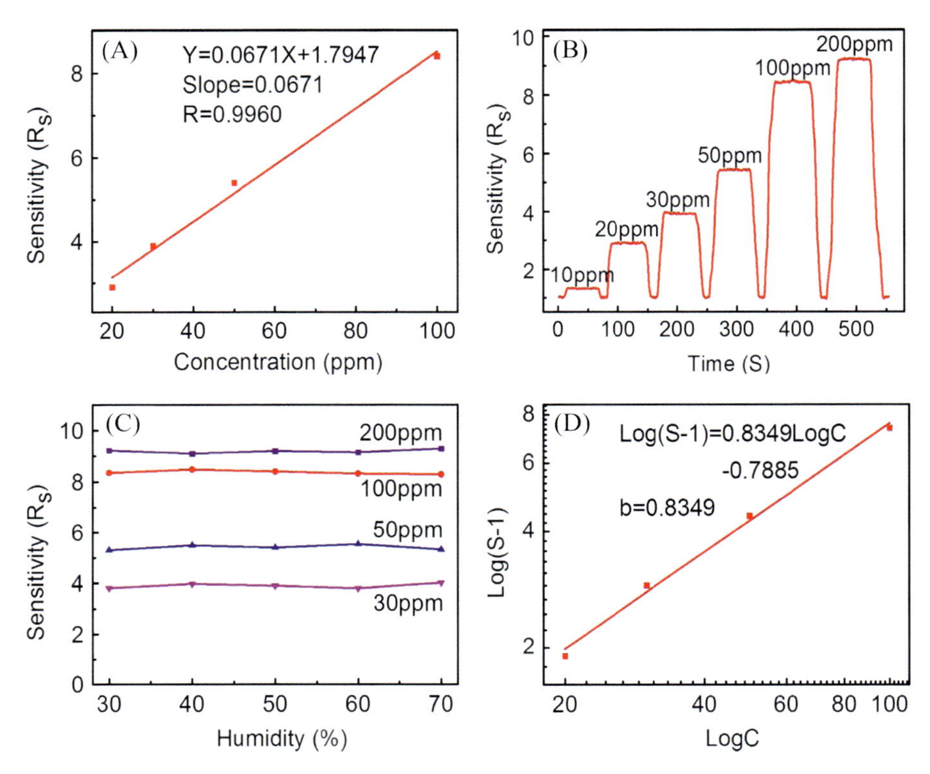

Figure 9.9 (A) Plot of sensitivity versus concentration; (B) response–recovery characteristics to different ethanol concentration; (C) the sensitivity to different concentration ethanol under different humidity; (D) image of Log (S-1) and Log C for Ce-SnO_2 sensor at 230°C.
Source: Reproduced with permission from Qin, Z., Liu, Y., Chen, W., Wu, Y., & Li, S. (2016). The highly promotive sensing performance of a single cerium doped SnO_2 nanobelt sensor to ethanol. *Materials Science in Semiconductor Processing*, *52*, 75–81.

showing that humidity has minimal effect on the response. Fig. 9.9D shows a correlation of Log (S-1) and Log C at 230°C for the cerium-doped tin oxide sensor. It showed that Log (S-1) had a sufficient linear correlation to Log C, and the slope was determined at 0.8349, which closes to 1, implying that the abovementioned oxygen ions at the surface are mostly made up of O^- (Qin et al., 2016).

Table 9.2 shows the application of several rare earth–doped SnO_2 ceramic nanomaterials in the detection of different gases.

9.3.2 Photocatalysis

Photocatalysis is the acceleration of a photoinduced reaction via a photocatalyst. The substrate absorbs light during the process. The ease with which the catalyst creates electron–hole pairs determines the activity of the catalyst and reaction kinetic.

Table 9.2 Application of rare earth–doped SnO_2 ceramic nanomaterials as gas sensor.

Materials	Target species	Reference
Yb-doped SnO_2	Liquefied petroleum gas	Deepa, Kumari, and Saidh (2019)
Yb-doped SnO_2	Ethanol	Wang et al. (2015)
Yb-doped SnO_2	Ethandiol	Chen et al. (2016)
Sm-doped SnO_2	Hydrogen	Singh, Hastir, and Singh (2016)
Sm-doped SnO_2	Humidity	Bakiya Lakshmi and Vimala Juliet (2019)
Sm-doped SnO_2	Isopropanol	Zhao, Li, Wan, Ren, and Zhao (2018)
Er-doped SnO_2	Hydrogen	Singh, Virpal, and Singh (2019)
Nd-doped SnO_2	Ethanol	Wu et al. (2010), Qin et al. (2016)
La-doped SnO_2	CO_2 gas	Kim, Yoon, Park, and Kim (2000), Marsal, Cornet, and Morante (2003)
Eu-doped SnO_2	Hydrogen	Singh, Kohli, and Singh (2017)

Via the reaction of the substrate and electron–hole pairs, a radical is produced. Then, other reactions occur when these radicals react with the reactant to form several beneficial products. Photocatalysts are dedicated to an essential role in the degradation of water pollutants (Liu & Chen, 2017; Ohtani, 2011; Ross, 2019). The use of nanomaterials in the photocatalyst process has been able to make a big revolution in this field. Nanomaterials can be very effective in photocatalysis due to their attractive properties that induced-high specific surface area and quantum confinement (Aamir, 2021; Al Farraj, Al-Mohaimeed, Alkufeidy, & Alkubaisi, 2021; Chinnathambi et al., 2021; Fan et al., 2019; Uma, Ananda, & Kumar, 2021; Wu et al., 2018).

Salavati-Niasari et al. prepared $Nd_2Sn_2O_7$ nanostructures and investigated its photocatalytic activity against methyl orange (MO) as a water pollutant. PL and Dr-UV–vis spectroscopy were applied to determine the optical properties of as-produced $Nd_2Sn_2O_7$. The results showed that the PL spectra had an emission band at roughly 397 nm. The Dr-UV–vis spectra of $Nd_2Sn_2O_7$ revealed an absorption peak at 357 nm. Tauc's equation was used to determine the bandgap, E_g based on Dr-UV–vis data. 3.25 eV has been established as the E_g quantity of $Nd_2Sn_2O_7$ ceramic nanomaterials. The E_g quantity found indicates that the as-produced $Nd_2Sn_2O_7$ nanomaterials might be used as a photocatalyst. Fig. 9.10 displays the outcome of photocatalytic activity of $Nd_2Sn_2O_7$. Without UV light irradiation or as-prepared nanomaterials, no MO dye was virtually removed after 50 minutes due to the minimal effect of self-decomposition. After 50 minutes of UV light exposure, the MO degradation efficiency was around 87%, according to photocatalytic calculation. The proposed degradation mechanism is given below (Morassaei et al., 2017):

$$Nd_2Sn_2O_7 + h\upsilon \rightarrow Nd_2Sn_2O_{7*} + e^- + h^+$$
$$h^+ + H_2O \rightarrow OH\bullet + H^+$$
$$e^- + O_2 \rightarrow O_2^-\bullet$$
$$OH\bullet + O_2^-\bullet + \text{organic pollutant} \rightarrow CO_2 + H_2O$$

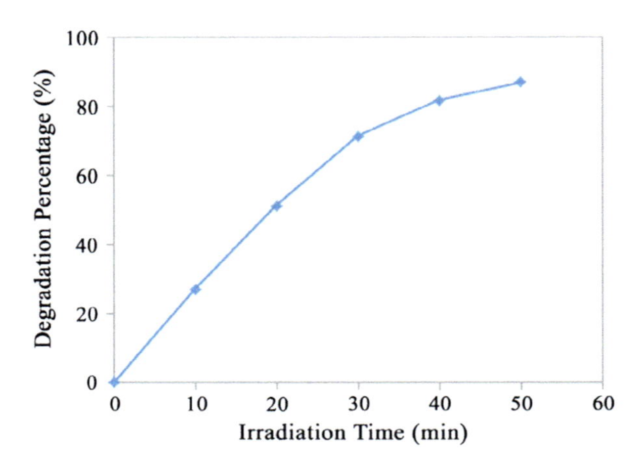

Figure 9.10 Photocatalytic methyl orange dye degradation of the as-synthesized neodymium stannate nanostructures (sample no. 8).
Source: Reproduced with permission from Morassaei, M. S., Zinatloo-Ajabshir, S., & Salavati-Niasari, M. (2017). $Nd_2Sn_2O_7$ nanostructures: New facile Pechini preparation, characterization, and investigation of their photocatalytic degradation of methyl orange dye. *Advanced Powder Technology*, *28*, 697–705.

Table 9.3 represents some rare earth stannate and rare earth–doped SnO_2 ceramic nanomaterials applied for photocatalytic degradation of organic pollutants.

9.3.3 Solar cells

A solar cell, also known as a photovoltaic cell, is a device that converts light energy into electrical energy using photovoltaic characteristics. Solar panels made from 150- to 200-μm-thick crystalline silicon wafers are used in traditional first-generation solar panels.

Thin film photovoltaics are a type of second-generation solar cell that is gaining popularity owing to lower production costs, greater environmental sustainability, and a broad range of architectural integration possibilities, including applications that need lightweight modules, flexibility, and semitransparency (Fuentes, Vega, Arias, & Morales, 2021; Li et al., 2021; Shin, Lim, Shin, Lee, & Kang, 2021). With the advent of nanoscience, nanostructures have been widely used in solar cell fields (Abdel-Galil, Hussien, & Yahia, 2020; In, Park, & Jung, 2020; Otoufi et al., 2020; Pan et al., 2020; Wang, Li, Xu, Yuan, & Yang, 2020). In this section, the application of rare earth stannates and rare earth–doped SnO_2 ceramic nanomaterials in the field of solar cells is discussed.

Ma et al. prepared La-doped SnO_2 in perovskite solar cells (PSCs). The doping of La leads to reducing tin oxide crystal agglomeration and resulting in coverage in its entirety and homogeneous film. Furthermore, La:SnO_2 may successfully lower the SnO_2 layer's band offset, resulting in a high open-circuit voltage (V_{oc}) of 1.11 V.

Table 9.3 Applied rare earth stannate and rare earth–doped SnO_2 ceramic nanomaterials as a photocatalyst for the degradation of organic pollutants.

Photocatalyst	Target pollutant	Reference
$Dy_2Sn_2O_7$	Acid Violet 7 and crystal violet	Zinatloo-Ajabshir, Morassaei, Amiri, and Salavati-Niasari (2020)
$Nd_2Sn_2O_7$-SnO_2	Rhodamine B and eosin Y	Zinatloo-Ajabshir, Morassaei, and Salavati-Niasari (2019)
La-doped SnO_2	Methyl orange	Arif et al. (2017)
Gd-doped SnO_2	Phenol	Al-Hamdi, Sillanpää, and Dutta (2015)
Nd-doped SnO_2	Phenol	Al-Hamdi, Sillanpää, and Dutta (2014)
NdOCl-$Nd_2Sn_2O_7$-SnO_2	Erythrosine	Morassaei, Zinatloo-Ajabshir, and Salavati-Niasari (2016)
$Dy_2Sn_2O_7$-SnO_2	Eosin Y, Eriochrome Black T, erythrosine	Zinatloo-Ajabshir, Morassaei, and Salavati-Niasari (2017)
Ce-doped SnO_2	Phenol	Al-Hamdi et al. (2014)

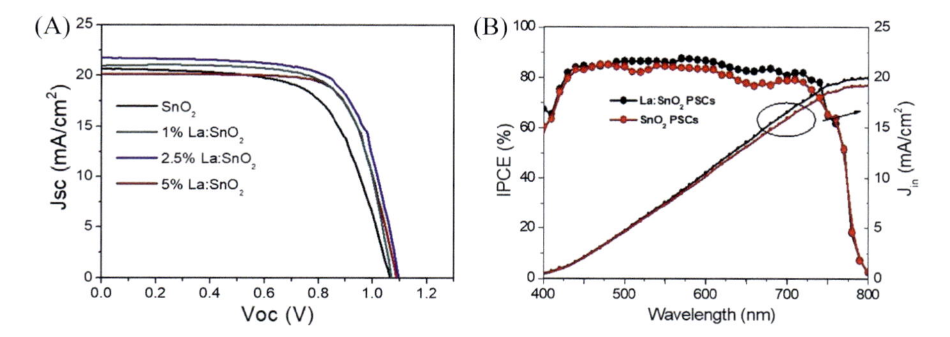

Figure 9.11 (A) The J – V curves of PSC devices were based on pristine SnO_2 and La:SnO_2 with different La doping contents. (B) IPCE of the PSCs was based on SnO_2 and La:SnO_2 ETLs, respectively.
Source: Reproduced with permission from Xu, Z., Teo, S. H., Gao, L., Guo, Z., Kamata, Y., Hayase, S., Ma, & T. (2019). La-doped SnO_2 as ETL for efficient planar-structure hybrid perovskite solar cells. *Organic Electronics*, *73*, 62–68.

Fig. 9.11A shows the photovoltaic performance of La:SnO_2-based PSCs. The power conversion efficiency was significantly enhanced to 17.08% from 14.24% when the amount of lanthanum molar ratio was adjusted to the optimal 2.5% (for pure tin oxide). However, the concentration of lanthanum being increased even further (5%) does not modify the power conversion efficiency, and the power conversion efficiency decreased to 15.92%. The IPCE and integrated current based on the different electron transport layers (ETLs) are shown in Fig. 9.11B. The IPCE-integrated current density for the tin oxide–based cell was measured at

19.31 mA/cm^2, and it rose to a higher amount for the device on lanthanum-doped based layer, which matches the J–V measurements (Xu et al., 2019).

9.3.4 Transistor

Radio-frequency magnetron sputtering was used via Shen et al. to synthesize Er-SnO_2 thin film transistors (TFTs). The Er-doped tin oxide thin films were used to manufacture bottom-gate top-contact TFTs (thickness: 5–10 nm). Fig. 9.12A shows typical transfer characteristic curves ($V_{DS} = 10$ V). With Er^{3+} concentration increasing from 0% to 4.6%, the off-state current reduces from 8.56×10^{-11} to 4.71×10^{-12} A, indicating erbium ion's evident carrier suppression impact. As a result, μ_{FE} drops, whereas *V*th amounts rise via the presence of erbium. Performance differences in devices are well aligned with film characteristics. Under testing settings, the Er-doped tin oxide TFTs with an erbium ion amount of 3.1% has a pretty high performance. The

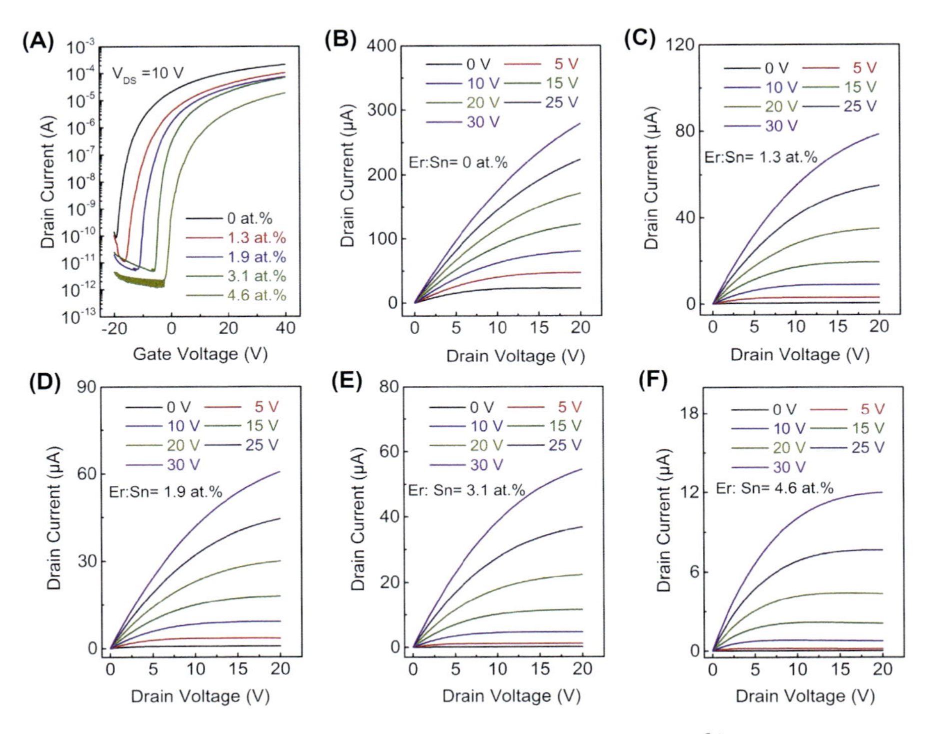

Figure 9.12 Electrical characteristics of ErSnO-TFTs with different Er^{3+} concentrations: (A) transfer curves of ErSnO-TFTs and (B)–(F) output curves of ErSnO-TFTs. All TFTs are swept with the gate voltage from −20 to 40 V.
Source: Reproduced with permission from Ren, J.-h., Li, K.-w., Shen, J., Sheng, C.-m., Huang, Y.-t., & Zhang, Q. (2019). Effects of rare-earth erbium doping on the electrical performance of tin-oxide thin film transistors. *Journal of Alloys and Compounds 791*, 11–18.

ability of Er-SnO_2 TFTs diminishes noticeably when erbium amount rises to 4.6%. The 4.6% has fewer oxygen defects, according to XPS data. Er-SnO_2 is a thin film that cannot ensure good performance in a sequential manner. Fig. 9.12B–F shows common output curves of Er-SnO_2 TFTs with various erbium ion concentrations. The ohmic contact between indium tin oxide electrodes and Er-doped SnO_2 channels is shown by all TFTs displaying a linear rise in I_D at low V_D. Furthermore, current crowding effects were not found in obtained results, demonstrating the Er-doped SnO_2 TFTs' dependability in this study. It can be concluded that erbium as a dopant is an effective carrier inhibitor, indicating that lanthanide oxides have a lot of promise in SnO_2-based TFTs (Ren et al., 2019).

9.4 Conclusion and outlook

In conclusion, rare earth stannate ceramic nanomaterials have been found more attention in recent years. The intrinsic physical and chemical properties of these materials, along with the characteristics of induced nanoscale, make them attractive in various applications. The properties of rare earth stannate ceramic nanomaterials depend intensively on their shape and size, which also depends strongly on the preparation methods. Various methods have been applied for the synthesis of rare earth stannate and rare earth–doped SnO_2 ceramic nanomaterials, such as coprecipitation, hydrothermal, sonochemical, microwave-assisted, solid-state, and sol–gel routes. The morphology of these ceramic nanomaterials can be engineered by controlling experimental conditions in preparation methods. The prepared rare earth stannate ceramic nanomaterials can be applied in catalysis, sensor, photocatalysis, solar cell, and transistor-based applications. Because of their high fusing point ($>2000^\circ C$), they may be employed as a high-temperature catalyst, and the use of uniqueness is arranged in the aspect of control in the processing of vehicle exhaust. The nanoscale particle size and higher specific surface area of $Re_2Sn_2O_7$ materials generated as nanostructures can result in altered phase transition temperatures, thereby enhancing catalytic activity.

References

Aamir, L. (2021). Novel p-type Ag-WO_3 nano-composite for low-cost electronics, photocatalysis, and sensing: Synthesis, characterization, and application. *Journal of Alloys and Compounds*, *864*, 158108.

Abdel-Galil, A., Hussien, M. S. A., & Yahia, I. S. (2020). Low cost preparation technique for conductive and transparent Sb doped SnO_2 nanocrystalline thin films for solar cell applications. *Superlattices and Microstructures*, *147*, 106697.

Ahmadnia-Feyzabad, S., Mortazavi, Y., Khodadadi, A. A., & Hemmati, S. (2013). Sm_2O_3 doped-SnO_2 nanoparticles, very selective and sensitive to volatile organic compounds. *Sensors and Actuators B: Chemical*, *181*, 910–918.

Ahmed, M. (2020). Chapter 16 - Nanomaterial synthesis. In R. Narain (Ed.), *Polymer science and nanotechnology* (pp. 361–399). Elsevier.

Al Farraj, D. A., Al-Mohaimeed, A. M., Alkufeidy, R. M., & Alkubaisi, N. A. (2021). Facile synthesis and characterization of CeO_2-Al_2O_3 nano-heterostructure for enhanced visible-light photocatalysis and bactericidal applications. *Colloid and Interface Science Communications*, *41*, 100375.

Alam, M. M., Asiri, A. M., & Rahman, M. M. (2021). Wet-chemically synthesis of SnO_2-doped Ag_2O nanostructured materials for sensitive detection of choline by an alternative electrochemical approach. *Microchemical Journal*, *165*, 106092.

Al-Hamdi, A. M., Sillanpää, M., & Dutta, J. (2014). Photocatalytic degradation of phenol in aqueous solution by rare earth-doped SnO_2 nanoparticles. *Journal of Materials Science*, *49*, 5151–5159.

Al-Hamdi, A. M., Sillanpää, M., & Dutta, J. (2015). Gadolinium doped tin dioxide nanoparticles: An efficient visible light active photocatalyst. *Journal of Rare Earths*, *33*, 1275–1283.

Aragón, F. H., Chitta, V. A., Coaquira, J. A. H., Hidalgo, P., & Brito, H. F. (2013). Long-range ferromagnetic order induced by a donor impurity band exchange in SnO_2:Er^{3+} nanoparticles. *Journal of Applied Physics*, *114*, 203902.

Arif, H. S., Murtaza, G., Hanif, H., Ali, H. S., Yaseen, M., & Khalid, N. R. (2017). Effect of La on structural and photocatalytic activity of SnO_2 nanoparticles under UV irradiation. *Journal of Environmental Chemical Engineering*, *5*, 3844–3851.

Asaithambi, S., Sakthivel, P., Karuppaiah, M., Balamurugan, K., Yuvakkumar, R., Thambidurai, M., & Ravi, G. (2021). Synthesis and characterization of various transition metals doped SnO_2@MoS_2 composites for supercapacitor and photocatalytic applications. *Journal of Alloys and Compounds*, *853*, 157060.

Bakiya Lakshmi, R., & Vimala Juliet, A. (2019). Effect of annealing on humidity sensing properties of Sm-doped SnO_2 thin films. *Journal of Materials Research and Technology*, *8*, 5862–5866.

Bakiya Lakshmi, R., Vimala Juliet, A., Sundararaj, A., Abhinav, E. M., & Chandrasekaran, G. (2018). Effect of Sm doping and annealing on magnetic and structural properties of sputter deposited SnO_2 thin films. *Materials Research Express*, *5*, 064004.

Ben Saber, N., Mezni, A., Alrooqi, A., & Altalhi, T. (2020). A review of ternary nanostructures based noble metal/semiconductor for environmental and renewable energy applications. *Journal of Materials Research and Technology*, *9*, 15233–15262.

Binetti, E., Longo, A., & Carotenuto, G. (2015). Colloidal synthesis of rare earth-based nanoparticles. In *2015 IEEE 15th International Conference on Nanotechnology (IEEE-NANO)*, pp. 366–369.

Boomashri, M., Perumal, P., Khan, A., El-Toni, A. M., Ansari, A. A., Gupta, R. K., ... Kumar, K. D. A. (2021). Zn influence on nanostructured tin oxide (SnO_2) films as ammonia sensor at room temperature. *Surfaces and Interfaces*, 101195.

Bouzigues, C., Gacoin, T., & Alexandrou, A. (2011). Biological applications of rare-earth based nanoparticles. *ACS Nano*, *5*, 8488–8505.

Chang, S.-S., & Jo, M. S. (2007). Luminescence properties of Eu-doped SnO_2. *Ceramics International*, *33*, 511–514.

Chatterjee, A., Kwatra, N., & Abraham, J. (2020). Chapter 8 - Nanoparticles fabrication by plant extracts. In N. Thajuddin, & S. Mathew (Eds.), *Phytonanotechnology* (pp. 143–157). Elsevier.

Chen, W., Liu, Y., Qin, Z., Wu, Y., Li, S., & Gong, N. (2016). Improved ethanediol sensing with single Yb ions doped SnO_2 nanobelt. *Ceramics International*, *42*, 10902–10907.

Chen, Z., Xu, Z., & Zhao, H. (2021). Flame spray pyrolysis synthesis and H2S sensing properties of CuO-doped SnO_2 nanoparticles. *Proceedings of the Combustion Institute*, *38*, 6743–6751.

Chinnathambi, A., Syed, A., Elgorban, A. M., Marraiki, N., Al-Rashed, S., & Yassin, M. T. (2021). Performance analysis of novel La_6WO_{12}/Ag_2WO_4 nano-system for efficient visible-light photocatalysis and antimicrobial activity. *Journal of Alloys and Compounds*, 160075.

Cojocaru, B., Avram, D., Kessler, V., Parvulescu, V., Seisenbaeva, G., & Tiseanu, C. (2017). Nanoscale insights into doping behavior, particle size and surface effects in trivalent metal doped SnO_2. *Scientific Reports*, *7*, 9598.

Das, S., Som, S., Yang, C.-Y., & Lu, C.-H. (2018). Optical temperature sensing properties of SnO_2: Eu^{3+} microspheres prepared via the microwave assisted solvothermal process. *Materials Research Bulletin*, *97*, 101–108.

Deepa, S., Kumari, K. P., & Saidh, A. (2019). Investigation of structural and optical properties of ytterbium modified tin oxide thin films for gas sensing application. *AIP Conference Proceedings*, *2162*, 020084.

Deng, X., Zhu, M., Ke, J., Li, W., Xiong, D., Feng, Z., & He, M. (2021). SnO_2-ZnO nanoparticles wrapped in graphite nanosheets as a large-capacity, high-rate and long-lifetime anode for lithium-ion batteries. *Chemical Physics Letters*, *769*, 138392.

Divya, J., Pramothkumar, A., Joshua Gnanamuthu, S., Bernice Victoria, D. C., & Jobe Prabakar, P. C. (2020). Structural, optical, electrical and magnetic properties of Cu and Ni doped SnO_2 nanoparticles prepared via co-precipitation approach. *Physica B: Condensed Matter*, *588*, 412169.

Dong Hyun, Y., Chul Han, K., Hyung-Ki, H., Hyun Woo, S., Seung-Ryeol, K., & Kyuchung, L. (1996). Abnormal current-voltage characteristics of WO_3-doped SnO_2 oxide semiconductors and their applications to gas sensors. *Sensors and Actuators B: Chemical*, *35*, 48–51.

Dung, C. T. M., Giang, L. T. T., Binh, D. H., Hieu, L. V., & Van, T. T. T. (2021). Understanding up and down-conversion luminescence for Er^{3+}/Yb^{3+} co-doped SiO_2-SnO_2 glass-ceramics. *Journal of Alloys and Compounds*, *870*, 159405.

Escudero, A., Becerro, A. I., Carrillo-Carrión, C., Núñez, N. O., Zyuzin, M. V., Laguna, M., ... Parak, W. J. (2017). Rare earth based nanostructured materials: Synthesis, functionalization, properties and bioimaging and biosensing applications. *Nanophotonics*, *6*, 881–921.

Fan, W., Zhou, Z., Wang, W., Huo, M., Zhang, L., Zhu, S., ... Wang, X. (2019). Environmentally friendly approach for advanced treatment of municipal secondary effluent by integration of micro-nano bubbles and photocatalysis. *Journal of Cleaner Production*, *237*, 117828.

Feng, J., Xiao, B., Qu, Z. X., Zhou, R., & Pan, W. (2011). Mechanical properties of rare earth stannate pyrochlores. *Applied Physics Letters*, *99*, 201909.

Feng, Y., Bai, C., Wu, K., Dong, H., Ke, J., Huang, X., ... He, M. (2020). Fluorine-doped porous SnO_2@C nanosheets as a high performance anode material for lithium ion batteries. *Journal of Alloys and Compounds*, *843*, 156085.

Fuentes, S., Vega, M., Arias, M., & Morales, P. (2021). Upconversion of $Bi_4Ti_3O_{12}$:Er and its evaluation in silicon solar cell yield. *Materials Letters*, *296*, 129889.

Ganose, A. M., & Scanlon, D. O. (2016). Band gap and work function tailoring of SnO_2 for improved transparent conducting ability in photovoltaics. *Journal of Materials Chemistry C*, *4*, 1467–1475.

Gao, Y., Hou, Q., & Liu, Q. (2020). First-principles study on the electronic structures and magneto-optical properties of $Fe^{2+/3+}$ doped SnO_2. *Solid State Communications*, *305*, 113764.

Gherardi, S., Zonta, G., Astolfi, M., & Malagù, C. (2021). Humidity effects on SnO_2 and $(SnTiNb)O_2$ sensors response to CO and two-dimensional calibration treatment. *Materials Science and Engineering: B*, *265*, 115013.

Hu, Y., Li, L., Xu, C., & Yang, P. (2021). Study of high metal doped SnO_2 for photovoltaic devices. *Materials Today Communications*, *27*, 102148.

In, S. J., Park, M., & Jung, J. W. (2020). Reduced interface energy loss in non-fullerene organic solar cells using room temperature-synthesized SnO_2 quantum dots. *Journal of Materials Science & Technology*, *52*, 12–19.

Irtyugo, L. A., Denisova, L. T., Kargin, Y. F., Beletskii, V. V., & Denisov, V. M. (2016). Synthesis and investigation of the heat capacity of $Sm_2Sn_2O_7$ in the 346–1050 K range. *Russian Journal of Inorganic Chemistry*, *61*, 701–703.

Jadhav, V. V., Shinde, P. V., Mane, R. S., & O'Dwyer, C. (2019). Chapter 11 - Shape-controlled hybrid nanostructures for cancer theranostics. In R. Ashok Bohara, & N. Thorat (Eds.), *Hybrid nanostructures for cancer theranostics* (pp. 209–227). Elsevier.

Jain, A., Fournier, P. G. J., Mendoza-Lavaniegos, V., Sengar, P., Guerra-Olvera, F. M., Iñiguez, E., ... Juárez, P. (2018). Functionalized rare earth-doped nanoparticles for breast cancer nanodiagnostic using fluorescence and CT imaging. *Journal of Nanobiotechnology*, *16*, 26.

Jia, L., Chen, R., Xu, J., Zhang, L., Chen, X., Bi, N., ... Zhao, T. (2021). A stick-like intelligent multicolor nano-sensor for the detection of tetracycline: The integration of nano-clay and carbon dots. *Journal of Hazardous Materials*, *413*, 125296.

Jiang, C., Tian, Z., & Yu, W. (2021). Displacement measurement technology of nano grating sensor based on HHT algorithm. *Microelectronics Journal*, *109*, 104986.

Jiang, Z., Zhao, R., Sun, B., Nie, G., Ji, H., Lei, J., & Wang, C. (2016). Highly sensitive acetone sensor based on Eu-doped SnO_2 electrospun nanofibers. *Ceramics International*, *42*, 15881–15888.

Kaliyaperumal, C., Jayabalan, S., Sankarakumar, A., & Paramasivam, T. (2020). Structural and electrical characteristics of nanocrystalline $La_2Sn_2O_7$ pyrochlore. *Solid State Sciences*, *105*, 106245.

Karmaoui, M., Jorge, A. B., McMillan, P. F., Aliev, A. E., Pullar, R. C., Labrincha, J. A., & Tobaldi, D. M. (2018). One-step synthesis, structure, and band gap properties of SnO_2 nanoparticles made by a low temperature nonaqueous sol–gel technique. *ACS Omega*, *3*, 13227–13238.

Kaur, H., Bhatti, H. S., & Singh, K. (2019). Europium doping effect on 3D flower-like SnO_2 nanostructures: Morphological changes, photocatalytic performance and fluorescence detection of heavy metal ion contamination in drinking water. *RSC Advances*, *9*, 37450–37466.

Kennedy, B. J., Hunter, B. A., & Howard, C. J. (1997). Structural and bonding trends in tin pyrochlore oxides. *Journal of Solid State Chemistry*, *130*, 58–65.

Kim, D. H., Yoon, J. Y., Park, H. C., & Kim, K. H. (2000). CO_2-sensing characteristics of SnO_2 thick film by coating lanthanum oxide. *Sensors and Actuators B: Chemical*, *62*, 61–66.

Li, G., Wang, Z., Fei, X., Li, J., Zheng, Y., Li, B., & Zhang, T. (2021). Identification and elimination of cancer cells by folate-conjugated CdTe/CdS quantum dots chiral nano-sensors. *Biochemical and Biophysical Research Communications*, *560*, 199–204.

Li, S., Pomaska, M., Lambertz, A., Duan, W., Bittkau, K., Qiu, D., ... Ding, K. (2021). Transparent-conductive-oxide-free front contacts for high-efficiency silicon heterojunction solar cells, Joule.

Li, S., Yu, L., Man, X., Zhong, J., Liao, X., & Sun, W. (2017). The synthesis and band gap changes induced by the doping with rare-earth ions in nano-SnO_2. *Materials Science in Semiconductor Processing*, *71*, 128—132.

Liao, J., Nie, L., Wang, Q., Liu, S., Fu, J., & Wen, H.-R. (2017). Microwave hydrothermal method and photoluminescence properties of $Gd_2Sn_2O_7$: Eu^{3+} reddish orange phosphors. *Journal of Luminescence*, *183*, 377—382.

Liu, Y., & Chen, X. (2017). Chapter eleven - black titanium dioxide for photocatalysis. In Z. Mi, L. Wang, & C. Jagadish (Eds.), *Semiconductors and semimetals* (pp. 393—428). Elsevier.

Luque, P. A., Garrafa-Gálvez, H. E., Nava, O., Olivas, A., Martínez-Rosas, M. E., Vilchis-Nestor, A. R., ... Chinchillas-Chinchillas, M. J. (2021). Efficient sunlight and UV photocatalytic degradation of methyl orange, methylene blue and rhodamine B, using Citrus × paradisi synthesized SnO_2 semiconductor nanoparticles. *Ceramics International*.

Ma, Y., Wang, X., You, X., Liu, J., Tian, J., Xu, X., ... Chen, X. (2014). Nickel-supported on $La_2Sn_2O_7$ and $La_2Zr_2O_7$ pyrochlores for methane steam reforming: Insight into the difference between tin and zirconium in the B site of the compound. *ChemCatChem*, *6*, 3366—3376.

Maiti, K., Kim, N. H., & Lee, J. H. (2021). Strongly stabilized integrated bimetallic oxide of Fe_2O_3-MoO_3 nano-crystal entrapped N-doped graphene as a superior oxygen reduction reaction electrocatalyst. *Chemical Engineering Journal*, *410*, 128358.

Marsal, A., Cornet, A., & Morante, J. R. (2003). Study of lanthanum compounds influence on tin oxide CO_2 gas sensors. *Sensors for Environmental Control*, 30—34.

Matussin, S. N., Harunsani, M. H., Tan, A. L., Cho, M. H., & Khan, M. M. (2020). Effect of Co^{2+} and Ni^{2+} co-doping on SnO_2 synthesized via phytogenic method for photoantioxidant studies and photoconversion of 4-nitrophenol. *Materials Today Communications*, *25*, 101677.

Morassaei, M. S., Zinatloo-Ajabshir, S., & Salavati-Niasari, M. (2016). New facile synthesis, structural and photocatalytic studies of $NdOCl$-$Nd_2Sn_2O_7$-SnO_2 nanocomposites. *Journal of Molecular Liquids*, *220*, 902—909.

Morassaei, M. S., Zinatloo-Ajabshir, S., & Salavati-Niasari, M. (2017). $Nd_2Sn_2O_7$ nanostructures: New facile Pechini preparation, characterization, and investigation of their photocatalytic degradation of methyl orange dye. *Advanced Powder Technology*, *28*, 697—705.

Mounkachi, O., Salmani, E., Lakhal, M., Ez-Zahraouy, H., Hamedoun, M., Benaissa, M., ... Benyoussef, A. (2016). Band-gap engineering of SnO_2. *Solar Energy Materials and Solar Cells*, *148*, 34—38.

Nasrollahzadeh, M., Issaabadi, Z., Sajjadi, M., Sajadi, S. M., & Atarod, M. (2019). Chapter 2 - Types of nanostructures. In M. Nasrollahzadeh, S. M. Sajadi, M. Sajjadi, Z. Issaabadi, & M. Atarod (Eds.), *Interface science and technology* (pp. 29—80). Elsevier.

Nunes, D., Pimentel, A., Santos, L., Barquinha, P., Pereira, L., Fortunato, E., & Martins, R. (2019). 2 - Synthesis, design, and morphology of metal oxide nanostructures. In D. Nunes, A. Pimentel, L. Santos, P. Barquinha, L. Pereira, E. Fortunato, & R. Martins (Eds.), *Metal oxide nanostructures* (pp. 21—57). Elsevier.

Ohtani, B. (2011). Chapter 10 - Photocatalysis by inorganic solid materials: Revisiting its definition, concepts, and experimental procedures. In Rv Eldik, & G. Stochel (Eds.), *Advances in inorganic chemistry* (pp. 395—430). Academic Press.

Oliveira, A. G., Andrade, J. d. L., Montanha, M. C., Ogawa, C. Y. L., de Souza Freitas, T. K. F., Moraes, J. C. G., ... de Oliveira, D. M. F. (2021). Wastewater treatment using Mg-doped ZnO nano-semiconductors: A study of their potential use in environmental remediation. *Journal of Photochemistry and Photobiology A: Chemistry*, *407*, 113078.

Otoufi, M. K., Ranjbar, M., Kermanpur, A., Taghavinia, N., Minbashi, M., Forouzandeh, M., & Ebadi, F. (2020). Enhanced performance of planar perovskite solar cells using TiO_2/SnO_2 and TiO_2/WO_3 bilayer structures: Roles of the interfacial layers. *Solar Energy*, *208*, 697–707.

Pan, B., Yang, Y., Bian, J., Hu, X., & Zhang, W. (2019). Quantum dot decorated nano-pyramid fiber tip for scanning near-field optical microscopy. *Optics Communications*, *445*, 273–276.

Pan, J., Li, S., Ou, W., Liu, Y., Li, H., Wang, J., ... Li, C. (2020). The photovoltaic conversion enhancement of NiO/Tm:CeO_2/SnO_2 transparent p-n junction device with dual-functional Tm:CeO_2 quantum dots. *Chemical Engineering Journal*, *393*, 124802.

Park, S., Hwang, H. J., & Moon, J. (2003). Catalytic combustion of methane over rare earth stannate pyrochlore. *Catalysis Letters*, *87*, 219–223.

Pascariu, P., Koudoumas, E., Dinca, V., Rusen, L., & Suchea, M. P. (2019). Chapter 14 - Applications of metallic nanostructures in biomedical field. In V. Dinca, & M. P. Suchea (Eds.), *Functional nanostructured interfaces for environmental and biomedical applications* (pp. 341–361). Elsevier.

Prabhakaran, D., Wang, S., & Boothroyd, A. T. (2017). Crystal growth of pyrochlore rare-earth stannates. *Journal of Crystal Growth*, *468*, 335–339.

Qin, G., Gao, F., Jiang, Q., Li, Y., Liu, Y., Luo, L., ... Zhao, H. (2016). Well-aligned Nd-doped SnO_2 nanorod layered arrays: Preparation, characterization and enhanced alcohol-gas sensing performance. *Physical Chemistry Chemical Physics*, *18*, 5537–5549.

Qin, Z., Liu, Y., Chen, W., Wu, Y., & Li, S. (2016). The highly promotive sensing performance of a single cerium doped SnO_2 nanobelt sensor to ethanol. *Materials Science in Semiconductor Processing*, *52*, 75–81.

Quader, A., Mustafa, G. M., Kumail Abbas, S., Ahmad, H., Riaz, S., Naseem, S., & Atiq, S. (2020). Efficient energy storage and fast switching capabilities in Nd-substituted $La_2Sn_2O_7$ pyrochlores. *Chemical Engineering Journal*, *396*, 125198.

Rafique, M., Tahir, M. B., Rafique, M. S., Safdar, N., & Tahir, R. (2020). Chapter 2 - Nanostructure materials and their classification by dimensionality. In M. B. Tahir, M. Rafique, & M. S. Rafique (Eds.), *Nanotechnology and photocatalysis for environmental applications* (pp. 27–44). Elsevier.

Ren, J.-h., Li, K.-w., Shen, J., Sheng, C.-m., Huang, Y.-t., & Zhang, Q. (2019). Effects of rare-earth erbium doping on the electrical performance of tin-oxide thin film transistors. *Journal of Alloys and Compounds*, *791*, 11–18.

Ross, J. R. H. (2019). Chapter 13 - Environmental catalysis. In J. R. H. Ross (Ed.), *Contemporary catalysis* (pp. 291–314). Amsterdam: Elsevier.

Ryumin, M. A., Nikiforova, G. E., Tyurin, A. V., Khoroshilov, A. V., Kondrat'eva, O. N., Guskov, V. N., & Gavrichev, K. S. (2020). Heat capacity and thermodynamic functions of $La_2Sn_2O_7$. *Inorganic Materials*, *56*, 97–104.

Shaikh, F. I., Chikhale, L. P., Patil, J. Y., Mulla, I. S., & Suryavanshi, S. S. (2017). Enhanced acetone sensing performance of nanostructured Sm_2O_3 doped SnO_2 thick films. *Journal of Rare Earths*, *35*, 813–823.

Shin, D.-Y., Lim, J. R., Shin, W.-G., Lee, C.-G., & Kang, G.-H. (2021). Layup-only modulization for low-stress fabrication of a silicon solar module with 100 μm thin silicon solar cells. *Solar Energy Materials and Solar Cells*, *221*, 110903.

Singh, A. K., Yadav, T. P., Pandey, B., Gupta, V., & Singh, S. P. (2019). Chapter 15 - Engineering nanomaterials for smart drug release: recent advances and challenges. In S. S. Mohapatra, S. Ranjan, N. Dasgupta, R. K. Mishra, & S. Thomas (Eds.), *Applications of targeted nano drugs and delivery systems* (pp. 411–449). Elsevier.

Singh, G., Hastir, A., & Singh, R. C. (2016). Hydrogen sensor based on Sm-doped SnO_2 nanostructures. *AIP Conference Proceedings*, *1731*, 050117.

Singh, G., Kohli, N., & Singh, R. C. (2017). Preparation and characterization of Eu-doped SnO_2 nanostructures for hydrogen gas sensing. *Journal of Materials Science: Materials in Electronics*, *28*, 2257–2266.

Singh, G., Virpal., & Singh, R. C. (2019). Highly sensitive gas sensor based on Er-doped SnO_2 nanostructures and its temperature dependent selectivity towards hydrogen and ethanol. *Sensors and Actuators B: Chemical*, *282*, 373–383.

Singh, J. P., Singh, V., Sharma, A., Pandey, G., Chae, K. H., & Lee, S. (2020). Approaches to synthesize MgO nanostructures for diverse applications. *Heliyon*, *6*, e04882.

Singh, S., Kumar, S., & Sharma, S. (2020). Room temperature high performance ammonia sensor using MoS_2/SnO_2 nanocomposite. *Materials Today: Proceedings*, *28*, 52–55.

Sivakumar, P., Akkera, H. S., Ranjeth Kumar Reddy, T., Srinivas Reddy, G., Kambhala, N., & Nanda, N. (2021). Kumar Reddy, Influence of Ga doping on structural, optical and electrical properties of transparent conducting SnO_2 thin films. *Optik*, *226*, 165859.

Soussi, L., Garmim, T., Karzazi, O., Rmili, A., El Bachiri, A., Louardi, A., & Erguig, H. (2020). Effect of (Co, Fe, Ni) doping on structural, optical and electrical properties of sprayed SnO_2 thin film. *Surfaces and Interfaces*, *19*, 100467.

Sun, Y., Guo, J., Qi, J., Liu, B., & Yu, T. (2020). Comparisons of SnO2 gas sensor degradation under elevated storage and working conditions. *Microelectronics Reliability*, *114*, 113808.

Tian, J., Peng, H., Xu, X., Liu, W., Ma, Y., Wang, X., & Yang, X. (2015). High surface area La2Sn2O7 pyrochlore as a novel, active and stable support for Pd for CO oxidation. *Catalysis Science & Technology*, *5*, 2270–2281.

Tong, Y., Zhao, S., Wang, X., & Lu, L. (2009). Synthesis and characterization of $Er_2Sn_2O_7$ nanocrystals by salt-assistant combustion method. *Journal of Alloys and Compounds*, *479*, 746–749.

Trujillano, R., Martín, J. A., & Rives, V. (2016). Hydrothermal synthesis of $Sm_2Sn_2O_7$ pyrochlore accelerated by microwave irradiation. A comparison with the solid state synthesis method. *Ceramics International*, *42*, 15950–15954.

Uma, H. B., Ananda, S., & Kumar, M. S. V. (2021). Electrochemical synthesis and characterization of CuO/ZnO/SnO nano photocatalyst: Evaluation of its application towards photocatalysis, photo-voltaic and antibacterial properties. *Chemical Data Collections*, *32*, 100658.

Wang, J., Li, D., Xu, C., Yuan, X., & Yang, P. (2020). Numerical study on photoelectric characteristics of Mo-doped SnO_2. *Superlattices and Microstructures*, *138*, 106387.

Wang, J., Xu, F., Wheatley, R. J., Choy, K.-L., Neate, N., & Hou, X. (2015). Investigation of La^{3+} Doped $Yb_2Sn_2O_7$ as new thermal barrier materials. *Materials & Design*, *85*, 423–430.

Wang, S. M., Lu, M. K., Zhou, G. J., Zhou, Y. Y., Zhang, H. P., Wang, S. F., & Yang, Z. S. (2006). Synthesis and luminescence properties of $La_{2-x}RE_xSn_2O_7$ (RE = Eu and Dy) phosphor nanoparticles. *Materials Science and Engineering: B*, *133*, 231–234.

Wang, S. M., Xiu, Z. L., Lü, M. K., Zhang, A. Y., Zhou, Y. Y., & Yang, Z. S. (2007). Combustion synthesis and luminescent properties of Dy^{3+}-doped $La_2Sn_2O_7$ nanocrystals. *Materials Science and Engineering: B*, *143*, 90–93.

Wang, S. M., Zhou, G. J., Lu, M. K., Zhou, Y. Y., & Yang, Z. S. (2006). Nanorods of $La_2Sn_2O_7$ synthesized in ethanol solvent. *Journal of Alloys and Compounds, 424*, L3–L5.

Wang, T. T., Ma, S. Y., Cheng, L., Luo, J., Jiang, X. H., & Jin, W. X. (2015). Preparation of Yb-doped SnO_2 hollow nanofibers with an enhanced ethanol–gas sensing performance by electrospinning. *Sensors and Actuators B: Chemical, 216*, 212–220.

Wu, S., Li, C., Wei, W., Wang, H., Song, Y., & Zhu, Y. (2010). l. Lu, Nd-doped SnO_2: Characterization and its gas sensing property. *Journal of Rare Earths, 28*, 171–173.

Wu, Y., Hu, W., Xie, R., Liu, X., Yang, D., Chen, P., ... Zhang, F. (2018). Composite of nano-goethite and natural organic luffa sponge as template: Synergy of high efficiency adsorption and visible-light photocatalysis. *Inorganic Chemistry Communications, 98*, 115–119.

Xu, Z., Teo, S. H., Gao, L., Guo, Z., Kamata, Y., Hayase, S., & Ma, T. (2019). La-doped SnO_2 as ETL for efficient planar-structure hybrid perovskite solar cells. *Organic Electronics, 73*, 62–68.

Yan, C.-H., Yan, Z.-G., Du, Y.-P., Shen, J., Zhang, C., & Feng, W. (2011). Chapter 251 - Controlled synthesis and properties of rare earth nanomaterials. In K. A. Gschneidner, J.-C. G. Bünzli, & V. K. Pecharsky (Eds.), *Handbook on the physics and chemistry of rare earths* (pp. 275–472). Elsevier.

Yang, J., & Su, Y. (2010). *Novel 3D octahedral $La_2Sn_2O_7$:Eu^{3+} microcrystals: Hydrothermal synthesis and photoluminescence properties, Materials Letters* (64, pp. 313–316).

Yang, J., Su, Y., Li, H., Liu, X., & Chen, Z. (2011). Hydrothermal synthesis and photoluminescence of Ce3 + and Tb3 + doped $La_2Sn_2O_7$ nanocrystals. *Journal of Alloys and Compounds, 509*, 8008–8012.

Yang, J.-y., Su, Y.-c., & Liu, X.-y. (2011). Hydrothermal synthesis, characterization and optical properties of $La_2Sn_2O_7$: Eu^{3+} micro-octahedra. *Transactions of Nonferrous Metals Society of China, 21*, 535–543.

Yu, Z., Eich, C., & Cruz, L. J. (2020). Recent advances in rare-earth-doped nanoparticles for NIR-II imaging and cancer. *Theranostics, Frontiers in Chemistry, 8*.

Zhao, X., He, S., & Tan, M. C. (2016). Design of infrared-emitting rare earth doped nanoparticles and nanostructured composites. *Journal of Materials Chemistry C, 4*, 8349–8372.

Zhao, Y., Li, Y., Wan, W., Ren, X., & Zhao, H. (2018). Surface defect and gas-sensing performance of the well-aligned Sm-doped SnO_2 nanoarrays. *Materials Letters, 218*, 22–26.

Zinatloo-Ajabshir, S., Morassaei, M. S., Amiri, O., & Salavati-Niasari, M. (2020). Green synthesis of dysprosium stannate nanoparticles using Ficus carica extract as photocatalyst for the degradation of organic pollutants under visible irradiation. *Ceramics International, 46*, 6095–6107.

Zinatloo-Ajabshir, S., Morassaei, M. S., & Salavati- Niasari, M. (2017). Facile fabrication of $Dy_2Sn_2O_7$-SnO_2 nanocomposites as an effective photocatalyst for degradation and removal of organic contaminants. *Journal of Colloid and Interface Science, 497*, 298–308.

Zinatloo-Ajabshir, S., Morassaei, M. S., & Salavati-Niasari, M. (2019). Facile synthesis of $Nd_2Sn_2O_7$-SnO_2 nanostructures by novel and environment-friendly approach for the photodegradation and removal of organic pollutants in water. *Journal of Environmental Management, 233*, 107–119.

Rare-earth molybdates ceramic nanomaterials

10

Hossein Safardoust-Hojaghan
Young Researchers and Elite Club, Marand Branch, Islamic Azad University, Marand, Iran

10.1 General introduction

Molybdates are compounds that contain oxymolybdenum ions that are negatively charged. Mineralogically, the basic tetraoxomolybdate(VI), or molybdate ion, is present for the most part (Goldberg & Suarez, 2006; Williams, 2005; Zhang & Lo, 2013). Molybdate ions polymerize in acidic environments, and this process may include other chemical entities. For their attractive chemical and physical properties, various researches have been focused on the preparation and application of molybdate-based compounds (Ansari, Shukla, & Ansari, 2021; Dridi, Zid, & Maczka, 2018; Kögler et al., 2021; Vats, Shafeeq, & Kesari, 2021; Wu & Shi, 2021; Yang & Wang, 2021; Zhu et al., 2021; Zolotova et al., 2021).

Rare-earth molybdates with the general formula, $Re_2(MoO_4)_3$ (Re = Sc, Y, and La–Lu) is found more and more attention for their fascinating, piezoelectric, fluorescence, narrow band gap, laser ferroelectric, pyroelectric, and ferroelastic properties (Bazarova, Tushinova, Bazarov, & Dorzhieva, 2017; Borchardt & Bierstedt, 1967; Brixner, Barkley, & Jeitschko, 1979; Forbes, Kong, & Cava, 2018; Kaczmarek & Van, 2013; Ponomarev & Zhukov, 2012; Tkachenko & Fedorov, 2003; Xiao, Schlenz, Bosbach, Suleimanov, & Alekseev, 2016). While rare-earth as such, especially in the form of oxides, have been known since the late 18th century, it was not until more than a century later that Hitchcock identified the first molybdates of these elements (1895) (Hitchcock, 1895). Before Borchardt and Bierstedt (1966) who reported members of the $RE_2(MoO_4)_3$ series in the orthorhombic crystal form, very few research work was done in this field (Borchardt & Bierstedt, 1967). By reporting the ferroelectric properties of this molybdate series, they put them at the center of attention. Therefore in recent decades, because of their distinctive chemical and physical features, many types of research have been conducted on their synthesis and application (Bharat, Raju, & Yu, 2017; Kato, Oishi, Shishido, Yamazaki, & Iida, 2005; Li & Van Deun, 2018; Pratap, Gaur, & Lal, 1987; Pu, Lin, Liu, Wang, & Wang, 2020; Roy, Choudhary, & Acharaya, 1989; Sha et al., 2021; Shlyakhtina et al., 2018; Sinha, Mahata, & Kumar, 2018; Sofich et al., 2018).

The rare-earth molybdates have a considerable melting point; 1045°C for $Pr_2(MoO_4)_3$ and 1172°C for $Tb_2(MoO_4)_3$. At 1222°C, $Dy_2(MoO_4)_3$ crystallizes into a cubic phase, and at 1030°C, it turns into β-structure. The compounds of this group perform a transition from the tetragonal β-phase to the monoclinic α-phase process

Advanced Rare Earth-Based Ceramic Nanomaterials. DOI: https://doi.org/10.1016/B978-0-323-89957-4.00001-3

as it cools. The β-α phase transition takes place at 987°C for $Pr_2(MoO_4)_3$ and 805°C for $Dy_2(MoO_4)_3$. As the thermodynamically metastable -phases of $Re_2(MoO_4)_3$ are cooled further, the second transition occurs, resulting in lower symmetry ferroelastic-ferroelectric. They are moreover metastable in terms of thermodynamics. $Pr_2(MoO_4)_3$ β′-β phase transition temperature is 235°C, while $Dy_2(MoO_4)_3$ is 145°C. It should be noted that the single-crystal structure of $Re_2(MoO_4)_3$ is transparent in the range of 400–700 nm wavelength. Interestingly, both β′ and β phases show piezoelectric characteristics. In addition, the β′ phase structures show ferroelectric properties. The orthorhombic β′-phases of $Re_2(MoO_4)_3$ are paramagnetic at temperatures under 1 K (Joukoff, Grimouille, Leroux, Daguet, & Pougnet, 1979; Logvinovich et al., 2010; Ponomarev & Zhukov, 2012; Savvin et al., 2015; Suzuki, Honma, & Komatsu, 2011; Trnovcová, Škubla, & Schultze, 2005; Yu et al., 2021).

The above-mentioned structural properties led to the application of $Re_2(MoO_4)_3$ in electrochemical sensors, photocatalysts, supercapacitors, luminescent material, magnetic material, and piezoelectric material (Bispo-Jr, Shinohara, Pires, & Cardoso, 2018; Chen, Bu, Zhang, & Shi, 2008; Freed & Haenisch, 1937; Gonzalez-rojas, Bonifacio-Martinez, Ordoñez-Regil, & Fernandez-Valverde, 1998; Laufer et al., 2013; Tian, Chen, Hua et al., 2011; Tian, Qi, Wu, Hua, & Chen, 2009; Wang et al., 2009, 2018).

When the size of the $Re_2(MoO_4)_3$ decreased to the nanoscale, the mentioned properties are greatly improved. It is well known that the higher specific surface and the possibility of quantum effects at the nanoscale are responsible for different material properties in the nanostructures (Asha & Narain, 2020; Cheng, 2014; Daniels-Race, 2014; Goswami et al., 2017; Khan, Saeed, & Khan, 2019; Mittal & Banerjee, 2016; Nasrollahzadeh, Issaabadi, Sajjadi, Sajadi, & Atarod, 2019; Nunes et al., 2019; Yang, Deng, Pan, Fu, & Bao, 2015; Zhu, Luo, & Liu, 2020). Because of the scientific and industrial significance of size-dependent properties, the study of size and shape effects on material properties has gotten a lot of attention (Benelmekki, Singh, Baughman, Bohra, & Kim, 2021; Chen & Suganuma, 2019; Fukui, Wang, & Kato, 2021; Goyal & Singh, 2020; Khademalrasool, Farbod, & Talebzadeh, 2021). The number of dimensions of materials that are outside the nanoscale (100 nm) is applied to classify it. As a result, all dimensions in zero-dimensional (0D) nanomaterials are determined on the nanoscale (there are no dimensions larger than 100 nm). Nanoparticles are the most popular 0D nanomaterials. One dimension is beyond the nanoscale in one-dimensional nanomaterials (1D). Nanotubes, nanofibers, and nanowires all fall under this category. Two dimensions are beyond the nanoscale of two-dimensional nanomaterials (2D). Graphene, nanofilms, and nanolayers are examples of this genre, which have plate-like shapes. Materials that are not limited to the nanoscale in any dimension are referred to as three-dimensional nanomaterials (3D). Bulk powders, nanoparticle dispersions, nanowire bundles, and nanotubes all fall into this category (Daulbayev, Sultanov, Bakbolat, & Daulbayev, 2020; Ding, Ma, Yue, Gao, & Jia, 2019; Dong, Li, Ashour, Dong, & Han, 2021; Garnett, Mai, & Yang, 2019; Li & Wang, 2020; Wang, Hu, Liang, & Wei, 2020; Xu, Bo, He, Tian, & Yan, 2020).

This chapter discusses the properties, preparation, and application of $Re_2(MoO_4)_3$ ceramic nanomaterials. Although, the effect of size and morphology on the characteristics and subsequent application of $Re_2(MoO_4)_3$ ceramic nanomaterials have been investigated.

10.2 Preparation methods of rare-earth molybdates ceramic nanomaterials

In general, various synthesis routes are applied to prepared nanostructures. All of these methods can be categorized under two approaches: Top-down approach, and bottom-up approach. The top-down approach is the destructive approach that used larger materials as the precursor and decomposed it into smaller units to reach the nanoscale. Bottom-up approaches, on the other hand, include putting together a very small structure in the material and allowing the precursors to self-assemble into a certain morphology (de Oliveira, Torresi, Emmerling, & Camargo, 2020; Gerberich, Jungk, & Mook, 2003; Ma et al., 2020; Ramesh, Vetrivel, Suresh, & Kaviarasan, 2020; Wang & Xia, 2004). The importance of synthesis methods is determined when it is noted that the main properties of $Re_2(MoO_4)_3$ ceramic nanomaterials highly dependent on their preparation method. In the following, different methods of synthesis of nanostructures and factors affecting them are investigated.

10.2.1 Coprecipitation route

Coprecipitation is one of the simple and low-cost pathways to synthesize $RE_2(MoO_4)_3$ ceramic nanomaterials. The characteristic of salts that have varying degrees of solubility in different solvents is exploited in the typical coprecipitation reaction. Various water-soluble salts that can react together are used as reagents. In the liquid process, one or more water-insoluble salts are formed. When the concentration of this substance in the reaction media exceeds the solubility product value, precipitation occurs. It is found that the experimental conditions such as the type of salts used, the reaction temperature, concentration, stirring speed, used capping agents, and the pH value of the solution have all been found to have a significant effect on the synthesized $Re_2(MoO_4)_3$ ceramic nanomaterials (Ashik, Kudo, & Hayashi, 2018; Ayni, Sabet, & Salavati-Niasari, 2016; Jeseentharani, Dayalan, & Nagaraja, 2018; Ravichandran, Praseetha, Arun, & Gobalakrishnan, 2018; Shahri, Sobhani, & Salavati-Niasari, 2013; Tian, Chen, Tian et al., 2011; Tian et al., 2009; Wu et al., 2016). By using the coprecipitation process, $La_{2-x}Sm_x(MoO_4)_3$, with different samarium ions amounts were prepared. The value of x was determined 0.1, 0.06, 0.02, 0.01, 0.005, which nominated, as Sm 5%, Sm 3, Sm 1%, Sm 0.5%, and Sm 25%. The morphological and structural features of synthesized nanostructures were studied via transmission electron microscopy (TEM), and X-ray diffraction (XRD) analysis respectively. Fig. 10.1 shows TEM images of $La_{1.998}Sm_{0.002}(MoO_4)_3$ nanocrystal. Tiny nanostructures with spherical shape stack together to form high aspect ratio nanorod-like morphology as well as shown in Fig. 10.1A. The in-depth

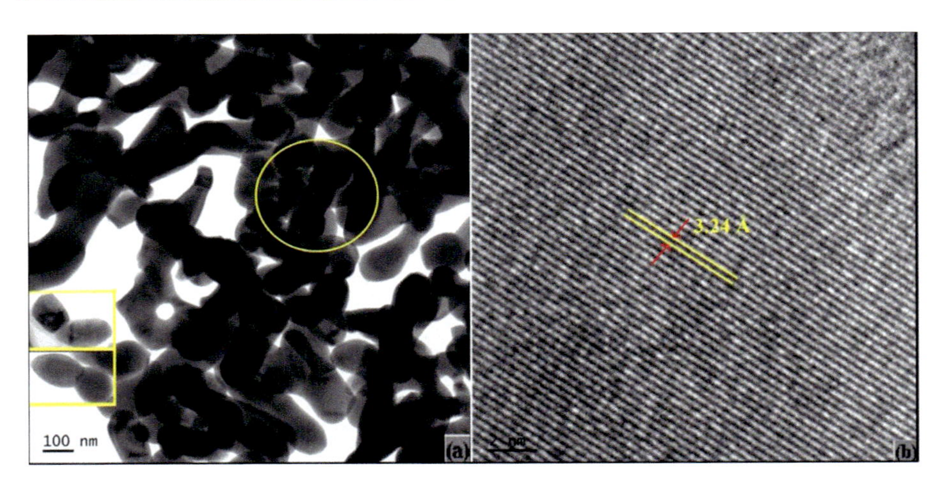

Figure 10.1 (A) Transmission electron microscopy image and (B) HRTEM image of $La_{1.998}Sm_{0.002}(MoO_4)_3$ [Sm 0.1%] nanocrystals.
Source: Reproduced with permissions from Thomas, K., Alexander, D., Sisira, S., Jacob, L. A., Biju, P. R., Unnikrishnan, N. V., . . ., Joseph, C., (2019). Sm^{3+} doped tetragonal lanthanum molybdate: A novel host sensitized reddish orange emitting nanophosphor, *Journal of Luminescence*, *211*, 284–291.
HRTEM, High-resolution transmission electron microscopy.

crystalline properties of the produced nanostructure was investigated by the high-resolution transmission electron microscopy (HRTEM) that confirms a d-spacing of 3.24 Å which is in good agreement with XRD-obtained results. The optical properties of prepared samples were investigated via UV-Vis analysis. The UV–Vis analysis of $La_{1.9}Sm_{0.1}(MoO_4)_3$ shows a wide band in the range of 200–400 nm accompanied by many intense peaks in the range of 400–700 nm. The strong band in the UV-Vis spectrum can be related to the combined effect of charge transfer transitions between O^{2-}-Mo^{6+}, and O^{2-}-Sm^{3+} (Thomas et al., 2019).

Mariappan et al. prepared $Cu_3Mo_2O_9$ (CMO) and $La_2Mo_3O_{12}$ (LMO) nanoparticles by facile coprecipitation and postannealing methods. For $La_2Mo_3O_{12}$ case, $La(NO_3)_3 \cdot 6H_2O$ and Na_2MoO_4 were applied as starting material for lanthanum and molybdenum respectively. The as-prepared precipitate was filtered and washed several times. In the end, the precursors were annealed for 4 hours at 900°C to form the crystalline structure of $La_2Mo_3O_{12}$. The preparation procedure has been illustrated in Fig. 10.2A schematically. Fig. 10.2B displays the XRD pattern of as-produced products. As well as shown in the XRD pattern of $La_2Mo_3O_{12}$, the 2theta positions of XRD patterns were attributed to 101, 112, 004, 200, 211, 204, 220, 116, 312, and 224 Bragg reflections (JCPDS No. 45–0407). The structure of tetragonal with the space group of $I4_1/a$ is approved for $La_2Mo_3O_{12}$ from the X-ray diffraction pattern. The SEM and TEM analyses were applied for in-depth investigation of morphological properties of prepared nanoparticles. The field emission scanning electron microscope (FE-SEM)

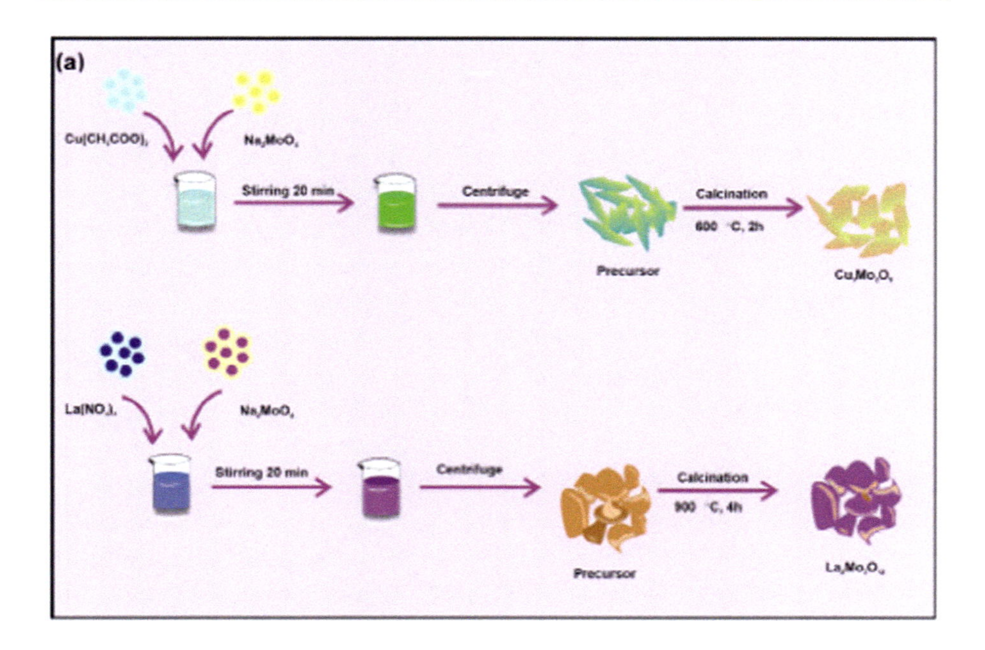

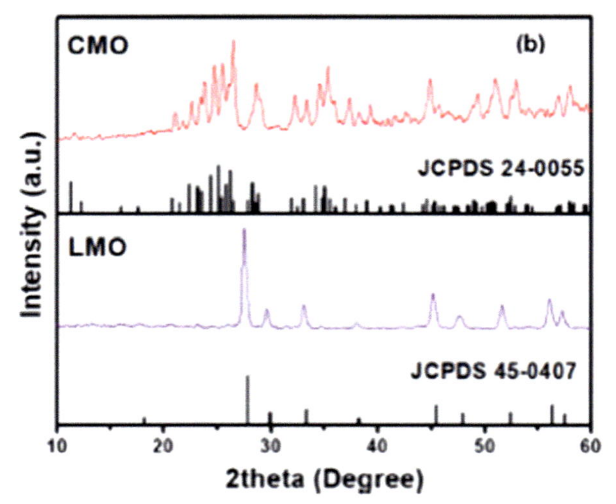

Figure 10.2 (A) Schematic illustration of preparation of $Cu_3Mo_2O_9$ (CMO) and $La_2Mo_3O_{12}$ (LMO). (B) X-ray diffraction patterns of CMO and LMO.
Source: Reproduced with permissions from Gajraj, V. & Mariappan, C. R., (2020). Electrochemical performances of asymmetric aqueous supercapacitor based on porous $Cu_3Mo_2O_9$ petals and $La_2Mo_3O_{12}$ nanoparticles fabricated through a simple co-precipitation method, *Applied Surface Science*, *512* 145648.

Figure 10.3 FE-SEM images of CMO (A,B) and LMO (C,D). (E) transmission electron microscopy (TEM) image of CMO (F) HRTEM image of CMO (G) TEM image of LMO, and (H) enlarged lattice pattern of LMO. Inset shows the SAED pattern of LMO.

(Continued)

images of CMO (Fig. 10.3A and B) reveal a large number of aggregated petals-like properties, while the FE-SEM images of LMO (Fig. 10.3C and D) reveal a large number of agglomerated particles. The agglomerated particles are visible in the TEM picture of the LMO compound (Fig. 10.3G). The bright spots in the electron diffraction pattern in Fig. 10.3H relate to the Millar indices (1 1 2), (0 0 4), (2 0 0), and (2 0 4) respectively. These findings are consistent with XRD results (Gajraj & Mariappan, 2020).

10.2.2 Sonochemical route

Sonochemistry is derived from the intense temporary state caused via ultrasound, that creates hot spots with temperatures exceeding 5000 K and pressures over 10^3 atm, and cooling and heating rates exceeding 10^{10} K/S. Ultrasonic cavitation provides conditions that enable advanced materials to be synthesized at lower temperature and pressure than other applied synthesis routes. Till now a wide range of ceramic nanomaterials were prepared via the ultrasonic method (Ferreira, Andrade Neto, Bomio, & Motta, 2019; Rabbani, Ghasemi, & Hosseini, 2020; Lou, Rajaji, Chen, & Chen, 2020; Ruíz-Baltazar, 2020).

The sonochemical method was used to make samarium molybdate via Tzyy-Jiann Wang et al. Briefly, in 50 mL of DI water, 0.1 mol/L of samarium (III) nitrate hexahydrate and 0.1 mol/L of sodium molybdate were mixed by stirring at 30°C. Following that, a small content of polyvinylpyrrolidone was gradually added to the as-prepared solution. The as-obtained mixture was then subjected to 1 hour of ultrasonic irradiation at a temperature that remains constant. Then, the as-prepared white-color sample was washed with deionized water and ethanol several times and finally dried for 12 hours at 50°C. The schematic ultrasonic-assisted preparation of samarium molybdate was illustrated in Fig. 10.4. The proposed formation mechanism is given below:

$$Na_2MoO_4.2H_2O \rightarrow 2Na^+ + MoO_4^{2-} + 2H_2O$$
$$Sm(NO_3)_3.6H_2O \rightarrow \left[Sm(H_2O)_X\right]^{3-} + 3NO_3^- + 6H_2O$$
$$2Na^+ + MoO_4^{2-} + \left[Sm(H_2O)_X\right]^{3-} + 3NO_3^- \rightarrow Sm_3(MoO_4)_3.H_2O + 2NaNO_3^- \text{(Ultrasonication)}$$
$$Sm_3(MoO_4)_3.H_2O \rightarrow Sm_3(MoO_4)_3 \text{(Heating)}$$

Fig. 10.5 depicts the prepared samarium molybdate's field emission electron microscope-provided images, HRTEM-based images, selected area electron diffraction (SAED), and FFT patterns (A to H). The FE-SEM images in Fig. 10.5 show a

◀ *Source*: Reproduced with permissions from Gajraj, V. & Mariappan, C. R., (2020). Electrochemical performances of asymmetric aqueous supercapacitor based on porous $Cu_3Mo_2O_9$ petals and $La_2Mo_3O_{12}$ nanoparticles fabricated through a simple co-precipitation method, *Applied Surface Science*, *512* (2020) 145648.
FE-SEM, Field emission scanning electron microscopy; *HRTEM*, high-resolution transmission electron microscope; *SAED*, selected area electron diffraction.

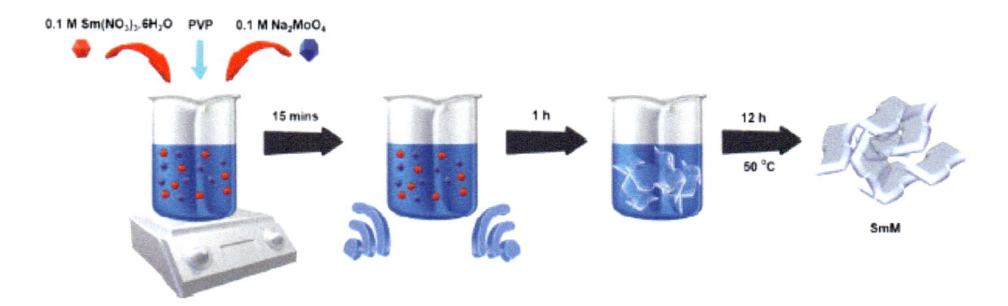

Figure 10.4 Schematic representation of sonochemical-assisted synthesis for samarium molybdate.
Source: Reproduced with permissions from Kokulnathan, T., Ashok Kumar, E., & Wang, T.-J., (2020). Synthesis of two-dimensional nanosheet like samarium molybdate with abundant active sites: Real-time carbendazimin analysis in environmental samples, *Microchemical Journal*, *158* 105227.

two-dimensional nanosheet morphology with nanometer thicknesses (A, B). In Fig. 10.5(C-E), the TEM images of samarium molybdate at various magnifications show an exclusive nanosheet-like structure, which is consistent with the field emission microscope images of samarium molybdate. The authors reported that these interesting structural and morphological characteristics make them an attractive option in the catalytic process (Kokulnathan, Ashok Kumar, & Wang, 2020).

Salavati-Niasari et al. prepared Pr_6MoO_{12} nanoparticles via sonochemical route according to the following procedure: 0.05 g of praseodymium (III) nitrate hexahydrate was solubilized in water and applied to the preprepared solution containing the polyvinylpyrrolidone, polyethylene glycol, and Triton X-100 under stirring via a typical procedure. The as-obtained solution was then placed in a sonication box, and a solution containing 0.02 g of $(NH_4)_6Mo_7O_{24}$. $4H_2O$ with a pH of 11 was progressively applied to the mixture under sonication at a specific power (30, 50, and 80 W). After that, ethanol and distilled water were used to wash the precipitate. It was then dried for 24 hours at 70°C under a vacuum. In the end, the sample was treated with heat for 4 hours at 800°C. It is investigated for the shape, size, crystallinity, and purity of prepared samples via SEM, TEM, XRD, and EDS analysis. Different capping agents were applied for engineering morphological properties of Pr_6MoO_{12} nanoparticles. The findings revealed that the size of particles grows as the ultrasound power is increased from 30 to 80 W. The authors reported that increasing ultrasonic irradiation power speeds up thermodynamic stabilization through the expanding in size of nuclei at the start and produces larger Pr_6MoO_{12} by providing higher energy in a shorter amount of time (Fig. 10.6). Fig. 10.6A, B, and C are respectively related to 30, 50, and 80 W (Namvar, Abass, Soofivand, Salavati-Niasari, & Moayedi, 2019).

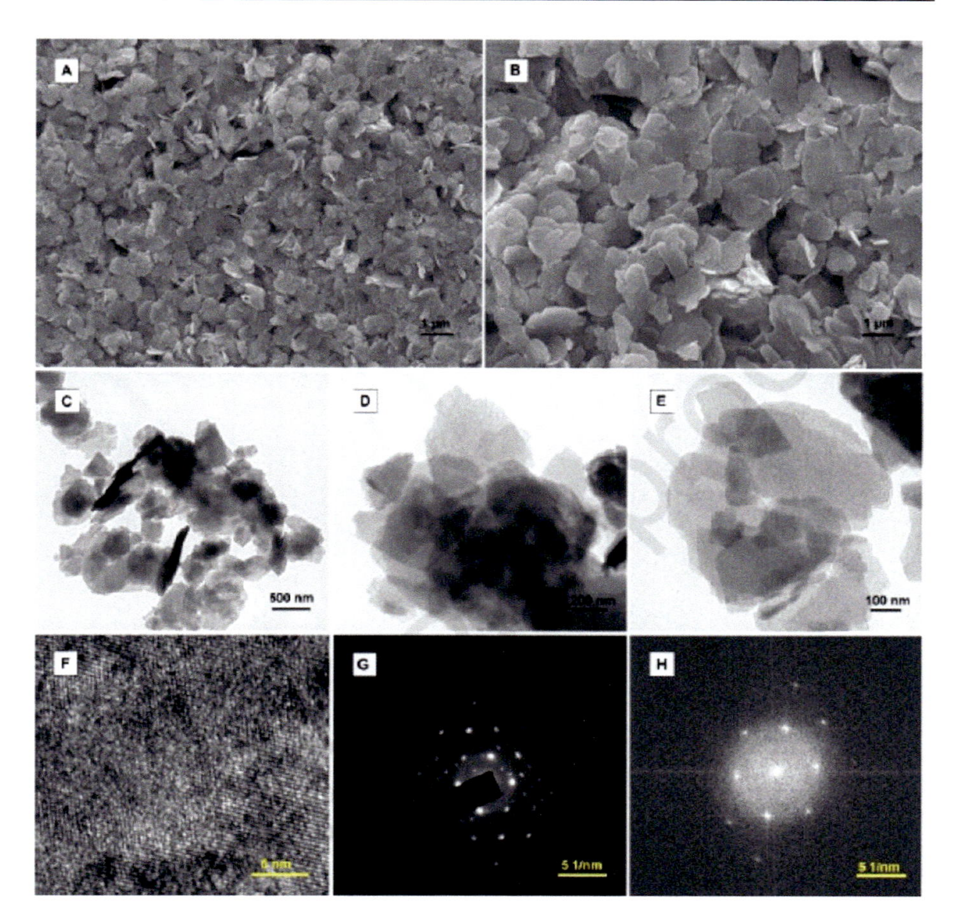

Figure 10.5 (A, B) FE-SEM images, (C−E) transmission electron microscopy images, (F) lattice fringes, (G) SAED pattern, and (H) FFT pattern of samarium molybdate.
Source: Reproduced with permissions from Kokulnathan, T., Ashok Kumar, E., & Wang, T.-J., (2020). Synthesis of two-dimensional nanosheet like samarium molybdate with abundant active sites: real-time carbendazimin analysis in environmental samples, *Microchemical Journal*, *158* (2020) 105227.
FE-SEM, Field emission scanning electron microscope; *SAED*, selected area electron diffraction.

In this work, Triton X-100, polyvinylpyrrolidone-40000, and polyethylene glycol-8000 were applied as morphological control agents and their effects on the shape and size of prepared Pr_6MoO_{12} were investigated. In the presence of PEG-8000, 2-dimensional dense structures were formed (Fig. 10.7A). Since the same structure of polyvinylpyrrolidone (Fig. 10.6A) and Triton X-100 (Fig. 10.7B), the morphology of prepared samples in the presence of PVP and Triton X-100 were the same.

Figure 10.6 Scanning electron microscopy images of prepared $Pr(NO_3)_{3}0.6H_2O$ in the presence of PVP-4000 and power of (A) 30 W, (B) 50 W and (C) 80 W.
Source: Reproduced with permissions from Namvar, F., Abass, S. K., Soofivand, F., Salavati-Niasari, M., & Moayedi, H., (2019). Sonochemical synthesis of Pr_6MoO_{12} nanostructures as an effective photocatalyst for waste-water treatment, *Ultrasonics Sonochemistry*, *58* 104687.

10.2.3 Solid-phase route

The solid-phase method is one of the most important preparation methods of ceramic nanomaterials (Ahmadpor and Milani-Hosseini, 2021; Dalafu et al., 2018; Lu et al., 2019; Morozova, Kalinina, Drozdova, & Shilova, 2018; Radeva, Blaskov, Klissurski, Mitov, & Toneva, 1997; Sakthivel & Prasanna Venkatesh, 2012). Radio

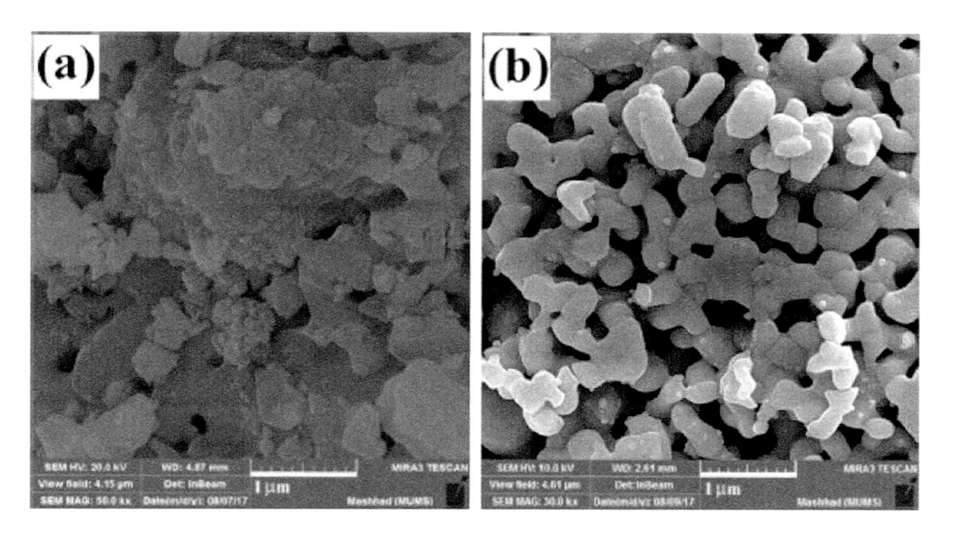

Figure 10.7 Scanning electron microscopy images of prepared $Pr(NO_3)_30.6H_2O$ with sonication power 30 W in the presence of (A) PEG-8000, and (B) Triton X-100. *Source*: Reproduced with permissions from Namvar, F., Abass, S. K., Soofivand, F., Salavati-Niasari, M., & Moayedi, H., (2019). Sonochemical synthesis of Pr_6MoO_{12} nanostructures as an effective photocatalyst for waste-water treatment, *Ultrasonics Sonochemistry*, *58* 104687.

et al. prepared $Nd_{5\text{-}x}Ln_xMo_3O_{16+y}$ ($y \sim 0.5$), where Ln = lanthanum, cerium, and praseodymium via solid-phase pathway. They considered the different value of x (0−5) for different rare-earth molybdates. The solid-phase approach was used to make samples from mixtures of neodymium oxide, lanthanum oxide, cerium(III) oxide, praseodymium oxide, and molybdenum trioxide. The sample weight had a total mass of 1.00 g. Initially, Nd_2O_3, La_2O_3, CeO_3, and Pr_6O_{11} were calcined for 1 hour at 1000°C, while MoO_3 was calcined for 4 hours at 500°C. The batch was then homogenized for 30 minutes with polyvinyl alcohol. Then, it was heated at 500°C in the air to stop molybdenum dioxide from sublimating and then heated for 10 and 20 hours at 800°C. After that, a 5% polyvinyl alcohol solution was put on, and the powders were pressed into tablets of 8 mm size. The synthesis was completed by calcining the tablets at 1050°C for 20 hours, after which they were triturated into powder. For relative volatility and consequently sublimation of molybdenum (VI) oxide during the preparation process, multistage calcination at 500, 800, and 1050°C was used to bind it and hinder sublimation. The findings of the elemental analysis indicated the absence of MoO_3 sublimation during the synthesis (Table 10.1). XRD analysis confirmed that only lines of solid solutions with the structure of fluorite-like were found in the X-ray patterns of $Nd_{5\text{-}x}Pr_xMo_3O_{16+y}$ (Get'man et al., 2016).

Yttrium molybdate and lanthanum molybdate were synthesized via Liang et al. through laser-assisted solid-phase route. Briefly, commercial rare-earth oxides

Table 10.1 Elemental composition of the samples, at.%.

$Nd_{4.5}La_{0.5}Mo_3O_{16+y}$			$Nd_3Pr_2Mo_3O_{16+y}$			$Nd_{4.9}Ce_{0.1}Mo_3O_{16+y}$		
Element	**Calcd.**	**Found**	**Element**	**Calcd.**	**Found**	**Element**	**Calcd.**	**Found**
Nd	18.8	18.8	Nd	12.5	11.2	Nd	20.4	19.3
La	2.1	2.1	Pr	8.3	7.6	Ce	0.4	0.3
Mo	12.5	13.3	Mo	12.5	12.5	Mo	12.5	13.7
O	66.7	65.7	O	66.7	68.7	O	66.7	65.9

Source: Reproduced with permissions from E.I. Get'man, K.A. Chebyshev, L.V. Pasechnik, L.I. Ardanova, N.I. Selikova, S.V. Radio, Isomorphous substitutions and conductivity in molybdates $Nd_{5-x}Ln_xMo_3O_{16+y}$ ($y \sim 0.5$), where Ln = La, Ce, Pr, *Journal of Alloys and Compounds* 686 (2016) 90–94.
Calcd, Calculated.

including; Y_2O_3, La_2O_3, and MoO_3 were combined in keeping with the exact ratios according to $La_2(MoO_4)_3$ and $Y_2(MoO_4)_3$ as destination products. After grounding it for 2 hours, it is pressed into certain shape pellets using a uni-axial cold press at 10 MPa. The pellets had a diameter of 18 mm and a thickness of 3 mm. They were dried in a baking oven for 2 hours at 373 K. A 5 kW CO_2 laser was used for the synthesis. The laser beam was focused on a pellet that was positioned at a distance of 12 cm with a defocus length of 12 cm. The beam spot on the specimen had a diameter of around 12 mm. The SEM analysis was applied for morphological investigation of prepared $La_2(MoO_4)_3$ and $Y_2(MoO_4)_3$ which were prepared at the power of 500 and 600 W and a scan rate of 1 mm/second. Results confirmed that both $La_2Mo_3O_{12}$ and $Y_2Mo_3O_{12}$ blocks have close shape and size that made up of nano dendrites or nanoparticles with particle sizes near 25 nm. It is found that the fine structures of $La_2Mo_3O_{12}$ and $Y_2Mo_3O_{12}$ prepared with CO_2 laser are particular for the laser-assisted methods (Liang, Huo, Wang, Chao, & Wang, 2009).

10.2.4 Hydrothermal method

Hydrothermal processing is a simple method for the synthesis of nanostructured ceramics (Beale et al., 2009; Kokulnathan et al., 2019; Kulkarni et al., 2020; Li, Li, & Wang, 2009; Martins, Coelho, Moreira, & Dias, 2018; Martins, Moreira, & Dias, 2020). In a brief, the hydrothermal synthesis process relies on almost all inorganic substances' solubility in water at high temperatures and pressures, followed by crystallization of the dissolved material from the solution (Huang, Lu, & Yang, 2019; Huo, 2011; Kumar & Nanda, 2019). Yan et al. prepared lanthanum molybdate, and lanthanum tungstate with different rare-earth ions (ytterbium/erbium and ytterbium/Thulium) by following process (Zhou, He, & Yan, 2014):

1. According to the molar ratio of La^{3+}:Yb^{3+}:Tm^{3+} = 77:20:3, 0.161 g of Gd_2O_3, 0.0473 g of Yb_2O_3, and 0.0069 g of Er_2O_3 were dissolved in dilute nitric acid with the molar ratio of Thulium:ytterbium:lanthanum = 3:20:77. For extraction of the residual HNO_3, the solution was evaporated to dryness, so the as-obtained residue solid was dissolved in 5 mL DI water with magnetic stirring.
2. 0.594 g of $Na_2MoO_40.2H_2O$ was dissolved in DI water via heating and then a sufficient amount of cetyltrimethylammonium bromide was solubilized and stirred for several minutes at 35°C. It should be noted that the ratio of rare-earth:molybdates was kept in 2:3. Then, under intense magnetic stirring, solution I was dropped into solution II drop by drop to observe white precipitates. The pH of the mixture was then changed to 8−9 by adding NaOH solution to it, and the as-obtained starting materials suspension was stirred for a time for establishing that all starting materials were uniformly distributed. The suspension was then considered for hydrothermal process in the autoclave at 180°C for 1−72 hours. The collected solids being allowed to cool to the temperature of the room. The products were centrifuged and washed several times with DI water and anhydrous ethanol. The precipitate was then collected and dried for 24 hours at 80°C.

They characterized the samples via X-ray powder diffraction, SEM, TEM, and EDS spectroscopy. The X-ray diffraction patterns were applied for the investigation of crystalline characteristics of the as-prepared products. The diffraction peaks of the rare-earth molybdate were in full accordance to the tetragonal-phase (I41/a) structure (JCPDS 45−0407). In addition, according to the Scherrer equation, $Dc = K\lambda/\beta Cos\theta$ observed broad peaks in X-ray diffraction patterns led to the nano-scale dimension of prepared rare-earth molybdate ceramic nanomaterials. Obtained information about scanning electron microscope, transmission electron microscope, and HRTEM images, selected area diffraction pattern, and energy-dispersive spectroscopy of ytterbium/erbium codoped yttrium molybdate ceramic nanomaterials are shown in Fig. 10.8. The SEM images (Fig. 10.8A) show an illustrative overview of the ytterbium/erbium codoped yttrium molybdate ceramic nanomaterials at low magnification. The sample is made up of a very homogenous pompon-like morphology. As shown in Fig. 10.8B, a close examination of an individual pompon-like structure reveals that its size is near the 2−3.5 micrometer and offers more details about its surface. Individual microstructures consist of various two-dimensional anisotropic nano flakes that observable from the edges of the pompon-like shape. The transmission electron microscope results were in good agreement with SEM images (Fig. 10.8C and D). Clear lattice fringes can be seen in the HRTEM images (Fig. 10.8E), and the d-spacing is 0.287 nanometer, which attributes to the (040) plane of ytterbium/erbium codoped yttrium molybdate ceramic nanomaterials. The single crystalline existence of the sample is indicated by the selected area diffraction pattern, which is carried out on a nano flake on the edge of the pompon-like shape. The presence of yttrium, molybdenum, ytterbium, and oxygen elements were confirmed via the EDS spectrum shown in Fig. 10.8G.

Xicheng et al. prepared Pr-doped $LiY_{1-x}Pr_x(MoO_4)_2$ which x was considered from 0.005 to 0.025. The hydrothermal method was applied as a preparation route. To make yttrium nitrate and praseodymium nitrate mixture, a sufficient concentration of $Pr(NO_3)_3$ and $Y(NO_3)_3$ were provided in nitric acid solution. To obtain yttrium nitrate solution (0.1 M) and praseodymium nitrate solution (0.5 M), the solutions were cooled to room temperature and deionized water was applied to each solution. In a 50 mL beaker, specific quantities of praseodymium nitrate and yttrium nitrate solutions were combined with stoichiometric ratio-based Pr-doped $LiY_{1-x}Pr_x(MoO_4)_2$. After that, solution I was prepared by magnetic stirring at room temperature for several minutes, and solution II was produced by dissolving the proper amount of sodium molybdate in DI water. Then, solution II was dropped into mixture I, until the mixed solution reached a pH of 6. The mixture was moved into a provided autoclave after 30 minutes of magnetic stirring. The autoclave was placed under hydrothermal condition for 24 hours at 180°C. The resulting solid was filtered before being dried at sufficient time and temperature, yielding the final sample. The X-ray diffraction patterns of praseodymium ions doped lithium yttrium molybdate ceramic nanomaterials with various dopant concentrations are shown in Fig. 10.9. The diffraction patterns for samples with different dopant concentrations display similar peaks, suggesting that doping does not affect the matrix's crystalline structure. All of the products have identical diffraction peaks, as shown in Fig. 10.9, and the peaks of $LiY_{1-x}Pr_x(MoO_4)_2$ ($x = 0.005-0.025$) can be indexed to scheelite-like form of lithium yttrium molybdate ceramic (JCPDS 17−0773). Scanning electron microscope images of Pr-doped $LiY_{1-x}Pr_x(MoO_4)_2$ with x value

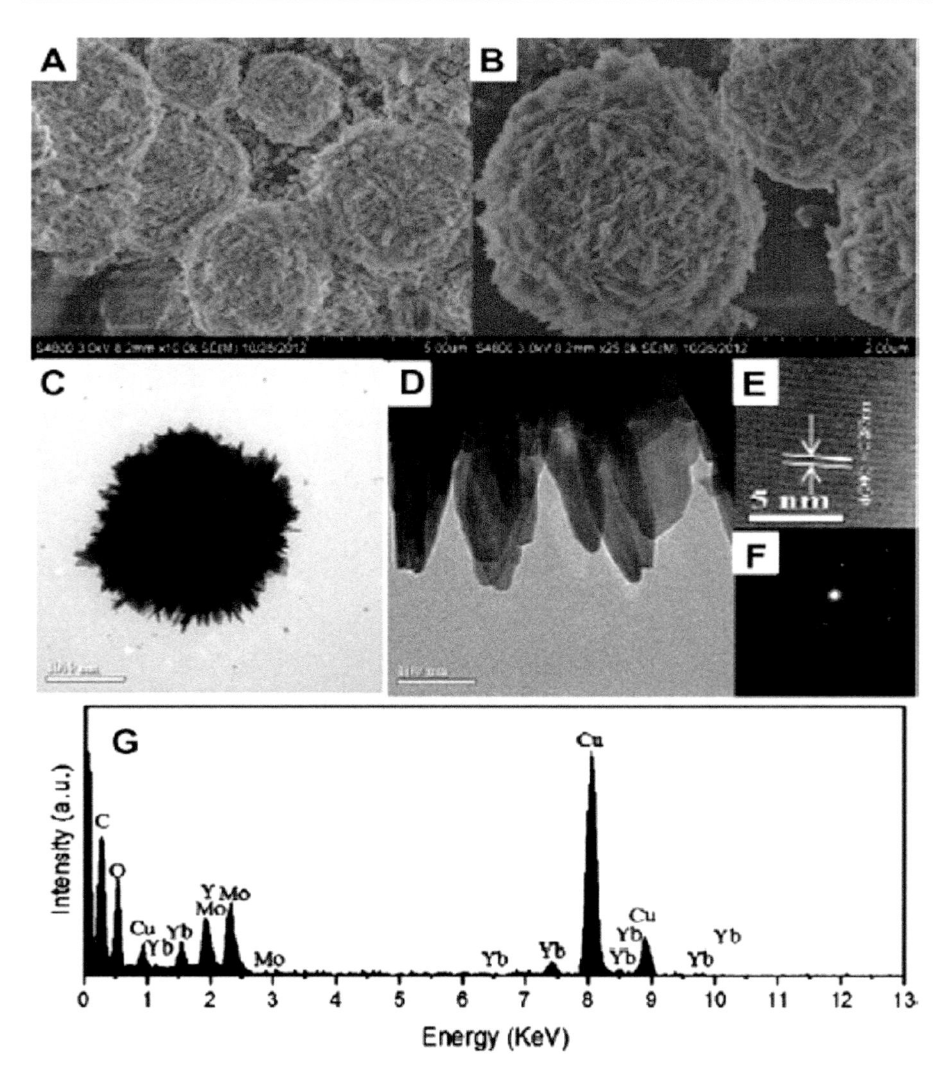

Figure 10.8 (A and B) Scanning electron microscopy (SEM), (C and D) Transmission electron microscopy (TEM), (E) HRTEM images, (F) SAED pattern and (G) energy-dispersive X-ray spectroscopy (EDS) spectrum of $Y_2(MoO_4)_3$:Yb^{3+}/Er^{3+} sample: (A, B) SEM images with different magnifications, (C) TEM image of an individual architecture, (D) TEM image of the fringe of an individual pumpon-like structure, (E) HRTEM image, (F) SAED pattern, (G) EDS spectrum.
Source: Reproduced with permissions from Zhou, Y., He, X.-H., & Yan, B., (2014). Self-assembled $RE_2(MO_4)_3$:Ln^{3+} (RE = Y, La, Gd, Lu; M = W, Mo; Ln = Yb/Er, Yb/Tm) hierarchical microcrystals: Hydrothermal synthesis and up-conversion luminescence, *Optical Materials*, *36* 602−607.
SAED, Selected area electron diffraction.

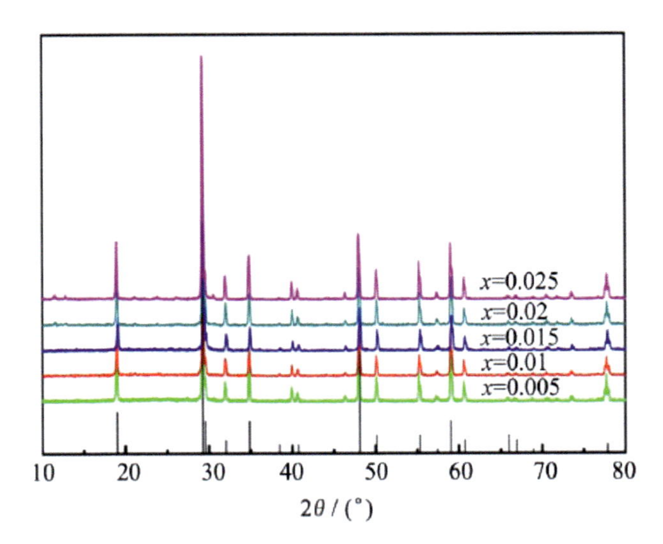

Figure 10.9 X-ray diffraction patterns of Pr^{3+} doped products of $LiY_{1-x}Pr_x(MoO_4)_2$ phosphor (x = 0.005–0.025).
Source: Reproduced with permissions from Li, Z., Zhao, X., & Jiang, Y., (2015). Hydrothermal preparation and photoluminescent property of $LiY(MoO_4)_2$:Pr^{3+} red phosphors for white light-emitting diodes, *Journal of Rare Earths*, *33* 33–36.

of 0.015 showed a lamellar structure with a smooth surface and particle size of approximately 3 mm. Investigations in the mentioned study revealed that Pr-doped $LiY_{1-x}Pr_x(MoO_4)_2$ ceramic (x = 0.015) with micro-lamellar shape could be produced using a hydrothermal process at high temperatures and pressures (Li, Zhao, & Jiang, 2015).

10.3 Applications methods of rare-earth molybdates ceramic nanomaterials

10.3.1 Electrocatalyst

Electrocatalysts are a form of catalyst that works at electrode surfaces or on the electrode itself to increase the rate of oxidation and reduction reactions in electrochemical reactions. Electrocatalysts can speed up the transfer of electrons between electrodes and reactants, as well as promote a chemical transformation that is characterized by an overall half-life (Nguyen, Kim, Hwang, & Won, 2020; Zeng and Li, 2015; Zhao et al., 2020; Zinola, Martins, Tejera, & Neves, 2012).

Nanostructures have been widely used in the design and preparation of electrodes of electrochemical sensor in recent decades to improve the S/N ratio and active surface (Maiti, Kim, & Lee, 2021; Mu & Wang, 2019; Nadar et al., 2020; Nejati, Davari, Akbari, Asadpour-Zeynali, & Rezvani, 2019). Rare-earth-based nanostructures have been employed for a variety of purposes, including fuel cells, sensors, electronics, heterogeneous catalysis, and the glass industry (Li, Wang, Lin, & Huang, 2021; Wang, Ma et al., 2021; Wang, Yu, Wei, Liu, & Zhao, 2021; Zhang et al., 2021). Rare-earth-based molybdates are a fascinating group of rare-earth-based materials with unique properties that make them suitable for use in the electrocatalytic process (Karthik et al., 2018; Karuppaiah, Ramachandran, Chen, Wan-Ling, & Wan, 2020; Vinoth, Govindasamy, Wang, Alothman, & Alshgari, 2021).

Chung et al. prepared Er_2MoO_6 nanoparticles via sonochemical route and subsequently prepared erbium molybdate/functionalized carbon black (f-CNF) nanocomposites. They investigated the morphological and structural properties of products comprehensively. After that, they designed the prepared erbium molybdate/functionalized carbon black-modified-printed electrode for the detection of phenothiazine (PTZ). Differential pulse voltammetry (DPV) is a common method for determining the sensitivity and selectivity of various modified electrodes used in electrochemical sensors and biosensors. The DPV tests were carried out at erbium molybdate/functionalized carbon black-modified-printed electrode with the addition of PTZ in 0.1 mol/L phosphate buffer for various concentrations (0.025–80 μM). The DPV curves display two oxidation curves at 0.375 and 0.81 V, as shown in Fig. 10.10A. With each successive addition of PTZ, the DPV response increased linearly. Fig. 10.10B shows the related calibration plot, which shows a linear correlation between peak currents and PTZ different amounts (0.025–80 μM). Ipa (μA) = 0.259 (μM) + 1.276 was found to be the linear regression equation, with $R^2 = 0.992$ as the regression coefficient. The sensitivity and limit of detection (LOD) were 3.707 $\mu A/\mu M/cm^2$ and 0.008 μM respectively. Furthermore, the designed sensor was found to be stable, selective, repeatable, and reusable (Liu, He, Sakthivel, & Chung, 2020).

10.3.2 Photocatalyst

Photocatalysts are materials that decompose harmful compounds in the presence of UV rays from the sun. In other words, photocatalysts absorb photons of a specific wavelength and are excited, forming an electron-hole pair on the catalyst's surface which lead to formation of hydroxyl radicals and subsequently degraded pollutants. Depending on the scope and specifications, photocatalysts are used as powders or thin films (Castillejos, Rodríguez-Ramos, & Guerrero-Ruiz, 2012; Kiwi & Rtimi, 2016; Liu & Chen, 2017; Ohtani, 2011; Ross, 2019). Due to the higher specific surface area, nanoparticles in powder form have been extensively studied as effective photocatalysts (Abdolmohammad-Zadeh & Zamani-Kalajahi, 2020; Bose, Barman, & Chakraborty, 2020; Cheng, Liu, Wang, Yang, & Ye, 2019; Ray & Hur, 2020; Wang, Guo, Yang, Chai, & Zhu, 2017; Yang et al., 2020). However, due to quick recombination losses and insufficient use of the solar spectrum, using nanoparticles for pollutant degradation has limitations. In recent times, rare-earth molybdates have been found more attention in the field of photocatalyst. This can be related to the unique optical properties of rare-earth molybdates (Fei & Gong, 2012; Karthik et al., 2017).

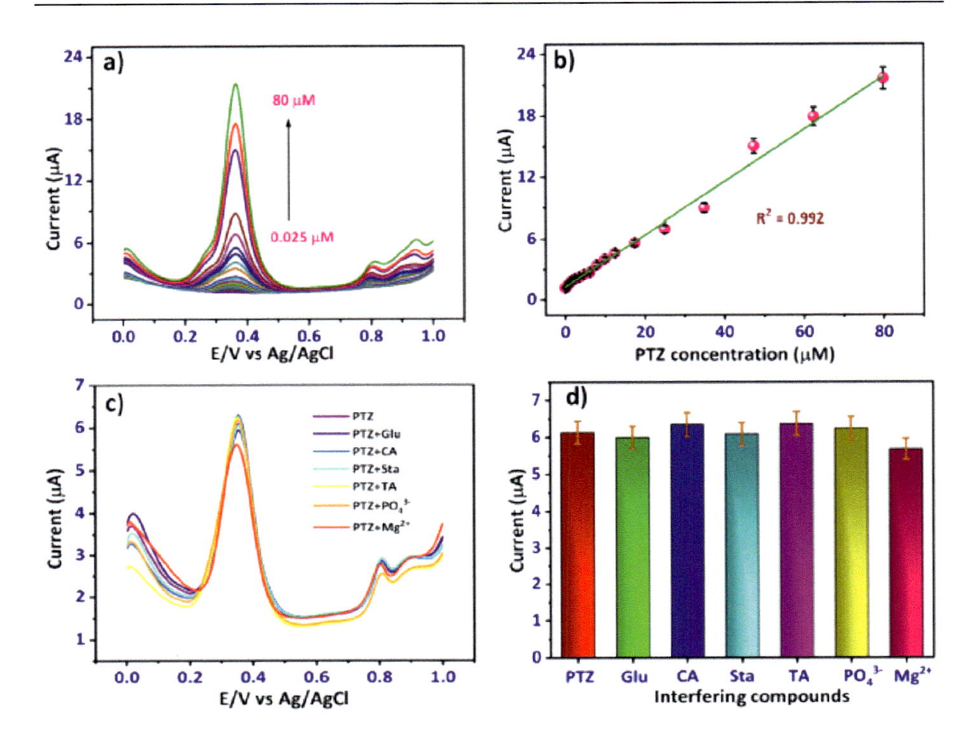

Figure 10.10 DPV curves of phenothiazine (PTZ) for the addition of various concentrations (0.025–80 μM) of f-CNF/Er_2MoO_6/SPE in 0.1 M phosphate buffer (pH = 5.0) (A). The calibration plot for the PTZ concentration versus anodic peak current (B). DPV curves of f-CNF/Er_2MoO_6/SPE in 0.1 M phosphate buffer (pH = 5) containing PTZ in the presence of inorganic substances and interfering compounds (C). The bar diagram for the interfering compounds with their current changes (D).
Source: Reproduced with permissions from Liu, X., He, J.-H., Sakthivel, R., & Chung, R.-J., (2020). Rare-earth erbium molybdate nanoflakes decorated functionalized carbon nanofibers: An affordable and potential catalytic platform for the electrooxidation of phenothiazine, *Electrochimica Acta*, *358* 136885.
DPV, Differential pulse voltammetry.

Niasari et al. prepared Pr_6MoO_{12} nanostructures via a fast and facile ultrasonic-assisted route. The photocatalytic degradation of prepared Pr_6MoO_{12} against acid red 92 and methylene blue under ultraviolet irradiation was investigated. The different scavengers for the determination of photocatalytic mechanism were applied. It is found that methylene blue degradation percent in the presence of benzoic acid, ethylenediaminetetraacetic acid, and argon gas were reduced to 67.04, 57.82, and 6.81, respectively. Argon gas, ethylenediaminetetraacetic acid, benzoic acid, and argon gas were considered as scavengers of superoxide anion radicals ($°O^{2-}$), holes (h^+), and hydroxyl radical (°OH), respectively. These findings suggested that $°O^{2-}$ were the most active photocatalyst for methylene blue degradation via of Pr_6MoO_{12}. Also, the photodegradation efficiency in the presence of benzoic acid, ethylenediaminetetraacetic acid, and argon as scavengers was determined to be about 62.98, 60.30, and 12.55, respectively, for acid blue 92 (Namvar et al., 2019).

It should be noted that in many types of research different rare-earth have been doped in molybdate-based ceramic nanomaterials to improve photocatalytic activity. Wohn Lee et al. prepared Er^{3+}/Yb^{3+} codoped Bi_2MoO_6 via microwave-hydrothermal route. The purity, crystallinity, optical features, shape and size, and specific surface area were all investigated in-depth. The infrared to visible upconversion luminescence of codoped products was studied under excitation at 532, 546, and 980 nm. The codoping of erbium/ytterbium ions into bismuth molybdate resulted in enhanced photodegradation performance for the removal of rhodamine B under visible light irradiation, according to the findings. The energy transfer between erbium/ytterbium ions and bismuth molybdate through IR to visible upconversion from erbium/ytterbium ions and size-dependent excellent specific surface area of the two-dimensional morphology of bismuth molybdate can be attributed to the reason for increased photocatalytic activity (Adhikari et al., 2014). The mechanism of the photocatalytic process of erbium/ytterbium ions codoped bismuth molybdate was provided in Fig. 10.11.

In another study, Hosseinpour-Mashkani et al. prepared cerium molybdate ceramic nanomaterials via the sonochemical method. A scanning electron microscope was used to examine the effects of experimental conditions such as the amount of glucose and power of ultrasonic irradiation on the shape and size of the as-obtained product. After 5 hours of UV light irradiation in the presence of cerium molybdate nanostructures, the best photocatalytic efficiency was measured 89% against methyl orange (MO) (Fig. 10.12A). Fig. 10.12B demonstrates spectrofluorimetric time-scans of MO solution irradiated at 450 nm with cerium molybdate

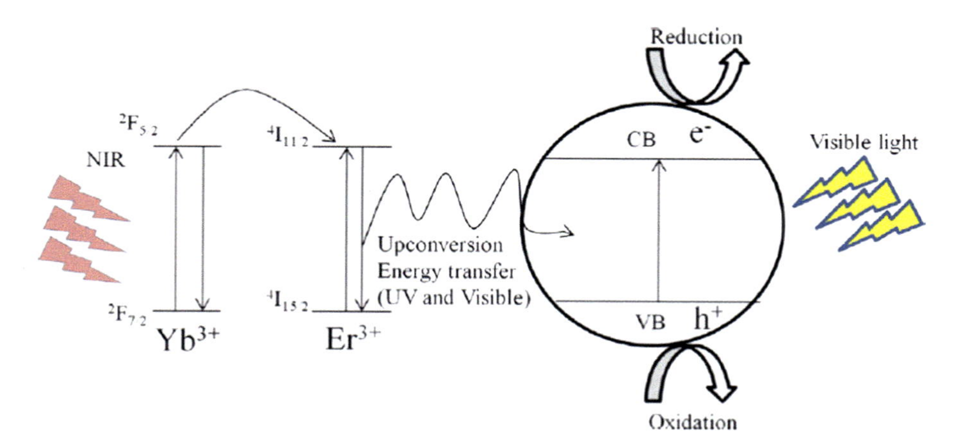

Figure 10.11 Schematic illustration of the upconversion photocatalysis.
Source: Reproduced with permissions from Adhikari, R., Gyawali, G., Cho, S. H., Narro-García, R., Sekino, T., & Lee, S.W. (2014). Er^{3+}/Yb^{3+} codoped bismuth molybdate nanosheets upconversion photocatalyst with enhanced photocatalytic activity, *Journal of Solid State Chemistry 209* 74–81.

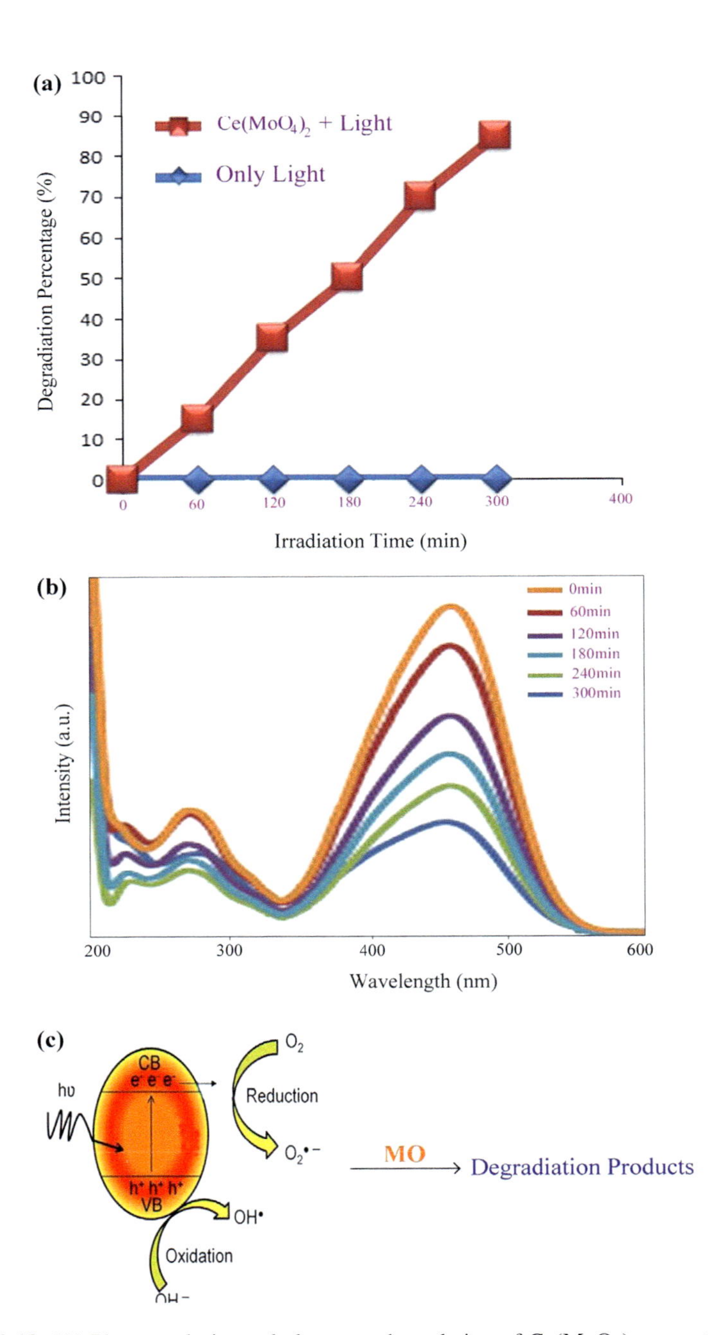

Figure 10.12 (A) Photocatalytic methyl orange degradation of $Ce(MoO_4)_2$ nanostructures under visible light; (B) fluorescence spectral time scan of methyl orange illuminated at 450 nm with $Ce(MoO_4)_2$ nanostructures; (C) reaction mechanism of methyl orange photodegradation over $Ce(MoO_4)_2$ under visible light irradiation.
Source: Reproduced with permissions from Sobhani-Nasab, A., Maddahfar, M., & Hosseinpour-Mashkani, S. M., (2016). $Ce(MoO_4)_2$ nanostructures: Synthesis, characterization, and its photocatalyst application through the ultrasonic method, *Journal of Molecular Liquids*, *216* 1–5.

ceramic nanomaterial, demonstrating continuous MO removal on cerium molybdate under visible light irradiation. Fig. 10.12C shows the mechanism of the photocatalytic process. The following is the exact mechanism of photodegradation of MO via cerium molybdate ceramic nanomaterial (Sobhani-Nasab, Maddahfar, & Hosseinpour-Mashkani, 2016):

$$Ce(MoO_4)_2 + h\upsilon \rightarrow Ce(MoO_4)_{2^\circ} + e^- + h^+$$
$$h^+ + H_2O \rightarrow OH$$
$$e^- + O_2 \rightarrow O_2^{-\circ}$$
$$OH_{UNDEFINEDring;} + O^-_{2^\circ} + MO \rightarrow H_2O + CO_2 + \text{degraded product}$$

10.3.3 Light-emitting diodes

The solid-state reaction technique was used via Xianghong et al. to synthesize europium ion-activated gadolinium molybdate stable phosphors with pseudo-pompon form. Its excitation band ranged from 348 to 425 nm, making it compatible with near-UV light-emitting diode (LED) chips' emission band. The $^7F0 \rightarrow {}^5L6$ transformation of Eu^{3+} was excited at 395 nm to produce pure and extreme red emission. Near-UV light could effectively excite the phosphors, and they emitted more red light than the commercial $Y_2O_2S:Eu^{3+}$ used in pc-white-LEDs. $Gd_2(MoO_4)_3:Eu^{3+}$ had CIE chromaticity coordinates that were similar to NTSC standard values to $Y_2O_2S:Eu^{3+}$. Combining this phosphor with a 395 nm InGaN chip resulted in the successful fabrication of an extreme red-emitting LED. Due to its broadened excitation band in the near-ultraviolet, smaller particle size, higher specific surface area, and intense red-emission with sufficient CIE chromaticity coordinates, the composition-optimized $Gd_2(MoO_4)_3:0.40Eu^{3+}$ phosphor could be a attractive option for solid-state lighting use (He et al., 2010).

10.3.4 Biosensor

Since the low-cost, rapid, fast, highly sensitive, and selective contribution of the biosensor to advances in novel medicines such as ultrasensitive point-of-care detection of disease markers, and individualized medicine, biosensor-based development, and research is becoming the most widely investigated field (Bouzigues, Gacoin, & Alexandrou, 2011; Esimbekova, Kalyabina, Kopylova, Torgashina, & Kratasyuk, 2021; Jones-Tabah, Mohammad, Clarke, & Hébert, 2021; Kotsiri, Vidic, & Vantarakis, 2022; Liu et al., 2021; Tian et al., 2021; Tran, Tran, Hwang, & Chang, 2021; Zhang et al., 2021). The rare-earth molybdate nanostructures can be applied in the field of biosensors interestingly (Li et al., 2020; Vinoth Kumar et al., 2018).

Chen et al. prepared Dy_2MoO_6 with the shape of flake-like ceramic nonmaterial and applied it for detection of the antibiotic drug metronidazole. To produce the electrode (Dy_2MoO_6/glassy carbon electrode), this nanostructure was applied to the surface of a glassy carbon electrode. The electrochemical properties were studied using cyclic and DPV techniques. The Dy_2MoO_6/glassy carbon electrode-based sensor had excellent selectivity and sensitivity for detecting the drug metronidazole, which could be due to Dy_2MoO_6-related strong affinity for the $-NO_2$ group in drug metronidazole, as well as Dy_2MoO_6-dependent good electrocatalytic activity and conductivity. The designed Dy_2MoO_6/glassy carbon electrode sensor had a better electrocatalytic activity response for drug metronidazole detection than untreated glassy carbon electrode and other $Re_2(MoO_4)_3$. In the dosage range of 0.01–2363 mm, the sensor showed a strong linear relationship. The detection and quantification limits, respectively, were measured to be 0.0030 and 0.010 mol/L (Karthik et al., 2019).

10.4 Conclusion and outlook

Today, molybdate-based ceramic nanomaterials are a widely used material in both academia and industry. For their unique physical and chemical properties, scientists pay more attention to these nanostructures. In recent years, rare-earth molybdate ceramic nanomaterials have found more application because they have exclusive, advantageous optical, magnetic, thermal, and mechanical properties. The applied procedure for the preparation of rare-earth molybdate ceramic nanomaterials plays a key role in the properties of obtained final products. In another word, the used preparation method intensively affects the physical and chemical properties of the rare-earth molybdate ceramic nanomaterials. Various methods have been widely applied for the preparation of these nanomaterials including; coprecipitation, sonochemical, hydrothermal, and solid-phase rout. For morphological and structural engineering, the experimental conditions such as the type of salts used, the reaction temperature, concentration, stirring speed, used capping agents, sonication power, the procedure time, and the pH value have all been controlled. X-ray diffraction (XRD), transmission electron microscopy (TEM), scanning electron microscopy (SEM), energy-dispersive X-ray spectroscopy (EDS) and UV-Vis, and photoluminescence (PL) analysis have been applied for the determination of chemical and physical properties of these nanomaterials. There are several important applications of rare-earth molybdate ceramic nanomaterials such as electrocatalyst, photocatalyst, LEDs, magnetic and magnetoelectric devices, and biosensor.

References

Abdolmohammad-Zadeh, H., & Zamani-Kalajahi, M. (2020). In situ generation of H_2O_2 by a layered double hydroxide as a visible light nano-photocatalyst: Application to bisphenol A quantification. *Microchemical Journal*, *158*, 105303.

Adhikari, R., Gyawali, G., Cho, S. H., Narro-García, R., Sekino, T., & Lee, S. W. (2014). Er^{3+}/Yb^{3+}co-doped bismuth molybdate nanosheets upconversion photocatalyst with enhanced photocatalytic activity. *Journal of Solid State Chemistry*, *209*, 74–81.

Ahmadpor, H., & Milani-Hosseini, S. M.-R. (2021). A comparative study of different techniques for quantum dot nano-particles immobilization on glass surface for fabrication of solid phase fluorescence opto-sensors. *Surfaces and Interfaces*, *23*, 100907.

Ansari, A., Shukla, A. K., & Ansari, M. A. (2021). Preparation, characterization and electrical conductance studies of inorganic precipitate parchment supported barium molybdate membrane. *Materials Today: Proceedings*.

Asha, A. B., & Narain, R. (2020). Chapter 15 - Nanomaterials properties. In R. Narain (Ed.), *Polymer science and nanotechnology* (pp. 343–359). Elsevier.

Ashik, U. P. M., Kudo, S., & Hayashi, J.-I. (2018). Chapter 2 - An overview of metal oxide nanostructures. In S. Mohan Bhagyaraj, O. S. Oluwafemi, N. Kalarikkal, & S. Thomas (Eds.), *Synthesis of inorganic nanomaterials* (pp. 19–57). Woodhead Publishing.

Ayni, S., Sabet, M., & Salavati-Niasari, M. (2016). Synthesis and characterization of lead molybdate nanostructures with high photocatalytic activity via simple co-precipitation method. *Journal of Cluster Science*, *27*, 315–326.

Bazarova, J. G., Tushinova, Y. L., Bazarov, B. G., & Dorzhieva, S. G. (2017). Double molybdates of rare earth elements and zirconium. *Russian Chemical Bulletin*, *66*, 587–592.

Beale, A. M., Jacques, S. D. M., Sacaliuc-Parvalescu, E., O'Brien, M. G., Barnes, P., & Weckhuysen, B. M. (2009). An iron molybdate catalyst for methanol to formaldehyde conversion prepared by a hydrothermal method and its characterization. *Applied Catalysis A: General*, *363*, 143–152.

Benelmekki, M., Singh, V., Baughman, K. W., Bohra, M., & Kim, J. H. (2021). On the potentiality of a cluster-beam source to produce free-standing one-dimensional and two-dimensional FeAg nanostructures: mechanisms underlying their shape genesis. *Materials Today Chemistry*, *19*, 100405.

Bharat, L. K., Raju, G. S. R., & Yu, J. S. (2017). Red and green colors emitting spherical-shaped calcium molybdate nanophosphors for enhanced latent fingerprint detection. *Scientific Reports*, *7*, 11571.

Bispo-Jr, A. G., Shinohara, G. M. M., Pires, A. M., & Cardoso, C. X. (2018). Red phosphor based on Eu^{3+}-doped Y2(MoO4)3 incorporated with Au NPs synthesized via Pechini's method. *Optical Materials*, *84*, 137–145.

Borchardt, H. J., & Bierstedt, P. E. (1967). Ferroelectric Rare-Earth Molybdates. *Journal of Applied Physics*, *38*, 2057–2060.

Bose, D., Barman, S., & Chakraborty, R. (2020). Sustainable development of inexpensive visible-range $CuOTiO_2$ nano-photocatalysts deploying in situ recovered glass fiber and $Cu(CH_3COO)_2$ from waste printed wiring board: Optimal lignin photo-degradation for valuable products. *Sustainable Materials and Technologies*, *24*, e00162.

Bouzigues, C., Gacoin, T., & Alexandrou, A. (2011). Biological applications of rare-earth based nanoparticles. *ACS Nano*, *5*, 8488–8505.

Brixner, L. H., Barkley, J. R., & Jeitschko, W. (1979). *Chapter 30 Rare earth molybdates (VI). HPCRE* (pp. 609–654). Elsevier.

Castillejos, E., Rodríguez-Ramos, I., & Guerrero-Ruiz, A. (2012). Chapter 16-Catalytic removal of water-solved aromatic compounds by carbon-based materials. In J. M. D. Tascón (Ed.), *Novel carbon adsorbents* (pp. 499–520). Oxford: Elsevier.

Chen, C., & Suganuma, K. (2019). Microstructure and mechanical properties of sintered Ag particles with flake and spherical shape from nano to micro size. *Materials & Design*, *162*, 311–321.

Chen, Z., Bu, W., Zhang, N., & Shi, J. (2008). Controlled construction of monodisperse $La_2(MoO_4)_3$:Yb,Tm microarchitectures with upconversion luminescent property. *The Journal of Physical Chemistry C*, *112*, 4378–4383.

Cheng, L., Liu, L., Wang, D., Yang, F., & Ye, J. (2019). Synthesis of bismuth molybdate photocatalysts for CO_2 photo-reduction. *Journal of CO_2 Utilization*, *29*, 196–204.

Cheng, X. (2014). 10 - Nanostructures: Fabrication and applications. In M. Feldman (Ed.), *Nanolithography* (pp. 348–375). Woodhead Publishing.

Dalafu, H. A., Rosa, N., James, D., Asuigui, D. R. C., McNamara, M., Kawashima, A., ... Stoll, S. L. (2018). Solid-state and nanoparticle synthesis of EuSxSe1−x solid solutions. *Chemistry of Materials: a Publication of the American Chemical Society*, *30*, 2954–2964.

Daniels-Race, T. (2014). 12 - Nanodevices: Fabrication, prospects for low dimensional devices and applications. In M. Feldman (Ed.), *Nanolithography* (pp. 399–423). Woodhead Publishing.

Daulbayev, C., Sultanov, F., Bakbolat, B., & Daulbayev, O. (2020). 0D, 1D and 2D nanomaterials for visible photoelectrochemical water splitting. A review. *International Journal of Hydrogen Energy*, *45*, 33325–33342.

de Oliveira, P. F. M., Torresi, R. M., Emmerling, F., & Camargo, P. H. C. (2020). Challenges and opportunities in the bottom-up mechanochemical synthesis of noble metal nanoparticles. *Journal of Materials Chemistry A*, *8*, 16114–16141.

Ding, H., Ma, J., Yue, F., Gao, P., & Jia, X. (2019). Size and morphology dependent gas-sensing selectivity towards acetone vapor based on controlled hematite nano/microstructure (0D to 3D). *Journal of Solid State Chemistry*, *276*, 30–36.

Dong, S., Li, L., Ashour, A., Dong, X., & Han, B. (2021). Self-assembled 0D/2D nano carbon materials engineered smart and multifunctional cement-based composites. *Construction and Building Materials*, *272*, 121632.

Dridi, W., Zid, M. F., & Maczka, M. (2018). Characterization of a sodium molybdate compound β-$Na_4Cu(MoO_4)_3$. *Journal of Alloys and Compounds*, *731*, 955–963.

Esimbekova, E. N., Kalyabina, V. P., Kopylova, K. V., Torgashina, I. G., & Kratasyuk, V. A. (2021). Design of bioluminescent biosensors for assessing contamination of complex matrices. *Talanta*, 122509.

Fei, Z. Y., & Gong, J. H. (2012). Preparation of lanthanum molybdate powers by sol-gel. *Advanced Materials Research*, *412*, 121–124.

Ferreira, E. A. C., Andrade Neto, N. F., Bomio, M. R. D., & Motta, F. V. (2019). Influence of solution pH on forming silver molybdates obtained by sonochemical method and its application for methylene blue degradation. *Ceramics International*, *45*, 11448–11456.

Forbes, S., Kong, T., & Cava, R. J. (2018). $RE_3Mo_{14}O_{30}$ and $RE_2Mo_9O_{19}$, two reduced rare-earth molybdates with honeycomb-related structures (RE = La–Pr). *Inorganic Chemistry*, *57*, 3873–3882.

Freed, S., & Haenisch, E. L. (1937). The absorption spectra of samarium tungstate and of samarium molybdate at low temperatures. *The Journal of Chemical Physics*, *5*, 26–29.

Fukui, S., Wang, Z., & Kato, M. (2021). Dependences of local density of state on temperature, size, and shape in two-dimensional nano-structured superconductors. *Physica C: Superconductivity and its Applications*, *580*, 1353785.

Gajraj, V., & Mariappan, C. R. (2020). Electrochemical performances of asymmetric aqueous supercapacitor based on porous Cu3Mo2O9 petals and $La_2Mo_3O_{12}$ nanoparticles fabricated through a simple co-precipitation method. *Applied Surface Science*, *512*, 145648.

Garnett, E., Mai, L., & Yang, P. (2019). Introduction: 1D nanomaterials/nanowires. *Chemical Reviews*, *119*, 8955–8957.

Gerberich, W. W., Jungk, J. M., & Mook, W. M. (2003). The bottom-up approach to materials by design. In M. A. Meyers, R. O. Ritchie, & M. Sarikaya (Eds.), *Nano and microstructural design of advanced materials* (pp. 211–220). Oxford: Elsevier Science Ltd.

Get'man, E. I., Chebyshev, K. A., Pasechnik, L. V., Ardanova, L. I., Selikova, N. I., & Radio, S. V. (2016). Isomorphous substitutions and conductivity in molybdates $Nd_{5-x}Ln_xMo_3O_{16+y}$ ($y \sim 0.5$), where Ln = La, Ce, Pr. *Journal of Alloys and Compounds*, *686*, 90–94.
Goldberg, S., & Suarez, D. L. (2006). Chapter 18 - Prediction of anion adsorption and transport in soil systems using the constant capacitance model. In J. Lützenkirchen (Ed.), *Interface science and technology* (pp. 491–517). Elsevier.
Gonzalez-Rojas, G. P., Bonifacio-Martinez, J., Ordoñez-Regil, E., & Fernandez-Valverde, S. M. (1998). Synthesis of $La_2(MoO_4)_3$: electrocatalytic activity for the oxygen evolution in alkaline media. *International Journal of Hydrogen Energy*, *23*, 999–1003.
Goswami, L., Kim, K.-H., Deep, A., Das, P., Bhattacharya, S. S., Kumar, S., & Adelodun, A. A. (2017). Engineered nano particles: Nature, behavior, and effect on the environment. *Journal of Environmental Management*, *196*, 297–315.
Goyal, M., & Singh, M. (2020). Size and shape dependence of optical properties of nanostructures. *Applied Physics A*, *126*, 176.
He, X., Guan, M., Li, Z., Shang, T., Lian, N., & Zhou, Q. (2010). Luminescent properties and application of Eu^{3+}-activated $Gd_2(MoO_4)_3$ red-emitting phosphor with pseudo-pompon shape for solid-state lighting. *Journal of Rare Earths*, *28*, 878–882.
Hitchcock, F. R. M. (1895). The tungstates and molybdates of the rare earths. *Journal of the American Chemical Society*, *17*, 483–494.
Huang, G., Lu, C.-H., & Yang, H.-H. (2019). Chapter 3 - Magnetic nanomaterials for magnetic bioanalysis. In X. Wang, & X. Chen (Eds.), *Novel nanomaterials for biomedical, environmental and energy applications* (pp. 89–109). Elsevier.
Huo, Q. (2011). Chapter 16 - Synthetic chemistry of the inorganic ordered porous materials. In R. Xu, W. Pang, & Q. Huo (Eds.), *Modern inorganic synthetic chemistry* (pp. 339–373). Amsterdam: Elsevier.
Jeseentharani, V., Dayalan, A., & Nagaraja, K. S. (2018). Nanocrystalline composites of transition metal molybdate ($Ni_{1-x}Co_xMoO_4$; x = 0, 0.3, 0.5, 0.7, 1) synthesized by a co-precipitation method as humidity sensors and their photoluminescence properties. *Journal of Physics and Chemistry of Solids*, *115*, 75–83.
Jones-Tabah, J., Mohammad, H., Clarke, P. B. S., & Hébert, T. E. (2021). In vivo detection of GPCR-dependent signaling using fiber photometry and FRET-based biosensors. *Methods (San Diego, Calif.)*.
Joukoff, B., Grimouille, G., Leroux, G., Daguet, C., & Pougnet, A. M. (1979). Crystal growth and crystallographic data of some $Ln_2(MoO_4)_3$ type mixed rare earth molybdates. *Journal of Crystal Growth*, *46*, 445–450.
Kaczmarek, A. M., & Van Deun, R. (2013). Rare earth tungstate and molybdate compounds – from 0D to 3D architectures. *Chemical Society Reviews*, *42*, 8835–8848.
Karthik, R., Kumar, J. V., Chen, S.-M., Karuppiah, C., Cheng, Y.-H., & Muthuraj, V. (2017). A study of electrocatalytic and photocatalytic activity of cerium molybdate nanocubes decorated graphene oxide for the sensing and degradation of antibiotic drug chloramphenicol. *ACS Applied Materials & Interfaces*, *9*, 6547–6559.
Karthik, R., Kumar, J. V., Chen, S.-M., Kokulnathan, T., Yang, H.-Y., & Muthuraj, V. (2018). Design of novel ytterbium molybdate nanoflakes anchored carbon nanofibers: Challenging sustainable catalyst for the detection and degradation of assassination weapon (Paraoxon-Ethyl). *ACS Sustainable Chemistry & Engineering*, *6*, 8615–8630.
Karthik, R., Mutharani, B., Chen, S.-M., Vinoth Kumar, J., Abinaya, M., Chen, T.-W., ... Hao, Q. (2019). Synthesis, characterization and catalytic performance of nanostructured

dysprosium molybdate catalyst for selective biomolecule detection in biological and pharmaceutical samples. *Journal of Materials Chemistry B*, *7*, 5065–5077.

Karuppaiah, B., Ramachandran, R., Chen, S.-M., Wan-Ling, S., & Wan, J. Y. (2020). Hierarchical construction and characterization of lanthanum molybdate nanospheres as an unassailable electrode material for electrocatalytic sensing of the antibiotic drug nitrofurantoin. *New Journal of Chemistry*, *44*, 46–54.

Kato, A., Oishi, S., Shishido, T., Yamazaki, M., & Iida, S. (2005). Evaluation of stoichiometric rare-earth molybdate and tungstate compounds as laser materials. *Journal of Physics and Chemistry of Solids*, *66*, 2079–2081.

Khademalrasool, M., Farbod, M., & Talebzadeh, M. D. (2021). Investigation of shape effect of silver nanostructures and governing physical mechanisms on photo-activity: Zinc oxide/silver plasmonic photocatalyst. *Advanced Powder Technology*.

Khan, I., Saeed, K., & Khan, I. (2019). Nanoparticles: Properties, applications and toxicities. *Arabian Journal of Chemistry*, *12*, 908–931.

Kiwi, J., & Rtimi, S. (2016). 3 - Environmentally mild self-cleaning processes on textile surfaces under daylight irradiation: Critical issues. In J. Hu (Ed.), *Active coatings for smart textiles* (pp. 35–54). Woodhead Publishing.

Kokulnathan, T., Ashok Kumar, E., & Wang, T.-J. (2020). Synthesis of two-dimensional nanosheet like samarium molybdate with abundant active sites: Real-time carbendazimin analysis in environmental samples. *Microchemical Journal*, *158*, 105227.

Kokulnathan, T., Chen, T.-W., Chen, S.-M., Kumar, J. V., Sakthinathan, S., & Nagarajan, E. R. (2019). Hydrothermal synthesis of silver molybdate/reduced graphene oxide hybrid composite: An efficient electrode material for the electrochemical detection of tryptophan in food and biological samples. *Composites Part B: Engineering*, *169*, 249–257.

Kotsiri, Z., Vidic, J., & Vantarakis, A. (2022). Applications of biosensors for bacteria and virus detection in food and water—A systematic review. *JEnvS*, *111*, 367–379.

Kulkarni, A. K., Tamboli, M. S., Nadargi, D. Y., Sethi, Y. A., Suryavanshi, S. S., Ghule, A. V., & Kale, B. B. (2020). Bismuth molybdate (α-$Bi_2Mo_3O_{12}$) nanoplates via facile hydrothermal and its gas sensing study. *Journal of Solid State Chemistry*, *281*, 121043.

Kumar, A., & Nanda, D. (2019). Chapter 3 - Methods and fabrication techniques of superhydrophobic surfaces. In S. K. Samal, S. Mohanty, & S. K. Nayak (Eds.), *Superhydrophobic polymer coatings* (pp. 43–75). Elsevier.

Kögler, F., Hartmann, F. S. F., Schulze-Makuch, D., Herold, A., Alkan, H., & Dopffel, N. (2021). Inhibition of microbial souring with molybdate and its application under reservoir conditions. *International Biodeterioration & Biodegradation*, *157*, 105158.

Laufer, S., Strobel, S., Schleid, T., Cybinska, J., Mudring, A.-V., & Hartenbach, I. (2013). Yttrium(iii) oxomolybdates(vi) as potential host materials for luminescence applications: An investigation of Eu^{3+}-doped $Y_2[MoO_4]_3$ and $Y_2[MoO_4]_2[Mo_2O_7]$. *New Journal of Chemistry*, *37*, 1919–1926.

Li, H., Li, K., & Wang, H. (2009). Hydrothermal synthesis and photocatalytic properties of bismuth molybdate materials. *Materials Chemistry and Physics*, *116*, 134–142.

Li, H., Wang, P., Lin, G., & Huang, J. (2021). The role of rare earth elements in biodegradable metals: A review. *Acta Biomaterialia*.

Li, K., & Van Deun, R. (2018). Photoluminescence and energy transfer properties of a novel molybdate $KBaY(MoO_4)_3$:Ln^{3+} ($Ln^{3+} = Tb^{3+}$, Eu^{3+}, Sm^{3+}, Tb^{3+}/Eu^{3+}, Tb^{3+}/Sm^{3+}) as a multi-color emitting phosphor for UV w-LEDs. *DTr*, *47*, 6995–7004.

Li, X., Gao, P., Li, J., Guan, L., Li, X., Wang, F., ... Li, X. (2020). Enhancing upconversion emission and temperature sensing modulation of the $La_2(MoO_4)_3$: Er^{3+}, Yb^{3+} phosphor by adding alkali metal ions. *Ceramics International*, *46*, 20664–20671.

Li, X., & Wang, J. (2020). One-dimensional and two-dimensional synergized nanostructures for high-performing energy storage and conversion. *InfoMat*, *2*, 3–32.

Li, Z., Zhao, X., & Jiang, Y. (2015). Hydrothermal preparation and photoluminescent property of $LiY(MoO_4)_2$:Pr^{3+} red phosphors for white light-emitting diodes. *Journal of Rare Earths*, *33*, 33–36.

Liang, E. J., Huo, H. L., Wang, Z., Chao, M. J., & Wang, J. P. (2009). Rapid synthesis of $A_2(MoO_4)_3$ (A = Y^{3+} and La^{3+})with a CO_2 laser. *Solid State Sciences*, *11*, 139–143.

Liu, N., Xiang, X., Fu, L., Cao, Q., Huang, R., & Wu, L. (2021). Regenerative field effect transistor biosensor for in vivo monitoring of dopamine in fish brains. *Biosensors and Bioelectronics*, 113340.

Liu, X., He, J.-H., Sakthivel, R., & Chung, R.-J. (2020). Rare earth erbium molybdate nanoflakes decorated functionalized carbon nanofibers: An affordable and potential catalytic platform for the electrooxidation of phenothiazine. *Electrochimica Acta*, *358*, 136885.

Liu, Y., & Chen, X. (2017). Chapter Eleven - black titanium dioxide for photocatalysis. In Z. Mi, L. Wang, & C. Jagadish (Eds.), *Semiconductors and semimetals* (pp. 393–428). Elsevier.

Logvinovich, D., Arakcheeva, A., Pattison, P., Eliseeva, S., Tomeš, P., Marozau, I., & Chapuis, G. (2010). Crystal structure and optical and magnetic properties of $Pr_2(MoO_4)_3$. *Inorganic Chemistry*, *49*, 1587–1594.

Lou, B.-S., Rajaji, U., Chen, S.-M., & Chen, T.-W. (2020). A simple sonochemical assisted synthesis of $NiMoO_4$/chitosan nanocomposite for electrochemical sensing of amlodipine in pharmaceutical and serum samples. *Ultrasonics Sonochemistry*, *64*, 104827.

Lu, X., Jia, W., Chai, H., Hu, J., Wang, S., & Cao, Y. (2019). Solid-state chemical fabrication of one-dimensional mesoporous β-nickel molybdate nanorods as remarkable electrode material for supercapacitors. *Journal of Colloid and Interface Science*, *534*, 322–331.

Ma, Y., Teng, A., Zhao, K., Zhang, K., Zhao, H., Duan, S., ... Wang, W. (2020). A top-down approach to improve collagen film's performance: The comparisons of macro, micro and nano sized fibers. *Food Chemistry*, *309*, 125624.

Maiti, K., Kim, N. H., & Lee, J. H. (2021). Strongly stabilized integrated bimetallic oxide of Fe_2O_3-MoO_3 nano-crystal entrapped N-doped graphene as a superior oxygen reduction reaction electrocatalyst. *Chemical Engineering Journal*, *410*, 128358.

Martins, G. M., Coelho, P. O., Moreira, R. L., & Dias, A. (2018). Hydrothermal synthesis and polarized micro-Raman spectroscopy of copper molybdates. *Ceramics International*, *44*, 12426–12434.

Martins, G. M., Moreira, R. L., & Dias, A. (2020). Microstructure and optical vibration features of complex cobalt molybdates synthesized by the microwave and conventional hydrothermal processes. *Vibrational Spectroscopy*, *109*, 103107.

Mittal, A. K., & Banerjee, U. C. (2016). Chapter 5 - Current status and future prospects of nanobiomaterials in drug delivery. In A. M. Grumezescu (Ed.), *Nanobiomaterials in drug delivery* (pp. 147–170). William Andrew Publishing.

Morozova, L. V., Kalinina, M. V., Drozdova, I. A., & Shilova, O. A. (2018). Preparation and characterization of nanoceramics for solid oxide fuel cells. *Inorganic Materials*, *54*, 79–86.

Mu, B., & Wang, A. (2019). 11 - Fabrication and applications of carbon/clay mineral nanocomposites. In A. Wang, & W. Wang (Eds.), *Nanomaterials from clay minerals* (pp. 537–587). Elsevier.

Nadar, A., Banerjee, A. M., Pai, M. R., Antony, R. P., Patra, A. K., Sastry, P. U., ... Tripathi, A. K. (2020). Effect of Mo content on hydrogen evolution reaction activity of Mo_2C/C electrocatalysts. *International Journal of Hydrogen Energy*, *45*, 12691–12701.

Namvar, F., Abass, S. K., Soofivand, F., Salavati-Niasari, M., & Moayedi, H. (2019). Sonochemical synthesis of Pr_6MoO_{12} nanostructures as an effective photocatalyst for waste-water treatment. *Ultrasonics Sonochemistry*, *58*, 104687.

Nasrollahzadeh, M., Issaabadi, Z., Sajjadi, M., Sajadi, S. M., & Atarod, M. (2019). Chapter 2 - Types of nanostructures. In M. Nasrollahzadeh, S. M. Sajadi, M. Sajjadi, Z. Issaabadi, & M. Atarod (Eds.), *Interface science and technology* (pp. 29–80). Elsevier.

Nejati, K., Davari, S., Akbari, A., Asadpour-Zeynali, K., & Rezvani, Z. (2019). A highly active oxygen evolution electrocatalyst: Ni-Fe-layered double hydroxide intercalated with the molybdate and vanadate anions. *International Journal of Hydrogen Energy*, *44*, 14842–14852.

Nguyen, D. L. T., Kim, Y., Hwang, Y. J., & Won, D. H. (2020). Progress in development of electrocatalyst for CO_2 conversion to selective CO production. *Carbon Energy*, *2*, 72–98.

Nunes, D., Pimentel, A., Santos, L., Barquinha, P., Pereira, L., Fortunato, E., & Martins, R. (2019). 2 - Synthesis, design, and morphology of metal oxide nanostructures. In D. Nunes, A. Pimentel, L. Santos, P. Barquinha, L. Pereira, E. Fortunato, & R. Martins (Eds.), *Metal oxide nanostructures* (pp. 21–57). Elsevier.

Ohtani, B. (2011). Chapter 10 - Photocatalysis by inorganic solid materials: Revisiting its definition, concepts, and experimental procedures. In R. V. Eldik, & G. Stochel (Eds.), *Advances in Inorganic Chemistry* (pp. 395–430). Academic Press.

Ponomarev, B. K., & Zhukov, A. (2012). Magnetic and magnetoelectric properties of rare earth molybdates. *Physics Research International*, *2012*, 276348.

Pratap, V., Gaur, K., & Lal, H. B. (1987). Electrical transport in heavy rare-earth molybdates. *Materials esearch Bulletin*, *22*, 1381–1393.

Pu, Y., Lin, L., Liu, J., Wang, J., & Wang, D. (2020). High-gravity-assisted green synthesis of rare-earth doped calcium molybdate colloidal nanophosphors. *Chinese Journal of Chemical Engineering*, *28*, 1744–1751.

Rabbani, O., Ghasemi, S., & Hosseini, S. R. (2020). Sonochemical assisted synthesis of manganese–nickel molybdate/reduced graphene oxide nanohybrid for energy storage. *Journal of Alloys and Compounds*, *840*, 155665.

Radeva, D. D., Blaskov, V., Klissurski, D., Mitov, I., & Toneva, A. (1997). Effect of the mechanical activation of the reagents on the solid phase synthesis of iron (III) molybdate. *Journal of Alloys and Compounds*, *256*, 108–111.

Ramesh, S., Vetrivel, S., Suresh, P., & Kaviarasan, V. (2020). Characterization techniques for nano particles: A practical top down approach to synthesize copper nano particles from copper chips and determination of its effect on planes. *Materials Today: Proceedings*, *33*, 2626–2630.

Ravichandran, K., Praseetha, P. K., Arun, T., & Gobalakrishnan, S. (2018). Chapter 6 - Synthesis of nanocomposites. In S. Mohan Bhagyaraj, O. S. Oluwafemi, N. Kalarikkal, & S. Thomas (Eds.), *Synthesis of inorganic nanomaterials* (pp. 141–168). Woodhead Publishing.

Ray, S. K., & Hur, J. (2020). Surface modifications, perspectives, and challenges of scheelite metal molybdate photocatalysts for removal of organic pollutants in wastewater. *Ceramics International*, *46*, 20608–20622.

Ross, J. R. H. (2019). Chapter 13 - Environmental catalysis. In J. R. H. Ross (Ed.), *Contemporary catalysis* (pp. 291–314). Amsterdam: Elsevier.

Roy, M., Choudhary, R. N. P., & Acharaya, H. N. (1989). Differential scanning calorimetric studies of ferroelectric rare-earth molybdates. *Thermochimica Acta*, *145*, 11–17.

Ruíz-Baltazar, Á (2020). Green synthesis assisted by sonochemical activation of Fe_3O_4-Ag nano-alloys: Structural characterization and studies of sorption of cationic dyes. *Inorganic Chemistry Communications*, *120*, 108148.

Sakthivel, S., & Prasanna Venkatesh, R. (2012). Solid state synthesis of nano-mineral particles. *International Journal of Mining Science and Technology*, *22*, 651–655.

Savvin, S. N., Shlyakhtina, A. V., Borunova, A. B., Shcherbakova, L. G., Ruiz-Morales, J. C., & Núñez, P. (2015). Crystal structure and proton conductivity of some Zr-doped rare-earth molybdates. *Solid State Ionics*, *271*, 91–97.

Sha, X., Chen, B., Zhang, X., Zhang, J., Xu, S., Li, X., ... Hua, R. (2021). Pre-assessments of optical transition, gain performance and temperature sensing of Er^{3+} in $NaLn(MoO_4)_2$ (Ln = Y, La, Gd and Lu) single crystals by using their powder-formed samples derived from traditional solid state reaction. *Optics & Laser Technology*, *140*, 107012.

Shahri, Z., Sobhani, A., & Salavati-Niasari, M. (2013). Controllable synthesis and characterization of cadmium molybdate octahedral nanocrystals by coprecipitation method. *Materials Research Bulletin*, *48*, 3901–3909.

Shlyakhtina, A. V., Kolbanev, I. V., Degtyarev, E. N., Lyskov, N. V., Karyagina, O. K., Chernyak, S. A., & Shcherbakova, L. G. (2018). Kinetic aspects of the synthesis of $Ln_{6-\square}MoO_{12-\delta}$ (Ln = Sm, Ho -Yb; x = 0, 0.5) rare-earth molybdates using mechanical activation of oxides. *Solid State Ionics*, *320*, 272–282.

Sinha, S., Mahata, M. K., & Kumar, K. (2018). Comparative thermometric properties of bi-functional Er^{3+}–Yb^{3+} doped rare earth (RE = Y, Gd and La) molybdates. *Materials Research Express*, *5*, 026201.

Sobhani-Nasab, A., Maddahfar, M., & Hosseinpour-Mashkani, S. M. (2016). $Ce(MoO_4)_2$ nanostructures: Synthesis, characterization, and its photocatalyst application through the ultrasonic method. *Journal of Molecular Liquids*, *216*, 1–5.

Sofich, D., Tushinova, Y. L., Shendrik, R., Bazarov, B. G., Dorzhieva, S. G., Chimitova, O. D., & Bazarova, J. G. (2018). Optical spectroscopy of molybdates with composition $Ln_2Zr_3(MoO_4)_9$ (Ln: Eu, Tb). *Optical Materials*, *81*, 71–77.

Suzuki, F., Honma, T., & Komatsu, T. (2011). Laser patterning and morphology of two-dimensional planar ferroelastic rare-earth molybdate crystals on the glass surface. *Materials Chemistry and Physics*, *125*, 377–381.

Thomas, K., Alexander, D., Sisira, S., Jacob, L. A., Biju, P. R., Unnikrishnan, N. V., ... Joseph, C. (2019). Sm3 + doped tetragonal lanthanum molybdate: A novel host sensitized reddish orange emitting nanophosphor. *Journal of Luminescence*, *211*, 284–291.

Tian, Y., Chen, B., Hua, R., Sun, J., Cheng, L., Zhong, H., ... Yu, H. (2011). Optical transition, electron-phonon coupling and fluorescent quenching of $La_2(MoO_4)_3$:Eu^{3+} phosphor. *Journal of Applied Physics*, *109*, 053511.

Tian, Y., Chen, B., Tian, B., Hua, R., Sun, J., Cheng, L., ... Meng, Q. (2011). Concentration-dependent luminescence and energy transfer of flower-like $Y_2(MoO_4)_3$: Dy^{3+} phosphor. *Journal of Alloys and Compounds*, *509*, 6096–6101.

Tian, Y., Du, L., Zhu, P., Chen, Y., Chen, W., Wu, C., & Wang, P. (2021). Recent progress in micro/nano biosensors for shellfish toxin detection. *Biosensors and Bioelectronics*, *176*, 112899.

Tian, Y., Qi, X., Wu, X.-w., Hua, R., & Chen, B. (2009). Luminescent properties of $Y_2(MoO_4)_3:Eu^{3+}$ red phosphors with flowerlike shape prepared via coprecipitation method. *Journal of Physical Chemistry C*, *113*, 10767–10772.

Tkachenko, E. A., & Fedorov, P. P. (2003). Lower rare-earth molybdates. *Inorganic Materials*, *39*, S25–S45.

Tran, V. V., Tran, N. H. T., Hwang, H. S., & Chang, M. (2021). Development strategies of conducting polymer-based electrochemical biosensors for virus biomarkers: Potential for rapid COVID-19 detection. *Biosensors and Bioelectronics*, *182*, 113192.

Trnovcová, V., Škubla, A., & Schultze, D. (2005). Anisotropy of the ionic conductivity in potassium bismuth/rare earth molybdate crystals. *Solid State Ionics*, *176*, 1739–1742.

Vats, B. G., Shafeeq, M., & Kesari, S. (2021). Triple molybdates and tungstates scheelite structures: Effect of cations on structure, band-gap and photoluminescence properties. *Journal of Alloys and Compounds*, *865*, 158818.

Vinoth, S., Govindasamy, M., Wang, S.-F., Alothman, A. A., & Alshgari, R. A. (2021). Surface engineering of roselike lanthanum molybdate electrocatalyst modified screen-printed carbon electrode for robust and highly sensitive sensing of antibiotic drug. *Microchemical Journal*, *164*, 106044.

Vinoth Kumar, J., Karthik, R., Chen, S.-M., Natarajan, K., Karuppiah, C., Yang, C.-C., & Muthuraj, V. (2018). 3D Flower-Like gadolinium molybdate catalyst for efficient detection and degradation of organophosphate pesticide (fenitrothion). *ACS Applied Materials & Interfaces*, *10*, 15652–15664.

Wang, C., Ma, R., Zhou, Y., Liu, Y., Daniel, E. F., Li, X., . . . Ke, W. (2021). Effects of rare earth modifying inclusions on the pitting corrosion of 13Cr4Ni martensitic stainless steel. *Journal of Materials Science & Technology*.

Wang, L., Cai, X., Li, J. I. A., Song, Q., Wang, X. U. E., Han, Y., & Jia, G. (2018). Hydrothermal synthesis of red phosphor $La_2(MoO_4)_3$ doped with Eu^{3+} and its application for ferric ions assay. *Surface Review and Letters*, *26*, 1950046.

Wang, L., Yu, X., Wei, Y., Liu, J., & Zhao, Z. (2021). Research advances of rare earth catalysts for the catalytic purification of vehicle exhausts. *Journal of Rare Earths*.

Wang, M., Guo, P., Yang, G., Chai, T., & Zhu, T. (2017). The honeycomb-like Eu^{3+}, Fe^{3+} doping bismuth molybdate photocatalyst with enhanced performance prepared by a citric acid complex process. *Materials Letters*, *192*, 96–100.

Wang, S.-F., Koteswara Rao, K., Wang, Y.-R., Hsu, Y.-F., Chen, S.-H., & Lu, Y.-C. (2009). Structural characterization and luminescent properties of a red phosphor series: $Y_{2-x}Eu_x(MoO_4)_3$ ($x = 0.4–2.0$). *Journal of the American Ceramic Society*, *92*, 1732–1738.

Wang, Y., & Xia, Y. (2004). Bottom-up and top-down approaches to the synthesis of monodispersed spherical colloids of low melting-point metals. *Nano Letters*, *4*, 2047–2050.

Wang, Z., Hu, T., Liang, R., & Wei, M. (2020). Application of zero-dimensional nanomaterials in biosensing. *Frontiers in Chemistry*, *8*.

Williams, P. A. (2005). Minerals|Molybdates. In R. C. Selley, L. R. M. Cocks, & I. R. Plimer (Eds.), *Encyclopedia of geology* (pp. 551–552). Oxford: Elsevier.

Wu, J. L., Cao, B. S., Lin, F., Chen, B. J., Sun, J. S., & Dong, B. (2016). A new molybdate host material: Synthesis, upconversion, temperature quenching and sensing properties. *Ceramics International*, *42*, 18666–18673.

Wu, M., & Shi, J. (2021). Beneficial and detrimental impacts of molybdate on corrosion resistance of steels in alkaline concrete pore solution with high chloride contamination. *Corrosion Science*, *183*, 109326.

Xiao, B., Schlenz, H., Bosbach, D., Suleimanov, E. V., & Alekseev, E. V. (2016). The structural effects of alkaline- and rare-earth element incorporation into thorium molybdates. *CrystEngComm*, *18*, 113–122.

Xu, X. L., Bo, G. X., He, X., Tian, X. K., & Yan, Y. J. (2020). Structural effects of dimensional nano-fillers on the properties of Sapium sebiferum oil-based polyurethane matrix: Experiments and molecular dynamics simulation. *Polymer*, *202*, 122709.

Yang, F., Deng, D., Pan, X., Fu, Q., & Bao, X. (2015). Understanding nano effects in catalysis. *National Science Review*, *2*, 183–201.

Yang, G., Liang, Y., Li, K., Yang, J., Wang, K., Xu, R., & Xie, X. (2020). Engineering the dimension and crystal structure of bismuth molybdate photocatalysts via a molten salt-assisted assembly approach. *Journal of Alloys and Compounds*, *844*, 156231.

Yang, P.-T., & Wang, S.-L. (2021). Sorption and speciation of molybdate in soils: Implications for molybdenum mobility and availability. *Journal of Hazardous Materials*, *408*, 124934.

Yu, Y., Shao, K., Zhu, X., Zhang, X., Qiu, H., Wu, S., & Wang, G. (2021). Research on a novel molybdate Er^{3+}:$KBaY(MoO_4)_3$ crystal as a prominent 1.55 μm laser medium. *Journal of Luminescence*, 118194.

Zeng, M., & Li, Y. (2015). Recent advances in heterogeneous electrocatalysts for the hydrogen evolution reaction. *Journal of Materials Chemistry A*, *3*, 14942–14962.

Zhang, J., Zhang, X., Wei, X., Xue, Y., Wan, H., & Wang, P. (2021). Recent advances in acoustic wave biosensors for the detection of disease-related biomarkers: A review. *Analytica Chimica Acta*, *1164*, 338321.

Zhang, K. Y., & Lo, K. K. W. (2013). 8.18 - Chemosensing and diagnostics. In J. Reedijk, & K. Poeppelmeier (Eds.), *Comprehensive inorganic chemistry II* (2nd Ed., pp. 657–732). Amsterdam: Elsevier.

Zhang, N., Yan, H., Li, L., Wu, R., Song, L., Zhang, G., ... He, H. (2021). Use of rare earth elements in single-atom site catalysis: A critical review—commemorating the 100th anniversary of the birth of Academician Guangxian Xu. *Journal of Rare Earths*, *39*, 233–242.

Zhao, B., Lei, H., Wang, N., Xu, G., Zhang, W., & Cao, R. (2020). Underevaluated solvent effects in electrocatalytic co_2 reduction by FeIII chloride tetrakis(pentafluorophenyl)porphyrin. *Chemistry – A European Journal*, *26*, 4007–4012.

Zhou, Y., He, X.-H., & Yan, B. (2014). Self-assembled $RE_2(MO_4)_3$:Ln^{3+} (RE = Y, La, Gd, Lu; M = W, Mo; Ln = Yb/Er, Yb/Tm) hierarchical microcrystals: Hydrothermal synthesis and up-conversion luminescence. *Optical Materials*, *36*, 602–607.

Zhu, L., Luo, D., & Liu, Y. (2020). Effect of the nano/microscale structure of biomaterial scaffolds on bone regeneration. *International Journal of Oral Science*, *12*, 6.

Zhu, W., Huang, X., Zhang, Y., Yin, Z., Yang, Z., & Yang, W. (2021). Renewable molybdate complexes encapsulated in anion exchange resin for selective and durable removal of phosphate. *Chinese Chemical Letters*.

Zinola, C. F., Martins, M. E., Tejera, E. P., & Neves, N. P. (2012). Electrocatalysis: Fundamentals and applications. *International Journal of Electrochemistry*, *2012*, 874687.

Zolotova, E. S., Solodovnikov, S. F., Solodovnikova, Z. A., Yudin, V. N., Uvarov, N. F., & Sukhikh, A. S. (2021). Selection of alkali polymolybdates as fluxes for crystallization of double molybdates of alkali metals, zirconium or hafnium, revisited crystal structures of $K_2Mo_2O_7$, $K_2Mo_3O_{10}$, $Rb_2Mo_3O_{10}$ and ionic conductivity of $A_2Mo_2O_7$ and $A_2Mo_3O_{10}$ (A = K, Rb, Cs). *Journal of Physics and Chemistry of Solids*, *154*, 110054.

Rare earth–doped semiconductor nanomaterials

*Noshin Mir**
Department of Mechanical and Nuclear Engineering, Virginia Commonwealth University, Richmond, VA, United States
*Corresponding author E-mail address: mir.n63@gmail.com

11.1 General introduction

11.1.1 Doping of semiconductor

Impurities are the heart bit of a semiconductor material and can determine the bottom line of its performance. By using different doping elements, we select to tune and moderate the properties of semiconductors in our desired fashion. The type of the doping elements and their concentrations can be both important in determining photonic, conductivity, magnetic, and catalytic activities of a material. One historical great challenge in introducing impurities into semiconductors is the limitation of incorporating doping agents in the system without expelling the guest element in the form of either an oxide or a secondary phase (Erwin et al., 2005). This limitation is considered as a bottleneck for the utilization of certain class of materials in different applications, especially electronics and photonics. The bandgap of a semiconductor plays an important role in success or failure of hosting external elements. With increasing the bandgap, for example from Si→GaAs→ZnSe→ZnO, symmetric doping in negatively- and positively-charged fashions becomes more challenging (Zunger, 2003).

Dopants are generally introduced into a semiconductor as form of defects, which could enact as donors or acceptors. Donor and acceptor impurities emit electrons to the conduction or holes to the valence bands, respectively (Dierolf & Ferguson, 2021). Semiconductors can be divided into two categories, *intrinsic* and *extrinsic*. Intrinsic is the pure form of the semiconductor in which the number of free electrons and holes are equal. The defects in this system are usually oxygen vacancies and can act as shallow donor states bellow the conduction band (Shukla & Sharma, 2020). When cations or anions are introduced to the semiconductor lattice, the final doped semiconductor is called extrinsic. If the dopant is a donor (n[1]-type), free electrons can be created without simultaneous creation of holes and thus create higher number of electrons. On the other hand, if the dopant is an acceptor (p[2]-type), the number of holes will exceed the number of electrons (Dasgupta et al., 2004).

[1] n stands for negatively charged electrons.
[2] p stands for positively charged holes.

Advanced Rare Earth-Based Ceramic Nanomaterials. DOI: https://doi.org/10.1016/B978-0-323-89957-4.00013-X

11.1.2 Rare earth elements

The rare earth elements (REEs or REs[4]) are referred to a chemically similar group of elements that their f orbitals is partially filled. Cerium through lutetium (atomic number of 57−71, respectively) as well as scandium and yttrium are included in this group (Atwood, 2013). Scandium and yttrium, with atomic numbers of 21 and 39, are included in this group because they have similar chemical properties and frequently occur with REEs in nature. It is noteworthy that scandium occurs in smaller quantity than yttrium because of its smaller atomic and ionic size (Shukla & Sharma, 2020). Although the nomenclature name associated with RE refers to their scarcity in the nature, their abundances in crustal rocks are not exceptionally lower than many other elements. However, an early paper on REEs "Dispersed and not-so-rare earths" (Atwood, 2013) resulted in this terminology.

The history of REEs goes back to 1787, when a Swedish mineralogist Lt. Carl Axel Arrhenius discovered Ytterby in the form of an unusual black mineral. It was the starting point of the journey of discovering many other REEs, most of which inherited their names from the location they were first found (Steckl & Zavada, 1999b). The commercial uses of REEs started right after World War II, and the main application until few decades ago was permanent magnets. As mentioned previously, REEs have extraordinary magnetic moment of all elements due to the unpaired 4f electrons. Many of the separation and purification techniques for REEs were developed during the Manhattan Project[3] during World War II (Steckl & Zavada, 1999b).

The common ionic charge that these elements take is " + 3", and many of the interesting properties of the RE^{3+} ions stem from the fact that $5s^2$ and $5p^6$ outer-shell electrons provide an energy levels for 4f shells, independent of their surroundings (Steckl & Zavada, 1999b). In fact, in europium with divalent and cerium with tetravalent possible ionic charge are the only exceptions; apart from these examples, in all geochemical systems, REEs are trivalent (Atwood, 2013). Based on their geochemical activities, REEs are subdivided in two groups called the light REs (LREs) and heavy RE elements (HREs) (La−Sm and Gd−Lu, respectively). Europium has a different redox geochemistry that makes it anomalous in RE family, but can be categorized as an LRE. In the earth's crust, HREs are naturally rarer than LREs (Atwood, 2013).

Same as their outstanding optical properties, their magnetic properties, which are originating from the spin and orbital movements of the single electrons in 4f orbitals, are also of great interest. Interestingly, some REEs can show the highest magnetic moments of any elements (Hite & Zavada, 2019; Steckl & Zavada, 1999b). Except the La^{3+} and Lu^{3+}, the rest of RE^{3+} ions show paramagnetic properties, meaning that the electrons order their spins in the direction of an external magnetic field. Since the La^{3+} has no 4f electrons and Lu^{3+} has only paired electrons in 4f

[4] Both terms as well as the singular forms of each will be used in this chapter

[3] The Manhattan Project was one of the ongoing projects during World War II for inventing the first nuclear weapons.

orbitals, they both show diamagnetic properties, meaning that they align their electron spins against the external field (Hite & Zavada, 2019). Results of Van Vleck theoretical model (Van Vleck, 1932), an approach for describing magnetic susceptibilities for the lanthanides, showed that the effective magnetic moment (μ_{eff}) is larger for some HREs (Gd^{3+} to Tm^{3+}) than LREs (from Ce^{3+} to Eu^{3+}). Interestingly, Gd^{3+} to Tm^{3+} become ferromagnetic at very low temperatures, which are of great interest for ferromagnetic semiconductors applications (Hite & Zavada, 2019).

11.2 Applications of RE-doped semiconductor nanomaterial

Fig. 11.1 shows the key applications of RE-doped semiconductors. The REE application in semiconductor industry was initially recognized for their excellent photoluminescence (PL) properties. There has been an increasing demand for the integration of compatible REEs with semiconductors specially the conventional fibers and microelectronics. Silicon, III-N compounds (GaN, InN, A1N, InGaN, AlGaN), and SiC are important materials that have been extensively used for the development of RE-doped materials for applications, such as laser, photonics, light-emitting diode (LED) devices, and optical fibers (Kane, Gupta, & Ferguson, 2016;

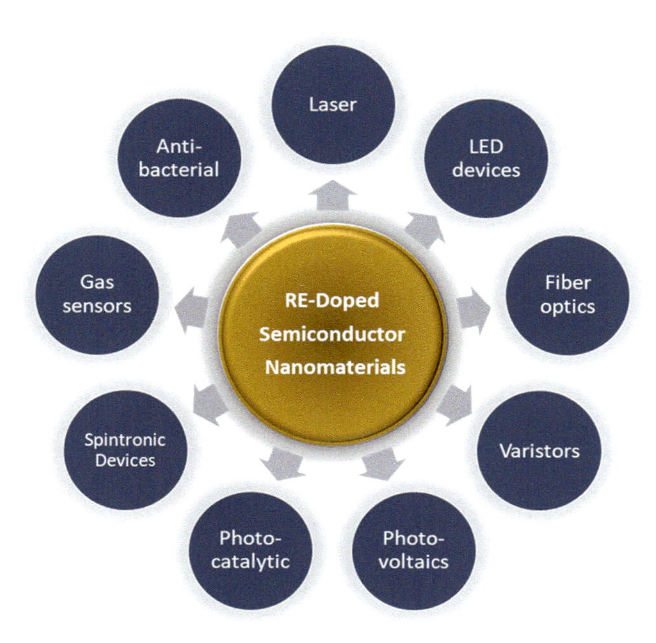

Figure 11.1 Common applications of RE-doped semiconductor nanomaterials.

Kränkel, Marzahl, Moglia, Huber, & Metz, 2016; Li, Li, Wang, & Zheng, 2018; O'Donnell, 2015; Steckl, Heikenfeld, Lee, & Garter, 2001; Zavada et al., 2000).

Another important application of RE-doped semiconductors is in spintronic devices. In these devices, in addition to the charges of the electrons, the spin of magnetic is also exploited. By replacing some of the cations in the host network by magnetic ions such as REEs, an emerging class of materials can be produced, namely diluted magnetic semiconductors (DMS). DMS that can be fabricated by using certain transition or REEs, such as Sc, Ti, V, Cr, Mn, Fe, Co, Ni, Sm, Eu, Gd, Tb, Dy, and Er, are used in spintronic devices (Bonanni, 2007). Some semiconductor host networks, including but not limited to III-N, ZnS, and ZnO could be used for making DMS specific for application in spintronic devices (Gupta et al., 2008; Kane et al., 2016; O'Donnell & Dierolf, 2010; Poornaprakash, Ramu, Park, Vijayalakshmi, & Reddy, 2016; Ungureanu et al., 2007). They are meant to exploit the spin of magnetic materials along with—like in standard electronics—the charge of electrons in semiconductors. It is generally expected that new functionalities for electronic and photonics will arise if the injection, transfer, and detection of carrier spins can be mastered above room temperature (RT).

11.3 RE-doped semiconductors

Doping semiconductors with REEs could be defined as part of the nanoengineering, in which by manipulating the properties of the materials at nanoscale, multifaceted applications to science and industry can be developed. These applications can be in a wide range including chemical, physical, biological, and engineering (Daksh & Agrawal, 2016). This technology can manipulate semiconductor properties in the atomic levels to enhance their current performance or even create a completely new application for them.

In this section, different categories of rare earth element-doped (RE-doped) semiconductors will be briefly reviewed. There are hundreds of research literatures and review papers discussing synthesis, applications, and mechanisms behind the observed phenomena associated with different types of RE-doped semiconductors. The RE-doped semiconductors are very diverse and it is highly challenging if one wants to divide them into categories without missing out some materials. Moreover, some of the reported RE-doped semiconductors can fit into different categories. The selected categories here are mainly based on the application of the final doped materials and do not necessarily focus on the inherent material type or crystal structure.

11.3.1 Silicon

Silicon is a widely available element with excellent thermal and mechanical properties. It can be processed to an extraordinary pure level and is easy to work with in order to fabricate various devices. Complementary metal–oxide–semiconductor

(CMOS) is a single technology used as a single dominating processing technology, which accounts for more than 95% of the semiconductor chips market (Clemens, 1997). Silicon has been ruling the semiconductors industry for a long time; however, because of its unique properties in conducting strong emissions instead of intersystem crossing, internal conversion, and other nonradiative energy transfer paths, it remained unsuitable for optoelectronic applications for decades. Its indirect bandgap luminescence at 1.1 μm does not match with the required wavelength of 1.55 μm for fiber-optic communications systems. Moreover, the nonradiative recombination pathways result in a very short nonradiative lifetime of any excited electron (Kenyon, 2005). With swift shift in the communication technology recently, many research groups sought the alternative methods for responding to this high demand by adjusting optoelectronic function of silicone, especially in electronic circuits. The main idea was to integrate optical and electronic knowledge on processes and devices to exploit the benefit of both sides (Franzò, Vinciguerra, & Priolo, 1999).

Canham (1990) did the initial observation of light emission in porous silicon in 1990s. Since then, a great amount of research works have been devoted towards the development of emitting silicon materials. Doping Si with REEs has been a successful approach for producing luminescent silicon structures. The main motivation for doping silicon with REEs was to synthesize a CMOS-compatible material for different optical devices (Kenyon, 2005). The selection of REEs for this application is highly crucial since the emitted light should be in a wavelength range suitable for telecommunication applications. Here, two important examples of REEs including Er^{3+} and Eu^{2+} ions will be discussed.

11.3.1.1 Er-doped silicon

Erbium (Er) is one of the most successful elements with emission at 1.54 μm that has been ever studied in this field (Franzò et al., 1999). Despite some limitations of Er emission in Si (such as Auger, thermal decay pathways, energy transfer to electron–hole pairs (Palm, Gan, Zheng, Michel, & Kimerling, 1996)) Er:Si devices were fabricated more than two decades ago. The efficiency of the initial room temperature (RT) devices was around 0.1% (Du, Ni, Joelsson, & Hansson, 1997; Franzo, Priolo, Coffa, Polman, & Carnera, 1994; Stimmer et al., 1996; Zheng et al., 1994).

Erbium has atomic number of 68 and shows some intense luminescence bands with narrow shapes in the visible and near-infrared (NIR) regions. As mentioned previously, Er has an abnormal geochemistry and can take both +2 and +3 ionic charges, the former of which is rarely detected in semiconductors. Having [Xe] $4f^{12}6s^2$ electronic structure as an atom, its triple charged ion forms by loss of two electrons from 6s and one electron from 4f levels. The protective 5s and 5p levels can act as a shield to 4f orbitals and make its luminescence activities independent from the surrounding lattice. Although the parity forbidden electron transitions between 4f orbitals can be permitted due to low symmetry of some crystal lattice sites, fortunately it is not an option in silicon. Therefore due to the same crystal splitting of silicon for erbium ions, if there is no other thermal decay pathway, an extended luminescence is expected (Kenyon, 2005).

With continuation of the research works, an effective material, namely Si nanocrystal−embedded silicone dioxide, was developed. The confined excitons can be recombined and result in light emission in visible range (Brongersma et al., 1998; Shimizu-Iwayama, Kurumado, Hole, & Townsend, 1998; Song & Bao, 1997). The addition of REEs to this system creates a new pathway for transferring energy from the Si nanocrystals and the new guest element resulting in radiative deexcitation (Brongersma et al., 1998; Song & Bao, 1997). It was shown that Si nanocrystals incorporated in Er-doped SiO_2 can result in twofold higher RT luminescence than the pure silica (Fujii, Yoshida, Kanzawa, Hayashi, & Yamamoto, 1997).

The typical absorption of Si nanocrystals depend on their average size but it is usually detected near 600-nm shift to the lower wavelenghts. One important factor that shows high probability of energy transfer from Si nanocrystals to Er^{3+} is shown with absorption cross sections. For Si nanocrystals, the value is around 10^{-16} cm^2, which is higher than absorption cross section of Er^{3+} in stoichiometric silica (1×10^{-21} cm^2 at 477 nm and 8×10^{-21} cm^2 at 488 nm). Accurate measurements of the amount of transferring energy from Si to Er^{3+} have shown greater than 60% quantum efficiencies by pumping at 488 nm with micron second scale transfer time (Khriachtchev, 2016). Moreover, Si nanocrystals are excellent candidates for electrically pumped optical amplifiers because of their ability to enhance the dielectric refractive index, trapping light, and electrical conductivity (Lo Savio et al., 2014).

Franzò et al. (1999) explained the plausible mechanism of this interaction, as shown in Fig. 11.2. Based on their proposal, pumping laser first creates the initial excitation of the confined excitons in Si structure. These excitons then perform energy transfer to the illuminating centers of SiO_2. When REEs are used as doping agents, second pathway is created for the transfer of energy. With increased concentration of REEs, this route becomes the main transfer means. The suggested evidence for this mechanism was a decrease in 0.85 μm related to the nanocrystal luminescence. It also shows that the revoked RE ions are embedded within SiO_2, not in nanolattice.

11.3.1.2 Eu-doped silicon

Eu, another member of RE family with unprecedented ionic charge of +2 and +3, is one interesting doping agent for silicone. Eu is capable of emitting light in visible range, which finds extensive applications in phosphors used in photonic devices. A unique signature of Eu originating from its different ionic charges is switching between Eu^{2+} and Eu^{3+} ions and creating both ultraviolet and visible (blue) emissions in an LED device. The main problem of Eu for large-scale applications is its low solid solubility that limits its extensive use in silicone market. Extensive research works have been devoted to increasing the solubility of Eu as much as possible. Another limiting factor in application of Eu as a semiconductor-doping agent is difficulty in selective doping of either Eu^{2+} or Eu^{3+} ions. The emission signature

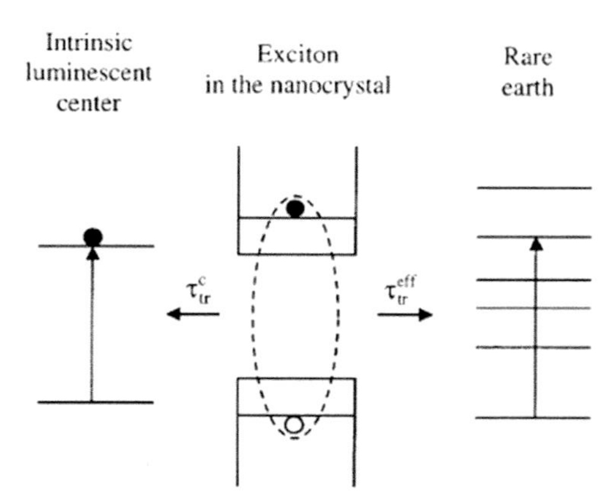

Figure 11.2 Proposed mechanism by Franzò et al. (1999) for Si nanocrystal–embedded Er-doped SiO_2 luminescence. This mechanism suggests that there are two plausible competitive routes for energy transfer after excitation of an exciton in Si nanocrystals, either to intrinsic luminescence centers in SiO_2 lattice or to the RE elements.
Source: Reprinted from Franzò, G., Vinciguerra, V., & Priolo, F. (1999). The excitation mechanism of rare-earth ions in silicon nanocrystals. *Applied Physics A: Materials Science and Processing 69*(1), 3–12, with the permission from Springer Nature.

of each ion is different from one another, Eu^{3+} shows weak dipole forbidden narrow transitions at about 615 nm, while Eu^{2+} emits a dipole-allowed wide wavelength in the range of 400–600 nm (Boninelli, Bellocchi, Franzò, Miritello, & Iacona, 2013).

The proposed approach for making Eu compatible with CMOS processing is the same as what was proposed for Er ions, namely incorporation of Eu in SiO_2 matrix. However, unlike Er, this approach is eminently limited by low solubility of Eu in solid matrix, resulting in massive clustering and precipitations of Eu ions (Li, Zhang, Jin, & Yang, 2010; Rebohle et al., 2008, 2009). This issue becomes even worse when high-temperature annealing is added in the process. Ion implantation method requires thermal annealing as a necessary step to introduce Eu ions into the network. This process removes defects and decreases the chance of nonradiative emissions, therefore provides a fresh active material for optical applications (Khriachtchev, 2016). Boninelli et al. (2013) studied Eu incorporation in SiO_2 network and tested annealing temperatures from 750°C to 1000°C. By doping 5.0×10^{20} Eu/cm^3 in a silica lattice, no precipitation was identified at 750°C or lower temperatures. However, bright field transmission electron microscopy (BF XTEM) images showed precipitate formation when samples was heat-treated at 900–1000°C in N_2. One remarkable observation was that at 900°C, although homogeneous clusters were all over the film, the measured scale of the particles was significantly smaller at the surface rather than at the substrate. Contrary, at 1000°C, the larger particles were accumulated at the surface. Fig. 11.3A–C shows this trend clearly. They confirmed and explained this phenomenon by investigating atomic

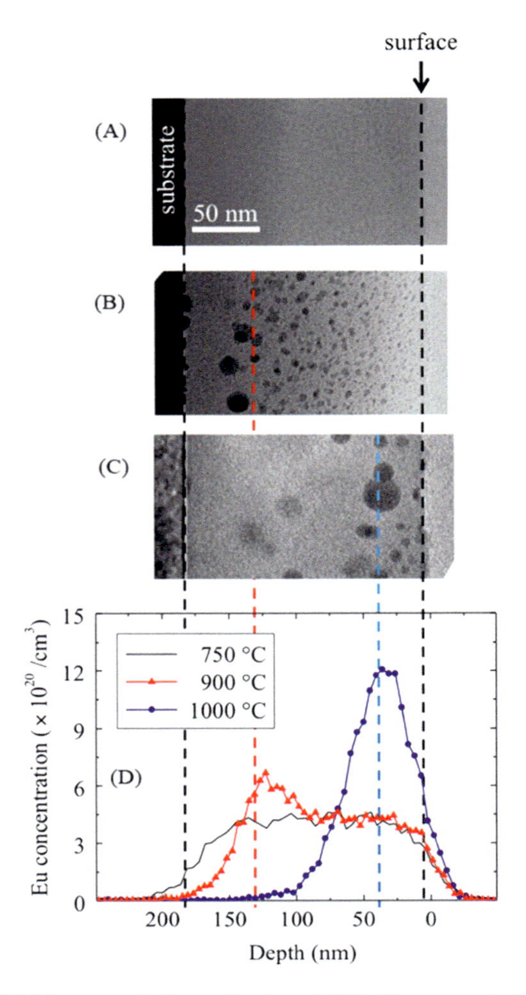

Figure 11.3 BF XTEM images relative to Eu-doped SiO_2 films annealed in N_2 ambient at (A) 750°C, (B) 900°C, and (C) 1000°C. Eu concentration is 5.0×10^{20} /cm^3 in all cases. (D) Eu concentration profiles, measured by RBS, relative to the samples are shown in panels (A)–(C).

Source: Reprinted from Boninelli, S., Bellocchi, G., Franzò, G., Miritello, M., & Iacona, F. (2013). New strategies to improve the luminescence efficiency of Eu ions embedded in Si-based matrices. *Journal of Applied Physics*, *113*(14), 143503, with the permission of AIP Publishing.

composition by Rutherford backscattering spectrometry (RBS) (Fig. 11.3D) and Eu diffusion toward the surface. They showed that the Eu peak depth is almost constant at 750°C while it decreases as the annealing temperature increases to 900°C and 1000°C, confirming the movement of the larger precipitates towards the surface (Boninelli et al., 2013).

This study as well as other research works (Jia et al., 2004) suggest that SiO_2 network is not a suitable host for Eu and an alternative material should be introduced. The desirable materials should possess two crucial factors, namely, being capable of accepting high concentrations of Eu and more importantly, being compatible with current silicone technology. There have been many reports on preparation of different host materials with high capabilities of dissolving Eu in their solid matrix. However, most of these materials, spanned from noncomplex oxides such as ytterbium or gadolinium oxide (Ahmadian, Al Hessari, & Arabi, 2019; Kumar, Ntwaeaborwa, Soga, Dutta, & Swart, 2017) to more complex materials such as $Ca_{18}Li_3Bi_{1-x}Eu_x$ $(PO_4)_{14}$ (Zhu et al., 2019) and $KCa_4(BO_3)_3{:}Ln^{3+}$ (Reddy et al., 2013) are very challenging to be compatible with CMOS technology (Boninelli et al., 2013).

Boninelli et al. (2013) suggested silicone oxycarbide (SiOC) matrix with low C content (around 5 at.%) as a potential candidate and explored the photonic performance of Eu in this material. SiOC materials exhibit inherent PL in the visible region when it is already activated at low temperatures under ultraviolet rays on pyrolysis. Blue light (457 nm), displaying white emission, can also excite SiOC (Stabler, Ionescu, Graczyk-Zajac, Gonzalo-Juan, & Riedel, 2018). SiOC is capable of hosting high concentrations of Eu and is compatible with Si technology (at low carbon contents). Additionally, owing to its chemical and structural properties, it can facilitate the reduction reaction of Eu^{3+} to Eu^{2+} and enhance the mobility of Eu ions, as important factors for decreasing precipitations. The promoted reduction of Eu can increase the chance for the presence of Eu^{2+} in the matrix and ensure visible light (440 mm), which is of interest of many photonic devices (Boninelli et al., 2013). Finally, the wide bandgap of SiC could match with both Eu^{3+} and Eu^{2+} (Khriachtchev, 2016). The RT PL spectra of the prepared Eu-doped SiOC in this study is shown in Fig. 11.4. All the samples were prepared by UHV magnetron-sputtering system, contained 1.5×10^{20} Eu/cm^3, and show very similar peak shapes, corresponding to Eu^{2+} emission. Eu-doped SiO_2 PL peak is also available in this figure and has lower intensity, by about more than two orders of magnitude, when compared with Eu-doped SiOC samples. Interestingly, Eu-doped SiOC showed the same PL emission at high temperatures. Two samples in this figure were annealed at 900°C and 1000°C while they were capped under a SiO_2 capping film. The role of this cap layer is decreasing oxygen and suppressing Eu precipitation. As can be seen, after adding the cap layer, an enhancement in PL intensity was obtained (Boninelli et al., 2013).

Lin et al. used magnetron-sputtering method followed by annealing to produce the sample containing both Eu^{3+} and Eu^{2+} that can produce intensified red and blue emissions from Eu-incorporated SiOC films. They showed that annealing temperature has a crucial role on enhancing light emission intensity of red and blue lights. Annealing temperature under 800°C increases the Eu^{3+}-emitting visible red light by threefold. On the other hand, at temperatures higher than 800°C, followed by a switch in dominant luminescence centers, the emission changes and increases the visible blue light intensity by 40 times. The energy transfer shift is from $^5D_0 \rightarrow {}^7F_2$ in Eu^{3+} to the $4f^6 5d \rightarrow 4f^7$ in Eu^{2+} (Lin et al., 2017).

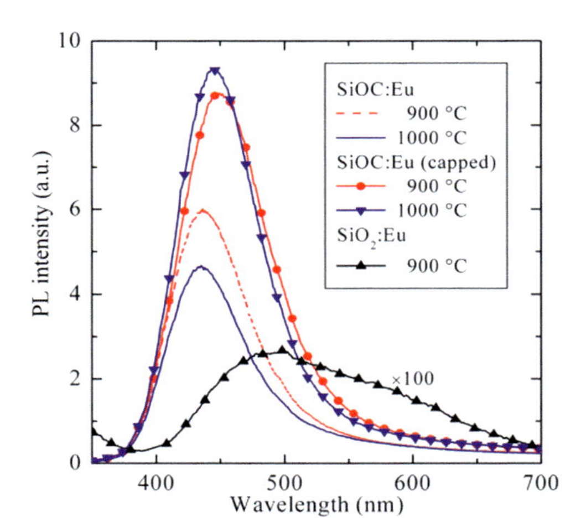

Figure 11.4 Room temperature PL spectra of Eu-doped SiOC and SiO_2 films annealed at 900°C and 1000°C in N_2 ambient. The PL spectrum of Eu-doped SiO_2 is multiplied by a factor of 100. In the case of SiOC, spectra of both SiO_2-capped and uncapped samples are shown. Eu concentration is $1.5 \times 10^{20}/cm^3$ in all of the samples.
Source: Reprinted from Boninelli, S., Bellocchi, G., Franzò, G., Miritello, M., & Iacona, F. (2013). New strategies to improve the luminescence efficiency of Eu ions embedded in Si-based matrices. *Journal of Applied Physics*, *113*(14), 143503, with the permission of AIP Publishing.

An important finding of this report was proposing a mechanism for the increased red and blue lights. The emission intensified because of the presence of $EuSiO_3$ nanoparticles and as the consequence of transferring energy from these nanostructures to either Eu^{3+} or Eu^{2+} ions. Fig. 11.5 schematically shows the proposed energy transfer mechanism from $EuSiO_3$ clusters to Eu^{3+}/Eu^{2+} ions in SiCO:Eu films as illustrated in Lin et al. (2017).

11.4 III–V RE-doped semiconductors

11.4.1 III-N

III-Nitride materials, consisting of nitride (N^{3-}) and group III elements, are parts of an attractive group of semiconductors. These materials include the binary, ternary, and quaternary alloys. GaN, InN, AlN, InGaN, AlGaN, and AlInGaN are some examples in this group, showing excellent optical and magnetic properties that are promising for modern optoelectronic applications. Historically, doping conventional semiconductors such as Si and GaAs with REEs has been always challenging because of the precipitation of REEs and sudden decrease in temperature. Favennec, L'haridon, Salvi, Moutonnet, and Le Guillou (1989) were able to

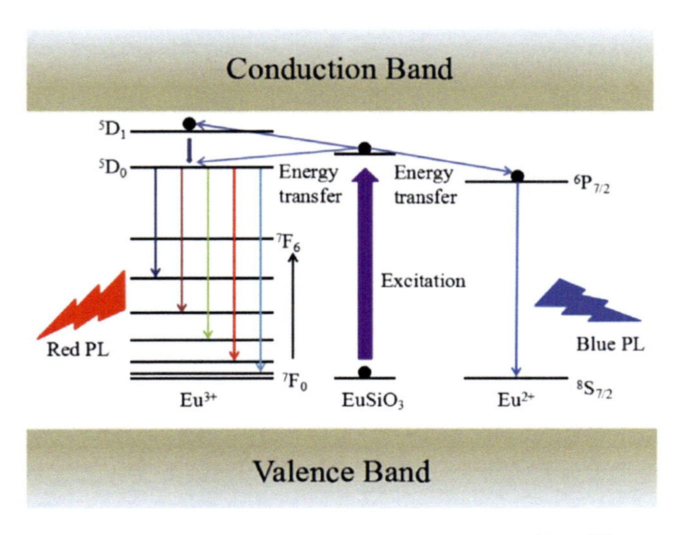

Figure 11.5 Energy transfer between $EuSiO_3$ nanoclusters and Eu^{3+}/Eu^{2+} ions in SiCO:Eu films.
Source: Reprinted from Lin, Z., et al. (2017). Dense nanosized europium silicate clusters induced light emission enhancement in Eu-doped silicon oxycarbide films. *Journal of Alloys and Compounds*, *694*, 946–951, with the permission from Elsevier.

mitigate the decreasing temperature in the semiconductors containing Er atoms through enhancing bandgap. Ever since, there was more interest in wide-bandgap semiconductors (WBGSs) for RE doping (Steckl et al., 2002). The arbitrary but accepted properties of WBGSs that seperate them from other semiconductors are their high bandgap energy (2 eV) (Steckl & Zavada, 1999a). III-N semiconductors are direct bandgap crystalline compounds with bandgaps between 0.7 and 6.2 eV. Some examples are InN 0.7 eV, GaN 6.2 eV, and AlN 6.2 eV (Hite & Zavada, 2019). Since these materials can provide various ranges of E_g, they are able to emit almost any desirable light wavelength. Although their typical crystal lattice is wurtzite, they sometime can form cubic structures, as well. To produce materials with large bandgaps with high bonding energy in the crystal lattice, synthetic methods with high processing temperatures are required. That is why the breakthrough in device applications of these materials did not occur until some advanced high-temperature methods such as metal–organic chemical vapor deposition (MOCVD) or molecular beam epitaxy (MBE) became available (Hite & Zavada, 2019). The optical emissions of these materials cover the whole light spectrum from infrared (IR) to visible (Feng, 2006). Since the start of these advanced synthetic methods, sapphire (Al_2O_3), and silicon carbide (SiC), and Si were used as some of the common substrates for epitaxy growth of the III-N films (Poust, 2003; Morita, 2006). Some examples of the remarkable usage of III-N materials in advanced applications are high-power blue laser diodes (LDs), LEDs (DenBaars et al., 2013) and circular-polarized laser diodes (DenBaars et al., 2013). III-Nitride-based novel LED applications have passed the breakthrough and are rising up. It is highly expected that this new application shape up photonic industry in 21st century, similar to Edison's electric light bulb in 1879 (Feng, 2006). The photonic and magnetic properties of RE-doped III-N materials were well reviewed in a number of excellent papers and books (Feng, 2006; O'Donnell & Dierolf, 2010; O'Donnell & Hourahine, 2006; Reshchikov & Morko, 2005; Steckl et al., 2001). Here, we aim to investigate the most recent advancements in this field.

11.4.1.1 Optical properties

The optical properties of RE-doped semiconductors come from the trivalent oxidation state RE^{3+}. Therefore in all different host lattices, there are always fixed numbers of electrons for a specific REE. For example, Er has single electron in f orbitals. As mentioned previously in this chapter (Section 11.1.2), since the 4f electrons are protected by 5s and 5p orbitals, the crystal field of the host has negligible influence on them. On the contrary, the REE defects can affect the crystal field orbitals of the host. Various factors such as RE ionic size, valence electrons, and their electronegativity might induce more energy gap level in the original crystal field of the host (O'Donnell & Dierolf, 2010). However, it is very interesting that in case of GaAs and GaN, REEs do not make bonding with excitons because they do not create new energy levels (Coutinho, Jones, Shaw, Briddon, & Öberg, 2004). In general, REEs incorporated in GaN need other dopants or defects to perform as effective exciton traps. A defect presence in the lattice structure of the semiconductor can pair with REE; therefore, high concentration of REE ($\sim 1\%$) is needed to be active and emit strong PL. In contrast, some other REEs, such as Er, Tm, and Eu when used in AlN, can introduce a deep donor level above the valence band (Vantomme et al., 2001). This result could be partly because of the wide bandgap (6.12 eV) due to ionic characteristic of the semiconductor (O'Donnell & Dierolf, 2010). Pr, Eu, Er, and Tm are being used as doping REEs in III-N semiconductors, frequently.

Reviewing milestones for developing GaN devices containing RE shows that the progress of GaN:RE in light-emitting materials has been significant (Steckl, Park, & Zavada, 2007). Fig. 11.6 shows some alloys of AlN, GaN, and InN from III-N compounds, which have a variety of bandgap energy values. They are desirable materials for laser application although some efficient wavelengths such as red and green are hard to be achieved by using these materials. Adding RE as dopant can help achieving a wider range of wavelengths in the spectrum. In addition, when codoping with multiple RE ions, we can tailor the final wavelength range.

One of the applications of RE in situ doping of III-N semiconductors is producing red, green, and blue (RGB) emissions for different colors and white light (Steckl et al., 2001). Thin-film electroluminescence (TFEL) of phosphors based on II–VI semiconductors that have uses in flat panel displays show multiple color capabilities. GaN:RE TFEL devices as another type of electroluminescence devices (ELDs) provide desirable colors with improved crystal quality and brightness, and superior durability (CN105932125B, 2021).

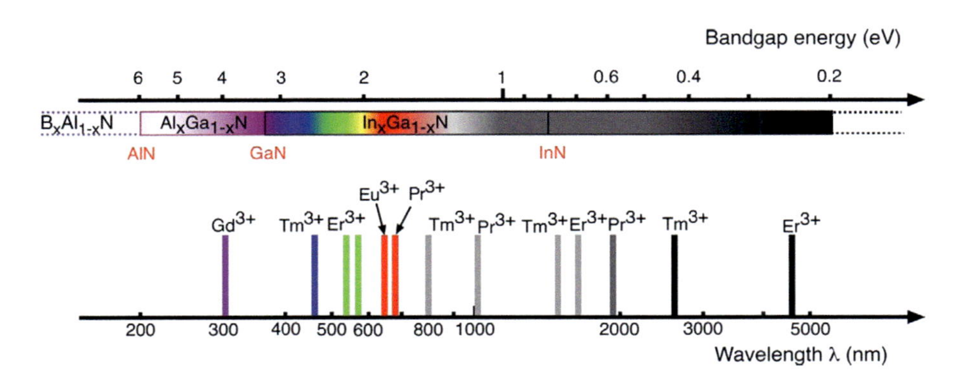

Figure 11.6 Emission wavelengths from selected transitions in rare earth ions and associated bandgap energies of alloys of III-N compound semiconductors.
Source: Reprinted from Steckl, A. J., Park, J. H., & Zavada, J. M. (2007). Prospects for rare earth doped GaN lasers on Si. *Materials Today*, *10*(7–8), 20–27 with permission from Elsevier.

There are two main types of ELDs, voltage-controlled ELDs and switchable color ELDs. In the former, increasing bias voltage is the cause of progress in red and green light, whereas in the latter polarity of applied bias is the origin of the color switch. Fig. 11.7 shows different color emissions for Eu, Er, and Tm (red, green, and blue, respectively) of GaN:RE ELDs. GaN:Er emission is at ~1550 nm, an important wavelength region in telecommunication applications (Steckl et al., 2001).

The main reason behind high brightness of GaN:Er ELDs (500–1000 cd/m^2 when biased with 50–100 V) is partly because hosting large concentrations of RE ions by GaN. However, when compared with other semiconductors that were not successful in incorporating such high RE concentrations without degradation of the host material (such as Si and GaAs), the ionic size mismatch between GaN and RE ions is more significant. Fig. 11.8A compares the ionic radii of Er-doped GaN, Si, ZnS, SrS, and AlN (Lide, 1997; Lozykowski & Jadwisienczak, 1997; Shannon, 1976) (the table) and depicts the experimental data for the Ga–N and Er–N bond lengths (Steckl et al., 2001). The high concentration of RE in GaN is attributed to partially covalent strong bonding nature of Ga–N and Er–N compared with the largely ionic weakly bonded II–VI compounds (largely ionic). In addition, due to the triple charges in GaN, it is capable to readily host trivalent RE.

NIR semiconductor lasers can be integrated with the CMOS silicon technology. They are also compatible with optical fiber–based devices. Er-doped semiconductors are prominent candidates since Er can provide light emission at the 1.5 μm via the $^4I_{13/2} \rightarrow {}^4I_{15/2}$ radiative transition (Kenyon, 2005). This wavelength is both within the minimum absorption band of optical fibers and is safe for eyes (Ho et al., 2020; Vinh, Ha, & Gregorkiewicz, 2009). GaN semiconductor has a direct bandgap of ~3.3 eV which makes it an attractive host material for RE doping. The low degree of nonphotonic quenching and the possibility of producing a strong

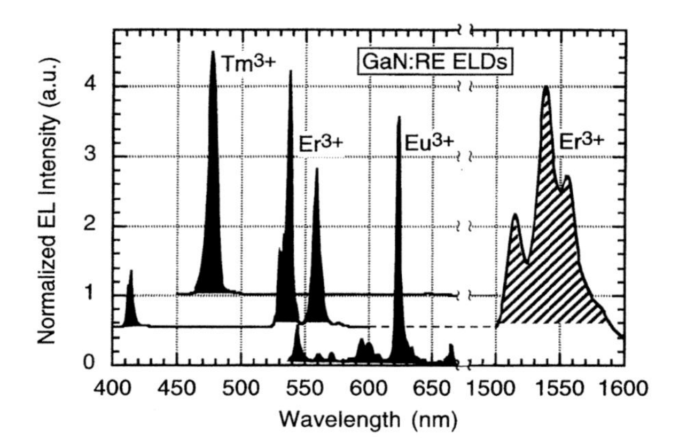

Figure 11.7 The emission spectra of Tm-, Er-, and Eu-doped GaN ELDs showing the visible and IR wavelengths of interest. All spectra are self-normalized and are not readily comparable to each other. The visible peaks for each RE correspond to saturated RGB colors, which can be utilized for single-, multiple-, mixed-, and full-color applications. The ~ 1550-nm emission from GaN:Er is of particular interest in telecommunication applications. *Source*: Reprinted from Steckl, A. J., Heikenfeld, J., Lee, D. S., & Garter, M. (2001). Multiple color capability from rare earth-doped gallium nitride, *Materials Science and Engineering B*, *81*(1), 97–101, with the permission from Elsevier.

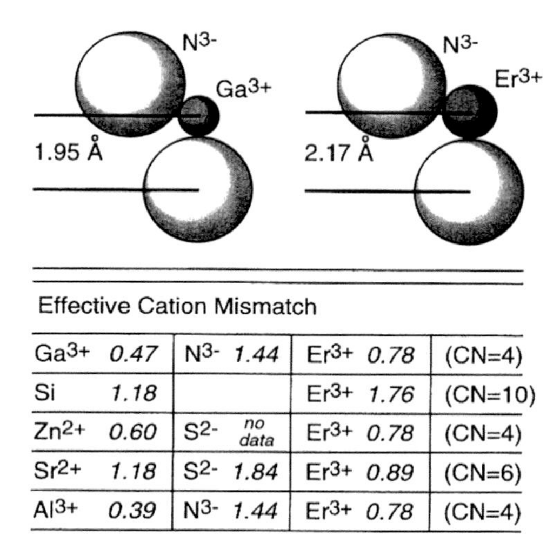

Effective Cation Mismatch

Ga^{3+}	*0.47*	N^{3-} *1.44*		Er^{3+}	*0.78*	(CN=4)
Si	*1.18*			Er^{3+}	*1.76*	(CN=10)
Zn^{2+}	*0.60*	S^{2-}	*no data*	Er^{3+}	*0.78*	(CN=4)
Sr^{2+}	*1.18*	S^{2-}	*1.84*	Er^{3+}	*0.89*	(CN=6)
Al^{3+}	*0.39*	N^{3-}	*1.44*	Er^{3+}	*0.78*	(CN=4)

Figure 11.8 Drawing (top) indicating experimental data for the Ga–N and Er–N bond length and atomic radii in wurtzite GaN:Er. Table (bottom) of effective ionic radii and coordination number (CN) for Er-doped GaN, Si, ZnS, SrS, and AlN.
Source: Reprinted from Steckl, A. J., Heikenfeld, J., Lee, D. S., & Garter, M. (2001). Multiple color capability from rare earth-doped gallium nitride, *Materials Science and Engineering B*, *81*(1), 97–101, with the permission from Elsevier.

emission at RT makes Er-doped GaN material an excellent candidate for photonic applications (Ho et al., 2018).

The first strong IR PL was observed in Er-implanted GaN thin films at 1.54 μm by Wilson et al. (1994). After that, many research works have been devoted towards developing Er-doped III-N semiconductors. In particular, GaN epilayers doped with Er ions showed a promising performance in terms of reduced thermal quenching and improved luminescence intensity at various temperatures.

Three mechanisms for optical excitation of III–V nitride semiconductors can be discussed (Zavada et al., 2000): (1) above-bandgap excitation, (2) below-bandgap excitation, and (3) direct excitation of the RE ions. In above-bandgap method, the created pair of electron-hole can do the energy transfer to RE ions and excite their 4f electrons to a higher energy level, which will emit luminescence in relaxation. This method is very common in case of Er-doped semiconductors and is usually performed by the use of a laser with higher energy than the bandgap of the host material. Energy transfer to Er ions in the second method is carried out by defects already present in the semiconductor host. Both first and second methods are representative of an indirect excitation of the Er ions. In the third method, however, Er ions are directly resonated by the laser and the 4f electrons receive equal energy that they require to excite to the next energy state (Kim et al., 1997; Thaik, Hömmerich, Schwartz, Wilson, & Zavada, 1997). George et al. (2015) reported direct evidence of resonant excitation through internal 4f orbitals and band-to-band excitation for Er^{3+} ions in GaN epilayers. The Er optical centers relating to defects showed different PL peak properties. Since 1.54-μm emission can be excited by either mechanism and is resulting from isolated Er centers, the band-to-band mechanism showed much higher excitation cross section. However, the 1.54 μm has the largest excitation value, which is originating from band-to-band excitation.

Temperature and defect concentration of semiconductors can have influence on the luminescence emission from Er-doped semiconductors. The ideal emission is the one that is not significantly influenced by these factors. Such materials can find widespread applications in laboratory and military systems. Zavada et al. (2000) reviewed the effect of temperature and oxygen and carbon impurities on PL emission of above-bandgap and below-bandgap excited Er-doped GaN semiconductors. When studied the effect of temperature, it was shown that in case of above-bandgap excitation, the PL intensity at 550 K decreased by only about 10% relative to its value at 15 K. In case of below-bandgap excitation, this value was about 50%. In addition, the effect of below-bandgap excitation was studied on GaN:Er samples with different impurities. The results showed that the GaN:Er sample with increased oxygen and carbon content has intensified PL emission than the one with low content of these elements. However, when used above-bandgap excitation, the influence of oxygen and carbon on emission was insignificant. Therefore when studying the effect of any external factor, it is important to take into the account the method of excitation (Zavada et al., 2000).

Since the discovery of promising optical properties associated with Er doping of GaN, research teams sought to explore possible ways for improving the emission intensities. Generally, in RE-doped semiconductors the probability of emission, which determines the final intensity, is affected by the host while the emission wavelength is not. One smart approach that has resulted in promising results is implementing quantum well structures around Er ions to increase the carrier density (Al Tahtamouni, Stachowicz, Li, Lin, & Jiang, 2015; Arai et al., 2015; Fragkos, Tan, Dierolf, Fujiwara, & Tansu, 2017; Li, Lin, & Jiang, 2016). In order to enhance excitation of the confined Er ions in the barriers, above-bandgap or near-edge-bandgap excitation mechanisms can be used. Using this approach, Al Tahtamouni et al. (2015) designed GaN/AlN multiple quantum well (MQW:Er) structures with an emission intensity nine times higher than that of the conventional GaN:Er epilayers. Fig. 11.9 shows layer structure of MQW:Er samples grown by MOCVD. GaN wells with Er incorporated and undoped AlN barriers grew on two AlN buffer layers (30 and 100 nm) deposited on a c-plane sapphire substrate, and a 1.0-μm AlN template. They showed that with engineering the well and the barrier widths, the PL emission at 1.54 μm could be adjustable. The well widths between 1.0 and 1.5 nm along with the largest possible barrier width (thin enough to allow penetration of the electron wave functions) were found to be the optimum values for achieving high PL intensity (Al Tahtamouni et al., 2015).

Er-doped GaN materials are one of the most appealing products in RE-doped III-N group. However, as mentioned previously (Section 11.4.1.1), application of

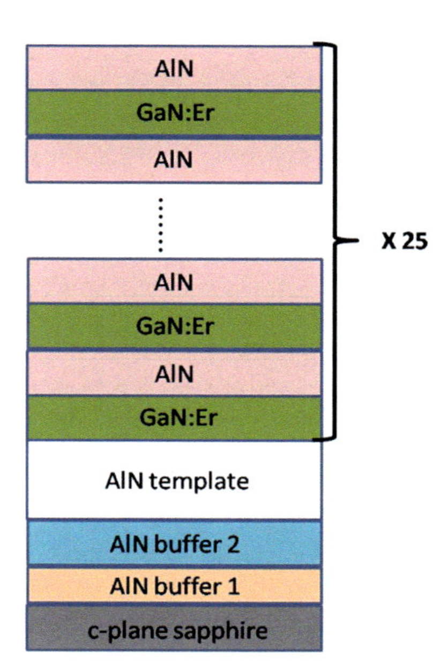

Figure 11.9 Schematic layer structure of Er-doped GaN/AlN multiple quantum wells (MQWs:Er) grown on AlN/sapphire template.
Source: Reprinted from Al Tahtamouni, T. M., Stachowicz, M., Li, J., Lin, J. Y., & Jiang, H. X. (2015). Dramatic enhancement of 1.54 μ m emission in Er doped GaN quantum well structures. *Applied Physics Letters*, *106*(12), 121106, with the permission of AIP Publishing.

some other REEs such as Eu and Tm is also of great interest in this field. Specifically, Eu-doped GaN are promising materials for red LEDs. InGaN/GaN is the most common material for covering green and blue colors in LEDs and is widely used in commercial applications. Much efforts have been devoted for developing InGaN/GaN-based red LEDs with successful outcomes (Mitchell, Dierolf, Gregorkiewicz, & Fujiwara, 2018). However, broad emission spectra, which are sensitive to current injection, largely limits their applications. Doping GaN with Eu^{3+} is an alternative solution to achieve an efficient red GaN-based LED. Since the first development of GaN:Eu LED, many advancements in terms of synthetic parameters and the final layer structures have been achieved (Heikenfeld, Garter, Lee, Birkhahn, & Steckl, 1999; Mitchell et al., 2017; Nishikawa, Kawasaki, Furukawa, Terai, & Fujiwara, 2009; Zhu et al., 2016, 2017). Zhu et al. (2017) successfully increased the emission rate by developing a multilayer structure of GaN: Eu and reducing the growth temperature to 960°C. They generated a peak value of 375 μW at 20 mA with an external quantum efficiency (EQE) of 4.6% by integrating 100 layers of varying 6-nm GaN and 3-nm GaN:Eu. Even higher EQE was achieved by further optimization of the processing condition and a device with the maximum output power of ~1.25 mW at 20 mA and a maximum EQE of 9.2% was developed (Mitchell et al., 2018).

The results of after-growth heat treatment and Eu insertion followed by crystallization on the AlGaN/GaN diode structure that contains Mg-doped GaN surface p-cap layers were investigated by Ben Sedrine et al. (2018). They established a high-temperature and high-pressure condition (1400°C in 1 GPa N_2) applied on a diode structure and showed that even at this harsh environment the crystalline quality of the diode can be stable. They also showed that the postannealing could partly remove the implantation defects and recover by optically activating the Eu^{3+} in the diode structure. They developed a model depending on the various excitation bands found in the host material and demonstrated that there is a transfer of energy between the superlattice excitons of AlGaN/GaN and the ions of Eu^{3+}. This will enhance and induce the red fluorescence of the ion by broadening the energy pathway of the excitons (Ben Sedrine et al., 2018).

More recently, Timmerman et al. studied the effect of optical and electrical excitation on Eu-doped GaN system. They reported that if the pump fluences are small, the amount of carrier traps that competes in energy transfer from the host material is the limiting factor. On the other hand, in case of large pump fluences, since the number of high-efficiency Eu^{3+}ions are limited and the excitation cross section is small, the quantum efficiency can be limited. They were also able to optimize the excitation parameters and show that at low temperatures, the EQE can reach 46%, showing the high potential for this material as an efficient light emitter (Timmerman et al., 2020).

11.4.1.2 Magnetic properties

A main group of the magnetic RE-doped semiconductors is called DMS, which stands for dilute magnetic semiconductors. DMS maerials are prepared from conventional semiconductors, but their magnetic properties are due to an impurity dopant. They are of interest because of their unique spintronics (short for "spin electronics") properties, which are potential for electronic-magnetic devices. These properties are suitable for using in spinotronic devices that in addition to charge of electrons in semiconductors exploit the spin of magnetic materials (Bonanni, 2007). These materials have been known for a long time; however, the DMS of GaAs and InAs were demonstrated not until 1989 (Munekata et al., 1989). Since then, the main barrier precluding real-world applications of this group of materials is their Curie temperature (TC), which was below RT and even after years of research is still limited to ~ 160 K (Nepal et al., 2016). After finding the theoretical p-type wide bandgap of Mn-doped DMS (Dietl, Ohno, Matsukura, Cibert, & Ferrand, 2000) and the follow on intense research in this field (Mukherji, Mathur, Samariya, & Mukherji, 2016; Viswanatha, Pietryga, Klimov, & Crooker, 2011), RE dopants received great attentions, as well. In addition to removing the temperature barriers, RE ions can provide the desired properties such as ferromagnetism at lower required concentrations than transition metals (Nepal et al., 2009). Five REEs (Er, Gd, Eu, Nd, and Tm) are the focus of the RT magnetic materials, mainly GaN thin films, functioning at RT.

Timmerman et al. (2020) reported the first study on magnetic properties of Er-doped GaN on a film grown on sapphire substrates (0001) by MBE method. The sample exhibited a predominant paramagnetic character with coexistence of ferromagnetic order. Zavada et al. used MOCVD method for producing GaN doped with Er with concentration up to $\sim 10^{21}$ cm^{-3} having saturation magnetization following a nearly linear fit to the Er concentration (Nepal et al., 2009). Based on their study and some other reports (Mitra & Lambrecht, 2009; Ugolini, Nepal, Lin, Jiang, & Zavada, 2006), MOCVD is the most effective in producing high-quality GaN:Er samples. Some evidence of coupling the magnetic states of Er ions to electronic states of the host under a magnetic field were reported by Woodward et al. (2011).

By combining some imaging techniques, namely atom probe tomography, PL spectroscopy, and atomic force microscopy, Mitchell et al. characterized GaN:Eu and GaN:Eu for their dopant distribution as well as surface and structural properties. The results showed that both Eu and Er accumulate within host GaN threading dislocations (Mitchell et al., 2020). Mitchell et al. (2017) identified two Eu centers in GaN:Eu, with dependency on temperature and the wavelength of the excitation laser, being related through the charge state of a local vacancy defect. They showed that this complex have a magnetic moment that varies with the number of captured carriers, explaining the magnetic moment observed in GaN:Eu systems.

Relevant magnetization of Gd-doped GaN was investigated and reported that ferromagnetism in Gd:GaN was due to N-site polarization or some potential extrinsic processes in these samples, such as magnetic polarization of defects or residue oxygen (Woodward et al., 2011). The study by Mitra and Lambrecht (2009) also confirmed that interstitial nitrogen and oxygen defects in GaN:Gd system could be likely the source of defect-induced magnetism. By using reactive MBE method, the formation and magnetic properties of GaN:Gd layers, developed directly on 6H-SiC (0001) surfaces, have been extensively investigated by Brandt, Dhar, Pérez, and

Sapega (2010), and with a Gd content smaller than 10^{16} cm^{-3}, the sample exhibited ferromagnetic properties above ambient temperature. A colossal Gd moment emerging from the Gd atom-induced deep spin polarization of the GaN network was detected, which was attributed to point defects or defect complexes. The work of Helbers, Mitchell, Woodward, and Dierolf (2015) supported these findings, as well. They reported the so-called anomalous asymmetry behavior in the RE-ion emission strength from Zeeman-split lines. They compared the emission spectra for applied fields parallel and antiparallel to the polar *c*-axis of a GaN:Nd sample and observed a difference that scaled with the degree of ferromagnetism of the samples on different substrates.

One of the strategies for increasing RT magnetic properties is codoping GaN:Er or GaN:Eu with Si. Magnetic behavior of the codoped GaN:(Eu,Si) is shown in Fig. 11.10 and clearly indicates the improved saturation magnetization of GaN:Eu by a factor of ~9 by adding Si as codopant (Wang, Steckl, Nepal, & Zavada, 2010). Kumar and Zavada (2016) reported ab initio pseudopotential calculations to find out how Si can facilitate the incorporation of RE ions in wurtzite lattice as well as in the $(GaN)_n$ structure. They reported that due to the energetically highly unfavorable deformation from oversized Eu doping, the local symmetry on a Ga site is lowered costing 1.84 eV. Adding Si can compensate this effect and lower the energy of strain by introducing undersized Si dopant to the lattice. Therefore the magnetic effects are enhanced due to the codoping of Si. They also reported a similar result when doped $(GaN)_n$ nanoclusters with either Gd or Nd.

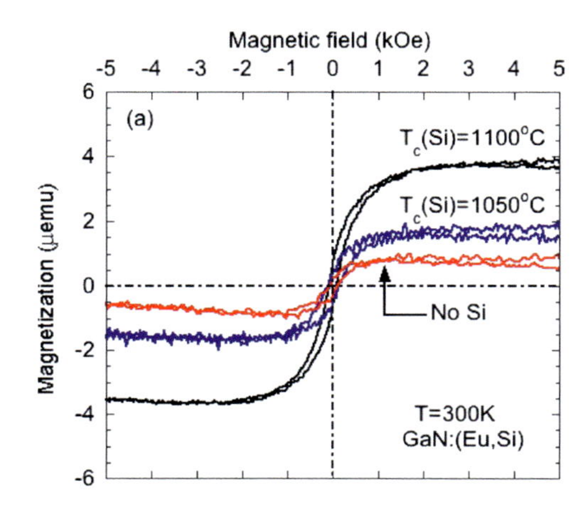

Figure 11.10 Hysteresis curves measured at RT for codoped Ga-N: (Eu,Si) films: magnetization measurements versus applied magnetic field data points for an undoped GaN film are also shown.
Source: Reprinted from Wang, R, Steckl, A. J., Nepal, N., & Zavada, J. M. (2010). Electrical and magnetic properties of GaN codoped with Eu and Si. *Journal of Applied Physics*, *107*(1), 013901, with the permission of AIP Publishing.

Hite and Zavada (2019) have recently reported a thorough review on magnetic properties of wide bandgap III-N semiconductors doped with REEs. Based on their report, MBE and MOCVD are the most successful methods for the synthesis of RE-doped III-N semiconductors. In addition to GaN, which has been the main and primary host material, other systems such as AlN, InGaN, and AlGaN alloys have shown promising results. They recognized the GaN:Gd as the most successful III-N:RE semiconductor in terms of magnetic properties, mainly because of the very low concentrations required for "colossal" values of μ_{eff} per Gd atom (Dhar, Brandt, Ramsteiner, Sapega, & Ploog, 2005; Dhar et al., 2005, 2006).

11.4.2 Other III–V

In addition to III-N semiconductors, other III–V materials show promising optical and magnetic properties for a variety of optoelectronic applications. Some examples are GaAs, InP, GaP, etc. Same as the III-N, other RE-doped III–V semiconductors can be applied in LEDs, solid-state lasers, and optical integration due to electroluminescence properties (Pomrenke, Ennen, & Haydl, 1986). One of the first reports on RE-doped III–V semiconductors was by Ennen and Haydi (Pomrenke et al., 1986). They investigated characteristic 1.54 μm emission in Er-doped GaAs, lnP, and GaP as a function of annealing temperature, time, and method. They used three different annealing methods in the range of 400°C–1000°C. Among the different tested materials, GaAs:Er showed the highest optical activation with optimum Er emissions between 650°C and 800°C. They also provided details of the most successful annealing method. After this report, an intense effort was devoted towards synthesis of RE-doped III–V semiconductors and their PL properties mainly by MBE and metal–organic vapor phase epitaxy (MOVPE) in the next decade (Ennen, Wagner, Müller, & Smith, 1987; Fang, Li, & Langer, 1993; Kozanecki & Gröetzschel, 1988; Raczyńska, Fronc, Langer, Lemańska, & Stapor, 1988; Redwing, Kuech, Gordon, Vaartstra, & Lau, 1994; Rzakuliev et al., 1988; Seghier et al., 1994; Taguchi, Kawashima, Takahei, & Horikoshi, 1993; Thonke, Hermann, & Schneider, 1988).

More recently, GaAs/transparent oxide heterojunctions such as ZnO (Du et al., 2007) and SnO_2 (Bueno, Scalvi, Saeki, & Li, 2015; Bueno, Ramos, Bailly, Mossang, & Scalvi, 2020; Pineiz, de Morais, Scalvi, & Bueno, 2013) have been used instead of the pure GaAs films. As one of the initiatives, Pineiz et al. (2013) proposed combining Eu^{3+}-doped SnO_2 with GaAs, to produce new heterojunction of SnO_2:Eu^{3+}/GaAs and GaAs/SnO_2:Eu^{3+} with high density of structural and interfacial defects. They showed that higher electrical conductivity could be achieved by the heterojunction SnO_2:2% Eu/GaAs compared with the individual films.

11.5 Re-doped metal oxides

One of the large categories of Re-doped semiconductors are metal oxides such as ZnO, TiO_2, SnO_2, ZrO_2, SiO_2, and Al_2O_3. The main applications of this group of materials are varistors (Jiang, Peng, Zang, & Fu, 2013), photovoltaics (Kumar et al., 2017), magneto-optical (Pearton et al., 2003), photocatalytic (Khaki, Shafeeyan, Raman, & Daud, 2017; us Saqib, Adnan, & Shah, 2016), antibacterial (Prathap Kumar, Suganya Josephine, Tamilarasan, Sivasamy, & Sridevi, 2018), and gas sensors (Dey, 2018).

ZnO is a wide direct bandgap (3.37 eV) semiconductor with 60-meV excitation energy at RT; therefore, it can be a suitable host for ions (Shukla & Sharma, 2020). The first reports on RE-doped ZnO was in 1991 when Kouyate et al. reported synthesis of Er-, Sm-, and Ho-doped ZnO and investigated their PL properties (Kouyate, Ronfard-Haret, & Kossanyi, 1991a; Kouyate, Ronfard-Haret, & Kossanyi, 1991b; Ronfard-Haret, Kouyate, & Kossanyi, 1991). The research on this subject continued to pick up in the following years (Bachir, Kossanyi, & Ronfard-Haret, 1994; Bachir, Kossanyi, Sandouly, Valat, & Ronfard-Haret, 1995; Bachir, Sandouly, Kossanyi, & Ronfard-Haret, 1996; Ohtake, Hijii, Sonoyama, & Sakata, 2006). A spectra centered around 550 nm is the commonly observed peak for doped or undoped ZnO that can be seen in electroluminescence, cathodoluminescence, and photoluminescene of this material (Kouyate, Ronfard-Haret, & Kossanyi, 1992). The origin of these emissions was reported to be from intrinsic self-activated centers of ZnO, emissions which can be reabsorbed partly by the dopant (Kossanyi et al., 1990; Kouyate et al., 1990).

The purpose of doping ZnO is preparing active phosphors with high fluorescence efficiency. Interestingly, the RE-doped ZnO nanostructures show ferromagnetism property, as well, which makes them even more attractive for different applications. One of the main challenges in the synthesis of RE-doped ZnO is low solubility of the dopant in the host lattice and formation of two separate phases (Pearton et al., 2003).

Eu-doped ZnO was of great interest at the initial stages of the research (Armelao et al., 2008; Du, Zhang, Sun, & Yan, 2008; Lima, Sigoli, Davolos, & Jafelicci, 2002; Trandafilović, Jovanović, Zhang, Ptasińska, & Dramićanin, 2017; Wang et al., 2011) for photonic and photocatalytic applications. Armelao et al. (2008) synthesized Eu^{3+}-doped zinc oxides by the sol–gel method and showed that after annealing at 800°C, cubic Eu_2O_3 formed. They also indicated that the luminescence properties of the Eu^{3+}-doped ZnO was strongly dependent on the annealing conditions. The undoped ZnO host and ZnO:Eu showed emission bands in the 300–420 nm and 390–550 nm wavelength ranges, respectively, demonstrating the presence of a spectral overlap between the two materials. This overlap provides a multicolored emission (green and red), a potential property for making this material as a candidate for light-emitting devices. Later on, Ahmed, Szymanski, El-Nadi, and El-Sayed (2014) confirmed this observation by conducting a more in-depth study on the ZnO annealing. They showed that in unannealed films the dynamics is consistent with energy transfer from O vacancies to the dopant, while in annealed film acceptor-type defects such as Zn vacancies play a role in energy transfer process. The schematic presentation of the energy levels involved in energy transfer from ZnO to Eu^{3+} and the final emission from there is shown in Fig. 11.11 (Ahmed et al., 2014). The defect states serve as intermediates and create the defect-level emissions. Energy transfer from ZnO host to a Eu^{3+} results in excitation. The electron within the initially unoccupied defect state relaxes back to the VB;

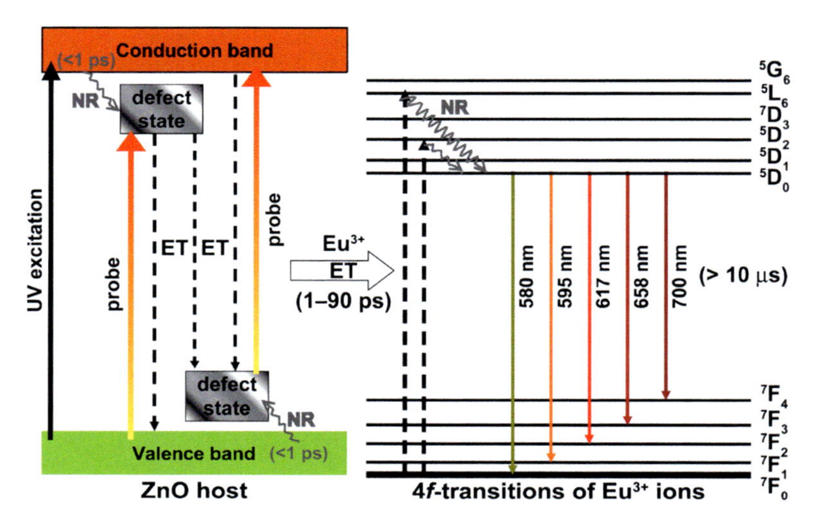

Figure 11.11 Energy-level diagram illustrating possible pathways involved in pump-probe spectroscopy of ZnO defects, nonradiative relaxation (NR), energy transfer (ET, dashed lines), and Eu^{+3}-radiative transitions. Time constants for major processes are shown, with ZnO excitonic and defect-level emission omitted because of quenching due to efficient ET. *Source*: Reprinted with permission from Ahmed, S. M., Szymanski, P., El-Nadi, L. M., & El-Sayed, M. A. (2014). Energy-transfer efficiency in Eu-doped ZnO thin films: The effects of oxidative annealing on the dynamics and the intermediate defect states. *ACS Applied Materials & Interfaces*, *6*(3), 1765–1772, Copyright (2014) American Chemical Society.

alternatively, if an electron in a conduction band undergoes relaxation to a deep hole trap, it could result in emission or energy transfer to Eu^{3+}. A green defect emission at ~0.88 eV is an example of such emission resulting from formation of a complex between two defects (Zn and O vacancies) (Cheng et al., 2010; Li et al., 2013).

Two other commonly used RE ions for doping ZnO are Ce^{3+} and Er^{3+}. Ce^{3+} is different from other RE ions since it only has one electron in the first shell of the 4f orbital and its luminescence transition emitted from the high-energy 5d orbitals is highly dependent on the host lattice. Ce-doped ZnO nanostructures has shown potential as photocatalysts for the degradation of dyes (Fangli, Ning, Zhang, & Yingzhong, 2010; Karunakaran, Gomathisankar, & Manikandan, 2010; Li et al., 2013). As discussed in previous sections of this capture, Er is a good candidate for addition to wide bandgap semiconductors to produce 1.54-μm IR emission. Not only is ZnO a wide bandgap semiconductor host (3.3 eV at 300 K) but also it has high electrical conductivity, which is required for current injection optoelectronic devices (Komuro et al., 2000; Shukla & Sharma, 2020). Er-doped ZnO films can be used in a variety of applications including gas sensors (Hastir, Kohli, & Singh, 2017), solar cells (Kumar et al., 2018), LEDs (Shi, Hu, Wang, Liao, & Ling, 2019), photo-induced piezooptics (Williams, Hunter, Pradhan, & Kityk, 2006), and nonlinear optics (Chen, Yao, & Hu, 2019).

Another RE used for doping ZnO is Dy^{3+}. The dysprosium ion is usually used when blue and yellow emissions are needed to be adjusted to produce a white emission. One of the characteristics of the Dy^{3+} ion is that unlike many other RE ions, its emission can be easily affected by crystal field of the host lattice since it is from transition of 5d→4f. Ho, Sm, Pr, Nd, and Gd ions are other REEs that have been added to ZnO as dopant that are extensively reviewed by Kumar, Nagpal, and Gupta (2017).

There are two main challenges for incorporating REs in ZnO network. First, in the ZnO network, low concentration of RE ions can be dissolved originating from difference in the ionic radius and charge between the dopant and host atoms. Second, the energy levels between RE ions and ZnO valence and conduction bands are not matching (Kumar et al., 2017; Zeng, Yuan, & Zhang, 2008).

It was briefly addressed in previous parts of this chapter (Section 11.1.2) that, due to the protecting impact of the 5s and 5p orbitals, the PL property of RE dopants were not based on their environment. There are two other distinctive processes in RE-doped ZnO, namely upconversion (UC) and downconversion (DC). These processes are especially important in dye-sensitized solar cells (DSSCs), perovskite solar cells (PSC), or organic solar cells and can use photon conversion processes to modify NIR/ultraviolet (UV) radiation into visible emission (Kumar et al., 2017; Yao et al., 2015). UC is a method of luminescence whereby two or even more low-energy infrared photons are absorbed and then converted into a high-energy visible photon. In the DC phase, on the other hand, the UV portion of the solar radiation is transformed into visible light (Yao et al., 2015).

Different synthetic methods have been used for synthesis of RE-doped ZnO, so far (Shukla & Sharma, 2020). Some of them are electrochemical deposition (Li, Lu, Zhao, Su, & Tong, 2008), sol–gel (Yang et al., 2008), hydrothermal (Yang et al., 2008), refluxing (George, Sharma, Chawla, Malik, & Qureshi, 2011), coprecipitation (Djaja & Saleh, 2013), hydrothermal method with no postannealing (Jung et al., 2012), solution combustion (Silambarasan, Saravanan, Ohtani, & Soga, 2014), and sputtering (Ahmed et al., 2014).

Another simple, low-lost, and commercially available metal oxide that has extensive photocatalytic applications if TiO_2. This material is of great interest since it shows attractive properties such as nontoxicity and chemical stability. When used as a photocatalyst, it can take the highest oxidation rate of the many photoactive metal oxides investigated. There are many reviews and books on synthesis and properties of TiO_2 that discuss its benefits in more details (Chandra, 2017; Fujishima, Hashimoto, & Watanabe, 1999; Ghosh, 2018; Hussain & Mishra, 2020; Yang, 2018). One of the limitation of TiO_2 photocatalyst though is their wide bandgap ($Eg = 3.2$ eV) that absorbs near-UV light in the solar spectrum. One of the approaches that could increase the visible light adsorption is doping with different elements. Different methods including sol–gel process (Gao, Liu, Lu, & Liu, 2012), hydrothermal (Gao et al., 2012), electrospinning (Gao et al., 2012), and magnetron sputtering (Gao et al., 2012) can be used for doping. Although metal-ion implantation is a good technique, it is not appropriate in TiO_2 case since RE ions are too large to be incorporated into the TiO_2 matrix (Bingham & Daoud, 2011). Doping elements can create new energy levels resulting in a red shift in the absorption spectra of TiO_2, which helps absorption of visible light of the solar spectrum and therefore results in more efficient photocatalytic activity. The new energy levels created by dopants can be a new energy state for the excited electrons from the VB, remaining a highly oxidizing hole that can readily react with water to produce reactive radicals (Bingham & Daoud, 2011). Xu, Gao, and Liu (2002) used different RE ions (RE = La^{3+}, Ce^{3+}, Er^{3+}, Pr^{3+}, Gd^{3+}, Nd^{3+}, Sm^{3+}) to prepare RE/TiO_2 photocatalysts by sol–gel method. The diffuse reflectance spectra of the doped samples showed red shifts in the bandgap transition in the order of $Gd^{3+} > Nd^{3+} > La^{3+} > Pr^{3+}$ $(Er^{3+}) >$ $Ce^{3+} > Sm^{3+}$. This shift to the longer wavelength was attributed to the charge-transfer transition between the RE-ion f electrons and the TiO_2 CB or VB. The highest enhancement in photocatalytic activity in the degradation of nitrite was obtained for Gd^{3+}-doped TiO_2, as was expected from the results of the red shift in the bandgap.

During the past two decades, a lot of research and review papers reported development of various types of RE-doped TiO_2 (El-Bahy, Ismail, & Mohamed, 2009; Hassan, Amna, Yang, Kim, & Khil, 2012; Liu, Yu, Chen, & Li, 2012; Reszczyńska et al., 2015; Štengl, Bakardjieva, & Murafa, 2009; Zinatloo-Ajabshir, Salavati-Niasari, Sobhani, & Zinatloo-Ajabshir, 2018). Bingham and Daoud (2011) reported a thorough review on RE-doped TiO_2 for visible-light active photocatalysts until 2010. In their excellent review, they reported the common synthetic techniques and RE atoms for obtaining the most efficient TiO_2 photocatalyst. They reported that large lanthanide ions are able to incorporate on TiO_2 surface and form oxide species that can make a red shift in the absorption edge of TiO_2. In addition to incorporation of new energy levels in the TiO_2 bandgap, in some elements such as gadolinium, scandium, and samarium a charge imbalance occurred that caused delaying in the recombination reaction. The best photocatalytic activity was observed for gadolinium-doped TiO_2 due to its stable half-filled electron configuration (Bingham & Daoud, 2011).

Another important application of RE doping in TiO_2 is UC/DC DSSCs. Due to their simplicity, availability, cost-effectiveness, and versatile application compatibility, DSSCs have attracted great attention since their first development (Amiri et al., 2018; Gholami, Mir, Masjedi-Arani, Noori, & Salavati-Niasari, 2014; Mir & Salavati-Niasari, 2012, 2013a, 2013b; Mir, Lee, Paramasivam, & Schmuki, 2012; Wang, Batentschuk, Osvet, Pinna, & Brabec, 2011). In DSSCs, the semiconductor substrate is sensitized with a highly visible light absorbing dye. A sandwich-structured TiO_2 DSSC consists of transparent conducting oxide (TCO) layer substrate glass, a thin dye-sensitized TiO_2 film, iodide/iodine (I_3^-/I^-) electrolyte, and Pt-coated TCO glass counter electrode (Fig. 11.12).

The materials of solar cells could react to a restricted window of solar photons with energy fitting the band gap of the material. In UC, the absorption of two or more photons to the emitted light of shorter wavelengths could allow photons with energy smaller than the bandgap to be used. The excess energy is usually wasted as other forms such as heat. This issue has been resolved to a great extent by using UC and DC materials. UC materials can convert 900–1600 nm light to another wavelength within the response range of solar cells. On the other hand, DC materials can convert the high-energy photons with frequencies <400 nm to the required energy for solar cells. UC converter can be coated at the back side of the solar cell while DC layer should be placed at front side of the solar cell (Rajeswari et al., 2020).

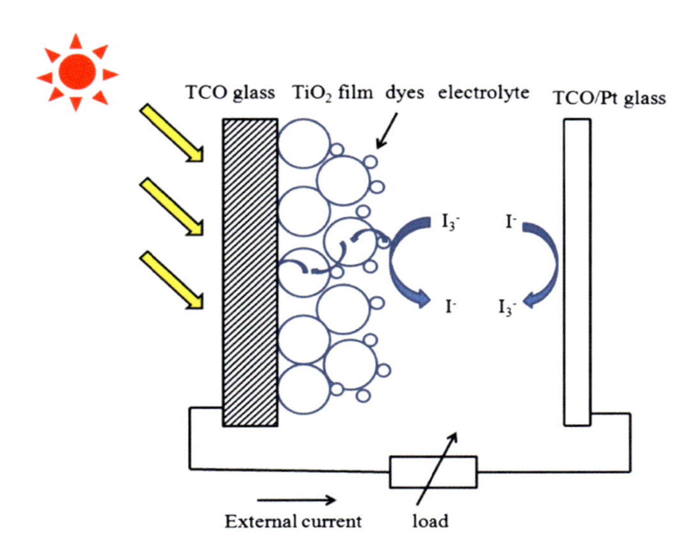

Figure 11.12 Schematic diagram of the dye-sensitized solar cell.
Source: Reprinted from Gong, J., Liang, J., & Sumathy, K. (2012). Review on dye-sensitized solar cells (DSSCs): Fundamental concepts and novel materials. *Renewable and Sustainable Energy Reviews*, *16*(8), 5848–5860, with the permission from Elsevier.

In UC/DC DSSCs, an additional layer is deposited on TiO_2 thin film. This can be done via different low-temperature solution coating approaches, such as spray, spin, dip coating, doctor blade, and screen printing (Rajeswari et al., 2020). Using aforementioned techniques, DC layer was deposited in two ways on DSSC and PSC devices. As shown in Fig. 11.13A and B, two methods were used to deposit DC layer on TiO_2 surface. One method is depositing the DC layer on the back surface of the fluorine-doped tin oxide (FTO) glass substrate and the other one is adding it to the TiO_2 paste and coated on the front side of the FTO glass substrate. However, depositing UC layer is only done on the TiO_2-coated FTO glass substrate, as shown in Fig. 11.13C. It has been shown in many recent studies that incorporation of DC/UC phosphor materials into the DCSs can improve the photovoltaic performance (Akman, Akin, Ozturk, Gulveren, & Sonmezoglu, 2020; Kaur, Mahajan, & Singh, 2020; Kharel, Zamborini, & Alphenaar, 2018; Li, Ågren, & Chen, 2018). Different REE including Eu, Sm, Nb, Pr, Nd, Sm, Gd, Er, and Yb have been investigated for this application, so far (Cavallo et al., 2015; Wu et al., 2010).

11.6 RE-doped perovskite

Three-dimensional (3D) semiconductor perovskite by a general formula of ABX_3, where A is an organic methylammonium (CH_3NH_3) or formamidinium ($NH = CHNH_3$) ion; B is Pb, Sn, or Cd ion; and X can be a halogen ion such as I^-,

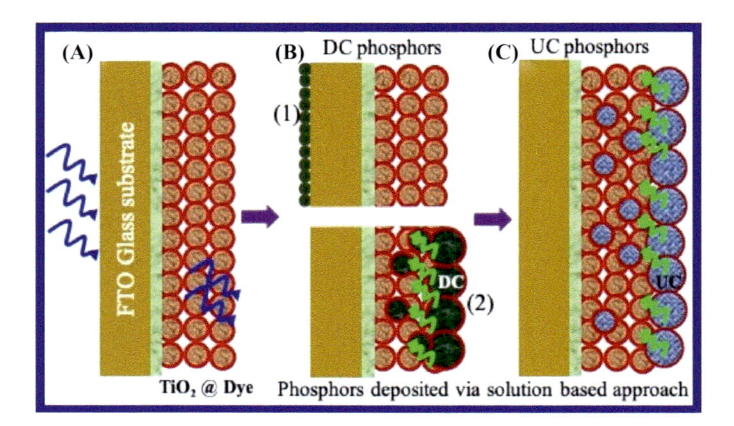

Figure 11.13 Schematic illustration of (A) pristine TiO_2 and (B, C) deposition of DC and UC phosphor layer–based DSSC devices.
Source: Reprinted from Rajeswari, R., Islavath, N., Raghavender, M., & Giribabu, L. (2020). Recent progress and emerging applications of rare earth doped phosphor materials for dye-sensitized and perovskite solar cells: A review. *The Chemical Record*, *20*(2), 65–68, with the permission from Wiley.

Br^-, or Cl^- have attracted significant interests (Ansari et al., 2020). They can be used in a variety of applications, such as solar cells, photodetectors, lasers, LEDs, etc. (Jung, Chueh, & Jen, 2015; Kovalenko, Protesescu, & Bodnarchuk, 2017). In recent years, the research on application of perovskite materials in solar cells has been greatly increased because of their high photovoltaic yield comparable with commercial silicon solar cells (Heikenfeld et al., 1999). Their light absorption window is very broad and covers the whole visible and NIR range (400–800 nm). The typical structure of PSCs, shown in Fig. 11.14, consists of a thin perovskite absorber layer ($\sim$300 nm) between an electron transport layer and a hole transporting layer (Ansari et al., 2018, 2020).

Although perovskite is an attractive material for the mentioned applications, it has some inherent shortcomings. The main issues are its low power conversion efficiency (PEC) and limited EQE as well as poor stability against oxygen, humidity, etc. One approach that has shown promising improvement is incorporating RE ions that can result in the desired properties for the target applications.

Similar to DSSCs, REEs can be added to PSCs as DC or UC materials. Some studies have reported RE-doped nanomaterials as UC phosphors for enhancing the efficiency of PSCs (Qiao et al., 2018; Zeng et al., 2020). One of the most efficient RE-doped UC nanomaterials is β-$NaYF_4$:Yb^{3+}, the Er^{3+} nanoparticle UC phosphor with bright green emission (Dyck & Demopoulos, 2014; Hu et al., 2017). The Yb^{3+} ion in this compound can absorb photon under 980-nm laser excitation and transfer it to Er^{3+} ion for UC (Qiao et al., 2018). In addition, the frequency of emitted light would be in the area of 410–654 nm that correlates mostly with PSC optical absorption. RE-doped UC product can thus be used to enhance the PCE of

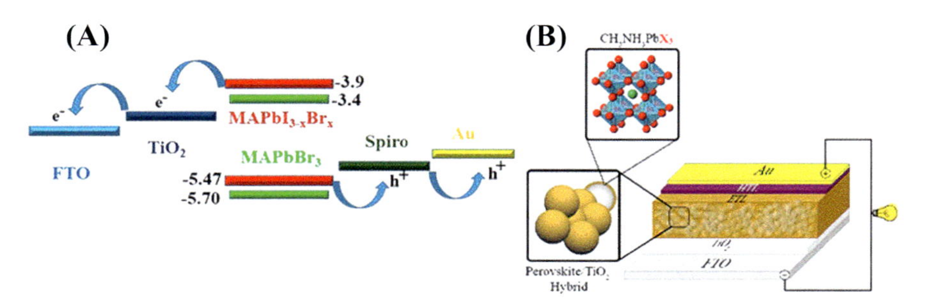

Figure 11.14 (A) Schematic illustration of energy-level diagrams of the FTO/TiO_2/ $MAPbI^{3-}xBrx$/MAPbBr3/Spiro-OMeTAD/Au. (B) Schematic illustration of the perovskite solar cell configuration, where a smooth and compact perovskite capping layer fully covers the mesoporous TiO_2 layer (mp-TiO_2) infiltrated with perovskite. *bl-TiO_2*, TiO_2 compact layer.
Source: Reprinted (adapted) with permission from Ansari, F., Salavati-Niasari, M., Nazari, P., Mir, N., Ahmadi, V., & Abdollahi Nejand, B. (2018). Long-term durability of bromide-incorporated perovskite solar cells via a modified vapor-assisted solution process. *ACS Applied Energy Materials*, *1*(11), 6018–6026, Copyright (2018) American Chemical Society.

PSCs, and TiO_2 or perovskite nanocrystal RE-ion doping is also an efficient way to boost PSC photovoltaic efficiency. Similarly, DC phosphors can be used in PCSs. In a study, UV degradation in TiO_2-based PSCs was mitigated using YVO4:Eu^{3+} (Chander et al., 2014). The obtained stability was more than 50% of its initial efficiency, while the PSC with no DC layer was only 35% stable.

In addition to solar cells, other applications of perovskite materials such as LEDs have also benefited from RE doping. In a very recent review, Chen et al. (2020) have summarized and discussed the coupling of lead halide perovskites and RE. Their discussion showed that to form a passivation film, RE ions are either distributed at the surface and grain boundaries of perovskite films or doped into the perovskite crystal structure to replace the Pb^{2+} ions partially. In addition, the B-site portion of lead-free perovskites such as $CsYbI_3$, $CsEuBr_3$, etc., may be employed as RE ions. Each of these possible areas cause RE ions to affect the optoelectronic performance of perovskites in various ways (Chen et al., 2020).

11.7 Synthesis methods of RE-doped semiconductors

11.7.1 Physical methods

11.7.1.1 Molecular beam epitaxy

For several years, the MBE technique has been introduced and currently the method can be successfully applied to the production of a broad variety of materials, notably group III–V semiconductor compounds (Feng, 2006). In an ultra-high vacuum

(UHV) setting, MBE uses concentrated beams of particles to provide the growing layer of a substrate surface with a supply of the components. The beams influence the crystal held at a sufficiently high temperatures that provides the incoming atoms with ample heat energy to transfer over the surface to lattice sites. The setting of UHV mitigates the risk of the surface contamination. The beam atoms and molecules move in almost collision-free pathways in the UHV environment before they enter either the substrate or the cooled edges of the reactor where they crystallize and therefore are efficiently removed from the device. An analogy close to MBE process could be "spray painting," in which atoms and molecules are like paints being deposited with a desired crystal lattice on the target substrate. Each layer that is deposited by a layer-by-layer process can be tailored in a required structure. All these great details in tailoring the structure comes with challenges that need high knowledge of surface processing and materials science (Arthur, 2002). Fig. 11.15 shows a schematic representation of an MBE system designed for the growth of heterostructured devices (Chen et al., 2020). Inside a vacuum chamber maintained at a pressure of $10^{-7}-10^{-10}$ Torr, the substrate is heated by a molybdenum heating block. Integrating a wide range of characterization into the MBE chamber, such as electron diffraction, mass spectrometer, and Auger analysis, allows the simultaneous identification of the growing epitaxial layers. Doping the growing semiconductor by RE or any other element can be done by putting the material source in independent liquid nitrogen surrounded by heated effusion ovens. By giving the required heating energy to the effusion ovens (containing either semiconductor or doping element), an adequate beam starts to flux and concentrate on the substrate surface. Adhesion coefficient of the substrate and flux atoms is an important factor determining the flux rate. For some materials such as AlGaAs on GaAs, the flux rate is in the range of $10^{12}-10^{14}$ atoms/(cm-second), which is due to the unity of

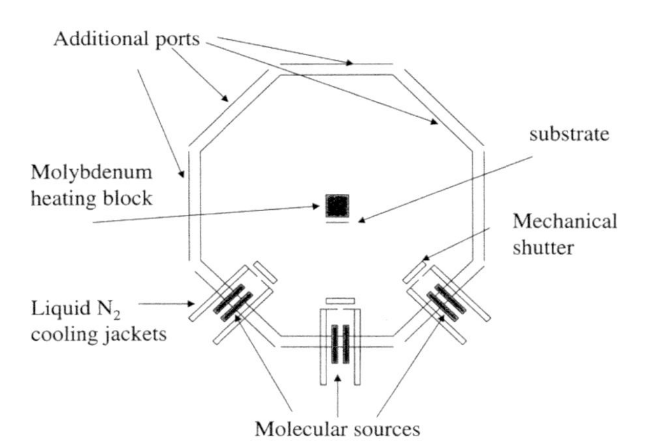

Figure 11.15 Schematic diagram of a molecular beam epitaxial growth system.
Source: Reprinted from Delfyett P. J. (2003). *Encyclopedia of physical science and technology* (3rd ed.), Lasers, Semiconductor, pp. 3443–3475, Copyright (2003), with the permission from Elsevier.

adhesion coefficient in the group III elements, specifically Al or Ga. However, it is not the case for group V elements and their adhesion coefficient differs considerably. Therefore the adhesion coefficient of the coating elements should be determined empirically to design the suitable reaction conditions (Chen et al., 2020).

11.7.1.2 Metal–organic vapor phase epitaxy

The motivation for using MOVPE for synthesis of semiconductors, especially III-nitrides started with Nakamura's work on nitride-based LEDs (Nakamura, Mukai, & Senoh, 1994). The commonly used precursors for the group III semiconductors nitrides and phosphides including Ga, Al, In, elements are mainly trimethyl-aluminum (TMA1), trimethyl-gallium (TMGa), and trimethyl-indium (TMIn). Ammonia (NH_3) and phosphine (PH_3) are usually used as source of nitrogen and phosphor, while sapphire and SiC are the most common selected substrates (Feng, 2006).

Growing INP through MOVPE process by TMIn and PH_3 precursor is shown in Fig. 11.16 (Bashir & Liu, 2015; Zhang et al., 2015). The metal–organic precursors reach out to the substrate and undergo pyrolysis on heated substrate surface. Methyl ligands are released as the metal–organic precursor decomposition at high temperature; they eventually leave the reactor as by-products. In this process, vacuum is not needed because a carrier gas (H_2 or N_2) at pressure of 2–100 kPa is purged throughout the whole process. Thermodynamically, the formation of devices

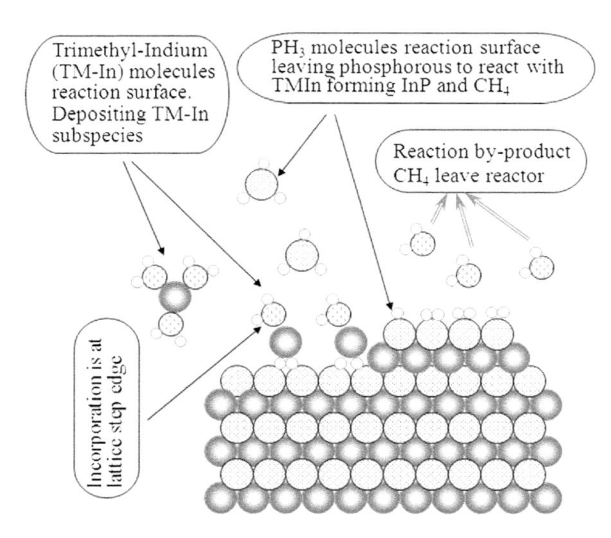

Figure 11.16 Schematic diagram of a metal–organic vapor phase epitaxy.
Source: Reprinted from Bashir, S., & Liu, J. (2015). Chapter 2 - overviews of synthesis of nanomaterials. In R. E. Bashir (Ed.), *Advanced nanomaterials and their applications in renewable energy* (pp. 51–115). Amsterdam: Elsevier, Copyright (2015), with permission from Elsevier.

incorporates metastable alloys. MOVPE method is a common approach in manufacturing different RE-doped semiconductors for a wide range of applications (LDs, solar cells, and LEDs) (Dierolf, Ferguson, & Zavada, 2016; Lian et al., 2013; O'Donnell & Dierolf, 2010).

11.7.1.3 Flash lamp annealing

FLA is one of the short-time annealing methods as part of the rapid thermal processing (RTP) technology, which includes thermal processes working in the range of several tens of seconds. These methods that encompasses annealing times over more than 10 orders of magnitude, namely between 10 ns and 100 s are divided into three groups, RTP in the range of 1–100 s, FLA in the range of 100 μs–100 Ms, and laser annealing in the nanosecond range and below (Prucnal, Rebohle, & Skorupa, 2017). The selection of method depends on the final application, materials type, doping solubility limit, etc. (Prucnal et al., 2017; Sedgwick, 1983).

Fig. 11.17 shows an FLA device consisting a process chamber, a sample holder, a series of Xenon flash lamps, a reflector for changing the light direction towards the substrate, and a preheating system (Prucnal et al., 2017). Heating inside the chamber can evaporate various types of elements that can damage the flash lamps. Therefore flash lamps are placed usually in a separate section by a suitable quartz window not in contact with the main chamber. Halogen lamps or a hot plate are usually used as the preheating system that can create an increased peak temperatures during the drop in the temperature gradient between front and back side (Smith, Seffen, McMahon, Voelskow, & Skorupa, 2006).

Application of FLA method for doping semiconductors with REEs comes with some challenges. The main challenge is clustering of RE atoms during high-temperature furnace annealing (FA), in particular for Er. One solution to this problem is using more moderate annealing temperatures, such as the "famous" FA

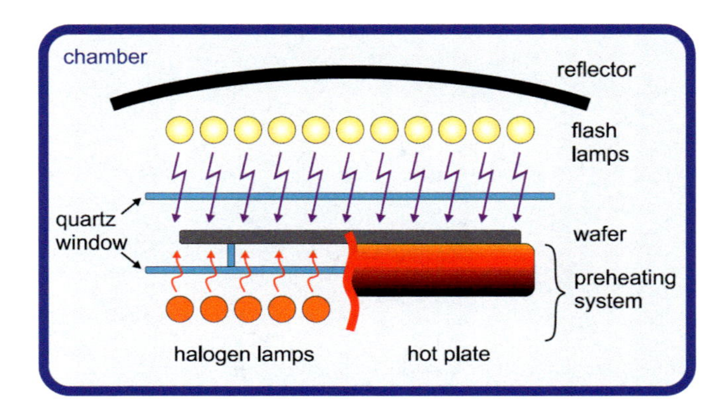

Figure 11.17 Basic scheme of an FLA tool as used for semiconductor wafer processing. *Source*: Reprinted from Prucnal, S., Rebohle, L., & Skorupa, W. (2017). Doping by flash lamp annealing. *Materials Science in Semiconductor Processing*, *62*, 115–127.

900°C in the case of Er. Although FLA is not a recommended method for doping Si-based LEDs, for the matrices that RE diffusivity is high, such as SiO_2, it is a superior method compared to other heat treatment approaches (Rebohle, Prucnal, & Skorupa, 2016).

11.7.1.4 Reactive magnetron cosputtering

Reactive sputtering is very common for the synthesis of thin films and doping them with a variety of elements. In this process, a gas with capability to react with the deposited material is used. High rate deposition systems are requested by industrial applications. It seems quite straightforward at the first sight; however, the interactions between the sputtered substance and the reacting gas can cause some production stability problems. The mixture of the elevated penetration depth of the prepared films and real compound stoichiometry turn out to be contradictory requirements. In certain cases, the primary explanation for this complication is that the compound formation will often take place on the surface of the sputtering target in addition to forming a compound of the deposited film. The sputtering yield of the compound material is usually considerably lower than that of the elemental target material. This triggers a decline in the deposition rate as the reactive gas supply rises. The relationship between the composition of the film and the reactive gas supply is nonlinear. For the deposition rate versus the availability of the reactive gas, this is also the case. Therefore the processes of reactive sputtering regulated by the reactive gas supply display very complex manufacturing actions (Berg & Nyberg, 2005).

11.7.2 Wet chemical methods

11.7.2.1 Chemical precipitation

The process of coprecipitation is an easy, cost-effective, and high-yielding method to obtain nanomaterials at the optimal ambient temperature in the laboratory. For the regulated synthesis of ZnO NPs, this method is stable, quick, and spontaneous by manipulating the variables of temperature, pH, solvent, and reaction time of precursors to form coprecipitation. Coprecipitation can be defined as if, using a basic solution of NaOH or NH_4OH, an inorganic metal substance dissolved in solvent is hydrolyzed, then these molecules condense to form a precipitate of metal oxide by increasing OH^- ion concentrations. To obtain the crystalline metal oxide material type, the obtained precipitate is then washed and dried (Daksh & Agrawal, 2016).

11.7.2.2 Sol–gel combustion

Sol–gel technique (Fig. 11.18) is a practical wet chemical source for the preparation of compounds based on metal alkoxide (M-O-M) or metallic inorganic compound (M-H-M) solutions that react to create colloidal particles (Bashir & Liu, 2015). Sol–gel method is often favored compared to all wet chemical methods because of its low cost, high yield, and fast processing time, as well as good ability

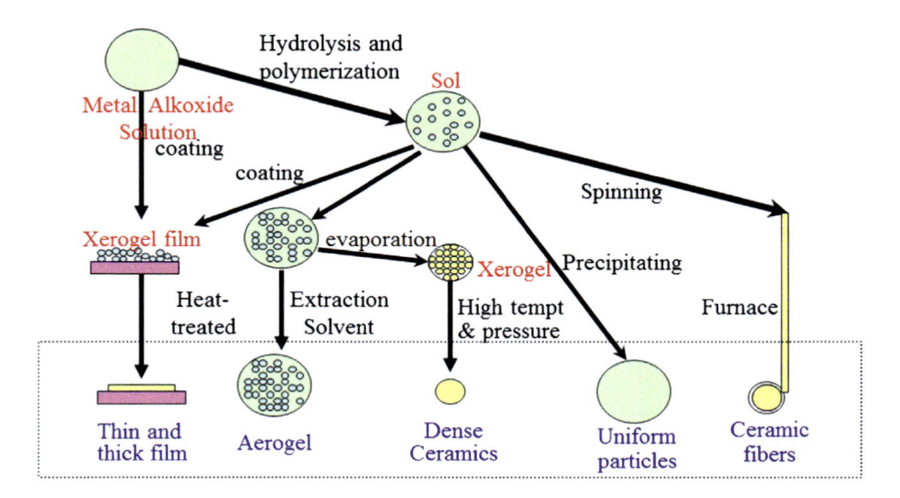

Figure 11.18 Diagram of sol–gel method used to demonstrate the formation of different types of materials.
Source: Reprinted from Bashir, S., & Liu, J. (2015). Chapter 2 - overviews of synthesis of nanomaterials. In R. E. Bashir (Ed.), *Advanced nanomaterials and their applications in renewable energy* (pp. 51–115). Amsterdam: Elsevier, Copyright (2015), with permission from Elsevier.

to achieve high purity in making complex oxide powders in single or multiphase. In addition, it offers the advantages of high purity and improved structural control with a uniform nanostructure achievable at low temperatures (Rajeswari et al., 2020). A gel consists of a steady 3D network that completely covers a liquid phase. In the colloidal gel phase, the system is formed from the aggregation of colloids. The concept behind sol–gel division is to "break" the compound in a solvent in a sequence to return in a defined order as solid (Bashir & Liu, 2015).

11.8 Rare earth elements resources and their recycling

Due to the limited resources and difficult extraction, recovery of REEs is of high importance. As discussed in this chapter, they have unique properties and are used in a wide range of applications in optics, magnetics, electronics, metallurgy, energy, etc. Specifically, there has been an increasing demand for the recovery of REEs for their use in hybrid cars, aerospace industry, fiber-optic technology, superconductors, sensors, and solar panels. In this scenario, not only does the recovery of REEs help us clean the water but it also minimizes the damage of mining, as a noneco-friendly process leaving behind harsh chemical by-products containing radioactive compounds.

The use of novel functionalized nanostructures and polymers is a promising strategy for the detection and removal of heavy metals. The key to a successful design of an efficient adsorbent is the right selection of functionalizing agents constituting the active sites for metal coordination. Extraction of heavy metals is an urgent issue, which demands to devise novel strategies. Different methods have been reported as efficient extraction techniques for heavy metal ions, including chemical precipitation, liquid–liquid extraction, ion exchange, supercritical fluid extraction, oxidation/reduction sedimentation, and solid-phase extraction. The last one though, abbreviated as SPE, is an emerging, effective, and economic technique in which solid substances such as carbon nanotubes, clays, metal–organic frameworks, graphene, biomaterials, natural wastes, and some other materials are used to adsorb a target ion from a complex matrix, and the adsorbent can be regenerated by simple procedures. The selected solid phase, being either highly porous or functionalized with organic molecules, should be sensitive and selective towards ions with specific sizes and charges (Behpour, Foulady-Dehaghi, & Mir, 2017; Dehaghi, Behpour, & Mir, 2018; Esmaeili, Mir, & Mohammadi, 2020; Mir, Heidari, Beyzaei, Mirkazehi-Rigi, & Karimi, 2017; Mir, Jalilian, Karimi, Nejati-Yazdinejad, & Khammarnia, 2018).

The world of chemistry has helped developing an increasing number of organic ligands for heavy metal adsorption. Many of these structures have shown to be high-performance sensors with extraordinary sensitivities towards a specific metal ion. However, their main drawback is the difficult separation of the adsorbent from the aqueous medium since most of them are either soluble or dispersed well in water. This problem limits their wide application due to the secondary pollution caused by the adsorbent itself. Therefore researchers have worked on innovative ways to introduce the organic ligands to the surface of a solid substrate to address this issue. The selected substrate could be chosen from nanocomposites, polymers, nanoparticles, carbon materials, and porous frameworks. In these approaches, the substrates are functionalized with chemical groups to provide further bonding to the ligand. The mesoporous substrates can also be attached to a magnetic core particle enabling a facile separation of the whole system from aqueous solution using a permanent magnet. Silicon dioxide, as a porous material with high numbers of hydroxyl sites available for anchoring functional groups, is one of the most used substrates in such applications. The obtained SiO_2-based SPE adsorbent is a cost-effective, easy-to-prepare, separable, safe, and portable probe that can be used for large-scale applications.

Another strategy for making separable adsorbents is using insoluble polymers or resins, an alternative that can be either easily separated from water or used in the form of an SPE column. Finding appropriate cost-effective chemicals and manipulating their structures for improving the coordinating properties for specific targets is an ongoing challenge in water treatment plants and, therefore, novel approaches in this field can benefit both environment and industry.

The development of a self-floating polymer for recovery of a group of REEs has recently been documented by our group (Mir et al., 2021). The polymer has been developed for the extraction and recovery of REEs, especially Yb^{3+}, with quite superior efficiency. Using a widely employed sizing product in the paper production, the formulated sorbents was quite cost-effective and easy to prepare. The designed polymer was used to remove La^{3+}, Ce^{3+}, Pr^{3+}, Nd^{3+}, Dy^{3+}, and Yb^{3+} from aqueous solutions with the highest adsorption of 191.87 mg/g (Mir et al., 2021) (Fig. 11.19).

11.9 Conclusion and outlook

RE-doped semiconductor nanomaterials are today the heart of modern electronics. The miniaturized electronic tools such as quantum computers, microprocessors, and tiny spinotronic devices require high-performance doped semiconductors with unique optical and electronic performance. Different semiconductor nanomaterials can host REEs including silicon, Si nanocrystal–embedded silicone dioxide, III-V-based materials, metal oxides (ZnO, TiO_2, SnO_2, ZrO_2, SiO_2, and Al_2O_3), perovskite, etc. Each of these groups of materials can have several potential applications including but not limited to optical fibers, lasers, LED devices, varistors, spintronic devices, photovoltaics, photocatalysts, gas sensors, and antimicrobials. Silicone as the dominating element in semiconductor industry suffers from its inefficiency for optoelectrical applications due to the mismatch of its bandgap with fiber-optic communication systems. Er is a key REEs in silicone industry especially for optoelectronic applications. A novel luminescent material known as Si nanocrystal–embedded silicone dioxide has shown even more promising RT performance than the pure silica. The main challenge in doping semiconductors with REEs is the capability of the host for accepting high concentration of doping elements without precipitation during the annealing process. Unfortunately, many suitable materials for hosting REEs have not found their places in semiconductor industry since they are not compatible with CMOS technology. RE-doped III-nitride-based semiconductors are other materials that are developing and rising up due to their excellent optical and magnetic properties. Among all the reported RE-doped metal oxide semiconductors with main applications in photovoltaics, photocatalysts, antimicrobial, and gas sensor applications,

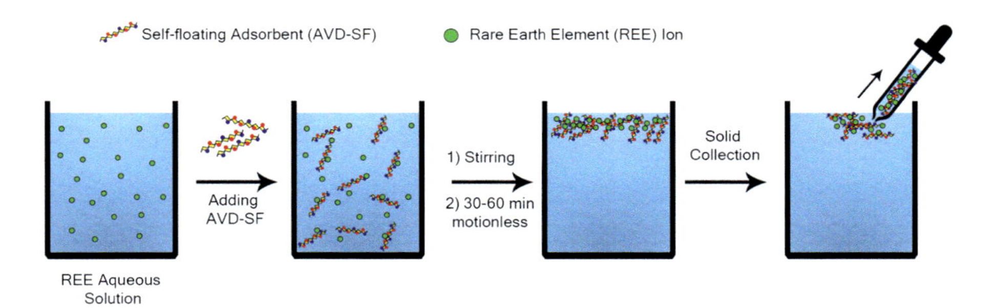

Figure 11.19 Application of a self-floating polymer for recovery of RE ions from water. *Source*: Reprinted from Mir, N., Castano, C. E., Rojas, J. V., Norouzi, N., Esmaeili, A. R., & Mohammadi, R. (2021). Self-separation of the adsorbent after recovery of rare-earth metals: Designing a novel non-wettable polymer. *Separation and Purification Technology*, *259*, 118152, with the permission from Elsevier.

ZnO and TiO_2 have been more promising. However, the champions in photovoltaic applications with high conversion efficiencies have been perovskite materials which have recently been doped with REEs to privilege from their UC and DC properties. Overall, REEs are the spark of the novel technologies in new era. Scarcity and mining challenges remain as the major limitations of RE applications. Seeking novel recycling methods could be a rescue scenario to overcome the challenges regarding the shortage of the RE resources as well as the geopolitical climates governing the precious metal extraction.

References

Ahmadian, H., Al Hessari, F., & Arabi, A. M. (2019). Preparation and characterization of Luminescent nanostructured Gd2O3-Y2O3: Eu synthesized by the solution combustion process. *Ceramics International*, *45*(15), 18778–18787.

Ahmed, S. M., Szymanski, P., El-Nadi, L. M., & El-Sayed, M. A. (2014). Energy-transfer efficiency in Eu-doped ZnO thin films: The effects of oxidative annealing on the dynamics and the intermediate defect states. *ACS Applied Materials & Interfaces*, *6*(3), 1765–1772. Available from https://doi.org/10.1021/am404662k.

Akman, E., Akin, S., Ozturk, T., Gulveren, B., & Sonmezoglu, S. (2020). Europium and terbium lanthanide ions co-doping in TiO2 photoanode to synchronously improve light-harvesting and open-circuit voltage for high-efficiency dye-sensitized solar cells. *Solar Energy*, *202*, 227–237.

Al Tahtamouni, T. M., Stachowicz, M., Li, J., Lin, J. Y., & Jiang, H. X. (2015). Dramatic enhancement of 1.54 μ m emission in Er doped GaN quantum well structures. *Applied Physics Letters*, *106*(12), 121106.

Amiri, O., Salavati-Niasari, M., Mir, N., Beshkar, F., Saadat, M., & Ansari, F. (2018). Plasmonic enhancement of dye-sensitized solar cells by using Au-decorated Ag dendrites as a morphology-engineered. *Renewable Energy*, *125*. Available from https://doi.org/10.1016/j.renene.2018.03.003.

Ansari, F., Salavati-Niasari, M., Amiri, O., Mir, N., Abdollahi Nejand, B., & Ahmadi, V. (2020). Magnetite as inorganic hole transport material for lead halide perovskite-based solar cells with enhanced stability. *Industrial & Engineering Chemistry Research*, *59*(2), 743–750. Available from https://doi.org/10.1021/acs.iecr.9b05173.

Ansari, F., Salavati-Niasari, M., Nazari, P., Mir, N., Ahmadi, V., & Abdollahi Nejand, B. (2018). Long-term durability of bromide-incorporated perovskite solar cells via a modified vapor-assisted solution process. *ACS Applied Energy Materials*, *1*(11). Available from https://doi.org/10.1021/acsaem.8b01075.

Arai, T., Timmerman, D., Wakamatsu, R., Lee, D., Koizumi, A., & Fujiwara, Y. (2015). Enhanced excitation efficiency of Eu ions in Eu-doped GaN/AlGaN multiple quantum well structures grown by organometallic vapor phase epitaxy. *Journal of Luminescence*, *158*, 70–74.

Armelao, L., et al. (2008). Structure – luminescence correlations in europium-doped sol – gel ZnO nan[opowders. *The Journal of Physical Chemistry C*, *112*(11), 4049–4054. Available from https://doi.org/10.1021/jp710207r.

Arthur, J. R. (2002). Molecular beam epitaxy. *Surface Science*, *500*(1–3), 189–217.

Atwood, D. A. (Ed.), (2013). *The rare earth elements: Fundamentals and applications*. Wiley.

Bachir, S., Kossanyi, J., & Ronfard-Haret, J. C. (1994). Electroluminescence of Ho3 + ions in a ZnO varistor-type structure. *Solid State Communications*, *89*(10), 859–863.

Bachir, S., Kossanyi, J., Sandouly, C., Valat, P., & Ronfard-Haret, J. C. (1995). Electroluminescence of Dy3 + and Sm3 + ions in polycrystalline semiconducting zinc oxide. *The Journal of Physical Chemistry*, *99*(15), 5674–5679.

Bachir, S., Sandouly, C., Kossanyi, J., & Ronfard-Haret, J. C. (1996). Rare earth-doped polycrystalline zinc oxide electroluminescent ceramics. *The Journal of Physics and Chemistry of Solids*, *57*(12), 1869–1879.

Bashir, S., & Liu, J. (2015). Chapter 2 - overviews of synthesis of nanomaterials. In: R. E. Bashir (Ed.), *Advanced nanomaterials and their applications in renewable energy* (pp. 51–115). Amsterdam: Elsevier.

Behpour, M., Foulady-Dehaghi, R., & Mir, N. (2017). Considering photocatalytic activity of N/F/S-doped TiO < inf > 2 < /inf> thin films in degradation of textile waste under visible and sunlight irradiation. *Solar Energy*, *158*. Available from https://doi.org/10.1016/j.solener.2017.10.034.

Ben Sedrine, N., et al. (2018). Eu-doped algan/gan superlattice-based diode structure for red lighting: Excitation mechanisms and active sites. *ACS Applied Nano Materials*, *1*(8), 3845–3858. Available from https://doi.org/10.1021/acsanm.8b00612.

Berg, S., & Nyberg, T. (2005). Fundamental understanding and modeling of reactive sputtering processes. *Thin Solid Films*, *476*(2), 215–230. Available from https://doi.org/10.1016/j.tsf.2004.10.051.

Bingham, S., & Daoud, W. A. (2011). Recent advances in making nano-sized TiO 2 visible-light active through rare-earth metal doping. *Journal of Materials Chemistry*, *21*(7), 2041–2050.

Bonanni, A. (2007). Ferromagnetic nitride-based semiconductors doped with transition metals and rare earths. *Semiconductor Science and Technology*, *22*(9), R41–R56. Available from https://doi.org/10.1088/0268-1242/2211.11./9/r01.

Boninelli, S., Bellocchi, G., Franzò, G., Miritello, M., & Iacona, F. (2013). New strategies to improve the luminescence efficiency of Eu ions embedded in Si-based matrices. *Journal of Applied Physics*, *113*(14), 143503. Available from https://doi.org/10.1063/1.4799407.

Brandt, O., Dhar, S., Pérez, L., & Sapega, V. (2010). High-temperature ferromagnetism in the super-dilute magnetic semiconductor GaN: Gd. *Topics in Applied Physics*, *124*, 309–342. Available from https://doi.org/10.1007/978-90-481-2877-8_10.

Brongersma, M. L., Polman, A., Min, K. S., Boer, E., Tambo, T., & Atwater, H. A. (1998). Tuning the emission wavelength of Si nanocrystals in SiO 2 by oxidation. *Applied Physics Letters*, *72*(20), 2577–2579.

Bueno, C. F., Ramos, A. Y., Bailly, A., Mossang, E., & Scalvi, L. V. A. (2020). X-ray absorption spectroscopy and Eu3 + -emission characteristics in GaAs/SnO2 heterostructure. *SN Applied Sciences*, *2*(9), 1–15. Available from https://doi.org/10.1007/s42452-020-03344-3.

Bueno, C. F., Scalvi, L. V. A., Saeki, M. J., & Li, M. S. (2015). Photoluminescence of the Eu-doped thin film heterojunction GaAs/SnO2 and rare-earth doping distribution. *IOP Conference Series: Materials Science and Engineering*, *76*, 12006. Available from https://doi.org/10.1088/1757-899x/76/1/012006.

Canham, L. T. (1990). Silicon quantum wire array fabrication by electrochemical and chemical dissolution of wafers. *Applied Physics Letters*, *57*(10), 1046–1048.

Cavallo, C., et al. (2015). Solid solutions of rare earth cations in mesoporous anatase beads and their performances in dye-sensitized solar cells. *Scientific Reports*, *5*(1), 1–15.

Chander, N., et al. (2014). Reduced ultraviolet light induced degradation and enhanced light harvesting using YVO4: Eu3 + down-shifting nano-phosphor layer in organometal halide perovskite solar cells. *Applied Physics Letters*, *105*(3), 33904.

Chandra, U. (2017). *Recent Applications in Sol-Gel Synthesis*. InTechOpen.

Chen, Y., et al. (2020). An overview of rare earth coupled lead halide perovskite and its application in photovoltaics and light emitting devices. *Progress in Materials Science*, 100737.

Chen, Z.-W., Yao, C.-B., & Hu, J.-Y. (2019). The nonlinear optical properties and optical transition dynamics of Er doped ZnO films. *Optical Laser Technology*, *119*, 105609.

Cheng, B., Zhang, Z., Liu, H., Han, Z., Xiao, Y., & Lei, S. (2010). Power-and energy-dependent photoluminescence of Eu3 + incorporated and segregated ZnO polycrystalline nanobelts synthesized by a facile combustion method followed by heat treatment. *Journal of Materials Chemistry*, *20*(36), 7821–7826.

Clemens, J. T. (1997). Silicon microelectronics technology. *Bell Labs Technical Journal*, *2* (4), 76–102.

CN105932125B - A kind of GaN base green light LED epitaxial structure and preparation method thereof. Google Patents. (2021). https://patents.google.com/patent/CN105932125B/en?q = GaN:RE + %2B + LED&oq = GaN:RE + %2B + LED (accessed 13.02.21).

Coutinho, J., Jones, R., Shaw, M. J., Briddon, P. R., & Öberg, S. (2004). Optically active erbium–oxygen complexes in GaAs. *Applied Physics Letters*, *84*(10), 1683–1685.

Dasgupta, N., & Dasgupta, A. (2004). *Semiconductor devices: Modelling and technology*. India: PHI Learning.

Daksh, D., & Agrawal, Y. K. (2016). Rare earth-doped zinc oxide nanostructures: A review. *Reviews in Nanoscience and Nanotechnology*, *5*(1), 1–27. Available from https://doi.org/10.1166/rnn.2016.1071.

Dehaghi, R. F., Behpour, M., & Mir, N. (2018). Purification of textile wastewater by using coated Sr/S/N doped TiO < inf > 2 < /inf > nanolayers on glass orbs. *Korean Journal Of Chemical Engineering*, *35*(7). Available from https://doi.org/10.1007/s11814-017-0176-0.

DenBaars, S. P., et al. (2013). Development of gallium-nitride-based light-emitting diodes (LEDs) and laser diodes for energy-efficient lighting and displays. *Acta Materialia*, *61* (3), 945–951.

Dey, A. (2018). Semiconductor metal oxide gas sensors: A review. *Materials Science and Engineering B: Solid-State Materials for Advanced Technology*, *229*, 206–217. Available from https://doi.org/10.1016/j.mseb.2017.12.036, no. November 2017.

Dhar, S., Brandt, O., Ramsteiner, M., Sapega, V. F., & Ploog, K. H. (2005). Colossal magnetic moment of Gd in GaN. *Physical Review Letters*, *94*(3), 37205. Available from https://doi.org/10.1103/PhysRevLett.94.037205.

Dhar, S., et al. (2005). Gd-doped GaN: A very dilute ferromagnetic semiconductor with a Curie temperature above 300 K. *Physical Review B*, *72*(24), 245203. Available from https://doi.org/10.1103/PhysRevB.72.245203.

Dhar, S., et al. (2006). Ferromagnetism and colossal magnetic moment in Gd-focused ion-beam-implanted GaN. *Applied Physics Letters*, *89*(6), 62503. Available from https://doi.org/10.1063/1.2267900.

Dierolf, J. M. Z. V., & Ferguson, I. T. (2021). *Doping in III-V semiconductors*. Google Books. https://www.google.com/books/edition/Doping_in_III_V_Semiconductors/Y6xhCgAAQBAJ?hl = en&gbpv = 0 (accessed 28.01.21).

Dierolf, V., Ferguson, I., & Zavada, J. M. (2016). *Rare earth and transition metal doping of semiconductor materials: Synthesis, magnetic properties and room temperature spintronics*. Elsevier Science.

Dietl, T., Ohno, H., Matsukura, F., Cibert, J., & Ferrand, D. (2000). Zener model description of ferromagnetism in zinc-blende magnetic semiconductors. *Science (80)*, *287*(5455), 1019–1022.

Djaja, N. F., & Saleh, R. (2013). Characteristics and photocatalytics activities of Ce-doped ZnO nanoparticles. *Materials Sciences and Applications*, *4*(02), 145.

Du, C.-X., Ni, W.-X., Joelsson, K. B., & Hansson, G. V. (1997). Room temperature 1.54 μ m light emission of erbium doped Si Schottky diodes prepared by molecular beam epitaxy. *Applied Physics Letters*, *71*(8), 1023–1025.

Du, G., et al. (2007). Visual-infrared electroluminescence emission from Zn O/ Ga As heterojunctions grown by metal-organic chemical vapor deposition. *Applied Physics Letters*, *90*(24), 243504.

Du, Y.-P., Zhang, Y.-W., Sun, L.-D., & Yan, C.-H. (2008). Efficient energy transfer in monodisperse Eu-doped ZnO nanocrystals synthesized from metal acetylacetonates in high-boiling solvents. *The Journal of Physical Chemistry C*, *112*(32), 12234–12241.

Dyck, N. C., & Demopoulos, G. P. (2014). Integration of upconverting β-NaYF 4: Yb 3 + , Er 3 + @ TiO 2 composites as light harvesting layers in dye-sensitized solar cells. *RSC Advances*, *4*(95), 52694–52701.

El-Bahy, Z. M., Ismail, A. A., & Mohamed, R. M. (2009). Enhancement of titania by doping rare earth for photodegradation of organic dye (Direct Blue). *Journal of Hazardous Materials*, *166*(1), 138–143.

Ennen, H., Wagner, J., Müller, H. D., & Smith, R. S. (1987). Photoluminescence excitation measurements on GaAs: Er grown by molecular-beam epitaxy. *Journal of Applied Physics*, *61*(10), 4877–4879.

Erwin, S. C., Zu, L., Haftel, M. I., Efros, A. L., Kennedy, T. A., & Norris, D. J. (2005). Doping semiconductor nanocrystals. *Nature*, *436*(7047), 91–94. Available from https://doi.org/10.1038/nature03832.

Esmaeili, A. R., Mir, N., & Mohammadi, R. (2020). A facile, fast, and low-cost method for fabrication of micro/nano-textured superhydrophobic surfaces. *Journal of Colloid and Interface Science*, *573*. Available from https://doi.org/10.1016/j.jcis.2020.04.027.

Fang, X. M., Li, Y., & Langer, D. W. (1993). Radiative and nonradiative transitions in GaAs: Er. *Journal of Applied Physics*, *74*(11), 6990–6992.

Fangli, D. U., Ning, W., Zhang, D., & Yingzhong, S. (2010). Preparation, characterization and infrared emissivity study of Ce-doped ZnO films. *Journal of Rare Earths*, *28*(3), 391–395.

Favennec, P. N., L'haridon, H., Salvi, M., Moutonnet, D., & Le Guillou, Y. (1989). Luminescence of erbium implanted in various semiconductors: IV, III-V and II-VI materials. *Electronics Letters*, *25*(11), 718–719.

Feng, Z. C. (2006). *III-Nitride semiconductor materials*. World Scientific.

Fragkos, I. E., Tan, C.-K., Dierolf, V., Fujiwara, Y., & Tansu, N. (2017). Pathway towards high-efficiency Eu-doped GaN light-emitting diodes. *Scientific Reports*, *7*(1), 1–13.

Franzo, G., Priolo, F., Coffa, S., Polman, A., & Carnera, A. (1994). Room-temperature electroluminescence from Er-doped crystalline Si. *Applied Physics Letters*, *64*(17), 2235–2237.

Franzò, G., Vinciguerra, V., & Priolo, F. (1999). The excitation mechanism of rare-earth ions in silicon nanocrystals. *Applied Physics A: Materials Science and Processing*, *69*(1), 3–12. Available from https://doi.org/10.1007/s003390050967.

Fujii, M., Yoshida, M., Kanzawa, Y., Hayashi, S., & Yamamoto, K. (1997). 1.54 μm photoluminescence of Er 3 + doped into SiO 2 films containing Si nanocrystals: Evidence for energy transfer from Si nanocrystals to Er 3 +. *Applied Physics Letters*, *71*(9), 1198–1200.

Fujishima, A., Hashimoto, K., & Watanabe, T. (1999). TiO_2 photocatalysis: Fundamentals and applications. BKC.

Gao, H., Liu, W., Lu, B., & Liu, F. (2012). Photocatalytic activity of La, Y co-doped TiO2 nanoparticles synthesized by ultrasonic assisted sol–gel method. *Journal of Nanoscience and Nanotechnology*, *12*(5), 3959–3965.

George, A., Sharma, S. K., Chawla, S., Malik, M. M., & Qureshi, M. S. (2011). Detailed of X-ray diffraction and photoluminescence studies of Ce doped ZnO nanocrystals. *Journal of Alloys and Compounds*, *509*(20), 5942–5946. Available from https://doi.org/10.1016/j.jallcom.2011.03.017.

George, D. K., et al. (2015). Excitation mechanisms of Er optical centers in GaN epilayers. *Applied Physics Letters*, *107*(17), 171105.

Gholami, T., Mir, N., Masjedi-Arani, M., Noori, E., & Salavati-Niasari, M. (2014). Investigating the role of a Schiff-base ligand in the characteristics of TiO<inf>2</inf> nano-particles: Particle size, optical properties, and photo-voltaic performance of dye-sensitised solar cells. *Materials Science in Semiconductor Processing*, *22*(1). Available from https://doi.org/10.1016/j.mssp.2014.01.013.

Ghosh, S. (2018). *Visible-light-active photocatalysis: Nanostructured catalyst design, mechanisms, and applications*. Wiley.

Gong, J., Liang, J., & Sumathy, K. (2012). Review on dye-sensitized solar cells (DSSCs): Fundamental concepts and novel materials. *Renewable and Sustainable Energy Reviews*, *16*(8), 5848–5860.

Gupta, S., et al. (2008). MOVPE growth of transition-metal-doped GaN and ZnO for spintronic applications. *Journal of Crystal Growth*, *310*(23), 5032–5038.

Hassan, M. S., Amna, T., Yang, O.-B., Kim, H.-C., & Khil, M.-S. (2012). TiO2 nanofibers doped with rare earth elements and their photocatalytic activity. *Ceramics International*, *38*(7), 5925–5930.

Hastir, A., Kohli, N., & Singh, R. C. (2017). Comparative study on gas sensing properties of rare earth (Tb, Dy and Er) doped ZnO sensor. *The Journal of Physics and Chemistry of Solids*, *105*, 23–34.

Heikenfeld, J., Garter, M., Lee, D. S., Birkhahn, R., & Steckl, A. J. (1999). Red light emission by photoluminescence and electroluminescence from Eu-doped GaN. *Applied Physics Letters*, *75*(9), 1189–1191. Available from https://doi.org/10.1063/1.124686.

Helbers, A., Mitchell, B., Woodward, N., & Dierolf, V. (2015). Anormolous magneto-optical behavior of rare earth doped gallium nitride. In *2015 European Conference on Lasers and Electro-Optics - European Quantum Electronics Conference*, p. CE_5b_2 [Online]. Available from: http://www.osapublishing.org/abstract.cfm?URI = CLEO_Europe-2015-CE_5b_2.

Higuchi, T., et al. (2011). R.; Sawatzky, GA; Hwang, HY, Applied Physics Letters, 98, 71902.

Hite, J. K., & Zavada, J. M. (2019). Dilute magnetic III-N semiconductors based on rare earth doping. *ECS Journal of Solid State Science and Technology*, *8*(9), P527–P535. Available from https://doi.org/10.1149/2.0261909jss.

Ho, V. X., Al Tahtamouni, T. M., Jiang, H. X., Lin, J. Y., Zavada, J. M., & Vinh, N. Q. (2018). Room-temperature lasing action in GaN quantum wells in the infrared 1.5 μm region. *ACS Photonics*, *5*(4), 1303–1309.

Ho, V. X., et al. (2020). Observation of optical gain in Er-doped GaN epilayers. *Journal of Luminescence, 221*, 117090. Available from https://doi.org/10.1016/j.jlumin.2020.117090.

Hu, J., et al. (2017). Enhanced performance of hole-conductor-free perovskite solar cells by utilization of core/shell-structured β-NaYF4: Yb3 + , Er3 + @ SiO2 nanoparticles in ambient air. *IEEE Journal of Photovoltaics, 8*(1), 132–136.

Hussain, C. M., & Mishra, A. K. (2020). *Handbook of smart photocatalytic materials: fundamentals, fabrications and water resources applications*. Elsevier.

Jia, C.-J., Sun, L.-D., Luo, F., Jiang, X.-C., Wei, L.-H., & Yan, C.-H. (2004). Structural transformation induced improved luminescent properties for LaVO 4: Eu nanocrystals. *Applied Physics Letters, 84*(26), 5305–5307.

Jiang, F., Peng, Z., Zang, Y., & Fu, X. (2013). Progress on rare-earth doped ZnO-based varistor materials. *Journal of Advanced Ceramics, 2*(3), 201–212. Available from https://doi.org/10.1007/s40145-013-0071-z, Tsinghua University Press.

Jung, J. W., Chueh, C., & Jen, A. K. (2015). High-performance semitransparent perovskite solar cells with 10% power conversion efficiency and 25% average visible transmittance based on transparent CuSCN as the hole-transporting material. *Advanced Energy Materials, 5*(17), 1500486.

Jung, Y.-I., Noh, B.-Y., Lee, Y.-S., Baek, S.-H., Kim, J. H., & Park, I.-K. (2012). Visible emission from Ce-doped ZnO nanorods grown by hydrothermal method without a post thermal annealing process. *Nanoscale Research Letters, 7*(1), 1–5.

Kane, M. H., Gupta, S., & Ferguson, I. T. (2016). *Transition metal and rare earth doping in GaN. Rare earth and transition metal doping of semiconductor materials: Synthesis, magnetic properties and room temperature spintronics* (pp. 315–370). Elsevier Inc.

Karunakaran, C., Gomathisankar, P., & Manikandan, G. (2010). Preparation and characterization of antimicrobial Ce-doped ZnO nanoparticles for photocatalytic detoxification of cyanide. *Materials Chemistry and Physics, 123*(2), 585–594. Available from https://doi.org/10.1016/j.matchemphys.2010.05.019.

Kaur, N., Mahajan, A., & Singh, D. P. (2020). Up-converting rare earth phosphor yttrium doped TiO2 material for dye sensitized solar cells application. *AIP Conference Proceedings, 2265*(1), 30621.

Kenyon, A. J. (2005). *Erbium in silicon* no. 12*Semiconductor Science and Technology* (vol. 20, p. R65). IOP Publishing. Available from http://doi.org/10.1088/0268-1242/20/12/R02.

Khaki, M. R. D., Shafeeyan, M. S., Raman, A. A. A., & Daud, W. M. A. W. (2017). Application of doped photocatalysts for organic pollutant degradation - A review. *Journal of Environmental Management, 198*, 78–94. Available from https://doi.org/10.1016/j.jenvman.2017.04.099, Academic Press.

Kharel, P. L., Zamborini, F. P., & Alphenaar, B. W. (2018). Enhancing the photovoltaic performance of dye-sensitized solar cells with rare-earth metal oxide nanoparticles. *Journal of the Electrochemical Society, 165*(3), H52.

Khriachtchev, L. (2016). *Silicon nanophotonics: Basic principles, present status, and perspectives*. CRC Press.

Kim, S., et al. (1997). Observation of multiple Er 3 + sites in Er-implanted GaN by site-selective photoluminescence excitation spectroscopy. *Applied Physics Letters, 71*(2), 231–233.

Komuro, S., Katsumata, T., Morikawa, T., Zhao, X., Isshiki, H., & Aoyagi, Y. (2000). 1.54 μm emission dynamics of erbium-doped zinc-oxide thin films. *Applied Physics Letters, 76*(26), 3935–3937.

Kossanyi, J., et al. (1990). Photoluminescence of semiconducting zinc oxide containing rare earth ions as impurities. *Journal of Luminescence*, *46*(1), 17–24.

Kouyate, D., Ronfard-Haret, J.-C., & Kossanyi, J. (1991a). Electroluminescent properties of polycrystalline zinc oxide electrodes doped with Er3 + ions. *Journal of Electroanalytical Chemistry and Interfacial Electrochemistry*, *319*(1–2), 145–160.

Kouyate, D., Ronfard-Haret, J.-C., & Kossanyi, J. (1991b). Photo-and electro-luminescence of rare earth-doped semiconducting zinc oxide electrodes: Emission from both the dopant and the support. *Journal of Luminescence*, *50*(4), 205–210.

Kouyate, D., Ronfard-Haret, J.-C., & Kossanyi, J. (1992). Electroluminescence of Sm3 + ions in semiconducting polycrystalline zinc oxide. *Journal of Materials Chemistry*, *2*(7), 727–732. Available from https://doi.org/10.1039/JM9920200727.

Kouyate, D., Ronfard-Haret, J. C., Valat, P., Kossanyi, J., Mammel, U., & Oelkrug, D. (1990). Quenching of zinc oxide photoluminescence by d-and f-transition metal ions. *Journal of Luminescence*, *46*(5), 329–337.

Kovalenko, M. V., Protesescu, L., & Bodnarchuk, M. I. (2017). Properties and potential optoelectronic applications of lead halide perovskite nanocrystals. *Science (80)*, *358* (6364), 745–750.

Kozanecki, A., & Gröetzschel, R. (1988). On the location of ytterbium in GaP and GaAs lattices. *Journal of Applied Physics*, *64*(6), 3315–3317.

Kränkel, C., Marzahl, D. T., Moglia, F., Huber, G., & Metz, P. W. (2016). Out of the blue: Semiconductor laser pumped visible rare-earth doped lasers. *Laser & Photonics Reviews*, *10*(4), 548–568. Available from https://doi.org/10.1002/lpor.201500290.

Kumar, P., Nagpal, K., & Gupta, B. K. (2017). Unclonable security codes designed from multicolor luminescent lanthanide-doped Y_2O_3 nanorods for anticounterfeiting. *ACS Applied Materials & Interfaces*, *9*(16), 14301–14308.

Kumar, V., Ntwaeaborwa, O. M., Soga, T., Dutta, V., & Swart, H. C. (2017). Rare earth doped zinc oxide nanophosphor powder: A future material for solid state lighting and solar cells. *ACS Photonics*, *4*(11), 2613–2637. Available from https://doi.org/10.1021/acsphotonics.7b00777, American Chemical Society.

Kumar, V., Pandey, A., Swami, S. K., Ntwaeaborwa, O. M., Swart, H. C., & Dutta, V. (2018). Synthesis and characterization of Er3 + -Yb3 + doped ZnO upconversion nanoparticles for solar cell application. *Journal of Alloys and Compounds*, *766*, 429–435.

Kumar, V., & Zavada, J. M. (2016). *Energetics, atomic structure, and magnetics of rare earth-doped GaN bulk and nanoparticles. Rare earth and transition metal doping of semiconductor materials: Synthesis, magnetic properties and room temperature spintronics* (pp. 103–126). Elsevier Inc.

Li, D., Ågren, H., & Chen, G. (2018). Near infrared harvesting dye-sensitized solar cells enabled by rare-earth upconversion materials. *Dalton Transactions*, *47*(26), 8526–8537.

Li, D., Zhang, X., Jin, L., & Yang, D. (2010). Structure and luminescence evolution of annealed Europium-doped silicon oxides films. *Optics Express*, *18*(26), 27191–27196.

Li, G. R., Lu, X. H., Zhao, W. X., Su, C. Y., & Tong, Y. X. (2008). Controllable electrochemical synthesis of Ce4 + -doped ZnO nanostructures from nanotubes to nanorods and nanocages. *Crystal Growth & Design*, *8*(4), 1276–1281. Available from https://doi.org/10.1021/cg7009995.

Li, J., Lin, J. Y., & Jiang, H. X. (2016). Current injection 1.54 μm light-emitting devices based on Er-doped GaN/AlGaN multiple quantum wells. *Optical Materials Express*, *6* (11), 3476–3481.

Li, M., et al. (2013). Origin of green emission and charge trapping dynamics in ZnO nanowires. *Physical Review B*, *87*(11), 115309.

Li, Y., Li, Y., Wang, R., & Zheng, W. (2018). Effect of silica surface coating on the luminescence lifetime and upconversion temperature sensing properties of semiconductor zinc oxide doped with gallium(III) and sensitized with rare earth ions Yb(III) and Tm(III). *Microchimica Acta, 185*(3), 1–9. Available from https://doi.org/10.1007/s00604-018-2733-6.

Lian, H., Hou, Z., Shang, M., Geng, D., Zhang, Y., & Lin, J. (2013). Rare earth ions doped phosphors for improving efficiencies of solar cells. *Energy, 57*, 270–283. Available from https://doi.org/10.1016/j.energy.2013.05.019.

Lide, D. R. (1997). *CRC handbook of chemistry and physics* (78th ed., pp. 4–50). Boca Raton, FL: Chem. Rubber Co. Press.

Lima, S. A. M., Sigoli, F. A., Davolos, M. R., & Jafelicci, M., Jr (2002). Europium (III)-containing zinc oxide from Pechini method. *Journal of Alloys and Compounds, 344*(1–2), 280–284.

Lin, Z., et al. (2017). Dense nanosized europium silicate clusters induced light emission enhancement in Eu-doped silicon oxycarbide films. *Journal of Alloys and Compounds, 694*, 946–951. Available from https://doi.org/10.1016/j.jallcom.2016.10.132, Feb.

Liu, H., Yu, L., Chen, W., & Li, Y. (2012). The progress of TiO2 nanocrystals doped with rare earth ions. *Journal of Nanomaterials*.

Lo Savio, R., et al. (2014). Photonic crystal light emitting diode based on Er and Si nanoclusters co-doped slot waveguide. *Applied Physics Letters, 104*(12), 121107.

Lozykowski, H. J., & Jadwisienczak, W. M. (1997). Luminescence properties of As, P and Bi as isoelectronic traps in GaN. *MRS Proceedings, 482*, 1033. doi:10.1557/PROC-482-1033.

Mir, N., Castano, C. E., Rojas, J. V., Norouzi, N., Esmaeili, A. R., & Mohammadi, R. (2021). Self-separation of the adsorbent after recovery of rare-earth metals: Designing a novel non-wettable polymer. *Separation and Purification Technology, 259*, 118152. Available from https://doi.org/10.1016/j.seppur.2020.118152, no. August.

Mir, N., Heidari, A., Beyzaei, H., Mirkazehi-Rigi, S., & Karimi, P. (2017). Detection of Hg2 + in aqueous solution by pyrazole derivative-functionalized Fe3O4@ SiO2 fluorescent probe. *Chemical Engineering Journal, 327*, 648–655.

Mir, N., Jalilian, S., Karimi, P., Nejati-Yazdinejad, M., & Khammarnia, S. (2018). 1, 3, 4-Thiadiazol derivative functionalized-Fe_3O_4@ SiO_2 nanocomposites as a fluorescent probe for detection of Hg^{2+} in water samples. *RSC Advances*.

Mir, N., Lee, K., Paramasivam, I., & Schmuki, P. (2012). Optimizing TiO2 nanotube top geometry for use in dye-sensitized solar cells. *Chemistry-A European Journal, 18*(38), 11862–11866.

Mir, N., & Salavati-Niasari, M. (2012). Photovoltaic properties of corresponding dye sensitized solar cells: Effect of active sites of growth controller on TiO2 nanostructures. *Solar Energy, 86*(11), 3397–3404.

Mir, N., & Salavati-Niasari, M. (2013a). Preparation of TiO < inf > 2 < /inf > nanoparticles by using tripodal tetraamine ligands as complexing agent via two-step sol-gel method and their application in dye-sensitized solar cells. *Materials Research Bulletin, 48*(4). Available from https://doi.org/10.1016/j.materresbull.2013.01.006.

Mir, N., & Salavati-Niasari, M. (2013b). Effect of tertiary amines on the synthesis and photovoltaic properties of TiO < inf > 2 < /inf > nanoparticles in dye sensitized solar cells. *Electrochimica Acta, 102*. Available from https://doi.org/10.1016/j.electacta.2013.03.141.

Mitchell, B., Dierolf, V., Gregorkiewicz, T., & Fujiwara, Y. (2018). Perspective: Toward efficient GaN-based red light emitting diodes using europium doping. *Journal of*

Applied Physics, *123*(16), 160901. Available from https://doi.org/10.1063/1.5010762, American Institute of Physics Inc.

Mitchell, B., Hernandez, N., Lee, D., Koizumi, A., Fujiwara, Y., & Dierolf, V. (2017). Charge state of vacancy defects in Eu-doped GaN. *Physical Review B*, *96*(6), 064308. Available from https://doi.org/10.1103/PhysRevB.96.064308.

Mitchell, B., et al. (2017). Synthesis and characterization of a liquid Eu precursor (EuCppm2) allowing for valence control of Eu ions doped into GaN by organometallic vapor phase epitaxy. *Materials Chemistry and Physics*, *193*, 140–146. Available from https://doi.org/10.1016/j.matchemphys.2017.02.021, Jun.

Mitchell, B., et al. (2020). Direct detection of rare earth ion distributions in gallium nitride and its influence on growth morphology. *Journal of Applied Physics*, *127*(1), 013102. Available from https://doi.org/10.1063/1.5134050.

Mitra, C., & Lambrecht, W. R. L. (2009). Interstitial-nitrogen- and oxygen-induced magnetism in Gd-doped GaN. *Physical Review B*, *80*(8), 81202. Available from https://doi.org/10.1103/PhysRevB.80.081202.

Morita, E. (2006). *Method of crystallizing a nitride III-V compound semiconductor layer on a sapphire substrate*. Google Patents, October 24, 2006.

Mukherji, R., Mathur, V., Samariya, A., & Mukherji, M. (2016). A review of transition metal-doped In 2 O 3-based diluted magnetic semiconductors. *IUP Journal of Electrical and Electronics Engineering*, *9*(4).

Munekata, H., Ohno, H., Von Molnar, S., Segmüller, A., Chang, L. L., & Esaki, L. (1989). Diluted magnetic III-V semiconductors. *Physical Review Letters*, *63*(17), 1849.

Nakamura, S., Mukai, T., & Senoh, M. (1994). Candela-class high-brightness InGaN/AlGaN double-heterostructure blue-light-emitting diodes. *Applied Physics Letters*, *64*(13), 1687–1689.

Nepal, N., Jiang, H. X., Lin, J. Y., Mitchell, B., Dierolf, V., & Zavada, J. M. (2016). *MOCVD growth of Er-doped III-N and optical-magnetic characterization. Rare earth and transition metal doping of semiconductor materials: Synthesis, magnetic properties and room temperature spintronics* (pp. 225–257). Elsevier Inc.

Nepal, N., et al. (2009). Optical enhancement of room temperature ferromagnetism in Er-doped GaN epilayers. *Applied Physics Letters*, *95*(2), 22510.

Nishikawa, A., Kawasaki, T., Furukawa, N., Terai, Y., & Fujiwara, Y. (2009). Room-temperature red emission from a p-type/europium-doped/n-type gallium nitride light-emitting diode under current injection. *Applied Physics Express*, *2*(7), 071004. Available from https://doi.org/10.1143/APEX.2.071004.

O'Donnell, K. P. (2015). The temperature dependence of the luminescence of rare-earth-doped semiconductors: 25 Years after Favennec. *Physica Status Solidi*, *12*(4–5), 466–468. Available from https://doi.org/10.1002/pssc.201400133.

O'Donnell, K. P., & Dierolf, V. (2010). *Rare-earth doped III-nitrides for optoelectronic and spintronic applications* (vol. 124). Springer Science & Business Media.

O'Donnell, K. P., & Hourahine, B. (2006). Rare earth doped III-nitrides for optoelectronics. *European Physical Journal Applied Physics*, *36*(2), 91–103. Available from https://doi.org/10.1051/epjap:2006122.

Ohtake, T., Hijii, S., Sonoyama, N., & Sakata, T. (2006). Electrochemical luminescence of n-type ZnO semiconductor electrodes doped with rare earth metals under the anodic polarization. *Applied Surface Science*, *253*(4), 1753–1757.

Palm, J., Gan, F., Zheng, B., Michel, J., & Kimerling, L. C. (1996). Electroluminescence of erbium-doped silicon. *Physical Review B*, *54*(24), 17603.

Pearton, S. J., et al. (2003). Wide band gap ferromagnetic semiconductors and oxides. *Journal of Applied Physics*, *93*(1), 1–13.

Pineiz, T. F., de Morais, E. A., Scalvi, L. V. A., & Bueno, C. F. (2013). Interface formation of nanostructured heterojunction SnO2:Eu/GaAs and electronic transport properties. *Applied Surface Science*, *267*, 200–205. Available from https://doi.org/10.1016/j.apsusc.2012.10.097.

Pomrenke, G. S., Ennen, H., & Haydl, W. (1986). Photoluminescence optimization and characteristics of the rare-earth element erbium implanted in GaAs, InP, and GaP. *Journal of Applied Physics*, *59*(2), 601–610. Available from https://doi.org/10.1063/1.336619.

Poornaprakash, B., Ramu, S., Park, S. H., Vijayalakshmi, R. P., & Reddy, B. K. (2016). Room temperature ferromagnetism in Nd doped ZnS diluted magnetic semiconductor nanoparticles. *Materials Letters*, *164*, 104–107. Available from https://doi.org/10.1016/j.matlet.2015.10.119.

Poust, B. D., et al. (2003). SiC substrate defects and III-N heteroepitaxy. *Journal of Physics D: Applied Physics*, *36*(10A), A102.

Prathap Kumar, M., Suganya Josephine, G. A., Tamilarasan, G., Sivasamy, A., & Sridevi, J. (2018). Rare earth doped semiconductor nanomaterials and its photocatalytic and antimicrobial activities. *Journal of Environmental Chemical Engineering*, *6*(4), 3907–3917. Available from https://doi.org/10.1016/j.jece.2018.05.046.

Prucnal, S., Rebohle, L., & Skorupa, W. (2017). Doping by flash lamp annealing. *Materials Science in Semiconductor Processing*, *62*, 115–127. Available from https://doi.org/10.1016/j.mssp.2016.10.040.

Qiao, Y., Li, S., Liu, W., Ran, M., Lu, H., & Yang, Y. (2018). Recent advances of rare-earth ion doped luminescent nanomaterials in perovskite solar cells. *Nanomaterials*, *8*(1), 43. Available from https://doi.org/10.3390/nano8010043.

Raczyńska, J., Fronc, K., Langer, J. M., Lemańska, A., & Stapor, A. (1988). Donor gettering in GaAs by rare-earth elements. *Applied Physics Letters*, *53*(9), 761–763.

Rajeswari, R., Islavath, N., Raghavender, M., & Giribabu, L. (2020). Recent progress and emerging applications of rare earth doped phosphor materials for dye-sensitized and perovskite solar cells: A review. *Chemical Record (New York, N.Y.)*, *20*(2), 65–88. Available from https://doi.org/10.1002/tcr.201900008.

Rebohle, L., Prucnal, S., & Skorupa, W. (2016). A review of thermal processing in the subsecond range: Semiconductors and beyond. *Semiconductor Science and Technology*, *31* (10), 103001. Available from https://doi.org/10.1088/0268-1242/31/10/103001, Institute of Physics Publishing.

Rebohle, L., et al. (2008). Blue and red electroluminescence of Europium-implanted metal-oxide-semiconductor structures as a probe for the dynamics of microstructure. *Applied Physics Letters*, *93*(7), 71908.

Rebohle, L., et al. (2009). Anomalous wear-out phenomena of europium-implanted light emitters based on a metal-oxide-semiconductor structure. *Journal of Applied Physics*, *106*(12), 123103.

Reddy, A. A., et al. (2013). KCa4 (BO3) 3: Ln3 + (Ln = Dy, Eu, Tb) phosphors for near UV excited white–light–emitting diodes. *AIP Advances*, *3*(2), 22126.

Redwing, J. M., Kuech, T. F., Gordon, D. C., Vaartstra, B. A., & Lau, S. S. (1994). Growth studies of erbium-doped GaAs deposited by metalorganic vapor phase epitaxy using novel cyclopentadienyl-based erbium sources. *Journal of Applied Physics*, *76*(3), 1585–1591.

Reshchikov, M. A., & Morko, H. (2005). Luminescence properties of defects in GaN. *Journal of Applied Physics*, *97*(6), 061301. Available from https://doi.org/10.1063/1.1868059.

Reszczyńska, J., et al. (2015). Visible light activity of rare earth metal doped (Er3 + , Yb3 + or Er3 + /Yb3 +) titania photocatalysts. *Applied Catalysis B: Environmental*, *163*, 40–49.

Ronfard-Haret, J.-C., Kouyate, D., & Kossanyi, J. (1991). Evidence for direct impact-excitation of luminescent rare-earth centers (Ho3 + and Sm3 +) in semiconducting zinc oxide. *Solid State Communications*, *79*(1), 85–88.

Rzakuliev, N. A., et al. (1988). Luminescence of Pr, Nd and Yb ions implanted in GaAs and GaP. *Czechoslovak Journal of Physics*, *38*(11), 1288–1293.

Sedgwick, T. O. (1983). Short time annealing. *Journal of the Electrochemical Society*, *130* (2), 484.

Seghier, D., et al. (1994). Optical and electrical properties of rare earth (Yb,Er) doped GaAs grown by molecular beam epitaxy. *Journal of Applied Physics*, *75*(8), 4171–4175. Available from https://doi.org/10.1063/1.356000.

Shannon, R. D. (1976). Revised effective ionic radii and systematic studies of interatomic distances in halides and chalcogenides. *Acta Crystallographica, Section A: Crystal Physics, Diffraction, Theoretical and General Crystallography*, *A32*, 751–767.

Shi, Y.-L., Hu, Y., Wang, S.-P., Liao, L.-S., & Ling, F. C.-C. (2019). High transmittance Er-doped ZnO thin films as electrodes for organic light-emitting diodes. *Applied Physics Letters*, *115*(25), 252102.

Shimizu-Iwayama, T., Kurumado, N., Hole, D. E., & Townsend, P. D. (1998). Optical properties of silicon nanoclusters fabricated by ion implantation. *Journal of Applied Physics*, *83*(11), 6018–6022.

Shukla, S., & Sharma, D. K. (2020). A review on rare earth (Ce and Er)-doped zinc oxide nanostructures. *Materials Today Proceedings*. Available from https://doi.org/10.1016/j.matpr.2020.05.264, no. xxxx.

Silambarasan, M., Saravanan, S., Ohtani, N., & Soga, T. (2014). Structural and optical studies of pure and Ni-doped ZnO nanoparticles synthesized by simple solution combustion method. *Japanese Journal of Applied Physics*, *53*(5S1), 05FB16.

Smith, M. P., Seffen, K. A., McMahon, R. A., Voelskow, M., & Skorupa, W. (2006). Analysis of wafer stresses during millisecond thermal processing. *Journal of Applied Physics*, *100*(6), 63515.

Song, H. Z., & Bao, X. M. (1997). Visible photoluminescence from silicon-ion-implanted SiO 2 sfilm and its multiple mechanisms. *Physical Review B*, *55*(11), 6988.

Stabler, C., Ionescu, E., Graczyk-Zajac, M., Gonzalo-Juan, I., & Riedel, R. (2018). Silicon oxycarbide glasses and glass-ceramics: 'All-Rounder' materials for advanced structural and functional applications. *Journal of the American Ceramic Society*, *101*(11), 4817–4856. Available from https://doi.org/10.1111/jace.15932.

Steckl, A. J., Heikenfeld, J., Lee, D. S., & Garter, M. (2001). Multiple color capability from rare earth-doped gallium nitride. *Materials Science and Engineering B*, *81*(1), 97–101. Available from https://doi.org/10.1016/S0921-5107(00)00745-5.

Steckl, A. J., & Zavada, J. M. (1999a). Optoelectronic properties and applications of rare-earth-doped GaN. *MRS Bulletin/Materials Research Society*, *24*(9), 33–38. Available from https://doi.org/10.1557/S0883769400053045.

Steckl, A. J., & Zavada, J. M. (1999b). Photonic applications of rare-earth-doped materials. *MRS Bulletin/Materials Research Society*, *24*(9), 16–17. Available from https://doi.org/10.1557/s0883769400053008.

Steckl, A. J., et al. (2002). Rare-earth-doped GaN: Growth, properties, and fabrication of electroluminescent devices. *IEEE Journal of Selected Topics in Quantum Electronics*, *8*(4), 749–766. Available from https://doi.org/10.1109/JSTQE.2002.801690.

Steckl, A. J., Park, J. H., & Zavada, J. M. (2007). Prospects for rare earth doped GaN lasers on Si. *Materials Today*, *10*(7–8), 20–27.

Štengl, V., Bakardjieva, S., & Murafa, N. (2009). Preparation and photocatalytic activity of rare earth doped TiO2 nanoparticles. *Materials Chemistry and Physics*, *114*(1), 217–226.

Stimmer, J., Reittinger, A., Nützel, J. F., Abstreiter, G., Holzbrecher, H., & Buchal, C. (1996). Electroluminescence of erbium–oxygen-doped silicon diodes grown by molecular beam epitaxy. *Applied Physics Letters*, *68*(23), 3290–3292.

Taguchi, A., Kawashima, M., Takahei, K., & Horikoshi, Y. (1993). Er luminescence centers in GaAs grown by migration-enhanced epitaxy. *Applied Physics Letters*, *63*(8), 1074–1076.

Thaik, M., Hömmerich, U., Schwartz, R. N., Wilson, R. G., & Zavada, J. M. (1997). Photoluminescence spectroscopy of erbium implanted gallium nitride. *Applied Physics Letters*, *71*(18), 2641–2643.

Thonke, K., Hermann, H. U., & Schneider, J. (1988). A Zeeman study of the 1.54 μm transition in molecular beam epitaxial GaAs: Er. *Journal of Physics C: Solid State Physics*, *21*(34), 5881.

Timmerman, D., Mitchell, B., Ichikawa, S., Tatebayashi, J., Ashida, M., & Fujiwara, Y. (2020). Excitation efficiency and limitations of the luminescence of Eu3 + Ions in Ga N. *Physical Review Applied*, *13*(1), 014044. Available from https://doi.org/10.1103/PhysRevApplied.13.014044.

Trandafilović, L. V., Jovanović, D. J., Zhang, X., Ptasińska, S., & Dramićanin, M. D. (2017). Enhanced photocatalytic degradation of methylene blue and methyl orange by ZnO: Eu nanoparticles. *Applied Catalysis B: Environmental*, *203*, 740–752.

Ugolini, C., Nepal, N., Lin, J. Y., Jiang, H. X., & Zavada, J. M. (2006). Erbium-doped GaN epilayers synthesized by metal-organic chemical vapor deposition. *Applied Physics Letters*, *89*(15), 151903. Available from https://doi.org/10.1063/1.2361196.

us Saqib, N., Adnan, R., & Shah, I. (2016). A mini-review on rare earth metal-doped TiO2for photocatalytic remediation of wastewater. *Environmental Science and Pollution Research*, *23*(16), 15941–15951. Available from https://doi.org/10.1007/s11356-016-6984-7.

Ungureanu, M., et al. (2007). Electrical and magnetic properties of RE-doped ZnO thin films (RE = Gd, Nd). *Superlattices and Microstructures*, *42*(1–6), 231–235.

Vantomme, A., et al. (2001). Suppression of rare-earth implantation-induced damage in GaN. *Nuclear Instruments and Methods in Physics Research Section B: Beam Interactions with Materials and Atoms*, *175*, 148–153.

Van Vleck, J. H. (1932). *Electric and magnetic susceptibilities*. Clarendon Press.

Vinh, N. Q., Ha, N. N., & Gregorkiewicz, T. (2009). Photonic properties of Er-doped crystalline silicon. *Proceedings of the IEEE*, *97*(7), 1269–1283.

Viswanatha, R., Pietryga, J. M., Klimov, V. I., & Crooker, S. A. (2011). Spin-polarized Mn2 + emission from Mn-doped colloidal nanocrystals. *Physical Review Letters*, *107*(6), 67402.

Wang, D., Xing, G., Gao, M., Yang, L., Yang, J., & Wu, T. (2011). Defects-mediated energy transfer in red-light-emitting Eu-doped ZnO nanowire arrays. *The Journal of Physical Chemistry C*, *115*(46), 22729–22735.

Wang, H. Q., Batentschuk, M., Osvet, A., Pinna, L., & Brabec, C. J. (2011). Rare-earth ion doped up-conversion materials for photovoltaic applications. *Advanced Materials*, *23* (22–23), 2675–2680. Available from https://doi.org/10.1002/adma.201100511.

Wang, R., Steckl, A. J., Nepal, N., & Zavada, J. M. (2010). Electrical and magnetic properties of GaN codoped with Eu and Si. *Journal of Applied Physics*, *107*(1), 13901. Available from https://doi.org/10.1063/1.3275508.

Williams, T. M., Hunter, D., Pradhan, A. K., & Kityk, I. V. (2006). Photoinduced piezo-optical effect in Er doped ZnO films. *Applied Physics Letters*, *89*(4), 43116.

Wilson, R. G., et al. (1994). 1.54-μm photoluminescence from Er-implanted GaN and AlN. *Applied Physics Letters*, *65*(8), 992–994.

Woodward, N. T., et al. (2011). Enhanced magnetization in erbium doped GaN thin films due to strain induced electric fields. *Applied Physics Letters*, *99*(12), 122506. Available from https://doi.org/10.1063/1.3643041.

Wu, J., Xie, G., Lin, J., Lan, Z., Huang, M., & Huang, Y. (2010). Enhancing photoelectrical performance of dye-sensitized solar cell by doping with europium-doped yttria rare-earth oxide. *Journal of Power Sources*, *195*(19), 6937–6940.

Xu, A. W., Gao, Y., & Liu, H. Q. (2002). The preparation, characterization, and their photocatalytic activities of rare-earth-doped TiO_2 nanoparticles. *Journal of Catalysis*, *207*(2), 151–157. Available from https://doi.org/10.1006/jcat.2002.3539.

Yang, D. (2018). *Titanium dioxide: Material for a sustainable environment*. IntechOpen.

Yang, J., et al. (2008). Low-temperature growth and optical properties of Ce-doped ZnO nanorods. *Applied Surface Science*, *255*(5), 2646–2650, Part 2. Available from https://doi.org/10.1016/j.apsusc.2008.08.001.

Yao, N., et al. (2015). Enhanced light harvesting of dye-sensitized solar cells with up/down conversion materials. *Electrochimica Acta*, *154*, 273–277.

Zavada, J. M., et al. (2000). Luminescence characteristics of Er-doped GaN semiconductor thin films. *Journal of Alloys and Compounds*, *300*, 207–213. Available from https://doi.org/10.1016/S0925-8388(99)00724-0.

Zeng, X., Yuan, J., & Zhang, L. (2008). Synthesis and photoluminescent properties of rare earth doped ZnO hierarchical microspheres. *The Journal of Physical Chemistry C*, *112* (10), 3503–3508.

Zeng, Z., et al. (2020). Rare-earth-containing perovskite nanomaterials: Design, synthesis, properties and applications. *Chemical Society Reviews*, *49*(4), 1109–1143.

Zhang, J., et al. (2015). Single-crystal indium phosphide nanowires grown on polycrystalline copper foils with an aluminum-doped zinc oxide template. *Journal of Materials Science*, *50*(14), 4926–4932.

Zheng, B., Michel, J., Ren, F. Y. G., Kimerling, L. C., Jacobson, D. C., & Poate, J. M. (1994). Room-temperature sharp line electroluminescence at $\lambda = 1.54$ μm from an erbium-doped, silicon light-emitting diode. *Applied Physics Letters*, *64*(21), 2842–2844.

Zhu, G., et al. (2019). Highly Eu 3 + ions doped novel red emission solid solution phosphors Ca 18 Li 3 (Bi, Eu)(PO 4) 14: Structure design, characteristic luminescence and abnormal thermal quenching behavior investigation. *Dalton Transactions*, *48*(5), 1624–1632.

Zhu, W., Mitchell, B., Timmerman, D., Koizumi, A., Gregorkiewicz, T., & Fujiwara, Y. (2017). High-power Eu-doped GaN red LED based on a multilayer structure grown at lower temperatures by organometallic vapor phase epitaxy. *MRS Advances*, *2*(3), 159–164. Available from https://doi.org/10.1557/adv.2017.67.

Zhu, W., Mitchell, B., Timmerman, D., Uedono, A., Koizumi, A., & Fujiwara, Y. (2016). Enhanced photo/electroluminescence properties of Eu-doped GaN through optimization of the growth temperature and Eu related defect environment. *APL Materials*, *4*(5), 056103. Available from https://doi.org/10.1063/1.4950826.

Zinatloo-Ajabshir, S., Salavati-Niasari, M., Sobhani, A., & Zinatloo-Ajabshir, Z. (2018). Rare earth zirconate nanostructures: Recent development on preparation and photocatalytic applications. *Journal of Alloys and Compounds*, *767*, 1164–1185.

Zunger, A. (2003). Practical doping principles. *Applied Physics Letters*, *83*(1), 57–59. Available from https://doi.org/10.1063/1.1584074.

Rare-earth-based nanocomposites 12

Razieh Razavi[1] *and Mahnaz Amiri*[2,3]
[1]Department of Chemistry, Faculty of Science, University of Jiroft, Jiroft, Iran,
[2]Neuroscience Research Center, Institute of Neuropharmacology, Kerman University of Medical Science, Kerman, Iran, [3]Department of Hematology and Medical Laboratory Sciences, Faculty of Allied Medicine, Kerman University of Medical Sciences, Kerman, Iran

12.1 General introduction

Today, nanomaterials have found extensive applications in both the applied and basic physics fields that cover extensive areas (e.g., health and biology, information communication technologies, systems of environment monitoring and structural engineering), due to the recent achievements in nanocomposite materials activated by the ions of rare-earth elements. During the recent decade, experimental attempts to improve and control the techniques and various methods for nanocomposite materials synthesis based on rare earth metals have increased significantly. The new achievements in the field of nanocomposites activated by the ions of rare earth metals have resulted in new opportunities in both the applied and basic physics fields that cover extensive areas (e.g., health and biology, information communication technologies, systems of environment monitoring and structural engineering). During the recent decades, RE (rare earth) elements (i.e., scandium, yttrium and the 14 elements found in the Ln (lanthanide) series) have attracted significant interest in both industries and academic research fields, which is attributable to their unique chemical and physical characteristics (Zepf, 2016). The increasing number of applications approves this claim, and therefore, they have become essential to some important technologies, which demonstrate the reason behind their increasing demands (Chakhmouradian & Wall, 2012; Du & Graedel, 2011). In fact, in terms of their abundance, most of the above elements are not rare, and the word "rare" recalls the difficult process of their extraction. This is due to the cooccurrence of these elements in nature, which are hardly separable from each other (Weller, Overton, Rourke, & Armstrong, 2014). Today, the design of materials with nanostructures containing rare earth elements, either as a dopant or as a major component, has facilitated the development of new applications. Particularly, NPs (nanoparticles) are of sizes (within the 1−100 nm range) in which the majority of the biomolecular interactions can happen (Hötzer, Medintz, & Hildebrandt, 2012; Mout, Moyano, Rana, & Rotello, 2012; Taeho & Taeghwan, 2014), thus, incorporating rare elements into nanoparticles makes their utilization possible in numerous biomedical applications, such as biosensing, bioimaging, drug delivery, targeting,

Advanced Rare Earth-Based Ceramic Nanomaterials. DOI: https://doi.org/10.1016/B978-0-323-89957-4.00005-0

and other therapies (Murthy, 2007). This study is aimed to present an overview of the advancements in nanocomposites systems based on rare earth elements, and also, of the most frequently applied strategies for the functionalization and synthesis of them. For a more detailed and more comprehensive description of the above issues, the readers are referred to some noticeable reviews available in the recent literature (Gnach & Bednarkiewicz, 2012). In the present work the goal is to explain which raw material and technique is most suited for processing of a particular nanocomposites based rare earth elements as well as application, advantages and drawbacks of these nanocomposites, we aimed to study the Rare-earth-based nanocomposites (recent advances related to preparation methods and applications) in order to give valuable information for readers and researchers.

12.2 Nanocomposite materials

12.2.1 Description

Nanocomposites are solid materials composed of various phases, and at least one of these phases is of one, two, or three nanometer dimensions. Nanocomposites proposed chances on totally new scales for resolving problems ranging from medical, pharmaceutical industry, food packaging, to electronics and energy industry. The synergy goal between various constituents is achievable via nanoscale phase procedure. Nanocomposites include nanoclays, nanoparticles, and nanofibers. In case the constituent size of the material is lower than a special level, called "critical size," the material characteristics are changed (Schmidt, Shah, & Giannelis, 2002). Lowering the dimensions of material to nanometer levels generates interactions phase interfaces, acting as important agents in improvement of materials characteristics. The surface area to the volume ratio of the reinforced material utilized during preparation of nanocomposites plays a direct role in perception of structure-property relations. The nanocomposites featuring at least one nanometric component (10 − 9 m) are composed of polymeric, nonmetallic, and metallic materials via special procedure and present extra advantages in maintaining primary characteristics to eliminate the defects and present some new features. These materials show multiphase crossover of the reinforcing materials and the matrix material. Whilst the reinforcing materials are dispersed phases, typically fibrous materials including organic fibers, glass fibers, etc., the material making the matrix is a continuous phase that is comprised of inorganic, metallic, polymer, and nonmetallic matrix materials (Wang et al., 2011). Nanocomposites act as a new alternative in overcoming the present constraints of monolithics and microcomposites, and thus, they can be deemed as materials with promising future. The major benefits of nanocomposites compared to other composites include:

- Higher surface/volume ratio allowing small distances between fillers as well as filler sizes;
- improved mechanical characteristics;

- Higher ductility with no strength loss, scratch resistance;
- Better optical characteristics (light transmission is dependent on particle sizes).
- The nanocomposite application has some disadvantages including:
- Toughness and impact performance related to incorporation of nanoparticles to the composite bulk-matrix;
- Insufficient perception of the property/ structure/ formulation relationships, requiring simpler particle dispersion and exfoliation;
- Cost-effectiveness.

Given their matrix materials, nanocomposites are categorized into three groups: 1-metal matrix nanocomposites, 2-polymer matrix nanocomposites, 3-ceramic matrix nanocomposites, in the following parts these three types of nanocomposite materials will be discussed in details.

12.2.2 Ceramic matrix nanocomposites

Composites with a ceramic matrix are those with intentionally-added one or more different ceramic phases, so that their chemical/thermal stability and wear resistance are improved. However, the major disadvantage of ceramics that prevents their vast industrial application is their lower toughness and being brittle. The above limitation can be overcome via CMNCs (ceramic-matrix nanocomposites) development. Examples of CMNC include matrices in which components with energy dissipating features (particles, fiber, or platelets) are incorporated within ceramic matrices to increase fracture durability and decrease brittleness (Becher, 1991; Harmer, Chan, & Miller, 1992; Lange, 1973). The raw materials used in CMNC matrices are SiN, SiC, Al_2O_3, etc. The materials used for crystalline reinforcement are Iron and other metal powders such as TiO_2, clays, and silica. The most common ones are layered silicates and clays due to the well-studied chemistry of their interactions and accessibility in very low particle sizes (Alexandre & Dubois, 2000; Fernando & Satyanarayana, 2005; Ogawa & Kuroda, 1997; Theng, 1974). Addition of even small quantities of layered silica and clay enhances the matrix characteristics (Noh & Lee, 1999). Numerous techniques have been developed for the CMNC synthesis (Long, Shao, Wang, & Wang, 2016; Sun, Jeurgensc, Burghardb, & Billb, 2017; Yu et al., 2017). The major modern methods are singlesource-precursor methods on the basis of melt spinning of hybrid precursors followed by pyrolysizing and curing of the fibers. Some former techniques are: polymer precursor route (He, Gao, Wang, Fang, & An, 2016; Yan, Sahimi, & Tsotsis, 2017); conventional powder technique (Ghasali, Yazdani-rad, Asadian, & Ebadzadeh, 2016); vapor techniques (Brooke et al., 2017) (CVD and PVD); spraypyrolysis (Choia, Yoonb, Hanb, Kima, & Othmanc, 2016). Chemical techniques include: the template synthesis; colloidal and precipitation method; sol-gel process (Camargo, Satyanarayana, & Wypych, 2009). The structures of nanocomposites are comprised of a matrix material which contains the nanosized components used for reinforcement (whiskers, particles, nanotubes, fibers). Easy fracturing due to propagation of cracks and high brittleness make ceramic an unsuitable material, but addition of ductile metal phases into the ceramics matrixes results in improved durability and consequently, improved mechanical characteristics (fracture toughness and hardness) due to relations between the various phases, reinforcements and matrix. The relationship between the volume of the reinforcement materials and the surface area acts as a crucial parameter in perception of the structure–property relation in ceramic-matrix nanocomposites. In this view, ceramic and metal nanocomposites may result in big impacts in various industries, ranging from military, automotive sector, aerospace, to electronics (Camargo et al., 2009). Some modern uses of ceramic nanocomposites are biomedical applications, acid fuchsine removal (Yua, Lia, Zhanga, Fenga, & Liua, 2016), and different photocurrent applications (Dezfuly, Yousefi, & Jamali-Sheini, 2016). The application of nanocrystalline ceramic materials is of the following benefits:

- Improved hardness/resistance
- Enhanced durability/toughness
- Lower ductility and elasticity
- Lower risk of rejection.

The combined characteristics of nanocomposites including enhanced hardness, creep resistance, stability, strength, and toughness may lead to the creation of a new generation of implants, prosthetic, and medical devices by combining mechanical characteristics with bioactive characteristics. Alumina/zirconia nanocomposites, CNT (ceramic/carbon nanotube) composites, alumina/silicon carbide nanocomposites, etc. are some examples of nanocomposite materials with ceramic base. According to experiments, given their hardness, excellent biocompatibility and lower wear rates, ceramics like alumina (Al_2O_3) and zirconia (ZrO_2) are suitable materials for the implants used in orthopedics. Zirconia is preferable for dental uses, in which aesthetic demands (translucency, color) are a must (Garmendia, Olalde, & Obieta, 2013). Gamal-Eldeen et al. demonstrated that nanocomposites made of ferromagnetic glass ceramic (CaO-ZnO-Fe_2O_3-SiO_2) are of anticancer impacts on bone cancerous Saos-2 cells (Gamal-Eldeena, Abdel-Hameedc, El-Dalya, Abo-Zeida, & Swellamb, 2017).

12.2.3 Metal matrix nanocomposites

MMNCs (metal matrix nanocomposites) are multiphase materials consisting of ductile alloy or metal matrices where some nanosized reinforcement materials are implanted. The main characteristics of MMNCs are their higher toughness, ductility, modulus and strength. Metal matrix nanocomposites are applicable in various industries such as automotive or aerospace industries (Lee, Choi, Anandhan, Baik, & Seo, 2004). The metal matrixes employed for production of metal matrix nanocomposites include Mg, Al, Sn, Pb, Fe, and W. Metal matrix nanocomposites have the same reinforcements as those of PMNC and CMNC. The common methods of MMNCs preparation include: liquid metal infiltration (Dezfuly et al., 2016), spray pyrolysis (Dermenci, Gencc, Ebinb, Olmez-Hanci, & Gürmen, 2014; Kobayashi, 2016), electrodeposition (Kashinath, Namratha, & Byrappa, 2016), vapor techniques (Abdelhamid, Talib, & Wu, 2016) (PVD, CVD), rapid solidification (Ren et al.,

2016), and chemical techniques including sol-gel (Abdelhamid et al., 2016) and colloidal (Ren et al., 2016) processes. Recent techniques used for metal matrix nanocomposites are the one-pot synthesis of the nanocomposite of carbon dots nanocomposite (CDs) and gold nanoparticles (AuNPs) (Królikowski & Rosłaniec, 2004) and melt falling-drop quenching technique (Zare & Shabani, 2015). In various industries, for example, aerospace, automotive, military, and electronics, metal nanocomposites play different roles. Polymer-metal nanocomposites (in which metal nanoparticles act as nanofillers and the polymer acts as the matrix,) are applicable in medical devices and in different biomedical uses (Spasówka & Rudnik, 2006).

12.2.4 Polymer matrix nanocomposites

The PMNC nanocomposites are in fact nanofillers that can be categorized as the following: 1D—linear (e.g., carbon nanotubes), 2D—layered (e.g., montmorillonite) and 3D—powder (e.g., silver nanoparticles) (Ogasawara, Ishida, Ishikawa, & Yokota, 2004). Owing to the interactions among the polymer matrix and nanofiller at molecular scale, the attraction impact among the nanocomposites is obviously observed. As a result, adding slight amounts of nanofiller with below 100 nm dimensions to matrix indexes modifies the composite material characteristics. The production of nanocomposites would be possible through the use of similar methods as for the production of typical composites, including in situ, solvent method or via mixing melted polymer matrix. The PMNCs characteristics are: excellent thermal stability enhanced mechanical features like good abrasion resistance and smaller gas permeability, that is greater barrier capacity (Alexandre & Dubois, 2000; Ren et al., 2016). Numerous techniques have been adopted so far for the production of polymer nanocomposites, among which the most frequently used, include: Polymer or prepolymer Intercalation from solution; In situ intercalative polymerization; Melt intercalation; and Template synthetization (Sol-gel method) (Anandhan & Bandyopadhyay, 2011; Hussain, Hojjati, Okamoto, & Gorga, 2006; Kornmann, Linderberg, & Bergund, 2001; Rehab & Salahuddin, 2005). Upon using traditional methods like injection molding and extrusion, a mechanical mixture of thermoplastic polymer and organophillic clay is composed at high temperature (Haraguchi, 2011). Compared with the traditional materials including metals, these composites exhibit the following properties:

High specific strength combined with high specific modulus provides a high performance and low density reinforcement fiber that would provide a glass fiber of relatively low modulus and high density. The glass fiber resin matrix nanocomposite is having a specific modulus relatively lower than the metallic materials (Wang et al., 2011). According to the literature, previous research conducted to achieve nonhalogenated flame retardants has resulted in the discovery of nanoclays. Research has revealed that an addition of as small amount of 5% nanosized clay particles would result in 63% decrease of nylon-6 flammability. Most recent studies have reported the polymer flame retardancy boost via dispersing the clay at molecular scale (Bai & Ho, 2008). Fiber and nanocomposite material matrix can enable good damping characteristics and incorporation of damping material enables absorption of vibrations. The vibrations may be ceased in a short while immediately after the occurrence (Wang et al., 2011). Improving the polymer nanocomposites' properties would make them suitable for wide variety of industrial applications. Some of the common applications are: power tool housing, fuel cells, solar power cell, plastic-based containers, fuel tanks, etc. One of the attractive fields of polymer composites include chemical protective applications and medical related products like surgical gloves usable for countering chemical warfare agents and medicine-based pollutants (Nayaka, Pabi, Kimb, & Murtyc, 2010; Smart, Cassady, Lu, & Martin, 2006). As an example, the incorporation of clay thermoset polymer in fuel cells is likely to increase proton conductivity, develop better ion exchange capacity, improved mechanical properties and higher rate of conductivity even at higher humidity (Fiorito, Serafino, Andreola, & Bernier, 2006; Hurt, Monthioux, & Kane, 2006).

12.3 Why dose rare-earth elements indicate many applications?

As it is stated in the previous chapter of the book, this section provides a brief description. Rare-earth-based nanostructures establish a type of functional materials broadly used and studied in the recent literature. Their presence in alloys, ceramic and glass compounds and in oxide compounds provide special material properties. The unusual physical and chemical properties of REEs are taken advantage of in several industrial and technological fields (Goonan, 2011). Typically, lighter REEs and yttrium are cheaper, are easier to mass produce, and employed more broadly when compared to heavier REEs. In contrast, the less ubiquitous and most expensive REEs (holmium to lutetium), are only employed in a few but very specialized scenarios (Long, 2011). For instance, cerium oxide is employed to produce glass that necessitates precision polish, such as flat panel displays screens. It is also used to as glass decorative. Two other elements that are used to greatly enhance the refractive index of optical glass are lanthanum and lutetium. The former is commonly used in camera lenses, while the latter (the more expensive one) is applied in immersion lithography which necessitates a high-refractive index. Erbium, holmium, neodymium, praseodymium, ytterbium, and yttrium are used as to apply color, and to supply glass with filtering and glare-reduction attributes. Europium is commonly used as a doping agent in optical fibers. Another avenue in which REEs see major use is catalysts. Lanthanum-based catalysts, for instance, are employed in the refining of petroleum, while cerium-based catalysts are used in automotive catalytic converters (Long, 2011). Of particular note is dysprosium, since substituting it for a small amount of neodymium enhances high-temperature performance, and increases demagnetization resistance. When heat stress is a constraint, samarium-cobalt magnets—who are less powerful, but have higher heat tolerance—are used instead of neodymium alternatives. Nickel-metal hydride batteries use a lanthanum-based alloy for anodes (Anderson & Patiño-Echeverri, 2009; Kopera, 2004). As nickel-metal hybrid batteries are phased out in favor of lithium ion alternatives, demands for REEs in batteries are expected to decrease accordingly (Anderson & Patiño-Echeverri, 2009). Lanthanum is an important component of nickel-metal hydride batteries in hybrid electric vehicles, using as much as 10–15 kg of the REE per vehicle. A mixed oxide containing cerium, lanthanum, neodymium, and praseodymium, are utilized in steel making to purifiers, as well as production of specific steel alloys. Each of these REEs as well as yttrium, are also utilized in several alloys of chromium, magnesium, molybdenum, tungsten, vanadium, and zirconium, either on their own, or in combination. A popular fertilizer uses neodymium as the active component. Some other fields in which REEs may become popular in the future are nanofillers and in memory devices, power converters, optical clocks, infrared decoy flares, and fusion energy. Lastly, in the category of HREEs, Tb is utilized in magnets and phosphors, ER in phosphor and glass industries, and Y is used in ceramics and phosphors. Though these elements are typically used in minor amounts, they are nevertheless crucial components in functional devices, and the need for them is likely to increase going forward (European Commission, 2014; European Rare Earths Competency Network ERECON, 2014; Pyrzynska, Kubiak, & Wysocka, 2016; Series, 2005). More uses are also likely to be found with future research. The following session indicates the biomedical applications of rare earth based nanocomposites.

12.4 Properties of rare earth elements based nanocomposites that leads to medical and biological applications

12.4.1 Fluorescence, CT, and MRI imaging

The various types of imaging methods each have differing strengths and weaknesses in sensitivity, resolution, penetration depth, and cost, brought about by the limitations of the equipment and light source of image modality. These limitations will further hamper the ability to produce effective and accurate information on target biological structures and physiological processes for the purpose of clinical diagnosis. As such, RE-doped NPs based bioimaging is evolving from single modality to multi-modality. NPs doped with more than one RE ions (Gd, Yb/Lu/Gd, and Er/Tm/Ho) (Chatterjee, Gnanasammandhan, & Zhang, 2010) exhibit performance in bioimaging, even more so in the fluorescence emission, MR, and X-ray attenuation. These NP properties resulted in their mass adoptions for use in the field of bioimaging as multi-modal imaging contrasting agents (Cheng et al., 2010; Cheng et al., 2011; Yang et al., 2011). Ren's research group has done great research in this field, synthesizing a small, homogeneous, strong NIR emission core-shell DC $NaYF_4$:5%Nd@$NaGdF_4$ nanoparticle with emission at 1060 nm (under excitation of 808 nm) via small nano-cluster medium method (Ren et al., 2018). The NPs were employed to portray the different tumor vasculature via NIR II fluorescence imaging of a breast tumor model, and as contrast agents for MR imaging and X-ray computed tomography imaging to create complementary anatomic structure of tumor tissue. The outcome supports the claim that NPs could be employed as a novel tri-modality imaging nanoplatform for the purpose of real-time biological diagnosis. Long has developed a novel step-wise synthetic process to synthesize a multifunctional bioimaging agent Fe_3O_4@$NaLuF_4$:Yb,Er/Tm (MUCNP) (Long, 2011). Lu^{3+}'s high X-ray absorption, makes it suitable for high-resolution CT imaging. MUNCP is supplied with high tissue-penetration for MRI through the superparamagnetism of Fe_3O_4 with saturation magnetism of 15 emu/g. Additionally, NIR excitation can help $NaLuF_4$:Yb,Er/Tm core achieve fluorescence imaging. MUCNP has been shown to be successful at multimode imaging via in-vivo experiments. In 2018, Liu et al. (Liu et al., 2019) reported a multimode imaging nanoparticle based on $NaLuF_4$:Yb/Er doped with PEGylated Mn^{2+} (PEG−UCNPs). PEG-UCNPs can generate bright fluorescence signals for use in fluorescence bioimaging, as well as be employed as T1-weighted MR imaging agents, thanks to its high longitudinal relaxation. The high X-ray absorption coefficient of Lu^{3+} supplies the CT imaging modality, making the PEG−UCNPs three imaging modalities for deep-tissue bioimaging (Deng, Huang, & Xu, 2018). Figs. 12.1 and 12.2 indicate the biomedical application of rare earth based nanocomposites.

12.4.2 Tumor therapy

In recent years, RE-doped NPs have been evolved from single function NPs used in cancer diagnosis to multi-purpose NPs used in diagnose and therapy (Liu et al., 2019). When lanthanide-doped NPs are emitted, the emissions are located at long wavelengths, that often convert the absorbed photons into hear energy or reactive oxygen species (ROS). The properties of lanthanide-doped NPs have been utilized to inhibit and eliminate tumor cells via photodynamic therapy (PDT), photothermal effect, and loading drug molecules (Lu, Yuan, & Zhang, 2018). The key step in diagnosing and treating different tumor cells is altering the NPs to grant them tumor-targeting capabilities. The process aims to enable attachment of tumors to the surface of NPs via chemical bonding or physical adsorption, while not suppressing the NP signals (whether fluorescence or other type). At present, FA, RGD, and various antibodies are widely used as target ligands (Abdukayum et al., 2013).

12.4.3 Drug delivery

Drug loaded NPs are a bridge on the gap between nanomaterials and modern medicine. NPs are smaller than cells by volume, and as such can be absorbed by tissues and their cells, allowing drug-loaded NPs to be transported to therapy's lesion area. Drug-loaded NPs are composed of carrier materials and therapeutic agents, as well as targeted ligands (Oliveira et al., 2019). It is vital for the carrier materials to have high enough loading capacity to deliver the cancer therapy's indicated volume of

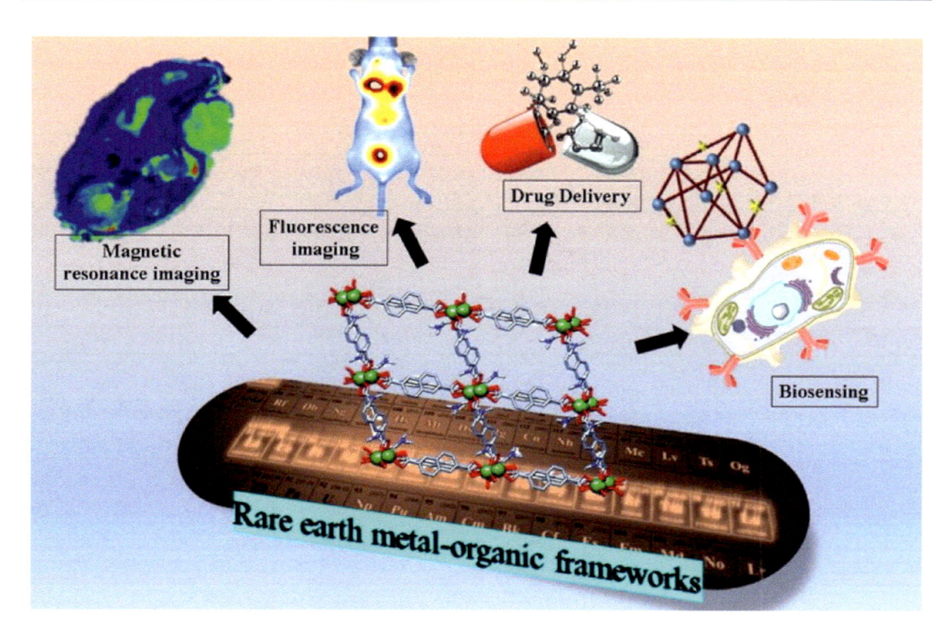

Figure 12.1 Biomedical applications of rare earth based metal organic frameworks.
Source: Reproduced with permissions from Younis, S. A., Bhardwaj, N., Bhardwaj, S. K., Kim, K., & Deep, A. (2021). Rare earth metal–organic frameworks (RE-MOFs): Synthesis, properties, and biomedical applications. *Coordination Chemistry Reviews*, *429*, 213620.

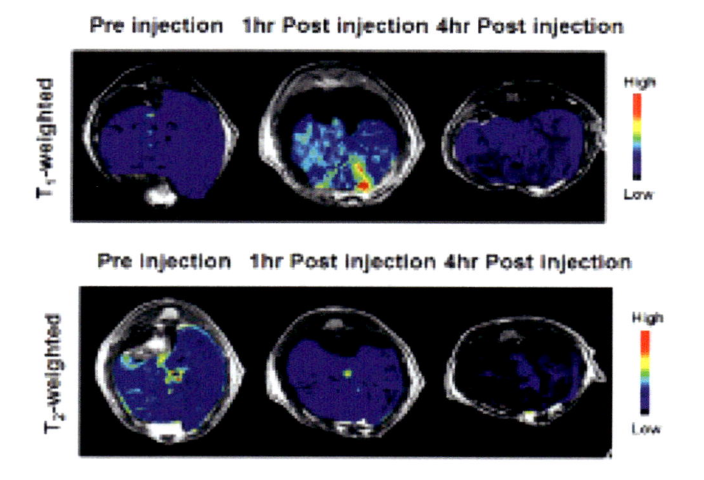

Figure 12.2 Axial T1- and T2-weighted MRI images of the liver before and after intravenous injection of Eu,Gd-NMOF@SiO_2NPs.
Source: Reproduced with permissions from Younis, S. A., Bhardwaj, N., Bhardwaj, S. K., Kim, K., & Deep, A. (2021). Rare earth metal–organic frameworks (RE-MOFs): Synthesis, properties, and biomedical applications. *Coordination Chemistry Reviews*, 429, 213620.

drug. For instance, mesoporous materials and polymeric micelles are popular choices when selecting drug-carrying material, due to their high porosities enabling them to carry large volumes of drugs. RE NPs nanoplatform are not only effective carriers, but also enable drug tracking via bioimaging, which could provide crucial information in clinical diagnosis. Additionally, drug release can be regulated efficiency, since the drug loading system could conditionally react to endogenous or exogenous stimuli such as the pH, light, and so on. pH-responsive drug-loaded NPs. pH-responsive drug-loaded NPs can greatly improve the drug utilization. The key advantage of these NPs is their delivery in physiological environments without leakage, absorption or corrosion. Tumor microenvironments are typically acidic, since the pH of endosomes and lysosome is 4–6. Therefore, pH-responsive drug-loaded NPs are widely studied for tumor treatment. Tawfik et al. (2018) have synthesized a stable and biocompatible polymer altered and have studied it for successful encapsulation, delivery and release of anticancer drug, doxorubicin (DOX). In-PBS encapsulation efficiency was found to be as high as 83%, while drug loading capacity was found to be 18.3% thanks to the large cavities of materials and the amphiphilic polymer shell. The DOX payload can be released efficiently in a highly-regulated and discriminating pH-responsive approach through FR-mediated endocytosis. The end result was improvements in DOX-loaded NPs ability to cause apoptosis and abrogation of KB cells. Additionally, flow cytometry quantifies the intake and aggregation of DOX within KB cancer cells. The results indicate the efficient cellular uptake of such NPs. In another case in 2018, Yang et al. (2018) have employed $GdPO_4$:Nd NPs as matrix materials. The egg yolk-shell structure $GdPO_4$:Nd was prepared via self-sacrificing template in which cetyltrimethylamine bromide is used as structure-directing agent. This shell can be used as drug carrier to responsive to change of pH. The NPs can be further linked to anticancer drug DOX by electrostatic interaction. $GdPO_4$'s aforementioned yolk-shell structure confers a large area for loading DOX. When in acidic environments such as those of tumors, the protonation of hydroxyl groups on the surface of NPs results in it becoming neutral, weakening the electrostatic interactions of NPs and DOX, ultimately resulting in controlled DOX release into tumor cells.

12.4.4 Tumor targeting of NPs

Crucial information such as the location of the tumor or its biological processes can be inferred from tumor-targeting NPs. Much information can be garnered from the state of the tumor as well. Some of these are protein kinase activity, hypoxia, angiogenesis, and apoptosis. This information can greatly assist clinicians when it comes to diagnosis, treatment, and prognosis of cancer patients (Blasberg, 2003; McDermott & Kilcoyne, 2015). Granting tumor targeting capabilities to NPs is done via functionalizing their surfaces. The function groups on the outermost layer are used to covalently bind targeting ligands such as FA, RGD, peptides, aptamers, antibodies, etc. (Xu, Fisher, & Juliano, 2011; Battogtokh & Ko, 2017). RGD receptor is commonly employed in cancer imaging since overexpression of certain receptors is a closely related to cancer progression. As such imaging based on receptor targets grants heightened tumor specificity, and quantitative information about the receptors. Because of this, a great deal of work has been done to create receptor-targeted techniques. RGD—a tumor-targeting penetration peptide—begins aggregation within tumors via targeting avb3 integrin receptors, which are specifically present in tumor cells and vasculature. It then binds to neuropilin-1 (NRP-1), consequently triggering tissue penetration (Desgrosellier & Cheresh, 2010; Ruoslahti, 2002; Sugahara et al., 2009; Teesalu et al., 2009; Wang et al., 2016; Xie et al., 2016). Physical binding, or chemical conjugation of NPs with iRGD enhances tumor penetration of the former, as well as improving the tumor-targeting via NRP-1-dependent fashion (Cheng et al., 2012; Gu et al., 2013; Jain & Stylianopoulos, 2010; Zako et al., 2009). In recent years, RGD has been broadly employed as surface modifiers of lanthanide NPs with the aim of enhancing tumor targeting. Yang's group have synthesized a sub-20 nm polyhedral cubic NIR nanoparticle coated with PAA by solvothermal process. By the ligand exchange with PAA, the NPs covalently linked to RGD. RGD-NPs can be utilized as tumor-targeting imaging agents of U87MG tumor cells. Additionally, RGD can be bound to various photosensitizers to grant the imaging agents tumor-targeting therapy capabilities. Zhou et al. (Zhou et al., 2012) have produced a novel kind of RGD-coated nanoparticle composed of $NaYF_4$:Yb,Er core and targeted RGD polymer shell. These RGD-coated rare-earth NPs have the advantages of defending the NPs from being caught by reticuloendothelial system, and targeting breast cancer cells. Furthermore, it can also be conjugated to photosensitizer to achieve tumor inhibition by photodynamics therapy (Wang et al., 2014).

12.5 Synthesis and functionalization of RE-based nanocomposites

There are many methods for the preparation of RE-based nanocomposites, including coprecipitation method, sol-gel processing, thermal decomposition, hydrothermal/solvothermal method, microemulsion method and so on. The purpose of this section is to provide a general and comprehensive overview of the current state of the synthesis methods of rare earth based nanoparticles as well as investigation of the advantages or disadvantages of various synthesis methods. Finally all of these items are summarized in the table in order to better understanding of readers.

12.5.1 Coprecipitation method

The coprecipitation method is extensively used for the preparation of CNPs through establishing precipitation reaction among positive and negative ions existing in the homogeneous solution to obtain a uniform precipitation. Wang et al. developed a synthetization of Lanthanide-oleate complex precursors through maintaining a reaction between lanthanide acetate and oleic acid at 150°C, and further addition of fluoride at low temperature followed by annealing in solvents boiling at high temperature to achieve monodispersed NaGdF4 nanocrystals, unsaturated fatty acids acting as surface ligands for controlling the growth of particles (Wang, Deng, Liu, 2014). Zhang et al. used coprecipitation method to produce Fe_3O_4@LaF_3:Yb^{3+}, Er^{3+} core/shell UCNPs. They initially made a synthetization of Fe_3O_4 nanoparticles (NPs), then made a mixture of NPs and $La(NO_3)_3.6H_2O$, $Yb(NO_3)_3 \cdot 6H_2O$, and Er $(NO_3)_3 \cdot 6H_2O$ Dropwise addition of NaF and stirring for 75°C for 2 hours was made to produce LaF_3:Yb^{3+},Er^{3+} NC shells. In the end, the magnetic/upconversion luminescent NPs were gained via warming up at 400°C for 1 hour under a N_2 atmosphere (Zhang et al., 2012). In another study, Guo et al. employed coprecipitation technique for synthetizing a set of $Lu_6O_5F_8$:20%Yb^{3+}, 1% Er^{3+}(Tm^{3+}) NPs with differing concentrations of Li^+. They reported that doping the UCNPs (Guo et al., 2013) with Li^+ could improve the UC, DC and CL emission intensities. Cascales et al. in their study synthesized the diamond-shaped Yb:Er:NaGd $(WO_4)_2$ nanoparticles (NPs) having a diagonal size of 5–7 and 10–12 nm using coprecipitation technique in oleic acid system (Cascales et al., 2017). Li et al. made a synthetization of $NaLnF_4$ UCNPs using modified high-temperature coprecipitation approach (305°C). They were able to simultaneously control the UCNPs size and phase by doping M^{2+} (Mg^{2+}, Co^{2+}). The small size UCNPs so produced with hexagonal phase structure demonstrated considerable improvement of the overall UCL (Li, Li, et al., 2017). Stochaj et al. made a synthetization of $NaLuF_4$:Tm^{3+}(0.5%)/Yb^{3+}(20%)/ Gd^{3+}(30%) UCNPs with 35 ± 2 nm sizing subjected to a set of reactions performed in a methanol solution combined with NaOH and NH_4F with 1-octadecene and oleic acid acting as surface stabilizers (Stochaj et al., 2018). The uneven size of products prepared in precipitation reaction makes the synthesized products to have a poor morphology. Moreover, the temperature of reaction, pH value, stoichiometry and the method of mixing have determinant impact on the synthesized UCNPs performance. Such factors can restrict the extensive application of co-preparation.

12.5.2 Sol-gel method

Sol-gel method often involves making use of metal compounds, metal or alkoxides or inorganic salts acting as substrate which gel them gradually through hydrolysis and polycondensation with the subsequent drying or sintering to produce CNPs. Park et al. used sol-gel process and soft lithography and produced $NaYF_4$ thin films and nanopatterns with success. They reported enhancement of light coupling output efficiency by nanopattern and an increase in the up-conversion intensity by 2–3 folds (Park et al., 2017). Liang et al. doped the β-NaYF4:Yb^{3+},Er^{3+} UCNPs with differing amounts of Li^+ and K^+ using sol-gel method without altering the form and size aiming at increasing the UC without changing the emission intensity (Liang, Wang, et al., 2017). All other materials but $NaYF_4$ may be utilized as the host metrics. In another study, synthesized Tm^{3+}/Yb^{3+}/Er^{3+}-doped Lu_3Ga using sol-gel process under nitric acid and citric acid (2 times of metal ion concentration) conditions (Mahalingam et al., 2008). The rare earth (Ho^{3+}, Er^{3+}, and Tm^{3+})-doped VYbOF nanoparticles was synthesized by Wen et al. with the application of sol-gel method and the subsequent use of PTFE as the fluorinating agent for fluorination treatment. The above study showed Yb-rich compounds having low-dimension structural characteristics are capable to work and UC host lattice in different applications (Wen et al., 2016).

12.5.3 Thermal decomposition method

The rare earth metal trifluoroacetate is often used in thermal decomposition method. Fixed proportions of materials of high purity are mixed and subsequently decomposed to produce UCNPs by applying high temperature and specific conditions. Despite minor alterations to conditions such as temperature, pressure, and additives having effects on the UCNPs, this is still a primary approach for synthesizing UCNPs. Boyer et al. introduced an appropriate quantity of sodium trifluoroacetate to the reaction vessel along with octadecene and oleic acid, heated it to 100°C in vacuum, and stirred the solution for 30 minutes to filter residual water and oxygen. Once heating under argon atmosphere to 300°C was carried out for 1 hour, $NaYF_4$:2%Er^{3+}, 20%Yb^{3+} were acquired via subsequent cooling, purification and drying (Boyer et al., 2006). The same team used thermal decomposition under octadecene and oleic acid conditions to synthesize Er^{3+}/Yb^{3+} and Tm^{3+}/Yb^{3+} doped $NaYF_4$ UCNPs as well (Boyer, Cuccia, & Capobianco, 2007). Thermal decomposition in the presence of oleylamine was carried out by Liu et al. to produce $NaLuF_4$ nanocrystals with sub-10 nm hexagonal phase co-doped with Gd^{3+}, Yb^{3+}, and Er^{3+} (or Tm^{3+}) (Liu et al., 2011). Fiorenzo et al. synthesized Er^{3+}, Yb^{3+}-doped $NaGdF_4$ nanoparticles in presence of 1-octadecene and oleic acid, through thermal decomposition approach. They devised and provided activecore/active-shell structured UCNPs. The UCL intensity of prepared UCNPs was larger in contrast to those of the active-core/inert-shell structured nanoparticles (Fiorenzo et al., 2009). Yb^{3+},Er^{3+}-doped $NaGdF_4$ UCNPs were synthesized in presence of oleic acid and 1-octadecene via two-pot thermal decomposition process as core structure by Nigoghossian et al. These UCNPs where produced under three different temperatures, 310°C, 315°C, and 320°C, employing a silica coating to alter the surface. Ultimately, cubic (310°C and 315°C) and hexagonal (320°C) monodisperse $NaGdF_4$:Yb^{3+}:Er^{3+} upconverting nanocrystals were synthesized (Nigoghossian et al., 2017). Core-shellshell heterostructures of α-$NaLuF_4$: Yb/Er@NaLuF = :Yb@MF_2 (M = Ca, Sr, Ba) heterostructures was fabricated via thermal decomposition by Su et al. The UCL intensity of the MF_2-shell coated UCNPs was significantly increased, when compared to core-only nanoparticles. Under 980 nm excitation, CaF_2-coated UCNPs have the strongest luminescence while BaF_2-coated samples have the longest lifetime (Su, Liu, & Lei, 2016). Zhang et al. fabricated high-quality Yb^{3+}/Er^{3+}-co-doped $LiYF_4$ nanoparticles via dropwise introduction of cation solution to a mixture of precursors, carried out at high temperature. The optimal rate of adding the LiF solution into the CF_3COORE mixture (Li/RE = 1:1−3:1) in a drop-wise manner was identified as 1−2 mL/min. The produced UCNPs depict high crystallinity, lowered agglomeration phenomenon, uniform tetragonal bipyramidal morphology, and excellent upconversion luminescence (Zhang et al., 2018). This method is extensively used, take note however that it has drawbacks, such as almost hard preparation conditions, complexity of reaction stages, high-priced reagents and severe toxicity.

12.5.4 Hydrothermal method

The hydrothermal synthesis method is the process of dissolving a material in water and recrystallizing the powder in a sealed pressure vessel. Using this method, it needs lower temperature, lower purity needed for the rear earth salt of the synthesis, easier controlling of the material formation process, good quality of the synthesized material crystal phase, uniformity of the particle size and higher yield of the product. The size and morphology of nanomaterials could easily be controlled via this method, and so making it an ideal synthesis method for UCNPs. Until now the researchers have been able to synthesize a number of up-conversion materials with success using the above method. Qian and Zhang have successfully employed the hydrothermal method in three steps to synthesize β-$NaYF_4$:Yb,Tm@β-$NaYF_4$:Yb, Er@β-$NaYF_4$:Yb,Tm nanoparticles. First, β-NaYF4:Yb,Tm core was prepared and used as seed crystals to induce shell growth. The core shell nanocrystals depicted multicolor MIR-to-visible luminescence under 980 nm excitation (Qian & Zhang, 2008). In another instance, the hydrothermal method in oleylamine system was applied to synthesize β-$NaYF_4$:Yb^{3+},Er^{3+}@β-$NaYF_4$ core/shell nanoparticles by Wang, Tu, et al. (2009). The hydrothermal method was applied by Du et al. (2017). In combination with synthesis temperature control with the aim of realizing the phase change and controlling the size of nanoparticles. Ultimately a series of Er^{3+}/ Yb^{3+} co-doped $NaYF_4$ upconverting nanoparticles were produced. In another instance, Ju et al. prepared core/shell microcrystals, using $NaYF_4$:Yb^{3+},Er^{3+} as core, and $NaYF_4$:Ce^{3+},Tb^{3+},Eu^{3+} as shell. The process utilized the hydrothermal method, known to confine the activator ions into a separate region, in turn minimizing the effect of surface quenching. The synthesized microcrystals emitted upconversion and downshifting emission, as well as being suitable for tunable emission of colors by adjusting the ratio of Tb^{3+}-Eu^{3+} of shells, and $Ce^{3+} \rightarrow Tb^{3+} \rightarrow Eu^{3+}$ energy transfer (Ju et al., 2018). The synthesization of $NaYF_4$:Yb,Er nanocrystals was made by Tang et al. through hydrothermal approach in oleic acid system. The analysis shows that red UCL was indiscriminately enhanced by $\sim$7 times through doping varying Fe^{3+} concentrations. A concentration of $\sim$5–20 mol.% Fe^{3+} resulted in the appearance of the pure hexagonal phase, while the size of $NaYF_4$: Yb,Er nanocrystals reached the maximum at 10 mol.% (Tang et al., 2015). In yet another case, Cao et al. produced high-quality water soluble surface-functionalized OA/AA-capped cylindrical LaF_3:Yb,Er/Tm (UCNPs-OAAA) through the hydrothermal synthesis process. The process involved use of OA with long alkyl chain and hydrophilic binary synthetic ligand with short alkyl chain (HR-BCL). And in order to control the crystal growth and nucleation of small nanoparticles, oleic acid and 6-aminocaproic acid were added (Cao et al., 2011). Lately, through capping with MA (malonic acid) via a facile one-step hydrothermal technique requiring heating at 200°C for 12 hours, Gerelkhuu et al. developed α-$NaLuGdF_4$:Yb^{3+}/Er^{3+}(Tm^{3+}) UCNPs soluble in water (Gerelkhuu et al., 2017). TiO_2 (Titanium dioxide; that can act as a potential photosensitizer) is of good chemical stability and photocurrent performance, however, given its narrow absorption band, it is of low utilization efficiency typically. To conquer this drawback, through hydrothermal reaction, a near-infrared ultraviolet-mediated photoelectrochemical (PEC) adapter platform on the basis of core-shell $NaYF_4$:Yb,Tm@TiO_2 upconverting microrods was synthesized by Qiu et al. And that proved to be satisfactory results (Qiu et al., 2018). Using hydrothermal technique at 180°C, Li et al. synthesized a series of Ba_2GdF_7:Yb^{3+}, Tm^{3+} nanophosphors (nanoparticles) successfully with up-conversion luminescence and down-conversion. A bright blue double-mode emission was created under the excitation by near-infrared 980 nm and UV 355 nm, the Ba_2GdF_7 phosphors co-doped with Tm^{3+} and Yb^{3+} ions (Li, Liu, et al., 2017). Using calcination and hydrothermal procedures, Liang et al acquired $GdVO_4$:Yb, which is a UCNP featuring water dispersibility and high crystallinity. The crude $GdVO_4$:Yb,Er NPs prepared through hydrothermal technique features a cubic morphology (approximate diameter of 45 nm). In order to prevent the aggregation and growth of particles throughout the heat treatment, a SiO_2 layer coating was applied on the nanoparticles surfaces. After etching the SiO_2 layer chemically in NaOH aqueous solution, not only the prepared highly crystalline $GdVO_4$:Yb,Er NPs preserve the characteristics of their nanostructure, but have good water dispersibility and high chemical stability as well (Liang, Noh, et al., 2017). Nonetheless, from another viewpoint, it is unfavorable to employ the hydrothermal reaction, because it takes a long time.

12.5.5 Solvothermal method

The solvothermal technique was developed based on the hydrothermal technique and refers to a technique where the primary mixture makes reactions in closed systems for example, an autoclave containing a nonaqueous or organic solvent as the solvent at the spontaneous pressure of the solution and a certain temperature. Since the solvothermal technique employs rare earth chlorides as precursors to decrease the pollution caused by fluorine in general, it is a more friendly synthesis technique for the environment. This technique employs the aqueous solution of a rare earth metal salt as the precursor, protected by oleylamine, octadecene, or oleic acid, a fluoride compound for example, ammonium fluoride or hydrogen fluoride is employed as a fluorine source that is heated in a vacuum so that a rare earth upconverting nanomaterial is formed. Simultaneously this method needs mild reaction conditions with highly active reaction. It can develop uniform size, monodisperse, morphologically controllable rare earth UCNPs. $NaYF_4$:Yb,Er NPs emitting bright green fluorescence was synthesized through solvothermal technique by Wang, Mi, et al. (2009). Zhou et al. presented solvothermal synthesis of $NaYF_4$:Yb,Er nanoparticles soluble in water. Through slow RE^{3+} ions release coordinated with branched PEI (polyethylene imine) molecules, they controlled the growth rate, modified the $NaYF_4$:Yb,Tm NCs size using various quantities of NH_4F, and the DLS average size synthesized was 28, 42 and 86 nm 3 kinds of NPs (Zhou et al., 2011). Through solvothermal technique, Yang et al. synthesized Lu_2O_3:Yb^{3+}/Er^{3+}/Tm^{3+} NPs that was heat treated at 800°C for 5 minutes so that green, blue, and red light-emitting NPs were obtained (Yang et al., 2009). For the synthesis of $BaYF_5$:Er^{3+},Yb^{3+} NPs via the solvothermal technique, Yao et al. employed $NaBF_4$ as a fluoride source, and discovered that the $BaYF_5$:Er^{3+} nanocrystals synthesis can be accelerated by addition of 5% of PEI as a surfactant in the solvent. Although the size of the nanoparticles and also the CIT (citrate) concentration are increased, the luminescence performance declines (Yao, Zhao, & Pan, 2017). Sodium oleate was employed by Shen et al. as the sodium source for synthesization of 11.0–21.4 nm sized hexagonal $NaYbF_4$:Er upconverted nanoparticles through the use of solvothermal method (Shen et al., 2017). Huang et al. synthesized the $Na_{0.52}YbF_{3.52}$:Er upconverted nanoparticles through using acial solvothermal method. They achieved a UC emission intensity three folds stronger compared with the green emission of hexagonal phase, 30 nm sized $NaYF_4$:Yb,Er UCNPs. This study reveals that nanoscale NaF and in situ generated HF play a significant role in the shaping the cubic monomers and UCNCs cubic to hexagonal phase conversion respectively. In this study 915 nm photo-triggered PDT was used for the first time and showed a considerably reduced overheating influence compared with the 980 nm therapeutic effects (Huang et al., 2016). In the study conducted by Kang, the lanthanide oleate compound ($Ln(OA)_3$) was utilized as precursor as a replacement for the conventional lanthanide chloride compound ($LnCl_3$) and finally the high purity $NaYF_4$ based UCNPs was produced in this study. In comparison to the conventional Ln-Cl approach by using $LnCl_3$ working as the precursor, the facial Ln-OA technique is able to prevent from producing NaCl by-product throughout the crystal growth phase of the $NaYF_4$-based UCNPs. Thus the method is capable of directly synthesizing high purity $NaYF_4$-based UCNPs without the need for additional purification (Kang et al., 2017). Mancic et al. used one-pot solvothermal synthesis to prepare $NaYF_4$:Yb,Er/PLGA [poly(lactic-*co*-glycolic acid)] nanoparticles. The PLGA ligands were added to the UCNPs to improve the cell uptake. The nanoparticles are able to reach good quality cell imaging when subjected to NIR light excitation. Additionally the $NaYF_4$:Yb, Er/PLGA UCNPs exhibited lower cytotoxicity as compared to HGC with 10–50 μg/mL concentration range (Mancic et al., 2018).

12.5.6 Microemulsion method

The microemulsion technique is a procedure where under the action of a surfactant, two immiscible solvents lead to formation of an emulsion, and after agglomeration, heat treatment, and nucleation in microbubbles, the NPs are obtained. Using a reverse microemulsion technique employing dual surfactants of 1-hexanol and polyoxyethylene (5) nonylphenylether, Qian et al. synthesized $NaYF_4$:Yb,Er/silica core/shell UCNPs (particle size of 11.1 ± 1.3 nm) (Qian et al., 2009). Using microemulsion techniques in the presence of amphiphilic block copolymer poly (styrene-block-allyl alcohol), Xu et al. combined ferrous IONPs (iron oxide nanoparticles) with hydrophobic UCNPs, and earned UC-IO multifunctional nanocomposites with triple-modal imaging capacity (UCL, FL, MR) (Xu et al., 2011). Campos-Gonçalves et al. recently synthesized a nanocomposite of multifunctional type (SPIONs@LaF_3:Yb^{3+}/Er^{3+}) successfully superparamagnetic core and UCL shell if performed through microemulsion and coprecipitation methods (Campos-Gonçalves et al., 2017). Table 12.1. summarizes the advantages and disadvantages of several

Table 12.1 Advantages and disadvantages of typical synthetic methods for CNPs.

Disadvantages	Advantages	Method
Requiring post heat treatment, lack of particle size control, considerable aggregation, typically need high temperature calcination, not applicable for large scale synthesis of UCNPs	Fast synthesis, low cost, does not require costly setup, or severe reaction conditions, simple procedures	**Coprecipitation method**
Post heat treatment is often needed, considerable particle aggregation, broad particle size distribution, irregular morphology, insoluble in water	Cheap precursors, can be used for large-scale production and the products usually offer high luminescence intensity due to the high crystallinity formed at high annealing temperature	**Sol-gel processing**
Temperature usually exceeding 300°C, use of expensive and air sensitive metal precursors, and the generation of toxic by-products, need the anaerobic and water-free reaction environment	High quality, monodispersed nanocrystals with strong upconversion emission	**Thermal decomposition**
Impossibility of observing the nanocrystal growth processes, hard to filter out the most optimized experimental conditions (the reaction temperature, surfactant type and concentration, reactant concentration, solvent and the composition of it, and reaction time), usually have large size distribution	High quality crystals with controllable particle size, shape and dopant ion concentrations, does not require stringent operation of the experimental process, the reaction temperature is much lower	**Hydro/ solvothermal method**
Usually generate small amount of products, difficulty of sample separation and narrow scope synthesis	Low-cost for equipment, easy operation, the small size of the UCNPs, and the controlled morphology of products by adjusting the dosage of the surfactant, solvent, as well as the aging time	**Microemulsion method**

Source: Reproduced with permissions from Hong, E., Liu, L., Bai, L., Xia, Ch., Gao, L., Zhang, L., Wang, B. (2019). Control synthesis, subtle surface modification of rare-earth-doped upconversion nanoparticles and their applications in cancer diagnosis and treatment. *Materials Science and Engineering: C*, *105*, 110097.

common synthetic methods. Nanotechnology is still in development and current limitations hinder global transition from macroscale to nanoscale.

12.6 Conclusion and outlook

In this chapter, a short brief of different nanocomposites based rare earth metals, their various properties and applications (industrial or medical uses) are discussed. Different synthesis methods along with various examples for the better understanding of readers are attached, from the synthesis point of view, among the synthesis method. The coprecipitation method is extensively used for the preparation of CNPs but in precipitation reaction makes the synthesized products to have a poor morphology. Sol-gel method often involves the use of metal compounds, metal or lakesides, or inorganic salts acting as substrate which gel them gradually through hydrolysis and polycondensation with the subsequent drying or sintering. The rare earth metal nanocomposites are synthesized via thermal decomposition method. This method is extensively used; however, it has some drawbacks, such as almost hard preparation conditions, complexity of reaction stages, high-priced reagents, and severe toxicity. For the use of hydrothermal method, it needs lower temperature, lower purity needed for the rear earth salt of the synthesis, easier controlling of the material formation process, good quality of the synthesized material crystal phase, uniformity of the particle size, and higher yield of the product. Despite the great advance achieved in the last years in the development of RE-based nanostructures for various applications, there is a lack of having many different compositions, morphology, size, properties and applications; therefore, some challenges and improvements still need to be addressed. Thus, from the synthetic point of view, efficient, large-scale, and cost-effective methods have to be developed. Regarding the persistent nanocomposites base rare earth materials, suitable methods for the synthesis of them are strongly demanded, and new routes for their synthesis at milder reaction conditions that are quicker and with a better control of the particle sizes should be developed.

References

Abdelhamid, H. N., Talib, A., & Wu, H. F. (2016). One pot synthesis of gold -carbon dots nanocomposite and its application for cytosensing ofmetals for cancer cells. *Talanta*. Available from https://doi.org/10.1016/j.talanta.2016.11.030.

Abdukayum, A., Chen, J. T., Zhao, Q., et al. (2013). Functional nearinfrared-emitting Cr3þ/Pr3þco-doped zinc gallogerma-nate persistent luminescent nanoparticles with super long after glow for in vivo targeted bioimaging. *Journal of the American Chemical Society*, *135*, 14125–14133.

Alexandre, M., & Dubois, P. (2000). Polymer-layered silicate nanocomposites:preparation, properties and uses of a new class of materials. *Materials Science and Engineering*, *28* (1–2), 1–63.

Anandhan, S., & Bandyopadhyay, S. (2011). *Polymer nanocomposites: fromsynthesis to applications*. Intech Open Access Publisher.

Anderson, D. L., & Patiño-Echeverri, D. (2009). *An evaluation of current and future costs for lithium-ion batteries for use in electrified vehicle powertrains* (44 p). Raleigh Durham, NC: Nicholas School of the Environment, Duke University master's thesis. Available at: http://dukespace.lib.duke.edu/dspace/bitstream/10161/1007/1/Li-Ion_Battery_costs_-_MP_Final.pdf. Accessed October 9, 2015.

Bai, H., & Ho, W. (2008). New sulfonated polybenzimidazole (SPBI)copolymer-based protonexchange membranes for fuel cells. *Journal of the Taiwan Institute of Chemical Engineers*, *40*, 260–267.

Battogtokh, G., & Ko, Y. T. (2017). Mitochondrial-targetedphotosensitizer-loaded folate-albumin nanoparticle forphotodynamic therapy of cancer. *Nanomedicine: Nanotechnology, Biology, and Medicine*, *13*, 733–743.

Becher, P. F. (1991). Microstructural design of toughened ceramics. *Journal of the American Ceramic Society*, *74*(2), 255–269.

Blasberg, R. G. (2003). Molecular imaging and cancer. *Molecular Cancer Therapeutics*, *2*, 335–343.

Boyer, J. C., Vetrone, F., Cuccia, L. A., et al. (2006). Synthesis of colloidal upconverting $NaYF_4$ nanocrystals doped with Er^{3+}, Yb^{3+} and Tm^{3+}, Yb^{3+} via thermal decomposition of lanthanide trifluoroacetate precursors. *Journal of the American Chemical Society*, *128*(23), 7444–7445.

Boyer, J. C., Cuccia, L. A., & Capobianco, J. A. (2007). Synthesis of colloidal upconverting $NaYF_4$:Er^{3+}/Yb^{3+} and Tm^{3+}/Yb^{3+} monodisperse nanocrystals. *Nano Letters*, *7*(3), 847–852.

Brooke, R., Fabretto, M., Murphy, P., Evans, D., Cottis, P., & Talemi, P. (2017). Recent advances in the synthesis of conducting polymers from thevapour phase. *Progress in Materials Science*. Available from https://doi.org/10.1016/j.pmatsci.2017.01.004.

Camargo, P. H. C., Satyanarayana, K. G., & Wypych, F. (2009). Nanocomposites:synthesis, structure, properties and new application opportunities. *Materials Research*, *12*(1), 1–39.

Campos-Gonçalves, I., Costa, B. F. O., Santos, R. F., et al. (2017). Superparamagnetic core-shell nanocomplexes doped with Yb^{3+}:Er^{3+}/Ho^{3+}, rare-earths for upconversionfluorescence. *Mater Design*, *130*, 263–274.

Cao, T., Yang, Y., Gao, Y., et al. (2011). High-quality water-soluble and surface-functionalized upconversion nanocrystals as luminescent probes for bioimaging. *Biomaterials*, *32* (11), 2959–2968.

Cascales, C., Paãno, C. L., Bazãn, E., et al. (2017). Ultrasmall, water dispersible, TWEEN80 modified Yb:Er:$NaGd(WO_4)_2$ nanoparticles with record upconversion ratiometric thermal sensitivity and their internalization by mesenchymal stem cells. *Nanotechnology*, *28* (18), 185101. Available from https://doi.org/10.1088/1361-6528/aa6834.

Chakhmouradian, A. R., & Wall, F. (2012). Rare earth elements: minerals, mines, magnets (and more). *Elements*, *8*, 333–340.

Chatterjee, D. K., Gnanasammandhan, M. K., & Zhang, Y. (2010). Small upconverting fluorescent nanoparticles for bio-medical applications. *Small (Weinheim an der Bergstrasse, Germany)*, *6*, 2781–2795.

Cheng, L., Yang, K., Zhang, S., et al. (2010). Highly-sensitive multiplexed in vivo imaging using PEGylated upconversion nanoparticles. *Nano Research*, *3*, 722–732.

Cheng, L., Yang, K., Shao, M., et al. (2011). Multicolor in vivoimaging of upconversion nanoparticles with emissionstuned by luminescence resonance energy transfer. *The Journal of Physical Chemistry C*, *115*, 2686–2692.

Cheng, Z., AlZaki, A., Hui, J. Z., et al. (2012). Multifunctionalnanoparticles: cost versus benefit of adding targetingand imaging capabilities. *Science (New York, N.Y.)*, *338*, 903–910.

Choia, H., Yoonb, S. P., Hanb, J., Kima, J., & Othmanc, M. R. (2016). Production of Al-SiC-TiC hybrid composites using pure and 1056 aluminum powders prepared through microwave and conventional heating methods. *Journal of Alloys and Compounds*. Available from https://doi.org/10.1016/j.jallcom.2016.08.145.

Deng, H., Huang, S., & Xu, C. (2018). Intensely red-emittingluminescent upconversion nanoparticles for deep-tissuemultimodal bioimaging. *Talanta*, *184*, 461–467.

Dermenci, K. B., Gencc, B., Ebinb, B., Olmez-Hanci, T., & Gürmen, S. (2014). Photocatalytic studies of Ag/ZnO nanocomposite particles pro-duced via ultrasonic spray pyrolysis method. *Journal of Alloys and Compounds*, *586*, 267–273.

Desgrosellier, J. S., & Cheresh, D. A. (2010). Integrins in cancer: Biological implications and therapeutic opportunities. *Nature Reviews Cancer*, *10*, 9.

Dezfuly, R. F., Yousefi, R., & Jamali-Sheini, F. (2016). Photocurrent applications of $Zn_{(1-x)}Cd_xO$/rGO nanocomposites. *Ceramics International*. Available from https://doi.org/10.1016/j.ceramint.2016.01.150.

Du, P., Deng, A. M., Luo, L., et al. (2017). Simultaneous phase and size manipulation inNaYF4:Er^{3+}/Yb^{3+} upconverting nanoparticles for non-invasion optical thermo-meter. *New Journal of Chemistry*, *41*(22), 13855–13861.

Du, X., & Graedel, T. E. (2011). Global in-use stocks of the rare earth elements: A first estimate. *Environmental Science & Technology*, *45*, 4096–4101.

European Commission. (2014). *Directorate general enterprise and industry EU critical raw materials profiles* (pp. 77–85). European Commission: Brussels, Belgium.

European Rare Earths Competency Network (ERECON). (2014). *Strengthening the European rare earths supply-chain challenges and policy options*. European Comission: Brussels, Belgium.

Fernando, W., & Satyanarayana, K. G. (2005). Functionalization of single layersand nanofibers: A new strategy to produce polymer nanocomposites with optimized properties. *Journal of Colloid and Interface Science*, *285*(1), 532–543.

Fiorenzo, V., Rafik, N., Venkataramanan, M., et al. (2009). The active-core/active-shellapproach: a strategy to enhance the upconversion luminescence in lanthanide-doped nanoparticles. *Advanced Functional Materials*, *19*(18), 2924–2929.

Fiorito, S., Serafino, A., Andreola, F., & Bernier, P. (2006). Effects of fullerenesand single-wall carbon nanotubes on murine and human macro-phages. *Carbon*, *44*(6), 1100–1105.

Gamal-Eldeena, A. M., Abdel-Hameedc, S. A. M., El-Dalya, S. M., Abo-Zeida, M. A. M., & Swellamb, M. M. (2017). Cytotoxic effect of ferrimagneticglass-ceramic nanocomposites on bone osteosarcoma cells. *Biomedicine & Pharmacotherapy*, *88*, 689–697.

Garmendia, N., Olalde, B., & Obieta, I. (2013). Biomedical applications of ceramic nanocomposites. *Ceramic Nanocomposites*, 530–547, A volume in Woodhead Publishing Series in Composites Science and Engineering.

Gerelkhuu, Z., Huy, B. T., Sharipov, M., et al. (2017). One-step synthesis of $NaLu_{80-x}Gd_xF_4$: ${Yb_{18}}^{3+}$/${Er_2}^{3+}$(Tm^{3+}) upconversion nanoparticles for in vitro cell imaging. *Materials Science & Engineering. C, Materials for Biological Applications*, *86*, 56–61.

Ghasali, E., Yazdani-rad, R., Asadian, K., & Ebadzadeh, T. (2016). Production ofAl-SiC-TiC hybrid composites using pure and 1056 aluminumpowders prepared through microwave and conventional heatingmethods. *Journal of Alloys and Compounds*. Available from https://doi.org/10.1016/j.jallcom.2016.08.145.

Gnach, A., & Bednarkiewicz, A. (2012). Lanthanide-doped up-converting nanoparticles: Merits and challenges. *Nano Today*, *7*, 532–563.

Goonan, T. G. (2011). *Rare earth elements—End use and recyclability: U.S. Geological Survey Scientific Investigations Report 2011–5094* (15 p.). Also available at http://pubs.usgs.gov/sir/2011/5094/.

Gu, G., Gao, X., Hu, Q., et al. (2013). The influence of the penetrating peptide iRGD on the effect of paclitaxel-loaded MT1-AF7p-conjugated nanoparticles on glioma cells. *Biomaterials*, *34*, 5138–5148.

Guo, L., Wang, Y., Wang, Y., et al. (2013). Structure, enhancement and white luminescenceof multifunctional $Lu_6O_5F_8$:20%Yb^{3+},1%Er^{3+}(Tm^{3+}) nanoparticles via furtherdoping with Li^+ under different excitation sources. *Nanoscale*, *5*(6), 2491–2504.

Haraguchi, K. (2011). Synthesis and properties of soft nanocomposite ma-terials with novel organic/inorganic network structures. *Polymer Journal*, *43*, 223–241, 287.

Harmer, M., Chan, H. M., & Miller, G. A. (1992). Unique opportunities for micro-structural engineering with duplex and laminar ceramic composites. *Journal of the American Ceramic Society*, *75*(2), 1715–1728.

He, J., Gao, Y., Wang, Y., Fang, J., & An, L. (2016). Synthesis of ZrB_2-SiC nano-composite powder via polymeric precursor route. *Ceramics International*. Available from https://doi.org/10.1016/j.ceramint.2016.10.073.

Huang, Y., Xiao, Q., Hu, H., et al. (2016). 915 nm light-triggered photodynamic therapy and MR/CT dual-modal imaging of tumor based on the nonstoichiometric Na0.52YbF3.52: Er upconversion nanoprobes. *Small (Weinheim an der Bergstrasse, Germany)*, *12*(31), 4200–4210.

Hurt, R. H., Monthioux, M., & Kane, A. (2006). Toxicology of carbonnanomaterials: status, trends, and perspectives on the special issue. *Carbon*, *44*(6), 1028–1033.

Hussain, F., Hojjati, M., Okamoto, M., & Gorga, R. E. (2006). Review article:polymer-matrix nanocomposites, processing, manufacturing, andapplication: an overview. *The Journal of Composite Materials*, *40*(17), 1511–1575.

Hötzer, B., Medintz, I. L., & Hildebrandt, N. (2012). Fluorescence in nanobiotechnology: Sophisticated fluorophores for novel applications. *Small (Weinheim an der Bergstrasse, Germany)*, *8*, 2297–2326.

Jain, R. K., & Stylianopoulos, T. (2010). Delivering nanomedicine to solid tumors. *Nature Reviews Clinical Oncology*, *7*, 653.

Ju, D. D., Song, F., Khan, A., et al. (2018). Simultaneous dual-mode emission and tunable-multicolor in the time domain from lanthanide-doped core-shell microcrystals. *Nanomaterials-Basel*, *8*(12). Available from https://doi.org/10.3390/nano8121023.

Kang, N., Ai, C. C., Zhou, Y. M., et al. (2017). Facile synthesis of upconversion nanoparti-cleswith high purity using lanthanide oleate compounds. *Nanotechnology*, *29*(7), 075601. Available from https://doi.org/10.1088/1361-6528/aa96ee.

Kashinath, L., Namratha, K., & Byrappa, K. (2016). Sol-gel assisted hydrother-mal synthesis and characterization of hybrid ZnS-RGO nanocom-posite for efficient photodegradation of dyes. *Journal of Alloys and Compounds*. Available from https://doi.org/10.1016/j.jallcom.2016.10.063.

Kobayashi, T. (2016). *Applied environmental materials science for sustain-ability*. IgI Global.

Kopera, J.J.C. (2004). *Inside the nickel metal hydride battery: Cobasys LLC Web page*. Available at http://www.cobasys.com/pdf/tutorial/InsideNimhBattery/inside_nimh_battery_technology.html. Accessed October 9, 2015.

Kornmann, X., Linderberg, H., & Bergund, L. A. (2001). Synthesis of epoxy–claynanocomposites: Influence of the nature of the curing agent onstructure. *Polymer*, *42*, 4493–4499.

Królikowski, W., & Rosłaniec, Z. (2004). Nanokompozyty polimerowe. Composites. Polish Ministry of Science. *Wydawnictwo Politechniki Czestochwskie*, *4*, 3–16.

Lange, F. F. (1973). Effect of microstructure on strength of Si3n4-SiC composite system. *Journal of the American Ceramic Society*, *56*(9), 445–450.

Lee, H. S., Choi, M. Y., Anandhan, S., Baik, D. H., & Seo, S. W. (2004). Microphasestructure and physical properties of polyurethane/organoclay nanocomposites. *ACS PMSE Preprints*, *91*, 638.

Li, H., Liu, G., Wang, J., et al. (2017). Dual-mode blue emission, enhanced up-conversion luminescence and paramagnetic properties of ytterbium and thulium-doped Ba_2GdF_7 multifunctional nanophosphors. *Journal of Colloid and Interface Science*, *501*, 215–221.

Li, Y., Li, X., Xue, Z., et al. (2017). M2 + doping induced simultaneous phase/size controland remarkable enhanced upconversion luminescence of $NaLnF_4$ probes for optical-guided tiny tumor diagnosis. *Advanced Healthcare Materials*, *6*(10), 1601231. Available from https://doi.org/10.1002/adhm.201601231.

Liang, Y. J., Noh, H. M., Xue, J. P., et al. (2017). High quality colloidal $GdVO_4$:Yb,Er upconversion nanoparticles synthesized via a proted calcination process for versatile applications. *Materials Design*. Available from https://doi.org/10.1016/j.matdes.2017.05.058.

Liang, Z., Wang, X., Wei, Z., et al. (2017). Upconversion nanocrystals mediated lateral-flow nanoplatform for in vitro detection. *ACS Applied Materials & Interfaces*, *9*(4), 3497–3504.

Liu, Q., Sun, Y., Yang, T., et al. (2011). Sub-10 nm hexagonal lanthanide-doped $NaLuF_4$ upconversion nanocrystals for sensitive bioimaging in vivo. *Journal of the American Chemical Society*, *133*(43), 17122–17125.

Liu, S., Li, W., Gai, S., et al. (2019). A smart tumor microenviron-ment responsive nanoplatform based on upconversion nanoparticles for efficient multimodal imaging guidedtherapy. *Biomaterials Science*, *7*, 951–962.

Long, K.R. (2011). *The future of rare earth elements—Will these high-tech industry elements continue in short supply?* (41 p.). U.S. Geological Survey Open-File Report 2011–1189. Also available at http://pubs.usgs.gov/of/2011/1189/.

Long, X., Shao, C., Wang, H., & Wang, J. (2016). Single-source-precursor syn-thesis of SiBNC-Zr ceramic nanocomposites fibers. *Ceramics International*, *42*(16), 19206–19211.

Lu, X., Yuan, P., Zhang, W., et al. (2018). A highly water-solubletriblock conjugated polymer for in vivo NIR-II imagingand photothermal therapy of cancer. *Polymer Chemistry*, *9*, 3118–3126.

Mahalingam, V., Mangiarini, F., Vetrone, F., et al. (2008). Bright white upconversion emission from Tm^{3+}/Yb^{3+}/Er^{3+}-doped $Lu_3Ga_5O_{12}$ nanocrystals. *The Journal of Physical Chemistry*, *C112*(46), 17745–17749.

Mancic, L., Djukic-Vukovic, A., Dinic, I., et al. (2018). NIR photo-driven upconversion in $NaYF_4$:Yb,Er/PLGA particles for in vitro, bioimaging of cancer cells. *Materials Science and Engineering: C*, *91*, 597–605.

McDermott, S., & Kilcoyne, A. (2015). Molecular imaging—Its current role in cancer. *QJM: Monthly Journal of the Association of Physicians*, *109*, 295–299.

Mout, R., Moyano, D. F., Rana, S., & Rotello, V. M. (2012). Surface functionalization of nanoparticles for nanomedicine. *Chemical Society Reviews*, *41*, 2539–2544.

Murthy, S. K. (2007). Nanoparticles in modern medicine: State of the art and future challenges. *International Journal of Nanomedicine*, *2*, 129–141.

Nayaka, S. S., Pabi, S. K., Kimb, D. H., & Murtyc, B. S. (2010). Microstructure-hardness relationship of Al–(L1$_2$)Al$_3$Ti nanocomposites preparedby rapid solidification processing. *Intermetallics*, *18*, 487–492.

Nigoghossian, K., Ouellet, S., Plain, J., et al. (2017). Upconversion nanoparticle-decoratedgold nanoshells for near-infrared induced heating and thermometry. *Journal of Materials Chemistry B*, *5*(34), 7109–7117.

Noh, M. W., & Lee, D. C. (1999). Synthesis and characterization of PS-clay nano-composite by emulsion polymerization. *Polymer Bulletin*, *42*(5), 619–626.

Ogasawara, T., Ishida, Y., Ishikawa, T., & Yokota, R. (2004). Characterization of multi-walled carbon nanotube/phenylethynyl terminated polyimidecomposites. *Composites Part A: Applied Science (New York, N.Y.)*, *35*(1), 67–74.

Ogawa, M., & Kuroda, K. (1997). Preparation of inorganic composites throughintercalation of organo ammoniumions into layered silicates. *Bulletin of the Chemical Society of Japan*, *70*(11), 2593–2618.

Oliveira, E., Santos, H. M., Jorge, S., et al. (2019). Sustainable syn-thesis of luminescent CdTe quantum dots coated withmodified silica mesoporous nanoparticles: Towards new-protein scavengers and smart drug delivery carriers. *Dye Pigment*, *161*, 360–369.

Park, H., Gang, Y. Y., Kim, M. S., et al. (2017). Thinfilm fabrication of upconversion lanthanide-doped NaYF$_4$ by a sol-gel method and soft lithographical nanopatterning. *Journal of Alloys and Compounds*, *728*, 927–935.

Pyrzynska, K., Kubiak, A., & Wysocka, I. (2016). Application of solid phase extraction procedures for rare earth elements determination in environmental samples. *Talanta*, *154*, 15–22.

Qian, H. S., & Zhang, Y. (2008). Synthesis of hexagonal-phase core-shell NaYF4nanocrystalswith tunable upconversionfluorescence. *Langmuir: The ACS Journal of Surfaces and Colloids*, *24*(21), 12123–12125.

Qian, L. P., Yuan, D., Shun Yi, G., et al. (2009). Critical shell thickness and emission enhancement of NaYF$_4$:Yb,Er/NaYF$_4$/silica core/shell/shell nanoparticles. *Journal of Materials Research*, *24*(12), 3559–3568.

Qiu, Z., Shu, J., & Tang, D. (2018). Near-infrared-to-ultraviolet light-mediated photoelec-tro-chemical aptasensing platform for cancer biomarker based on core-shellNaYF$_4$:Yb, Tm@TiO$_2$ upconversion microrods. *Analytical Chemistry*, *90*(1), 1021–1028.

Rehab, A., & Salahuddin, N. (2005). Nanocomposite materials based on poly-urethane intercalated into montmorillonite clay. *Materials Science and Engineering*, *399*, 368–376.

Ren, F., Ding, L., Liu, H., et al. (2018). Ultra-small nanoclustermediated synthesis of Nd^{3+}-doped core-shell nanocrystals with emission in the second near-infrared windowfor multimodal imaging of tumor vasculature. *Biomaterials*, *175*, 30–43.

Ren, Q., Su, H., Zhang, J., Ma, W., Yao, B., Liu, L., et al. (2016). Rapid eutectic growth of Al$_2$O$_3$/Er$_3$Al$_5$O$_{12}$ nanocomposite prepared by a new method: Melt falling-drop quenching. *Scripta Materialia.*, *125*, 39–43.

Ruoslahti, E. (2002). Specialization of tumour vasculature. *Nature Reviews Cancer*, *2*, 83.

Schmidt, D., Shah, D., & Giannelis, E. P. (2002). New advances in polymer/layered silicate nanocomposites. *Current Opinion in Solid State & Materials Science*, *6*(3), 205–212.

Series, S. (2005). In G. Liu, & B. Jacquier (Eds.), *Spectroscopic properties of rare earths in optical materials*. Berlin, Germany: Springer Science & Business Media.

Shen, J. W., Wang, Z., Liu, J., et al. (2017). Nano-sized NaF inspired intrinsic solvothermal-growth mechanism of rare-earth nanocrystals for facile control synthesis of high-quality and small-sized hexagonal $NaYbF_4$:Er. *Journal of Materials Chemistry*, *5*(37), 9579−9587.

Smart, S. K., Cassady, A. I., Lu, G. Q., & Martin, D. J. (2006). The biocompatibility ofcarbon nanotubes. *Carbon*, *44*(6), 1034−1047.

Spasówka, E., Rudnik, E., & Kijeński, J. (2006). Biodegradowalne nanokompozyty polimerowe. Cz. I. Metody otrzymywania. *Polimery*, *51*, 617−626.

Stochaj, U., Burbano, D. C. R., Cooper, D. R., et al. (2018). The effects of lanthanide-dopedupconverting nanoparticles on cancer cell biomarkers. *Nanoscale*, *10*(30), 14464−14471.

Su, Y., Liu, X., Lei, P., et al. (2016). Core-shell-shell heterostructures of α-$NaLuF_4$:Yb/Er@$NaLuF_4$:Yb@ MF_2(M = Ca, Sr, Ba) with remarkably enhanced upconversion luminescence. *Dalton T*, *45*(27), 11129−11136.

Sugahara, K. N., Teesalu, T., Karmali, P. P., et al. (2009). Tissue-penetrating delivery of compounds and nanoparticles into tumors. *Cancer Cell*, *16*, 510−520.

Sun, N., Jeurgensc, L. P. H., Burghardb, Z., & Billb, J. (2017). Ionic liquid assisted fabrication of high performance SWNTs reinforced ceramic matrix nano-composites. *Ceramics International*, *43*(2), 2297−2304.

Taeho, K., & Taeghwan, H. (2014). Applications of inorganic nanoparticles as therapeutic agents. *Nanotechnology*, *25*, 012001.

Tang, J., Chen, L., Li, J., et al. (2015). Selectively enhanced red upconversion luminescence and phase/size manipulation via Fe^{3+} doping in $NaYF_4$:Yb,Er nanocrystals. *Nanoscale*, *7*(35), 14752−14759.

Tawfik, S. M., Sharipov, M., Huy, B. T., et al. (2018). Naturally modified nonionic alginate functionalized upconversion nanoparticles for the highly efficient targeted pH-responsive drug delivery and enhancement of NIR-imaging. *Journal of Industrial and Engineering Chemistry*, *57*, 424−435.

Teesalu, T., Sugahara, K. N., Kotamraju, V. R., et al. (2009). C-end rule peptides mediate neuropilin-1-dependent cell, vascular, and tissue penetration. *Proceedings of the National Academy of Sciences of the United States of America*, *106*, 16157−16162.

Theng, B. K. G. (1974). *The chemistry of clay-organic reactions*. New York: Wiley.

Wang, F., Deng, R., & Liu, X. (2014). Preparation of core-shell $NaGdF_4$ nanoparticles doped with luminescent lanthanide ions to be used as upconversion-based probes. *Nature Protocols*, *9*(7), 1634−1644.

Wang, H., Dong, C., Zhao, P., et al. (2014). Lipid coated upconverting nanoparticles as NIR remote controlled transducer for simultaneous photodynamic therapy and cell imaging. *International Journal of Pharmaceutics*, *466*, 307−313.

Wang, J., Wang, H., Li, J., et al. (2016). iRGD-decorated polymeric nanoparticles for the efficient delivery of vandetanib to hepatocellular carcinoma: Preparation and in vitro and in vivo evaluation. *ACS Applied Materials & Interfaces*, *8*, 19228−19237.

Wang, M., Mi, C. C., Wang, W. X., et al. (2009). Immunolabeling and NIR-excited fluorescent imaging of HeLa cells by using $NaYF_4$:Yb,Er upconversion nanoparticles. *ACS Nano*, *3*(6), 1580−1586.

Wang, R. M., Zheng, S. R., & Zheng, Y. P. (2011). *Polymer matrix composites and technology*. Woodhead Publishing Limited and Science Press Limited.

Wang, Y., Tu, L. P., Zhao, J. W., et al. (2009). Upconversion luminescence of β-$NaYF_4$: Yb^{3+},Er^{3+}@β-$NaYF_4$ core/shell nanoparticles: Excitation power density and surface dependence. *The Journal of Physical Chemistry C*, *113*(17), 7164−7169.

Weller, M., Overton, T., Rourke, J., & Armstrong, F. (2014). *Inorganic chemistry*. Oxford: Oxford University Press.
Wen, T., Zhou, Y. N., Guo, Y. Z., et al. (2016). Color-tunable and single-band red upconversion luminescence form rare-earth doped vernier phase ytterbium oxyfluoride nanoparticles. *Journal of Materials Chemistry C*, *4*(4), 684–690.
Xie, H., Xu, X., Chen, J., et al. (2016). Rational design of multifunctional small-molecule prodrugs for simultaneous suppression of cancer cell growth and metastasis in vitro and in vivo. *Chemical Communications (Cambridge, England)*, *52*, 5601–5604.
Xu, H., Cheng, L., Wang, C., et al. (2011). Polymer encapsulated upconversion nanoparticle/iron oxide nanocomposites for multimodal imaging and magnetic targeted drug delivery. *Biomaterials*, *32*(35), 9364–9373.
Xu, R., Fisher, M., & Juliano, R. L. (2011). Targeted albumin-based nanoparticles for delivery of amphipathic drugs. *Bioconjugate Chemistry*, *22*, 870–878.
Yan, X., Sahimi, M., & Tsotsis, T. (2017). Fabrication of high-surface area nanoporous SiOC ceramics using pre-ceramic polymer precursorsand a sacrificial template: Precursor effects. *Microporous and Mesoporous Materials*. Available from https://doi.org/10.1016/j.micromeso.2016.12.027.
Yang, J., Zhang, C., Peng, C., et al. (2009). Controllable red, green, blue (RGB) and bright-white upconversion luminescence of Lu_2O_3:Yb^{3+}/Er^{3+}/Tm^{3+} nanocrystalsthrough single laser excitation at 980 nm. *Chemistry—A European Journal*, *15*, 4649–4655.
Yang, Q., Li, X., Xue, Z., et al. (2018). Short-wave near-infrared emissive $GdPO_4$:Nd^{3+} theranostic probe for in vivo bioimaging beyond 1300 nm. *RSC Advances*, *8*, 12832–12840.
Yang, X., Hong, H., Grailer, J. J., et al. (2011). cRGD-functionalized, DOX-conjugated, and ^{64}Cu-labeled superparamagnetic iron oxide nanoparticles for targeted anticancer drug delivery and PET/MR imaging. *Biomaterials*, *32*, 4151–4160.
Yao, J., Zhao, F., Pan, C., et al. (2017). Controlled synthesis of $BaYF_5$:Er^{3+},Yb^{3+} with different morphology for the enhancement of upconversion luminescence. *Nanoscale Research Letters*, *12*(1), 633. Available from https://doi.org/10.1186/s11671-017-2390-4.
Yu, Z., Pei, Y., Lai, S., Li, S., Feng, Y., & Liu, X. (2017). Single-source-precursorsynthesis, microstructure and high temperature behavior of TiC-TiB2-SiC ceramic nanocomposites. *Ceramics International*. Available from https://doi.org/10.1016/j.ceramint.2017.01.117.
Yua, Z., Lia, S., Zhanga, P., Fenga, Y., & Liua, X. (2016). Polymer-derived mesoporous Ni/SiOC(H) ceramic nanocomposites for efficient removalof acid fuchsin. *Ceramics International*. Available from https://doi.org/10.1016/j.ceramint.2016.12.104.
Zako, T., Nagata, H., Terada, N., et al. (2009). Cyclic RGD peptide-labeled upconversion nanophosphors for tumor cell-targeted imaging. *Biochemical and Biophysical Research Communications*, *381*, 54–58.
Zare, Y., & Shabani, I. (2015). Polymer/metal nanocomposites for biomedical applications. *Materials Science and Engineering*. Available from https://doi.org/10.1016/j.msec.2015.11.023.
Zepf, V. (2016). Chapter 1—An overview of the usefulness and strategic value of rare earth metals A2—Lima, Ismar Borges, De. In W. L. Filho (Ed.), *Rare earths industry* (pp. 3–17). Boston, MA: Elsevier.
Zhang, D., De, G., Zi, L., et al. (2018). Dropwise addition of cation solution: An approach forgrowing high-quality upconversion nanoparticles. *Journal of Colloid and Interface Science*, *512*, 141–150.

Zhang, L., Wang, Y. S., Yang, Y., et al. (2012). Magnetic/upconversion luminescent mesoparticles of Fe_3O_4@LaF_3:Yb^{3+},Er^{3+} for dual-modal bioimaging. *Chemical Communications*, *48*(91), 11238–11240.

Zhou, A., Wei, Y., Wu, B., et al. (2012). Pyropheophorbide A and c(RGDyK) comodified chitosan-wrapped upconversion nanoparticle for targeted near-infrared photodynamic therapy. *Molecular Pharmaceutics*, *9*, 1580–1589.

Zhou, J. C., Yang, Z. L., Dong, W., et al. (2011). Bioimaging and toxicity assessments of near-infrared upconversion luminescent $NaYF_4$:Yb, Tm nanocrystals. *Biomaterials*, *32*(34), 9059–9067.

Rare earth–based compounds for solar cells

13

Mahdiyeh Esmaeili-Zare[1] *and Omid Amiri*[2,3]
[1]Institute of Nano Science and Nano Technology, University of Kashan, Kashan, Islamic Republic of Iran, [2]Faculty of Chemistry, Razi University, Kermanshah, Iran, [3]Department of Chemistry, College of Science, University of Raparin, Rania, Iraq

13.1 General information

In the periodic table of elements, rare earth elements (REE) include 15 elements which extend from lanthanum to lutetium or in other words from atomic number (Z) of 57 to 71, and are evidently mentioned as the lanthanoids, although they are generally mentioned as the lanthanides. Also, Y ($Z = 39$) and Sc ($Z = 21$) due to their chemical and physical similarities are considered as an REE (Alonso et al., 2012; Massari & Ruberti, 2013; Van Gosen, Verplanck, Long, Gambogi, & Seal, 2014). The valence electronic configurations of the outermost shell for these elements are as [Xe] $4f^nd^0$, $16s^2$ ($n = 1-14$). Most REE are not as rare as the group's name offers. There are plenty of REE in the earth crust and their plentiness is close to that of chromium, tungsten, copper, niobium, and tin. REE are used in the optics, catalysis, and magnetic fields because of their unique 4f electronic structure (Mansour et al., 1984; Xu, Chen, Xu, Du, & Yan, 2020; Zhao et al., 2019). For example, in order to build high-performance laser, the presence of yttrium and neodymium elements is indispensable (Ikesue, Kinoshita, Kamata, & Yoshida, 1995). Vital components of strong permanent magnets include neodymium and samarium (Brown, Ma, & Chen, 2002; Hou et al., 2007). Also, scandium is a critical element for the production of high-strength alloys (Röyset & Ryum, 2005). Ceria is one of the most important material for use in decomposition, fuel cells, and photocatalytic water splitting (Fu, Saltsburg, & Flytzani-Stephanopoulos, 2003; Murray, Tsai, & Barnett, 1999; Primo, Marino, Corma, Molinari, & Garcia, 2011). Furthermore, ceria particles have a great polishing properties, so they can be used in advanced optical glass (Feng et al., 2006). Luminescence is one of the most important applications for REE and RE-based phosphors which are readily utilized in down-conversion luminescence, upconversion luminescence, and imaging in biomedicine (Wang et al., 2010; Wegh, Donker, Oskam, & Meijerink, 1999; Zhou, Shi, Jin, & Liu, 2015).

In this chapter, we summarize the application of RE-based compounds in various kinds of solar cells and also mentioned their preparation methods. Specific considerations have been paid to the use of these materials in solar cells as electrical devices. The chapter will not prepare a full review of the application of RE-doped

Advanced Rare Earth-Based Ceramic Nanomaterials. DOI: https://doi.org/10.1016/B978-0-323-89957-4.00006-2

material in solar cell and will not go into depth about their synthesis methods, but the purpose is to provide a summary of recent developments related to preparation methods and solar cell applications.

13.2 Application of RE-based compounds in solar cells

In the era of energy crisis and global warming, solar cells are considered as the top most choices for clean and economical energy generation (Kim et al., 2014). The solar radiation spectrum is composed of ultraviolet (UV), visible, and infrared (IR) lights. However, solar spectrum includes an enormous amount of energy, but solar cells can absorb a small amount of the narrow visible light area. The solar cell efficiency is limited due to the inefficient use of IR and UV spectra. Furthermore, efficiency of solar cells is restricted due to the damaging effect of UV light on them. One of the most important ways to overcome this problem is spectral conversion (Huang, Zhou, Liu, & Wu, 2003; Paschotta et al., 1997; Pillai & Green, 2010; Richards, 2006; Song, Zhang, & Zhu, 2015; Van Der Ende, Aarts, & Meijerink, 2009; van Sark, Korte, & Roca, 2012; Yang, Chen, Pun, Zhai, & Lin, 2013; Zhou, Teng, Ye, Lin, & Qiu, 2012). The downconversion luminescence and upconversion luminescence are utilized to transmute photons in the UV area. Upconversion is a procedure in which the successive absorption of two or more low-energy (long wavelength) photons results in their conversion into one high-energy photon. Therefore solar cell bandgap can be reduced by absorbing this upconverted photon with enough energy. Downconversion is the process that transforms one photon with high energy into two or more lower energy photons. This process occurs due to weak absorption of photon by photovoltaic cell. The photovoltaic cell can absorb all these photons with sufficient energy (Huang, Han, Huang, & Liu, 2013; Li, Zhang, Zhang, Hao, & Luo, 2014; Oskam, Wegh, Donker, Van Loef, & Meijerink, 2000; Qu et al., 2016; Richards, 2006; Trupke, Green, & Würfel, 2002; Wang, Jiang, Wan, & Zhai, 2015; Wang, Lei, Gao, & Mao, 2015; Wegh et al., 1999; Zhang & Huang, 2010). The down/upconversion (DC/UC) systems in composite materials including RE ions are widely considered in the literature in order to explain and to increase their luminescence for applications in photonics such as white-light generation, optical waveguides, and solar cells (Carvalho, Kassab, Del Cacho, da Silva, & Alayo, 2018; Florêncio et al., 2016; Sivakumar, van Veggel, & Raudsepp, 2005; Yuhua et al., 2015). In the following sections, we exhibit a broad overview of the suitable choice of the RE ions to modify the photoelectric conversion efficiency of different kinds of solar cells. Also, crystal structure of perovskite and dye-sensitized solar cells (DSSCs) is investigated.

13.2.1 Perovskite solar cells

Recently, organometallic-halide perovskite solar cells (PSCs) got attention due to their beneficial properties such as high conversion efficiency and minor environmental impact (Abate, Correa-Baena, Saliba, Su'ait, & Bella, 2018; Hodes, 2013; Service, 2014; Wu et al., 2017). In 2009, Kojima et al. achieved a power conversion efficiency (PEC) of 3.8% in DSSCs utilizing nanocrystalline perovskite semiconductors instead of conventional organic dyes (Kojima, Teshima, Shirai, & Miyasaka, 2009). Currently, the PSCs have noticeable specifications including a high charge carrier diffusion length (Stranks et al., 2013; Xing et al., 2013), high absorption factor (Green, Ho-Baillie, & Snaith, 2014), and easy solution process ability (Jeon et al., 2015), so they succeeded to attain an average PCE of 23.3%. Till now, several investigations have been performed on improving the performance and efficiency of PSCs through several approaches containing managing the crystallization of perovskite thin films as absorber (Burschka et al., 2013; Xiao et al., 2014), adjusting the energy level alignment at device joints (Zhou et al., 2014), composition engineering of perovskite materials (Jeon et al., 2015; Saliba et al., 2016; Saliba et al., 2016), and utilization of metallic nanostructures in solar cells (Luo et al., 2017; Zhang et al., 2013; Zhang et al., 2017). The PSCs waste a large part of sunlight energy because the spectral answer of PSCs is restricted to a small part of the solar spectrum from wavelength of 400–800 nm (Huang et al., 2013). Specially, in the matter of IR or near infrared (NIR) light, the subbandgap photons of sunlight have no contribution to make electricity for the PV cell since those have no adequate energy to produce electron–hole (e–h) pairs; in fact, they are either transported by the solar cell or absorbed to create warmth (Da, Xuan, & Li, 2018). The REE-codoped nanophosphors, as inorganic nanocrystals, are generally employed to decline the subbandgap loss. These elements can attract IR or NIR light and upconvert it to photons with superior visible energy through a nonlinear optical method (Goldschmidt & Fischer, 2015; Zhou, Liu, Feng, Sun, & Li, 2015).

13.2.1.1 Crystal structure of lead halide perovskites

Perovskites were named in honor of Russian mineralogist, Lev Perovski after a sample was discovered in the Ural mountains. The chemical formula of the first mineral perovskites was calcium titanate, combined of octahedron parts of TiO_6 (Tanaka & Misono, 2001). Commonly, the general chemical formula for perovskite compounds is like crystal structure with ABX_3 formula, in which A and B, respectively, are monovalent cations [Cs^+, MA^+ (methylammonium), FA^+ (formamidinium)] and divalent metal cation (lead$^{(2+)}$ ion), and X is a halide monovalent anion (chloride, bromide, and iodide ions). Furthermore, the researchers have prepared and studied a series of derivatives like ABX_3, $A_3B_2X_9$, $A_2BB'X_6$, A_2BX_6, with BX_6 units. Among them, the most important materials in solar cell applications were lead halide perovskites. Generally, the ideal structure of perovskite lattice is cubic, where 12 X anions surround the A cation and both of them fill the face-centers and vertexes, respectively, and the center of the cube is occupied by B cation, which is surrounded by six X anions (Fig. 13.1A). A three-dimensional (3D) perovskite structure is as BX_6 octahedron that it is constructed using the coordinates of central B ion with 6 X atoms. An ideal 3D perovskite structure with cubic phase has been shown in Fig. 13.1B (space group Pm3m). Nevertheless, perovskites is possible contorted away from the perfect cubic structure and transformed into an orthorhombic phase with less symmetry (Fig. 13.1C), because of the tilting of the PbX_6 octahedra. Nevertheless, the mechanisms that lead to the deviation from the perfect cubic structure are: (1) deformation of the octahedra, (2) distortions of the octahedra, and (3) shift of the cations (Woodward, 1997). The electronic fluctuations of the cation are the reasons for the first two mechanisms [namely (1) and (2)]. For instance, the Jahn–Teller distortion in $LaMnO_3$ (Lufaso & Woodward, 2004) and the shift of Ti in ferroelectric $BaTiO_3$ (Shirane, Danner, & Pepinsky, 1957) are the examples of electronic fluctuations. In the third case, perovskites are possibly contorted away from the perfect cubic structure while maintaining their corner-sharing connectivity and transformed into an orthorhombic phase with less symmetry

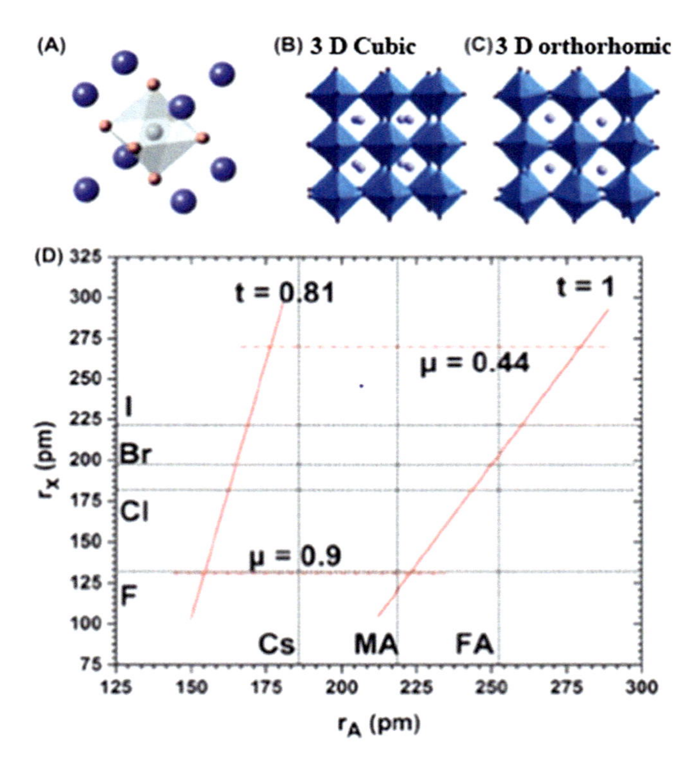

Figure 13.1 (A) Depiction of $APbX_3$ perovskites with cubic structure, and crystal structures of lead halide perovskites with (B) cubic and (C) orthorhombic phase. (D) Formability of 3D lead halide perovskites as a function of A-site cation and halide anion radii. The solid and dashed lines mark the bounds of the tolerance and octahedral factors, respectively.
Source: Reprinted with permission from Chen, Y., Liu, S., Zhou, N., Li, N., Zhou, H., Sun, L.-D., & Yan, C.-H. (2020). An overview of rare earth coupled lead halide perovskite and its application in photovoltaics and light emitting devices. *Progress in Materials Science*, *120*, 100737.

(Fig. 13.1C), because of the tilting of the PbX_6 octahedra (Stoumpos, Malliakas, & Kanatzidis, 2013; Wei, Cheng, & Lin, 2019; Yamada et al., 1998). These structural deviations lead to the formation of triclinic, rhombohedra, monoclinic, orthorhombic, and tetragonal crystal structures in perovskites (Woodward, 1996). The heterogeneous ions can be used in order to modify the structure stability, and enhanced Photoluminescence Quantum Yield (PLQY) of halide perovskites. In recent years, the RE-coupled lead halide perovskites have obtained vigorous consideration due to their unparalleled optical, magnetic, and electronic features. The incorporation RE ions into the lead halide perovskite is still difficult due to the influence of different agents containing the unbalanced electric charge, type of the host, and the temperature and duration of reaction.

13.2.1.2 RE-doped lead halide perovskites

Over the past few years, different types of RE-coupled perovskite precursors have been produced via many researchers and great optoelectronic and luminescence efficiencies of perovskite-based systems have been earned using advancements on morphology adjustment, defects passivation, and quantum yield betterment (Pan et al., 2017; Wang et al., 2019; Yin, Ahmed, Bakr, Brédas, & Mohammed, 2019).

The single and uniform Yb^{3+}- and Er^{3+}-doped Li (Gd, Y)F_4 upconversion nanophosphors with tetragonal phase were produced through thermal decomposition procedure by Deng et al. (2019) and also they developed a new PSC device with high efficiency through the addition of Yb^{3+}- and Er^{3+}-doped Li (Gd,Y)F_4 upconversion nanocrystals (UCNCs) to the hole transfer layer (HTL). They studied the effect of weight ratio of UCNCs on the PV fulfillment of the UC-PSC device. They found that the PV fulfillment of the UC-PSC system is affected using the weight ratio of UCNCs grown into the Spiro-OMeTAD-based HTL. Through optimizing the quantity of the UCNCs in the UC-PSC devices, these group succeeded to achieve the average PCE of 18.34% with Yb^{3+}- and Er^{3+}-doped Li (Gd,Y) UCNCs under sunlight (AM1.5G). Compared to PSC solar cells without UCNCs, a significant increase of more than 25% was obtained in PCE (Fig. 13.2). This research presents a probable way to effectively raise the efficiency of PSC systems and expand concentrator photovoltaic systems fabricated using low-cost UC nanomaterials.

SnO_2, because of its unique optical and electrical attributes, is a very hopeful electron transport layer (ETL) for extremely efficient PSCs. However, the presence of holes in the surface after an annealing action reduces its utilization in PSCs. To solve this issue, Xu et al. doped lanthanum (La) into the SnO_2 layer, which is able to reduce the aggregation of SnO_2 crystal and make full coverage and a monotonous film (Xu et al., 2019). This layer was synthesized through a low-temperature method at 180°C. They discovered that the charge injection and electron transfer are facilitated after doping the lanthanide into SnO_2 layer. Furthermore, the band deflection at the boundary of La:SnO_2 and perovskite was reduced using La:SnO_2, which leads to the $V_{oc} = 1.11$ V (Fig. 13.3). This study showed that usage of a novel effective dopant material, that is, lanthanide doping within the SnO_2 layer can be a

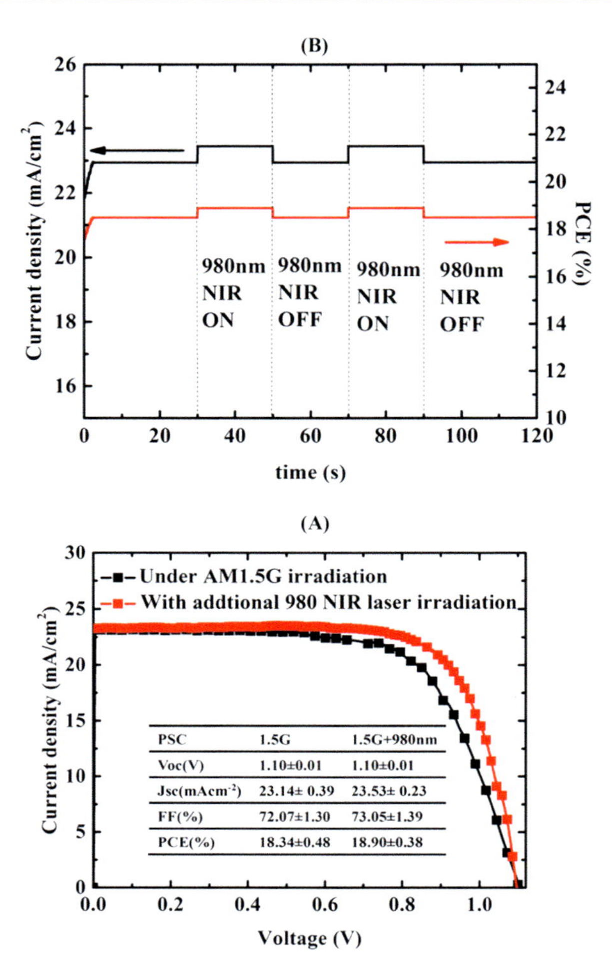

PSC	1.5G	1.5G+980nm
Voc(V)	1.10±0.01	1.10±0.01
Jsc(mAcm^{-2})	23.14± 0.39	23.53± 0.23
FF(%)	72.07±1.30	73.05±1.39
PCE(%)	18.34±0.48	18.90±0.38

Figure 13.2 (A) The current density–voltage (*J–V*) characteristics of CH3NH3PbI3 solar cells with 0.10 wt.% Li(Gd,Y)F4:Yb,Er UCNCs under AM1.5G and additional 980-nm laser irradiance. (B) The photocurrent density–time (*J–t*) curves measured at 0.80 V under AM1.5G and 980-nm laser irradiance. The laser was turned on and off every 20 s. *Source*: Reprinted with permission from Deng, X., Zhang, C., Zheng, J., Zhou, X., Yu, M., Chen, X., & Huang, S. (2019). Highly bright Li (Gd, Y) F4: Yb, Er upconverting nanocrystals incorporated hole transport layer for efficient perovskite solar cells. *Applied Surface Science 485*, 332–341.

significant method to optimize the efficiency of PSCs with an ETL processed at low temperature.

Utilization of trivalent RE metal cations, like neodymium^{3+}, europium^{3+}, and samarium^{3+}, in order to decrease the impurity phases (Kazhugasalamoorthy et al.,

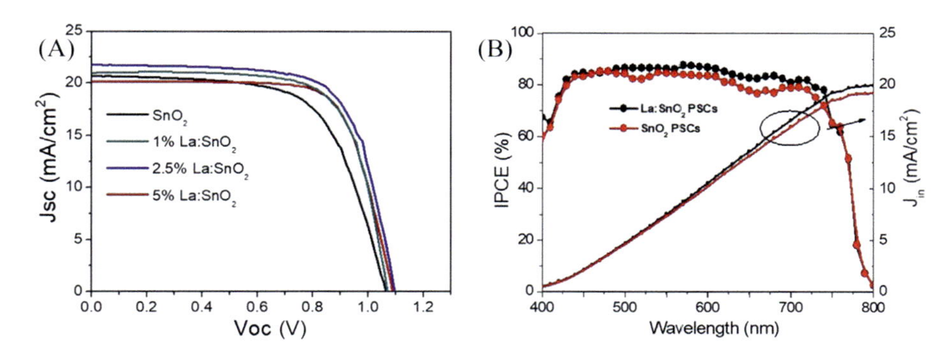

Figure 13.3 (A) The $J - V$ curves of PSC devices based on pristine SnO_2 and La:SnO_2 with different La doping contents. (B) IPCE of the PSCs based on SnO_2 and La:SnO_2 ETLs, respectively.
Source: Reprinted with permission from Xu, Z., Teo, S. H., Gao, L., Guo, Z., Kamata, Y., Hayase, S., & Ma, T. (2019) La-doped SnO_2 as ETL for efficient planar-structure hybrid perovskite solar cells. *Organic Electronics*, *73*, 62–68.

2010; Remya, Amirthapandian, Manivel Raja, Viswanathan, & Ponpandian, 2016), remove the deep defects (De Figueiredo et al., 2012; Yamamoto, Okamoto, & Kobayashi, 2002), upgrade the film state of superconductors and multiferroic substances (Singh, 2013), and hinder the chemical fluctuation (Wang & Nan, 2008) were rarely reported. For example, Nd^{3+} ions were utilized in inorganic perovskite materials to adjust the optoelectronic properties of them (Liu et al., 2011). Production of x mol.% Nd^{3+}-doped $MAPbI_3$ (where $x = 0.1$, 0.5, 1.0, and 5.0 mol. % as a formal ratio) as a new hybrid perovskite substance, where lead (II) ion at the B-site was replaced by heterovalent Nd^{3+} cations, has been reported by Wang et al. (2019). They found that utilizing thin films of x mol.% Nd^{3+}-doped $MAPbI_3$ as compared to the previous $MAPbI_3$ hybrid perovskite lead to make optimum crystallinity with decreased nonsimilar power, and also, better film quality with decreased trap states and pinholes. They came to this conclusion that planar heterojunction PSCs based on these recently improved x mol.% Nd^{3+}-doped $MAPbI_3$ thin films display an average PCE = 21.15% and significantly decreased photocurrent hysteresis. These results open a novel horizon for adjusting the electronic properties of hybrid perovskite substances and improving the equipment efficiency of PSCs.

The Yb^{3+}/Er^{3+}/Sc^{3+}-doped β-$NaYF_4$@$NaYF_4$ UCNPs with core shell construction have been synthesized by Guo et al. (2019). They found that the β-$NaYF_4$ with Sc^{3+} doping shows a total UC intensity of 10 and 16 times higher for red and green emissions than sample without Sc^{3+}as a doping agent. The synthesized UCNPs as optimal PSCs incorporated into TiO_2 mesoporous layer showed a high PCE = 20.19% (Fig. 13.4A), superior to those expressed formerly for UCNP incorporated PSCs. This research offers a useable way to provide the minimalization of nonabsorption loss of solar photons with modified energy adjustment for PSCs or other kinds of solar cells.

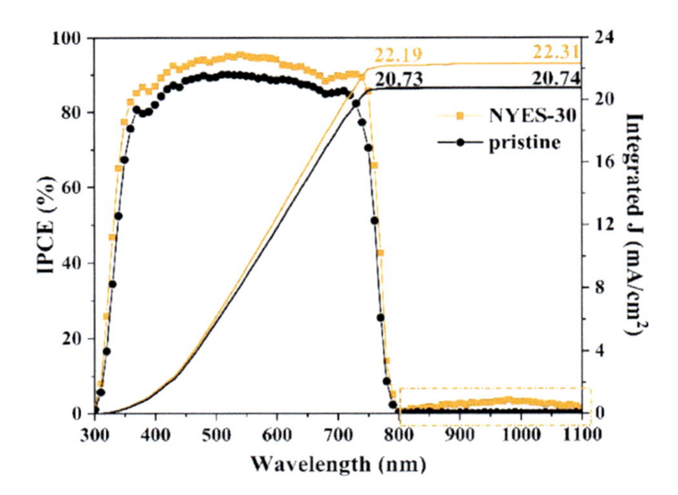

Figure 13.4 IPCE spectra of NYES-30 and pristine devices.
Source: Reprinted with permission from Guo, Q., Wu, J., Yang, Y., Liu, X., Jia, J., Dong, J., Lan, Z., Lin, J., Huang, M., & Wei, Y. (2019). High performance perovskite solar cells based on β-$NaYF_4$: Yb^{3+}/Er^{3+}/Sc^{3+}@ $NaYF_4$ core-shell upconversion nanoparticles. *Journal of Power Sources*, *426*, 178–187.

Until now, the champion PCE of the PSCs has achieved about 22.7% by using conventional n-i-p structures (Arora et al., 2017; Noh & Seok, 2015; Tsai et al., 2016; Yang et al., 2015; Yang et al., 2017; Zhang et al., 2019). TiO_2 nanostructures have been widely used in PSCs due to its well clearness, environmental stability, carrier divorce efficiency, and adjustability (Li et al., 2016). It makes a significant contribution as most common ETL material, in raising electron gathering and forwarding from the light absorbing layer to the related electrodes (Wang et al., 2014). Metal doping, especially RE ions, has been considered to be an effective way to increase the properties of PSC device. Accordingly, Zhang et al. synthesized the Sm^{3+}, Eu^{3+} @ TiO_2 ETLs in order to utilizing in n-i-p PSCs. In this codoped layer, Eu^{3+} ions act as the DC doping agent to enhance the UV light application as well as support the decadence of perovskite layer from UV radiation. Besides, a sensitizer (i.e., doped Sm^{3+} ions) was applied to raise the DC sufficiency of Eu^{3+} cations through extending the agitation wavelength area for Eu^{3+} publication because of the energy transition from Sm^{3+} to Eu^{3+} (Li, Huang, & Chen, 2011). This group successfully achieved a PCE = 19.01% with an fill factor (FF) = 76.9%, a V_{OC} = 1.10 V, and a J_{SC} = 22.47 mA/cm^2 by optimizing the concentration of Sm^{3+} and Eu^{3+} cations. They found that the TiO_2:REs ETL can lead to high electron elicitation and lower interfacial recombination. This work, in contrast with previous reports (Gao et al., 2016; Jiang et al., 2017; Xiang et al., 2017), prepares an applied procedure to modify the radiation stability and system efficiency of PSCs.

13.2.2 Dye-sensitized solar cells

In recent decade, renewable, clean, and low-cost energy fountains have been suggested as each of the elementary possible answers for universal warming (Ludin et al., 2018). So far, three generations of PV cell technologies have been introduced: (1) first-class PV cells are controlling the PV bazaar and include mono- and polycrystalline silicon. (2) Second-class PV cells are made up of noncrystalline silicon, copper gallium indium diselenide, and cadmium telluride. (3) Third-class PV cells contain quantum dots (QDs) and DSSCs (Byrne, Ahmad, Surolia, Gun'ko, & Thampi, 2014; Chiba et al., 2006; Grätzel, 2003; O'regan & Grätze, 1991; Wang, Yanagida, Sayama, & Sugihara, 2006), organic solar cells and PSCs. Among them, DSSCs as a new reusable and pure solar-to-electricity transformation device are employed instead of the conventional silicon-based photovoltaic devices.

13.2.2.1 Structure of dye-sensitized solar cells

Fig. 13.5 indicates a schematic presentation of DSSCs (Gong, Sumathy, Qiao, & Zhou, 2017). The system is composed of four principal parts: (1) a mesoporous oxide layer (generally, TiO_2) deposited on a transparent conductive glass substrate as a photoanode; (2) a monolayer of dye sensitizer covalently bonded to the surface of the TiO_2 layer to gather light and produce photon-excited electrons; (3) redox couple (generally, I^-/I^{3-}) in an organic solvent as an electrolyte to collect electrons at the counterelectrode and affecting dye-regeneration; and (4) a platinum-coated

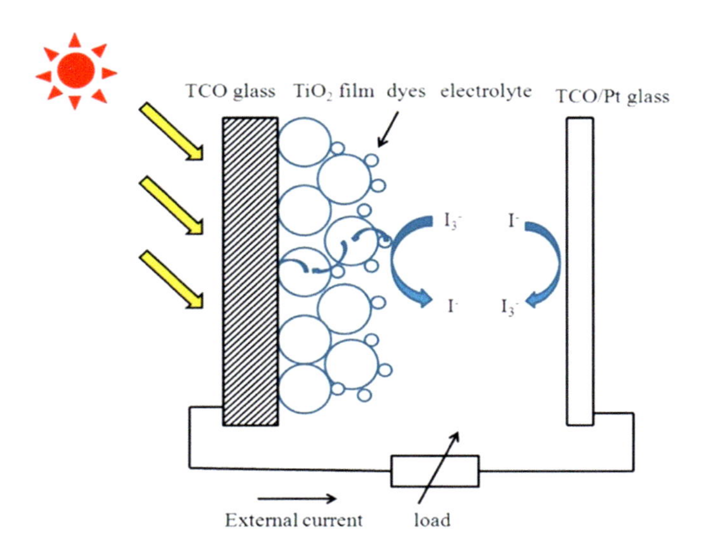

Figure 13.5 Schematic diagram of a dye-sensitized solar cell.
Source: Reprinted with permission from Gong, J., Sumathy, K., Qiao, Q., & Zhou, Z. (2017). Review on dye-sensitized solar cells (DSSCs): Advanced techniques and research trends. *Renewable and Sustainable Energy Reviews*, *68*, 234–246.

conductive glass substrate as a counterelectrode. When the sun radiation hits the solar cell, the electrons get injected into the conduction band of TiO_2 due to excited dye sensitizers on the surface of TiO_2 film. Inside the TiO_2 film, the infused electrons spread completely during the mesoporous film to the anode and are employed to do applied work at the outer burden. Eventually, to fulfill the cycle, an electrolyte is used to gather these electrons at counterelectrode that in turn are absorbed to reproduce the dye sensitizer (Akın, Açıkgöz, Gülen, Akyürek, & Sönmezoğlu, 2016; Akın et al., 2016; Carbas, Gulen, Tolu, & Sonmezoglu, 2017; Gulen, Sarilmaz, Patir, Ozel, & Sonmezoglu, 2018; Kaya, Akin, Akyildiz, & Sonmezoglu, 2018; Taş, Gülen, Can, & Sönmezoğlu, 2016; O'regan & Grätze, 1991).

13.2.2.2 RE codoped for utilization in dye-sensitized solar cells

TiO_2 nanostructures are one of the principal segments of the DSSCs, and it plays a pivotal role in cell efficiency (Akin, Ulusu, Waller, Lakey, & Sonmezoglu, 2018; Li et al., 2003; Sönmezoğlu, Çankaya, & Serin, 2012; Wei-Wei, Song-Yuan, Lin-Hua, Lin-Yun, & Kong-Jia, 2006; Yao, Liu, Peng, Wang, & Li, 2006). In fact, high PCE in DSSCs is achieved using an ideal photoanode with rapid charge forwarding and small electron recombination. In this way, the doping procedure is one of the practical methods for the isolation of electron-gap sets by creating crystal deformities or vacancy situations in substance or matrix and thus electron transport and quantum yield are improved (Akin, Erol, & Sonmezoglu, 2017; Fan, Yu, & Ho, 2017; Ozturk, Gulveren, Gulen, Akman, & Sonmezoglu, 2017; Sengupta, Das, Mondal, & Mukherjee, 2016). Among these doping elements, REE have extraordinary potential (Bingham & Daoud, 2011; Hafez, Saif, & Abdel-Mottaleb, 2011; Wang, Batentschuk, Osvet, Pinna, & Brabec, 2011; Wu et al., 2010; Zhang et al., 2009). Akmana et al. synthesized the TiO_2:Eu^{3+}, Tb^{3+} (RE^{3+} ion) nanostructures and used as photoanodes for application in DSSCs (Akman, Akin, Ozturk, Gulveren, & Sonmezoglu, 2020). As authors mentioned, it is the first efficiency presented for DSSCs using TiO_2:Eu^{3+}, Tb^{3+} as photoanode. In this research, the photocurrent density of DSSC for bare exhibited a $\sim$14.6% reduction after 3600 seconds testing, whereas in the same condition, the DSSC fabricated using photoanode: 1.0% Eu^{3+}/Tb^{3+} indicated slower destruction of about 2.7%. In another study, Wang et al. synthesized Eu^{3+}-doped $SrAl_2O_4$ DCNCs using a sol−gel procedure and next doped into TiO_2 just as a photoanode (Wang et al., 2016). The results showed that DSSCs with 3 wt.% $SrAl_2O_4$:Eu^{3+} @ TiO_2 photoanodes show that the photoelectric conversion efficiency is 20% superior to pure TiO_2 photoanode (Fig. 13.6). Table 13.1 shows photovoltaic criterions of DSSCs constructed with various kind of photoanodes.

In DSSCs, a number of studies have indicated that incorporation of an RE dopant in the photoanode can raise the cell efficiency (Hafez et al., 2011; Li & Gu, 2012; Wu et al., 2010; Yao et al., 2006; Zalas & Klein, 2012; Zhang et al., 2011; Zhang, Peng, Chen, Chen, & Han, 2012; Zhang et al., 2013). On the other hand, other semiconductor oxides such as TiO_2, ZnO, SnO_2, Nb_2O_5, $SrTiO_3$, Fe_2O_3, NiO, In_2O_3, $SrTiO_3$, $CaTiO_3$, and $BaTiO_3$ (Asbury, Wang, & Lian, 1999; Bauer, Boschloo, Mukhtar, & Hagfeldt, 2001; Chou, Zhang, Fryxell, & Cao, 2007; Dou et al., 2015; Du Pasquier, Chen, & Lu, 2006; Hsu, Xi, Yip, Djurišić, & Chan, 2008;

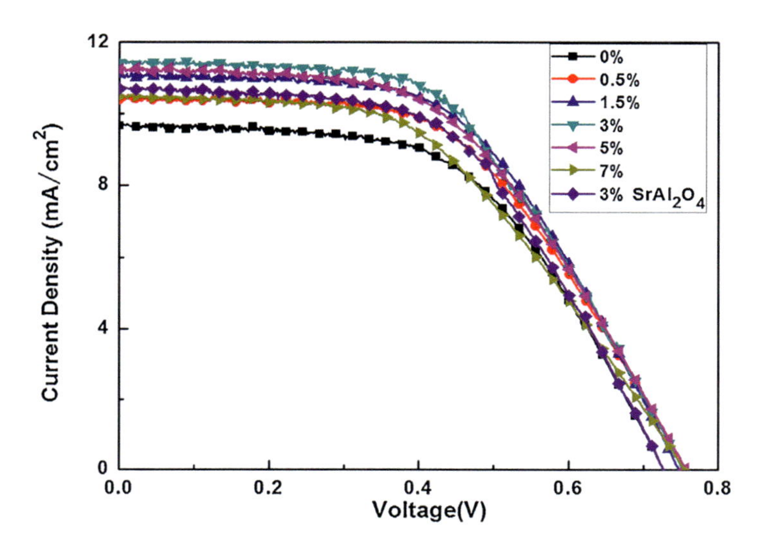

Figure 13.6 Current–voltage characteristics of Gd:CdTe QD-sensitized solar cells for different doping concentrations.
Source: Reprinted with permission from Wang, L., Guo, W., Hao, H., Su, Q., Jin, S., Li, H., Hu, X., Qin, L., Gao, W., & Liu, G. (2016). Enhancing photovoltaic performance of dye-sensitized solar cells by rare-earth doped oxide of $SrAl_2O_4$: Eu^{3+}. *Materials Research Bulletin*, *76*, 459–465.

Table 13.1 Photovoltaic parameters of DSSCs fabricated with different type of photoanodes.

$SrAl_2O_4$:Eu^{3+} (wt.%)	V_{oc} (V)	J_{sc} (mA/cm^2)	FF	η (%)
0%	0.73	9.66	0.55	3.86
0.5%	0.75	10.38	0.54	4.15
1.5%	0.75	11.09	0.54	4.49
3%	0.75	11.42	0.54	4.64
5%	0.76	11.23	0.51	4.38
7%	0.76	10.47	0.49	3.87
3% $SrAl_2O_4$	0.73	10.69	0.54	4.23

Source: Reprinted with permission from Wang, L., Guo, W., Hao, H., Su, Q., Jin, S., Li, H., Hu, X., Qin, L., Gao, W., & Liu, G. (2016). Enhancing photovoltaic performance of dye-sensitized solar cells by rare-earth doped oxide of $SrAl_2O_4$: Eu^{3+}. *Materials Research Bulletin*, 76, 459–465.

Katoh et al., 2004; Okamoto & Suzuki, 2014; Sayama, Sugihara, & Arakawa, 1998; Sheng, Zhao, Zhai, Jiang, & Zhu, 2007; Somdee & Osotchan, 2019; Xia, Li, Yang, Li, & Huang, 2007; Xie et al., 2018; Yang, Kou, Wang, Cheng, & Wang, 2009; Zhang, Chou, Russo, Jenekhe, & Cao, 2008; Zhang et al., 2008) have been applied as photoanode materials to develop great performance DSSCs. Chamanzadeh et al.

synthesized the ZnO:La nanorods and dye @ ZnO nanorods using two steps: (1) use of spin coating system in order to create seed layer and (2) use of hydrothermal method for nanorod growth (Chamanzadeh, Ansari, & Zahedifar, 2021). They grow ZnO nanorods on the surface of porous TiO_2 nanoparticles film (TiO_2 nanoparticles/ZnO nanorods) as the photoanodes and achieved to this conclusion that the PCE of DSSCs:La, Dy can be enhanced to 63.6% and 71.5%, respectively, in comparison with the formal TiO_2 nanoparticles photoanode. Lu et al. studied the effects of different RE (La, Nd, Ce, Gd, and Sm) ions on the photoelectrochemical properties of film electrodes built with the homemade ZnO nanoparticles and investigated the whole efficiencies of DSSCs built with ZnO films corrected with various RE (Ce, La, Sm, Nd, and Gd) ions (Lu, Li, Peng, Fan, & Dai, 2011). They found that the open-circuit photovoltage and fill factor of the ZnO-based solar cell can be increased by the Nd-, Sm- and Gd-ion modifications; while the La-, Ce-, Nd-, and Sm-ion modifications result in a reduced short-circuit photocurrent. Table 13.2 indicates the middle photovoltaic consequences of the parallel cells with the respective standard deviation of this research team.

13.2.3 Application of REs in other kinds of solar cells

Quantum dot sensitized solar cells (QDSSCs) are used as a novel advancement in the scope of DSSC. The single physical variation between DSSC and QDSSC is related to their sensitizers. All the other active components stay same. The DSSC was named as QDSSC due to the use of QDs as a sensitizer. QDSSCs increase the efficiency of solar cell at low cost. Arivarasan et al. fabricated Gd:CdTe QDs @ TiO_2 photoelectrodes and studied its photovoltaic efficiency through the making of sandwich-kind QDSSC (Arivarasan et al., 2020). They studied the photovoltaic response of Gd-doped CdTe QDs and achieved to the higher solar cell efficiency of about 2.24% for Gd 10% as doping agent. Totally, they found that the CdTe QDs with Gd 10% as doping agent can be employed as the prospective sensitizer for QDSSC (Fig. 13.7).

The crystalline silicon (c-Si) solar cells require effective light absorption to attain great efficiencies. Hence, an antireflection coating (ARC) is used to cover the top surface of the solar cells which can increase the efficiency of c-Si solar cells through enhancing light trapping in the active region. Therefore porous silicon (PS) layer was applied as a supreme ARC for c-Si solar cells. However, spontaneous oxidation of PS layer in ambient atmosphere leads to the reduction of the surface structures, photoluminescence, and electrical properties. To defeat this issue, many researchers have presented that filling the pores in the PS layer with the desired amount of conducting materials results in a growth of its photoluminescence and electrical properties. Atyaoui et al. improved the efficiency of silicon solar cells by creating a porous silicon layer modified with RE (Ce, La). They achieved a conversion efficiency of 7.7% and 9.3% in cell with La and Ce-doped PS layers, respectively (Atyaoui et al., 2013). Also, Yao et al. successfully prepared the bifunctional SiO_2 films including $NaYF_4$:Tb^{3+}-Yb^{3+} phosphors (Yao & Tang, 2020). They reported that the combining of film with silicon solar cells improves the J_{sc} and enhances the η of solar cells from 11.57% to 12.35% (Fig. 13.8).

Table 13.2 Effects of rare earth ion modifications on the photoelectrochemical properties of ZnO-based cells.

Sample	ZnO	La-ZnO	Ce-ZnO	Nd-ZnO	Sm-ZnO	Gd-ZnO
V_{oc} (mV)	494 ± 8.5	499 ± 9.7	450 ± 10.1	527 ± 11.5	530 ± 10.3	536 ± 9.3
I_{sc} (mA/cm^2)	520 ± 0.26	4.27 ± 0.27	0.66 ± 0.05	4.39 ± 0.29	5.02 ± 0.31	5.64 ± 0.33
FF	0.54 ± 0.01	0.57 ± 0.02	0.28 ± 0.01	0.58 ± 0.02	0.57 ± 0.02	0.65 ± 0.02
η (%)	1.37 ± 0.12	1.22 ± 0.11	0.09 ± 0.01	1.31 ± 0.11	1.52 ± 0.13	1.98 ± 0.17

Source: Reprinted with permission from Lu, L., Li, R., Peng, T., Fan, K., & Dai, K. (2011). Effects of rare earth ion modifications on the photoelectrochemical properties of ZnO-based dye-sensitized solar cells. *Renewable Energy*, 36, 3386–3393.

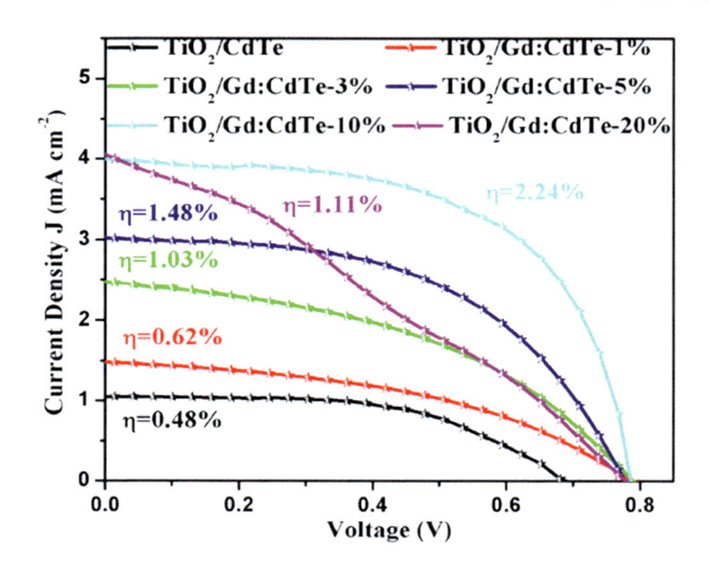

Figure 13.7 Current–voltage characteristics of Gd:CdTe QD-sensitized solar cells for different doping concentrations.
Source: Reprinted with permission from Arivarasan, A., Bharathi, S., Arasi, S. E., Arunpandiyan, S., Revathy, M., & Jayavel, R. (2020). Investigations of rare earth doped CdTe QDs as sensitizers for quantum dots sensitized solar cells. *Journal of Luminescence*, *219*, 116881.

Chalcopyrite Cu(InGa)Se_2 (CIGS) thin films are the other kinds of photovoltaic cells. Over the last few decades, this PV cells have improved the cell conversion efficiency record, which is currently at 23.35%—higher than polycrystalline Si (22.3%) and CdTe thin film (21.0%). Glass/Mo/Cu(InGa)Se_2/CdS/ZnO/ZnO:Al is a typical structure of CIGS cells. REE have been considered as a success applicant for a photon transformer (Bai et al., 2008; Balestrieri et al., 2013; Du, Zhang, Sun, & Yan, 2008; Gu et al., 2004; Li, Luo, Xia, & Wang, 2014; Liu, Luo, Li, Zhu, & Chen, 2009; Park et al., 2017). Park et al. examined the features of an ytterbium Yb: zinc tin oxide thin film and its utilization just as a capable DC of CIGS as thin film solar cells (Park et al., 2020). Their study showed that the greatest operation Yb-doped zinc tin oxide-CIGS solar cell has been built by selecting Yb: zinc tin oxide deposited at Sn sputter power and substrate temperature 15 W and 100°C, respectively.

13.3 Synthesis procedures

Various synthesis procedures are employed for the fabrication of RE-doped oxide nanophophors. An outline of the fabrication methods is described in the following sections.

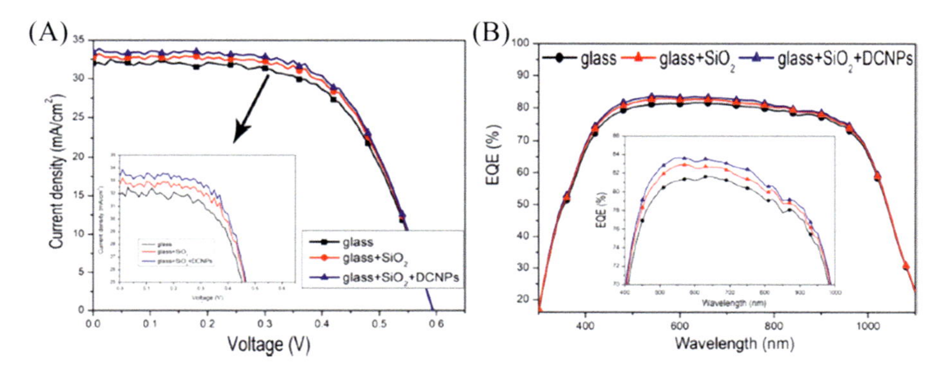

Figure 13.8 (A) J−V and (B) External Quantum Efficiency (EQE) curves of coatings with Si solar cells: pure glass, glass with SiO_2 coatings, and glass with SiO_2 coatings containing $NaYF_4$:Tb-Yb DCNPs (down-conversion Nanoparticles).
Source: Reprinted with permission from Yao, H., & Tang, Q. (2020). Luminescent anti-reflection coatings based on down-conversion emission of Tb^{3+}-Yb^{3+} co-doped $NaYF_4$ nanoparticles for silicon solar cells applications. *Solar Energy*, *211*, 446−452.

13.3.1 Solution combustion procedure

The solution combustion synthesis process is a simple and flexible methodology used to produce ceramic powders for various technologies, containing solid oxide fuel cells, actuators, UV absorbents, catalysts, gas sensors, and coloring agents (ceramic pigments). In this method, the solution is heated on a hot plate to evaporate the surplus amount of water, and spontaneous combustion happens. Initially, the solution releases a huge amount of gas, therefore it becomes dehydration. At the spontaneous ignition spot, the heat is released due to burning the solution. At the end of the combustion process, a burning solid is obtained when the solution evaporates immediately upon the release of carbon and nitrogen oxide gas products. This process is used in the production of different nanoscale materials, containing oxides, sulfides, metals, and alloys (Cho, Kumar, Holloway, & Singh, 1998; Luo & Cao, 2008; Shea, McKittrick, Lopez, & Sluzky, 1996; Ye, Gao, Xia, Nie, & Zhuang, 2010). Ogugua et al. synthesized the powders of dysprosium (Dy^{3+})-doped lanthanum gadolinium oxyorthosilicate, lanthanum yttrium oxyorthosilicate, and gadolinium yttrium oxyorthosilicate ($LaGdSiO_5$, $LaYSiO_5$ and $GdYSiO_5$, respectively) phosphors with urea-assisted ignition procedure (Ogugua, Shaat, Swart, & Ntwaeaborwa, 2015). Fig. 13.9 illustrates the method for the preparation of their powder phosphors schematically. Also, Araichimania et al. synthesized RE ion-doped silica nanoparticles via rice husk biomass using an easy and convenient microwave-assisted ignition method for the construction of bioimaging contrast factors (Araichimani et al., 2020).

13.3.2 Sol−gel procedure

The sol−gel procedure is a wet-chemical way that leads to the formation of a solid product during the reaction of an organometallic composition, which is usually a

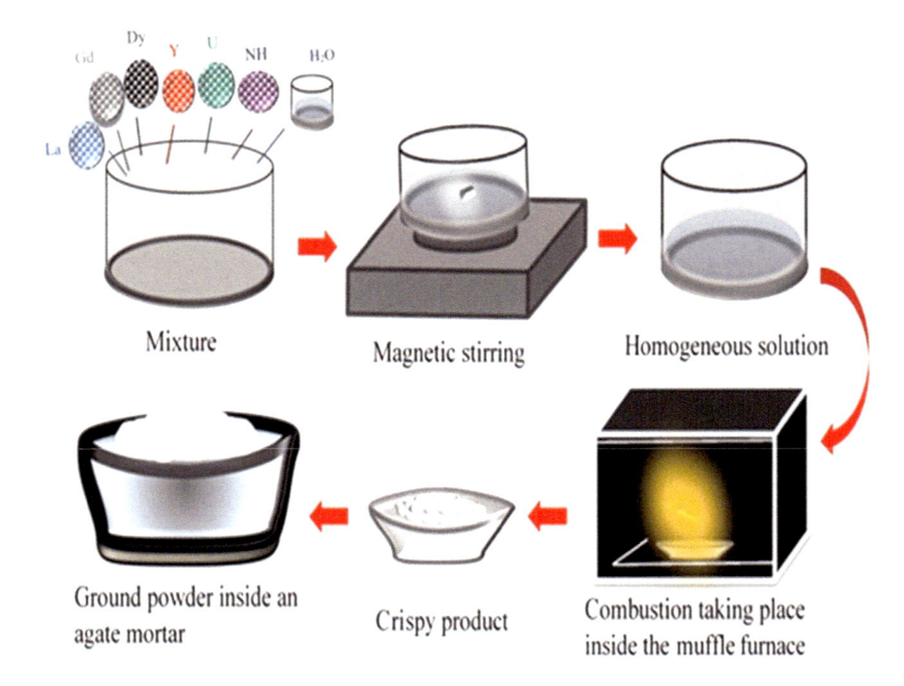

Figure 13.9 A schematic illustration of the synthesis procedure of powders (La = La$(NO_3)_3 0.6H_2O$, Gd = $Gd(NO_3)_3 0.6H_2O$, Dy = $Dy(NO_3)_3 0.6H_2O$, Y = $Y(NO_3)_3 0.6H_2O$, U = $CO(NH_4)_2$, and NH = NH_4NO_3).
Source: Reprinted with permission from Ogugua, S. N., Shaat, S., Swart, H. C., & Ntwaeaborwa, O. M. (2015). Optical properties and chemical composition analyses of mixed rare earth oxyorthosilicate (R_2SiO_5, R = La, Gd and Y) doped Dy^{3+} phosphors prepared by urea-assisted solution combustion method. *Journal of Physics and Chemistry of Solids*, *83*, 109–116.

nitrate, chloride, or alkoxide under aqueous conditions. This product can be an aerogel in the form of a metal oxide, an aggregation glass monolith, a nitride coating, a molecular filter with high surface area, or nanoparticle. This reaction is started by the formation of a gel using a mixture of reactants in a liquid phase through hydrolysis and condensation of a starting substance, followed by annealing at high temperature until aging, solvent evaporation, and eventually drying. An aqueous sol–gel procedure was used to prepare the nanosized magnetic RE iron garnets [$(RE)_3Fe_5O_{12}$, where RE = Sm, Eu, Gd, Tb, Dy, Ho, Er, Tm, Yb, Lu] by Opuchovic, Kareiva, Mazeika, & Baltrunas (2017). Moreover, RE (La^{3+}, Nd^{3+}, Sm^{3+}, Gd^{3+})-doped TiO_2-SiO_2 nanostructures, $NiFe_{1.98}(RE)_{0.02}O_4$ (where RE = La, Sm, Gd, and Dy) spinel ferrites, and (Eu^{3+}, Gd^{3+}, and Ho^{3+})-doped In_2O_3 gas sensor were constructed using sol–gel procedure (Mohamed & Mkhalid, 2010; Niu, Zhong, Wang, & Jiang, 2006; Samoila et al., 2017).

13.3.3 Hydrothermal method

Dharmadhikari and Athawale synthesized a series of RE cup rates (RE_2CuO_4, RE: Ce, Pr, Nd, Sm, and Gd) using template-free hydrothermal method (Dharmadhikari & Athawale, 2018) (Fig. 13.10). RE_2CuO_4 were fabricated through mixing the nitrate precursors of REE and copper in 2:1 molar ratio, followed by heating on hot plate at 80°C and adding KOH as a hydrolyzing agent. Finally, the mixture was transferred to a Teflon-lined stainless steel autoclave and subjected to hydrothermal activation at 140°C for 6 hours. In another work, Ba(RE)GeO_4 (OH) and Ba $(RE)_{10}(GeO_4)_4O_8$ (RE = holmium $^{3+}$ and erbium^{3+}) were synthesized using the reaction of BaO, RE_2O_3, and GeO_2 materials through high-pressure hydrothermal technique (Fulle et al., 2019).

13.3.4 Coprecipitation method

Mekala et al. fabricated 3 mol.% (Dy and Ce) @ ZrO_2 NPs via coprecipitation method (Mekala, Deepa, & Rajendran, 2018). In this method, the nitrate type of starting substances and urea are dissolved into methanol separately. The solutions are added together under stirring. After creating a homogeneous blending, the clear solution of the mixture was annealed at 120°C for 24 hours till a white powder was achieved. Then, the obtained products were calcined for 6 hours at 600°C. Furthermore, RE replaced cobalt ferrites $Co(RE)_xFe_{2-x}O_4$ (where RE = neodymium, gadolinium, and Samarium as well as $x = 0.1$ and 0.2) and $Mg_{1-x}Cd_xFe_2O_4 + 5\%Sm^{3+}$ ($x = 0$, 0.2, 0.4, 0.6, 0.8, and 1.0) were synthesized by this method (Gadkari, Shinde, & Vasambekar, 2010; Nikumbh et al., 2014).

13.3.5 Solid-state method

In a typical synthesis procedure, $Ca_5(PO_4)_3Cl$:(Eu^{3+} and Sm^{3+}) phosphors were produced by Kim et al. The phosphor material can be synthesized by solid-state reaction system as a high temperature method (Kim, Lee, Yang, Ha, & Hong, 2016). They mixed the starting materials through the grind in deionized water for 24 hours to prepare a uniform mixture, and so totally dried at 80°C on a hot plate. Then, they sintered the materials at 1100°C or 1300°C for 6 hours at room temperature.

13.4 Conclusion and outlook

REE are today a central material in both academia and industry. REE contain Sc and Y plus the series of lanthanide elements (La, Ce, Pr, Nd, Pm, Sm, Eu, Gd, Tb, Dy, Ho, Er, Tm, Yb, and Lu). Various synthesis methods are used to prepare RE-doped oxide nanophophors such as solution combustion, sol−gel, hydrothermal, coprecipitation, and solid-state methods. These ions show electronic, optical, catalytic, and magnetic properties because of the unique 4f electronic structure, which

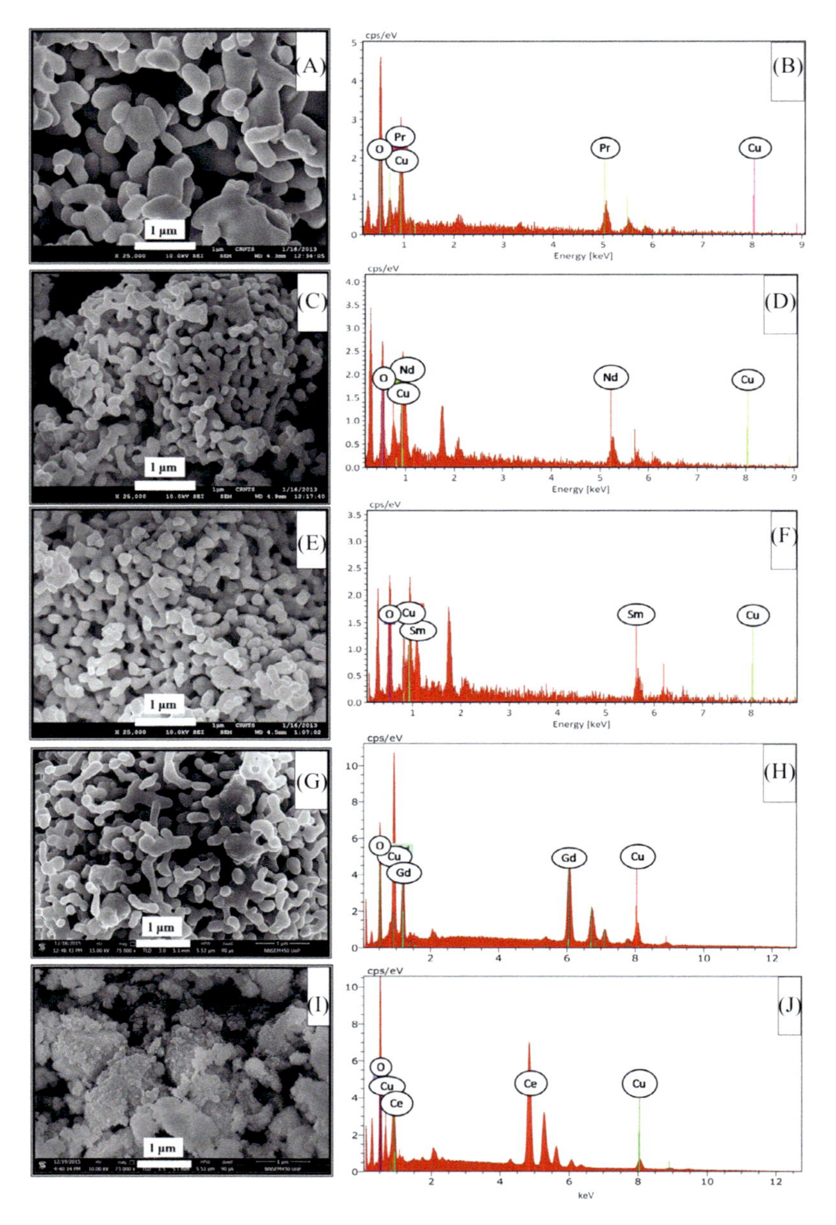

Figure 13.10 Scanning electron micrographs (A, C, E, G, and I) and energy-dispersive X-ray spectra (B, D, F, H, and J) of Pr_2CuO_4, Nd_2CuO_4, Sm_2CuO_4, Gd_2CuO_4/CuO, and CeO_2/CuO.

Source: Reprinted with permission from Dharmadhikari, D. V., & Athawale, A. A. (2018). Studies on structural and optical properties of rare earth copper oxides synthesized by template free hydrothermal method. *Materials Science and Engineering: B 229*, 70–78.

make excellent potential for application in the photovoltaic-based devices such as solar cells. An increase in the yield of photovoltaics is because of the rise of light production through DC of UV radiation to visible radiation and through UC of IR to visible radiation. Because of the importance of this subject, many attempts are done to enhance the efficiency of each component of the photovoltaic device. Therefore the research of novel REE-based solar cells with efficiencies higher than the real systems is an important research topic for the academic community and the solar industry. Furthermore, REE have application in the other fields such as electronics, manufacturing, medical science, and renewable energy. For example, in order to build a high-performance laser, the presence of yttrium and neodymium elements is indispensable. Vital components of strong permanent magnets include neodymium and samarium. Also, scandium is a critical element for the production of high-strength alloys. Ceria is one of the most important materials for use in decomposition, fuel cells, and photocatalytic water splitting. Furthermore, ceria particles have a great polishing properties, so they can be used in advanced optical glass.

References

Abate, A., Correa-Baena, J. P., Saliba, M., Su'ait, M. S., & Bella, F. (2018). Perovskite solar cells: From the laboratory to the assembly line. *Chemistry–A European Journal*, *24*, 3083–3100.

Akin, S., Erol, E., & Sonmezoglu, S. (2017). Enhancing the electron transfer and band potential tuning with long-term stability of ZnO based dye-sensitized solar cells by gallium and tellurium as dual-doping. *Electrochimica Acta*, *225*, 243–254.

Akin, S., Ulusu, Y., Waller, H., Lakey, J. H., & Sonmezoglu, S. (2018). Insight into interface engineering at TiO2/dye through molecularly functionalized caf1 biopolymer. *ACS Sustainable Chemistry & Engineering*, *6*, 1825–1836.

Akman, E., Akin, S., Ozturk, T., Gulveren, B., & Sonmezoglu, S. (2020). Europium and terbium lanthanide ions co-doping in TiO2 photoanode to synchronously improve light-harvesting and open-circuit voltage for high-efficiency dye-sensitized solar cells. *Solar Energy*, *202*, 227–237.

Akın, S., Gülen, M., Sayın, S., Azak, H., Yıldız, H. B., & Sönmezoğlu, S. (2016). Modification of photoelectrode with thiol-functionalized Calix [4] arenes as interface energy barrier for high efficiency in dye-sensitized solar cells. *Journal of Power Sources*, *307*, 796–805.

Akın, S., Açıkgöz, S., Gülen, M., Akyürek, C., & Sönmezoğlu, S. (2016). Investigation of the photoinduced electron injection processes for natural dye-sensitized solar cells: The impact of anchoring groups. *RSC Advances*, *6*, 85125–85134.

Alonso, E., Sherman, A. M., Wallington, T. J., Everson, M. P., Field, F. R., Roth, R., & Kirchain, R. E. (2012). Evaluating rare earth element availability: A case with revolutionary demand from clean technologies. *Environmental Science & Technology*, *46*, 3406–3414.

Araichimani, P., Prabu, K., Kumar, G. S., Karunakaran, G., Van Minh, N., Karthi, S., ... Kolesnikov, E. (2020). Rare-earth ions integrated silica nanoparticles derived from rice husk via microwave-assisted combustion method for bioimaging applications. *Ceramics International*, *46*, 18366–18372.

Arivarasan, A., Bharathi, S., Arasi, S. E., Arunpandiyan, S., Revathy, M., & Jayavel, R. (2020). Investigations of rare earth doped CdTe QDs as sensitizers for quantum dots sensitized solar cells. *Journal of Luminescence*, *219*, 116881.

Arora, N., Dar, M. I., Hinderhofer, A., Pellet, N., Schreiber, F., Zakeeruddin, S. M., & Grätzel, M. (2017). Perovskite solar cells with CuSCN hole extraction layers yield stabilized efficiencies greater than 20%. *Science (New York, N.Y.)*, *358*, 768–771.

Asbury, J. B., Wang, Y., & Lian, T. (1999). Multiple-exponential electron injection in Ru (dcbpy) 2 (SCN) 2 sensitized ZnO nanocrystalline thin films. *The Journal of Physical Chemistry. B*, *103*, 6643–6647.

Atyaoui, M., Dimassi, W., Atyaoui, A., Elyagoubi, J., Ouertani, R., & Ezzaouia, H. (2013). Improvement in photovoltaic properties of silicon solar cells with a doped porous silicon layer with rare earth (Ce, La) as antireflection coatings. *Journal of Luminescence*, *141*, 1–5.

Bai, Y., Wang, Y., Yang, K., Zhang, X., Song, Y., & Wang, C. (2008). Enhanced upconverted photoluminescence in Er3 + and Yb3 + codoped ZnO nanocrystals with and without Li + ions. *Optics Communications*, *281*, 5448–5452.

Balestrieri, M., Ferblantier, G., Colis, S., Schmerber, G., Ulhaq-Bouillet, C., Muller, D., ... Dinia, A. (2013). Structural and optical properties of Yb-doped ZnO films deposited by magnetron reactive sputtering for photon conversion. *Solar Energy Materials and Solar Cells*, *117*, 363–371.

Bauer, C., Boschloo, G., Mukhtar, E., & Hagfeldt, A. (2001). Electron injection and recombination in Ru (dcbpy) 2 (NCS) 2 sensitized nanostructured ZnO. *The Journal of Physical Chemistry B*, *105*, 5585–5588.

Bingham, S., & Daoud, W. A. (2011). Recent advances in making nano-sized TiO 2 visible-light active through rare-earth metal doping. *Journal of Materials Chemistry*, *21*, 2041–2050.

Brown, D., Ma, B.-M., & Chen, Z. (2002). Developments in the processing and properties of NdFeb-type permanent magnets. *Journal of Magnetism and Magnetic Materials*, *248*, 432–440.

Burschka, J., Pellet, N., Moon, S.-J., Humphry-Baker, R., Gao, P., Nazeeruddin, M. K., & Grätzel, M. (2013). Sequential deposition as a route to high-performance perovskite-sensitized solar cells. *Nature*, *499*, 316–319.

Byrne, O., Ahmad, I., Surolia, P. K., Gun'ko, Y. K., & Thampi, K. R. (2014). The optimisation of dye sensitised solar cell working electrodes for graphene and SWCNTs containing quasi-solid state electrolytes. *Solar Energy*, *110*, 239–246.

Carbas, B. B., Gulen, M., Tolu, M. C., & Sonmezoglu, S. (2017). Hydrogen sulphate-based ionic liquid-assisted electro-polymerization of PEDOT catalyst material for high-efficiency photoelectrochemical solar cells. *Scientific Reports*, *7*, 1–15.

Carvalho, D. O., Kassab, L. R., Del Cacho, V. D., da Silva, D. M., & Alayo, M. I. (2018). A review on pedestal waveguides for low loss optical guiding, optical amplifiers and nonlinear optics applications. *Journal of Luminescence*, *203*, 135–144.

Chamanzadeh, Z., Ansari, V., & Zahedifar, M. (2021). Investigation on the properties of La-doped and Dy-doped ZnO nanorods and their enhanced photovoltaic performance of dye-sensitized solar cells. *Optical Materials*, *112*, 110735.

Chen, Y., Liu, S., Zhou, N., Li, N., Zhou, H., Sun, L.-D., & Yan, C.-H. (2020). An overview of rare earth coupled lead halide perovskite and its application in photovoltaics and light emitting devices. *Progress in Materials Science*, *120*, 100737.

Chiba, Y., Islam, A., Watanabe, Y., Komiya, R., Koide, N., & Han, L. (2006). Dye-sensitized solar cells with conversion efficiency of 11.1%. *Japanese Journal of Applied Physics*, *45*, L638.

Cho, K., Kumar, D., Holloway, P., & Singh, R. K. (1998). Luminescence behavior of pulsed laser deposited Eu: Y 2 O 3 thin film phosphors on sapphire substrates. *Applied Physics Letters*, *73*, 3058–3060.

Chou, T. P., Zhang, Q., Fryxell, G. E., & Cao, G. (2007). Hierarchically structured ZnO film for dye-sensitized solar cells with enhanced energy conversion efficiency. *Advanced Materials*, *19*, 2588–2592.

Da, Y., Xuan, Y., & Li, Q. (2018). Quantifying energy losses in planar perovskite solar cells. *Solar Energy Materials and Solar Cells*, *174*, 206–213.

De Figueiredo, A., Longo, V., Da Silva, R., Mastelaro, V., Mesquita, A., Franco, R., ... Longo, E. (2012). Structural XANES characterization of Ca0. 99Sm0. 01TiO3 perovskite and correlation with photoluminescence emission. *Chemical Physics Letters*, *544*, 43–48.

Deng, X., Zhang, C., Zheng, J., Zhou, X., Yu, M., Chen, X., & Huang, S. (2019). Highly bright Li (Gd, Y) F4: Yb, Er upconverting nanocrystals incorporated hole transport layer for efficient perovskite solar cells. *Applied Surface Science*, *485*, 332–341.

Dharmadhikari, D. V., & Athawale, A. A. (2018). Studies on structural and optical properties of rare earth copper oxides synthesized by template free hydrothermal method. *Materials Science and Engineering: B*, *229*, 70–78.

Dou, Y., Wu, F., Mao, C., Fang, L., Guo, S., & Zhou, M. (2015). Enhanced photovoltaic performance of ZnO nanorod-based dye-sensitized solar cells by using Ga doped ZnO seed layer. *Journal of Alloys and Compounds*, *633*, 408–414.

Du, Y.-P., Zhang, Y.-W., Sun, L.-D., & Yan, C.-H. (2008). Efficient energy transfer in monodisperse Eu-doped ZnO nanocrystals synthesized from metal acetylacetonates in high-boiling solvents. *The Journal of Physical Chemistry C*, *112*, 12234–12241.

Du Pasquier, A., Chen, H., & Lu, Y. (2006). Dye sensitized solar cells using well-aligned zinc oxide nanotip arrays. *Applied Physics Letters*, *89*, 253513.

Fan, K., Yu, J., & Ho, W. (2017). Improving photoanodes to obtain highly efficient dye-sensitized solar cells: A brief review. *Materials Horizons*, *4*, 319–344.

Feng, X., Sayle, D. C., Wang, Z. L., Paras, M. S., Santora, B., Sutorik, A. C., ... Wang, X. (2006). Converting ceria polyhedral nanoparticles into single-crystal nanospheres. *Science (New York, N.Y.)*, *312*, 1504–1508.

Florêncio, L. d. A., Gómez-Malagón, L. A., Lima, B. C., Gomes, A. S., Garcia, J., & Kassab, L. R. (2016). Efficiency enhancement in solar cells using photon down-conversion in Tb/Yb-doped tellurite glass. *Solar Energy Materials and Solar Cells*, *157*, 468–475.

Fu, Q., Saltsburg, H., & Flytzani-Stephanopoulos, M. (2003). Active nonmetallic Au and Pt species on ceria-based water-gas shift catalysts. *Science (New York, N.Y.)*, *301*, 935–938.

Fulle, K., Sanjeewa, L. D., McMillen, C. D., De Silva, C. R., Ruehl, K., Wen, Y., ... Kolis, J. W. (2019). Hydrothermal crystal growth of 2-D and 3-D barium rare earth germanates: BaREGeO4 (OH) and BaRE10 (GeO4) 4O8 (RE = Ho, Er). *Journal of Alloys and Compounds*, *786*, 489–497.

Gadkari, A. B., Shinde, T. J., & Vasambekar, P. N. (2010). Magnetic properties of rare earth ion (Sm3 +) added nanocrystalline Mg–Cd ferrites, prepared by oxalate co-precipitation method. *Journal of Magnetism and Magnetic Materials*, *322*, 3823–3827.

Gao, X.-X., Ge, Q.-Q., Xue, D.-J., Ding, J., Ma, J.-Y., Chen, Y.-X., ... Hu, J.-S. (2016). Tuning the fermi-level of TiO 2 mesoporous layer by lanthanum doping towards efficient perovskite solar cells. *Nanoscale*, *8*, 16881–16885.

Goldschmidt, J. C., & Fischer, S. (2015). Upconversion for photovoltaics—a review of materials, devices and concepts for performance enhancement. *Advanced Optical Materials*, *3*, 510—535.

Gong, J., Sumathy, K., Qiao, Q., & Zhou, Z. (2017). Review on dye-sensitized solar cells (DSSCs): Advanced techniques and research trends. *Renewable and Sustainable Energy Reviews*, *68*, 234—246.

Green, M. A., Ho-Baillie, A., & Snaith, H. J. (2014). The emergence of perovskite solar cells. *Nature Photonics*, *8*, 506—514.

Grätzel, M. (2003). Dye-sensitized solar cells. *Journal of Photochemistry and Photobiology C: Photochemistry Reviews*, *4*, 145—153.

Gu, F., Wang, S. F., Lü, M. K., Zhou, G. J., Xu, D., & Yuan, D. R. (2004). Structure evaluation and highly enhanced luminescence of Dy3 + -doped ZnO nanocrystals by Li + doping via combustion method. *Langmuir: The ACS Journal of Surfaces and Colloids*, *20*, 3528—3531.

Gulen, M., Sarilmaz, A., Patir, I. H., Ozel, F., & Sonmezoglu, S. (2018). Ternary copper-tungsten-disulfide nanocube inks as catalyst for highly efficient dye-sensitized solar cells. *Electrochimica Acta*, *269*, 119—127.

Guo, Q., Wu, J., Yang, Y., Liu, X., Jia, J., Dong, J., ... Wei, Y. (2019). High performance perovskite solar cells based on β-NaYF4: Yb3 + /Er3 + /Sc3 + @ NaYF4 core-shell upconversion nanoparticles. *Journal of Power Sources*, *426*, 178—187.

Hafez, H., Saif, M., & Abdel-Mottaleb, M. (2011). Down-converting lanthanide doped TiO2 photoelectrodes for efficiency enhancement of dye-sensitized solar cells. *Journal of Power Sources*, *196*, 5792—5796.

Hodes, G. (2013). Perovskite-based solar cells. *Science (New York, N.Y.)*, *342*, 317—318.

Hou, Y., Xu, Z., Peng, S., Rong, C., Liu, J. P., & Sun, S. (2007). A facile synthesis of SmCo5 magnets from core/shell Co/Sm2O3 nanoparticles. *Advanced Materials*, *19*, 3349—3352.

Hsu, Y., Xi, Y., Yip, C., Djurišić, A., & Chan, W. (2008). Dye-sensitized solar cells using ZnO tetrapods. *Journal of Applied Physics*, *103*, 083114.

Huang, D., Zhou, S.-h., Liu, J., & Wu, C.-p. (2003). Preparation and properties of rare-earth doped polyurethane coating light conversion film. *Polymer Materials Science and Engineering*, *19*, 173—176.

Huang, X., Han, S., Huang, W., & Liu, X. (2013). Enhancing solar cell efficiency: The search for luminescent materials as spectral converters. *Chemical Society Reviews*, *42*, 173—201.

Ikesue, A., Kinoshita, T., Kamata, K., & Yoshida, K. (1995). Fabrication and optical properties of high-performance polycrystalline Nd: YAG ceramics for solid-state lasers. *Journal of the American Ceramic Society*, *78*, 1033—1040.

Jeon, N. J., Noh, J. H., Yang, W. S., Kim, Y. C., Ryu, S., Seo, J., & Seok, S. I. (2015). Compositional engineering of perovskite materials for high-performance solar cells. *Nature*, *517*, 476—480.

Jiang, L., Zheng, J., Chen, W., Huang, Y., Hu, L., Hayat, T., ... Dai, S. (2017). High-performance perovskite solar cells with a weak covalent TiO2: Eu3 + mesoporous structure. *ACS Applied Energy Materials*, *1*, 93—102.

Katoh, R., Furube, A., Yoshihara, T., Hara, K., Fujihashi, G., Takano, S., ... Tachiya, M. (2004). Efficiencies of electron injection from excited N3 dye into nanocrystalline semiconductor (ZrO2, TiO2, ZnO, Nb2O5, SnO2, In2O3) films, The. *Journal of Physical Chemistry B*, *108*, 4818—4822.

Kaya, İ. C., Akin, S., Akyildiz, H., & Sonmezoglu, S. (2018). Highly efficient tandem photoelectrochemical solar cells using coumarin6 dye-sensitized CuCrO2 delafossite oxide as photocathode. *Solar Energy*, *169*, 196–205.

Kazhugasalamoorthy, S., Jegatheesan, P., Mohandoss, R., Giridharan, N., Karthikeyan, B., Joseyphus, R. J., & Dhanuskodi, S. (2010). Investigations on the properties of pure and rare earth modified bismuth ferrite ceramics. *Journal of Alloys and Compounds*, *493*, 569–572.

Kim, H., Nam, S., Jeong, J., Lee, S., Seo, J., Han, H., & Kim, Y. (2014). Organic solar cells based on conjugated polymers: History and recent advances. *Korean Journal of Chemical Engineering*, *31*, 1095–1104.

Kim, Y.-K., Lee, M., Yang, H.-S., Ha, M., & Hong, K. (2016). Optical characteristics of the rare-earth-ions-doped calcium chlorapatite phosphors prepared by using the solid–state reaction method. *Current Applied Physics*, *16*, 357–360.

Kojima, A., Teshima, K., Shirai, Y., & Miyasaka, T. (2009). Organo metal halide perovskites as visible-light sensitizers for photovoltaic cells. *Journal of the American Chemical Society*, *131*, 6050–6051.

Li, F., & Gu, Y. (2012). Improvement of performance of dye-sensitized solar cells by doping Er2O3 into TiO2 electrodes. *Materials Science in Semiconductor Processing*, *15*, 11–14.

Li, H., Luo, K., Xia, M., & Wang, P. W. (2014). Synthesis and optical properties of Pr3 + -doped ZnO quantum dots. *Journal of non-crystalline solids*, *383*, 176–180.

Li, J., Zhang, J., Zhang, X., Hao, Z., & Luo, Y. (2014). Cooperative downconversion and near infrared luminescence of Tm3 + /Yb3 + codoped calcium scandate phosphor. *Journal of Alloys and Compounds*, *583*, 96–99.

Li, Q., Huang, J., & Chen, D. (2011). A novel red-emitting phosphors K2Ba (MoO4) 2: Eu3 + , Sm3 + and improvement of luminescent properties for light emitting diodes. *Journal of Alloys and Compounds*, *509*, 1007–1010.

Li, W., Wang, Y., Lin, H., Ismat Shah, S., Huang, C., Doren, D., ... Barteau, M. (2003). Band gap tailoring of Nd 3 + -doped TiO 2 nanoparticles. *Applied Physics Letters*, *83*, 4143–4145.

Li, W., Zhang, W., Van Reenen, S., Sutton, R. J., Fan, J., Haghighirad, A. A., ... Snaith, H. J. (2016). Enhanced UV-light stability of planar heterojunction perovskite solar cells with caesium bromide interface modification. *Energy & Environmental Science*, *9*, 490–498.

Liu, Q., Dai, J., Zhang, X., Zhu, G., Liu, Z., & Ding, G. (2011). Perovskite-type transparent and conductive oxide films: Sb-and Nd-doped SrSnO3. *Thin Solid Films*, *519*, 6059–6063.

Liu, Y., Luo, W., Li, R., Zhu, H., & Chen, X. (2009). Near-infrared luminescence of Nd 3 + and Tm 3 + ions doped ZnO nanocrystals. *Optics Express*, *17*, 9748–9753.

Lu, L., Li, R., Peng, T., Fan, K., & Dai, K. (2011). Effects of rare earth ion modifications on the photoelectrochemical properties of ZnO-based dye-sensitized solar cells. *Renewable Energy*, *36*, 3386–3393.

Ludin, N. A., Mustafa, N. I., Hanafiah, M. M., Ibrahim, M. A., Teridi, M. A. M., Sepeai, S., ... Sopian, K. (2018). Prospects of life cycle assessment of renewable energy from solar photovoltaic technologies: A review. *Renewable and Sustainable Energy Reviews*, *96*, 11–28.

Lufaso, M. W., & Woodward, P. M. (2004). Jahn–Teller distortions, cation ordering and octahedral tilting in perovskites. *Acta Crystallographica Section B: Structural Science*, *60*, 10–20.

Luo, Q., Zhang, C., Deng, X., Zhu, H., Li, Z., Wang, Z., ... Huang, S. (2017). Plasmonic effects of metallic nanoparticles on enhancing performance of perovskite solar cells. *ACS Applied Materials & Interfaces*, *9*, 34821–34832.

Luo, X.-x., & Cao, W.-h. (2008). Ethanol-assistant solution combustion method to prepare La2O2S: Yb, Pr nanometer phosphor. *Journal of Alloys and Compounds*, *460*, 529–534.

Mansour, A., Sayers, D., Cook, J., Jr, Short, D., Shannon, R., & Katzer, J. (1984). X-ray absorption studies of some platinum oxides. *The Journal of Physical Chemistry*, *88*, 1778–1781.

Massari, S., & Ruberti, M. (2013). Rare earth elements as critical raw materials: Focus on international markets and future strategies. *Resources Policy*, *38*, 36–43.

Mekala, R., Deepa, B., & Rajendran, V. (2018). Preparation, characterization and antibacterial property of rare earth (Dy and Ce) doping on ZrO2 nanoparticles prepared by co-precipitation method. *Materials Today: Proceedings*, *5*, 8837–8843.

Mohamed, R., & Mkhalid, I. (2010). The effect of rare earth dopants on the structure, surface texture and photocatalytic properties of TiO2–SiO2 prepared by sol–gel method. *Journal of Alloys and Compounds*, *501*, 143–147.

Murray, E. P., Tsai, T., & Barnett, S. A. (1999). A direct-methane fuel cell with a ceria-based anode. *Nature*, *400*, 649–651.

Nikumbh, A., Pawar, R., Nighot, D., Gugale, G., Sangale, M., Khanvilkar, M., & Nagawade, A. (2014). Structural, electrical, magnetic and dielectric properties of rare-earth substituted cobalt ferrites nanoparticles synthesized by the co-precipitation method. *Journal of Magnetism and Magnetic Materials*, *355*, 201–209.

Niu, X., Zhong, H., Wang, X., & Jiang, K. (2006). Sensing properties of rare earth oxide doped In2O3 by a sol–gel method. *Sensors and Actuators B: Chemical*, *115*, 434–438.

Noh, J. H., & Seok, S. I. (2015). Steps toward efficient inorganic–organic hybrid perovskite solar cells. *MRS Bulletin*, *40*, 648–653.

Ogugua, S. N., Shaat, S., Swart, H. C., & Ntwaeaborwa, O. M. (2015). Optical properties and chemical composition analyses of mixed rare earth oxyorthosilicate (R2SiO5, R = La, Gd and Y) doped Dy3 + phosphors prepared by urea-assisted solution combustion method. *Journal of Physics and Chemistry of Solids*, *83*, 109–116.

Okamoto, Y., & Suzuki, Y. (2014). Perovskite-type SrTiO3, CaTiO3 and BaTiO3 porous film electrodes for dye-sensitized solar cells. *Journal of the Ceramic Society of Japan*, *122*, 728–731.

Opuchovic, O., Kareiva, A., Mazeika, K., & Baltrunas, D. (2017). Magnetic nanosized rare earth iron garnets R3Fe5O12: Sol–gel fabrication, characterization and reinspection. *Journal of Magnetism and Magnetic Materials*, *422*, 425–433.

Oskam, K., Wegh, R., Donker, H., Van Loef, E., & Meijerink, A. (2000). Downconversion: A new route to visible quantum cutting. *Journal of Alloys and Compounds*, *300*, 421–425.

Ozturk, T., Gulveren, B., Gulen, M., Akman, E., & Sonmezoglu, S. (2017). An insight into titania nanopowders modifying with manganese ions: A promising route for highly efficient and stable photoelectrochemical solar cells. *Solar Energy*, *157*, 47–57.

O'regan, B., & Grätzel, M. (1991). A low-cost, high-efficiency solar cell based on dye-sensitized colloidal TiO 2 films. *Nature*, *353*, 737–740.

Pan, G., Bai, X., Yang, D., Chen, X., Jing, P., Qu, S., ... Xu, W. (2017). Doping lanthanide into perovskite nanocrystals: Highly improved and expanded optical properties. *Nano Letters*, *17*, 8005–8011.

Park, H., Alhammadi, S., Bouras, K., Schmerber, G., Ferblantier, G., Dinia, A., ... Kim, W. K. (2017). Nd-doped SnO2 and ZnO for application in Cu (InGa) Se2 solar cells. *Science of Advanced Materials*, *9*, 2114–2120.

Park, Y., Ferblantier, G., Slaoui, A., Dinia, A., Park, H., Alhammadi, S., & Kim, W. K. (2020). Yb-doped zinc tin oxide thin film and its application to Cu (InGa) Se2 solar cells. *Journal of Alloys and Compounds*, *815*, 152360.

Paschotta, R., Nilsson, J., Barber, P., Caplen, J., Tropper, A. C., & Hanna, D. C. (1997). Lifetime quenching in Yb-doped fibres. *Optics Communications*, *136*, 375–378.

Pillai, S., & Green, M. (2010). Plasmonics for photovoltaic applications. *Solar Energy Materials and Solar Cells*, *94*, 1481–1486.

Primo, A., Marino, T., Corma, A., Molinari, R., & Garcia, H. (2011). Efficient visible-light photocatalytic water splitting by minute amounts of gold supported on nanoparticulate CeO2 obtained by a biopolymer templating method. *Journal of the American Chemical Society*, *133*, 6930–6933.

Qu, B., Jiao, Y., He, S., Zhu, Y., Liu, P., Sun, J., ... Zhang, X. (2016). Improved performance of a-Si: H solar cell by using up-conversion phosphors. *Journal of Alloys and Compounds*, *658*, 848–853.

Remya, K., Amirthapandian, S., Manivel Raja, M., Viswanathan, C., & Ponpandian, N. (2016). Effect of Yb substitution on room temperature magnetic and dielectric properties of bismuth ferrite nanoparticles. *Journal of Applied Physics*, *120*, 134304.

Richards, B. (2006). Enhancing the performance of silicon solar cells via the application of passive luminescence conversion layers. *Solar Energy Materials and Solar Cells*, *90*, 2329–2337.

Richards, B. (2006). Luminescent layers for enhanced silicon solar cell performance: Down-conversion. *Solar Energy Materials and Solar Cells*, *90*, 1189–1207.

Röyset, J., & Ryum, N. (2005). Scandium in aluminium alloys. *International Materials Reviews*, *50*, 19–44.

Saliba, M., Matsui, T., Seo, J.-Y., Domanski, K., Correa-Baena, J.-P., Nazeeruddin, M. K., ... Hagfeldt, A. (2016). Cesium-containing triple cation perovskite solar cells: Improved stability, reproducibility and high efficiency. *Energy & Environmental Science*, *9*, 1989–1997.

Saliba, M., Matsui, T., Domanski, K., Seo, J.-Y., Ummadisingu, A., Zakeeruddin, S. M., ... Hagfeldt, A. (2016). Incorporation of rubidium cations into perovskite solar cells improves photovoltaic performance. *Science (New York, N.Y.)*, *354*, 206–209.

Samoila, P., Cojocaru, C., Sacarescu, L., Dorneanu, P. P., Domocos, A.-A., & Rotaru, A. (2017). Remarkable catalytic properties of rare-earth doped nickel ferrites synthesized by sol-gel auto-combustion with maleic acid as fuel for CWPO of dyes. *Applied Catalysis B: Environmental*, *202*, 21–32.

van Sark, W., Korte, L., & Roca, F. (2012). *Introduction–Physics and technology of amorphous-crystalline heterostructure silicon solar cells, physics and technology of amorphous-crystalline heterostructure silicon solar cells* (pp. 1–12). Springer.

Sayama, K., Sugihara, H., & Arakawa, H. (1998). Photoelectrochemical properties of a porous Nb2O5 electrode sensitized by a ruthenium dye. *Chemistry of Materials*, *10*, 3825–3832.

Sengupta, D., Das, P., Mondal, B., & Mukherjee, K. (2016). Effects of doping, morphology and film-thickness of photo-anode materials for dye sensitized solar cell application–A review. *Renewable and Sustainable Energy Reviews*, *60*, 356–376.

Service, R. F. (2014). Perovskite solar cells keep on surging. *Science*, *344*, 458.

Shea, L. E., McKittrick, J., Lopez, O. A., & Sluzky, E. (1996). Synthesis of red-emitting, small particle size luminescent oxides using an optimized combustion process. *Journal of the American Ceramic Society*, *79*, 3257–3265.

Sheng, X., Zhao, Y., Zhai, J., Jiang, L., & Zhu, D. (2007). Electro-hydrodynamic fabrication of ZnO-based dye sensitized solar cells. *Applied Physics A*, *87*, 715–719.

Shirane, G., Danner, H., & Pepinsky, R. (1957). Neutron diffraction study of orthorhombic BaTi O 3. *Physical Review*, *105*, 856.

Singh, S. K. (2013). Structural and electrical properties of Sm-substituted BiFeO3 thin films prepared by chemical solution deposition. *Thin Solid Films*, *527*, 126–132.

Sivakumar, S., van Veggel, F. C. M., & Raudsepp, M. (2005). Bright white light through up-conversion of a single NIR source from sol – gel-derived thin film made with Ln3 + -doped LaF3 nanoparticles. *Journal of the American Chemical Society*, *127*, 12464–12465.

Somdee, A., & Osotchan, T. (2019). Effect of precipitating agent NaOH on the synthesis of SrTiO3/TiO2 heterostructure for dye-sensitized solar cells. *Materials Chemistry and Physics*, *229*, 210–214.

Song, P., Zhang, C., & Zhu, P. (2015). Research phosphate glass in combination with Eu/Tb elements on turning sunlight into red/green light as photovoltaic precursors. *IEEE Journal of Quantum Electronics*, *51*, 1–5.

Stoumpos, C. C., Malliakas, C. D., & Kanatzidis, M. G. (2013). Semiconducting tin and lead iodide perovskites with organic cations: Phase transitions, high mobilities, and near-infrared photoluminescent properties. *Inorganic Chemistry*, *52*, 9019–9038.

Stranks, S. D., Eperon, G. E., Grancini, G., Menelaou, C., Alcocer, M. J., Leijtens, T., ... Snaith, H. J. (2013). Electron-hole diffusion lengths exceeding 1 micrometer in an organometal trihalide perovskite absorber. *Science (New York, N.Y.)*, *342*, 341–344.

Sönmezoğlu, S., Çankaya, G., & Serin, N. (2012). Influence of annealing temperature on structural, morphological and optical properties of nanostructured TiO2 thin films. *Materials Technology*, *27*, 251–256.

Tanaka, H., & Misono, M. (2001). Advances in designing perovskite catalysts. *Current Opinion in Solid State and Materials Science*, *5*, 381–387.

Taş, R., Gülen, M., Can, M., & Sönmezoğlu, S. (2016). Effects of solvent and copper-doping on polyaniline conducting polymer and its application as a counter electrode for efficient and cost-effective dye-sensitized solar cells. *Synthetic Metals*, *212*, 75–83.

Trupke, T., Green, M., & Würfel, P. (2002). Improving solar cell efficiencies by down-conversion of high-energy photons. *Journal of Applied Physics*, *92*, 1668–1674.

Tsai, H., Nie, W., Blancon, J.-C., Stoumpos, C. C., Asadpour, R., Harutyunyan, B., ... Tretiak, S. (2016). High-efficiency two-dimensional Ruddlesden–Popper perovskite solar cells. *Nature*, *536*, 312–316.

Van Der Ende, B. M., Aarts, L., & Meijerink, A. (2009). Near-infrared quantum cutting for photovoltaics. *Advanced Materials*, *21*, 3073–3077.

Van Gosen, B. S., Verplanck, P. L., Long, K. R., Gambogi, J., & Seal, II R. R. (2014). *The rare-earth elements: Vital to modern technologies and lifestyles*. United States Geological Survey.

Wang, F., Han, Y., Lim, C. S., Lu, Y., Wang, J., Xu, J., ... Liu, X. (2010). Simultaneous phase and size control of upconversion nanocrystals through lanthanide doping. *Nature*, *463*, 1061–1065.

Wang, H. Q., Batentschuk, M., Osvet, A., Pinna, L., & Brabec, C. J. (2011). Rare-earth ion doped up-conversion materials for photovoltaic applications. *Advanced Materials*, *23*, 2675–2680.

Wang, J. T.-W., Ball, J. M., Barea, E. M., Abate, A., Alexander-Webber, J. A., Huang, J., ... Snaith, H. J. (2014). Low-temperature processed electron collection layers of graphene/TiO2 nanocomposites in thin film perovskite solar cells. *Nano Letters*, *14*, 724–730.

Wang, K., Jiang, J., Wan, S., & Zhai, J. (2015). Upconversion enhancement of lanthanide-doped NaYF4 for quantum dot-sensitized solar cells. *Electrochimica Acta*, *155*, 357–363.

Wang, K., Zheng, L., Zhu, T., Yao, X., Yi, C., Zhang, X., ... Gong, X. (2019). Efficient perovskite solar cells by hybrid perovskites incorporated with heterovalent neodymium cations. *Nano Energy*, *61*, 352–360.

Wang, L., Guo, W., Hao, H., Su, Q., Jin, S., Li, H., ... Liu, G. (2016). Enhancing photovoltaic performance of dye-sensitized solar cells by rare-earth doped oxide of SrAl2O4: Eu3 + . *Materials Research Bulletin*, *76*, 459–465.

Wang, L., Zhou, H., Hu, J., Huang, B., Sun, M., Dong, B., ... Li, L. (2019). A Eu3 + -Eu2 + ion redox shuttle imparts operational durability to Pb-I perovskite solar cells. *Science (New York, N.Y.)*, *363*, 265–270.

Wang, W., Lei, X., Gao, H., & Mao, Y. (2015). Near-infrared quantum cutting platform in transparent oxyfluoride glass–ceramics for solar sells. *Optical Materials*, *47*, 270–275.

Wang, Y., & Nan, C.-W. (2008). Effect of Tb doping on electric and magnetic behavior of Bi Fe O 3 thin films. *Journal of Applied Physics*, *103*, 024103.

Wang, Z.-S., Yanagida, M., Sayama, K., & Sugihara, H. (2006). Electronic-insulating coating of CaCO3 on TiO2 electrode in dye-sensitized solar cells: Improvement of electron lifetime and efficiency. *Chemistry of Materials*, *18*, 2912–2916.

Wegh, R. T., Donker, H., Oskam, K. D., & Meijerink, A. (1999). Visible quantum cutting in LiGdF4: Eu3 + through downconversion. *Science (New York, N.Y.)*, *283*, 663–666.

Wei, Y., Cheng, Z., & Lin, J. (2019). An overview on enhancing the stability of lead halide perovskite quantum dots and their applications in phosphor-converted LEDs. *Chemical Society Reviews*, *48*, 310–350.

Wei-Wei, X., Song-Yuan, D., Lin-Hua, H., Lin-Yun, L., & Kong-Jia, W. (2006). Influence of Yb-doped nanoporous TiO2 films on photovoltaic performance of dye-sensitized solar cells. *Chinese Physics Letters*, *23*, 2288.

Woodward, P. M. (1996). Structural distortions, phase transitions, and cation ordering in the perovskite and tungsten trioxide structures.

Woodward, P. M. (1997). Octahedral tilting in perovskites. I. Geometrical considerations. *Acta Crystallographica Section B: Structural Science*, *53*, 32–43.

Wu, J., Xie, G., Lin, J., Lan, Z., Huang, M., & Huang, Y. (2010). Enhancing photoelectrical performance of dye-sensitized solar cell by doping with europium-doped yttria rare-earth oxide. *Journal of Power Sources*, *195*, 6937–6940.

Wu, J., Lan, Z., Lin, J., Huang, M., Huang, Y., Fan, L., ... Wei, Y. (2017). Counter electrodes in dye-sensitized solar cells. *Chemical Society Reviews*, *46*, 5975–6023.

Xia, J.-B., Li, F.-Y., Yang, H., Li, X.-H., & Huang, C.-H. (2007). A novel quasi-solid-state dye-sensitized solar cell based on monolayer capped nanoparticles framework materials. *Journal of Materials Science*, *42*, 6412–6416.

Xiang, Y., Ma, Z., Zhuang, J., Lu, H., Jia, C., Luo, J., ... Cheng, X. (2017). Enhanced performance for planar perovskite solar cells with samarium-doped TiO2 compact electron transport layers. *The Journal of Physical Chemistry C*, *121*, 20150–20157.

Xiao, M., Huang, F., Huang, W., Dkhissi, Y., Zhu, Y., Etheridge, J., ... Spiccia, L. (2014). A fast deposition-crystallization procedure for highly efficient lead iodide perovskite thin-film solar cells. *Angewandte Chemie International Edition*, *53*, 9898–9903.

Xie, Y., Zhou, X., Mi, H., Ma, J., Yang, J., & Cheng, J. (2018). High efficiency ZnO-based dye-sensitized solar cells with a 1H, 1H, 2H, 2H-perfluorodecyltriethoxysilane chain barrier for cutting on interfacial recombination. *Applied Surface Science*, *434*, 1144–1152.

Xing, G., Mathews, N., Sun, S., Lim, S. S., Lam, Y. M., Grätzel, M., ... Sum, T. C. (2013). Long-range balanced electron-and hole-transport lengths in organic-inorganic CH3NH3PbI3. *Science (New York, N.Y.)*, *342*, 344–347.

Xu, J., Chen, X., Xu, Y., Du, Y., & Yan, C. (2020). Ultrathin 2D rare-earth nanomaterials: Compositions, syntheses, and applications. *Advanced Materials*, *32*, 1806461.

Xu, Z., Teo, S. H., Gao, L., Guo, Z., Kamata, Y., Hayase, S., & Ma, T. (2019). La-doped SnO2 as ETL for efficient planar-structure hybrid perovskite solar cells. *Organic Electronics*, *73*, 62–68.

Yamada, K., Kuranaga, Y., Ueda, K., Goto, S., Okuda, T., & Furukawa, Y. (1998). Phase transition and electric conductivity of ASnCl3 (A = Cs and CH3NH3). *Bulletin of the Chemical Society of Japan*, *71*, 127–134.

Yamamoto, H., Okamoto, S., & Kobayashi, H. (2002). Luminescence of rare-earth ions in perovskite-type oxides: From basic research to applications. *Journal of luminescence*, *100*, 325–332.

Yang, J., Chen, B., Pun, E. Y. B., Zhai, B., & Lin, H. (2013). Excitation wavelength-sensitive multi-colour fluorescence in Eu/Tb ions doped yttrium aluminium garnet glass ceramics. *Journal of Luminescence*, *134*, 622–628.

Yang, S., Kou, H., Wang, H., Cheng, K., & Wang, J. (2009). The photoelectrochemical properties of N3 sensitized CaTiO3 modified TiO2 nanocrystalline electrodes. *Electrochimica Acta*, *55*, 305–310.

Yang, W. S., Noh, J. H., Jeon, N. J., Kim, Y. C., Ryu, S., Seo, J., & Seok, S. I. (2015). High-performance photovoltaic perovskite layers fabricated through intramolecular exchange. *Science (New York, N.Y.)*, *348*, 1234–1237.

Yang, W. S., Park, B.-W., Jung, E. H., Jeon, N. J., Kim, Y. C., Lee, D. U., ... Noh, J. H. (2017). Iodide management in formamidinium-lead-halide–based perovskite layers for efficient solar cells. *Science (New York, N.Y.)*, *356*, 1376–1379.

Yao, H., & Tang, Q. (2020). Luminescent anti-reflection coatings based on down-conversion emission of Tb3 + -Yb3 + co-doped NaYF4 nanoparticles for silicon solar cells applications. *Solar Energy*, *211*, 446–452.

Yao, Q., Liu, J., Peng, Q., Wang, X., & Li, Y. (2006). Nd-doped TiO2 nanorods: Preparation and application in dye-sensitized solar cells. *Chemistry–An Asian Journal*, *1*, 737–741.

Ye, X., Gao, W., Xia, L., Nie, H., & Zhuang, W. (2010). A modified solution combustion method to superfine Gd2O3: Eu3 + phosphor: Preparation, phase transformation and optical properties. *Journal of Rare Earths*, *28*, 345–350.

Yin, J., Ahmed, G. H., Bakr, O. M., Brédas, J.-L., & Mohammed, O. F. (2019). Unlocking the effect of trivalent metal doping in all-inorganic CsPbBr3 perovskite. *ACS Energy Letters*, *4*, 789–795.

Yuhua, W., Ge, Z., Shuangyu, X., Qian, W., Yanyan, L., Quansheng, W., ... Wanying, G. (2015). Recent development in rare earth doped phosphors for white light emitting diodes. *Journal of Rare Earths*, *33*, 1–12.

Zalas, M., & Klein, M. (2012). The influence of titania electrode modification with lanthanide ions containing thin layer on the performance of dye-sensitized solar cells. *International Journal of Photoenergy* (2012).

Zhang, B., Song, Z., Jin, J., Bi, W., Li, H., Chen, C., ... Song, H. (2019). Efficient rare earth co-doped TiO2 electron transport layer for high-performance perovskite solar cells. *Journal of Colloid and Interface Science*, *553*, 14–21.

Zhang, C., Luo, Q., Shi, J., Yue, L., Wang, Z., Chen, X., & Huang, S. (2017). Efficient perovskite solar cells by combination use of Au nanoparticles and insulating metal oxide. *Nanoscale*, *9*, 2852–2864.

Zhang, J., Han, Z., Li, Q., Yang, X., Yu, Y., Cao, W., & S-doped, N. (2011). TiO_2 anode effect on performance of dye-sensitized solar cells. *Journal of Physics and Chemistry of Solids*, *72*, 1239–1244.

Zhang, J., Peng, W., Chen, Z., Chen, H., & Han, L. (2012). Effect of cerium doping in the TiO_2 photoanode on the electron transport of dye-sensitized solar cells. *The Journal of Physical Chemistry C*, *116*, 19182–19190.

Zhang, J., Shen, H., Guo, W., Wang, S., Zhu, C., Xue, F., ... Yuan, Z. (2013). An upconversion NaYF4: Yb3 + , Er3 + /TiO2 core–shell nanoparticle photoelectrode for improved efficiencies of dye-sensitized solar cells. *Journal of Power Sources*, *226*, 47–53.

Zhang, L., Shi, Y., Peng, S., Liang, J., Tao, Z., & Chen, J. (2008). Dye-sensitized solar cells made from BaTiO3-coated TiO2 nanoporous electrodes. *Journal of Photochemistry and Photobiology A: Chemistry*, *197*, 260–265.

Zhang, Q., & Huang, X. (2010). Recent progress in quantum cutting phosphors. *Progress in Materials Science*, *55*, 353–427.

Zhang, Q., Chou, T. P., Russo, B., Jenekhe, S. A., & Cao, G. (2008). Aggregation of ZnO nanocrystallites for high conversion efficiency in dye-sensitized solar cells. *Angewandte Chemie*, *120*, 2436–2440.

Zhang, W., Wang, K., Zhu, S., Li, Y., Wang, F., & He, H. (2009). Yttrium-doped TiO2 films prepared by means of DC reactive magnetron sputtering. *Chemical Engineering Journal*, *155*, 83–87.

Zhang, W., Saliba, M., Stranks, S. D., Sun, Y., Shi, X., Wiesner, U., & Snaith, H. J. (2013). Enhancement of perovskite-based solar cells employing core–shell metal nanoparticles. *Nano Letters*, *13*, 4505–4510.

Zhao, H., Xia, J., Yin, D., Luo, M., Yan, C., & Du, Y. (2019). Rare earth incorporated electrode materials for advanced energy storage. *Coordination Chemistry Reviews*, *390*, 32–49.

Zhou, B., Shi, B., Jin, D., & Liu, X. (2015). Controlling upconversion nanocrystals for emerging applications. *Nature Nanotechnology*, *10*, 924–936.

Zhou, H., Chen, Q., Li, G., Luo, S., Song, T.-b., Duan, H.-S., ... Yang, Y. (2014). Interface engineering of highly efficient perovskite solar cells. *Science (New York, N.Y.)*, *345*, 542–546.

Zhou, J., Teng, Y., Ye, S., Lin, G., & Qiu, J. (2012). A discussion on spectral modification from visible to near-infrared based on energy transfer for silicon solar cells. *Optical Materials*, *34*, 901–905.

Zhou, J., Liu, Q., Feng, W., Sun, Y., & Li, F. (2015). Upconversion luminescent materials: Advances and applications. *Chemical Reviews*, *115*, 395–465.

Index

Note: Page numbers followed by "*f*" and "*t*" refer to figures and tables, respectively.

Printed in the United States
by Baker & Taylor Publisher Services